T0331619

p-ADIC DIFFERENTIAL EQUATIONS

Now in its second edition, this volume provides a uniquely detailed study of *p*-adic differential equations. Assuming only a graduate-level background in number theory, the text builds the theory from first principles all the way to the frontiers of current research, highlighting analogies and links with the classical theory of ordinary differential equations. The author includes many original results which play a key role in the study of *p*-adic geometry, crystalline cohomology, *p*-adic Hodge theory, perfectoid spaces, and algorithms for *L*-functions of arithmetic varieties. This updated edition contains five new chapters, which revisit the theory of convergence of solutions of *p*-adic differential equations from a more global viewpoint, introducing the Berkovich analytification of the projective line, defining convergence polygons as functions on the projective line, and deriving a global index theorem in terms of the Laplacian of the convergence polygon.

Kiran S. Kedlaya is the Stefan E. Warschawski Professor of Mathematics at University of California San Diego. He has published over 100 research articles in number theory, algebraic geometry, and theoretical computer science as well as several books, including two on the Putnam competition. He has received a Presidential Early Career Award, a Sloan Fellowship, and a Guggenheim Fellowship, and been named an ICM invited speaker and a Fellow of the American Mathematical Society.

p-adic Differential Equations

Second Edition

KIRAN S. KEDLAYA
University of California San Diego

CAMBRIDGE
UNIVERSITY PRESS

CAMBRIDGE
UNIVERSITY PRESS

University Printing House, Cambridge CB2 8BS, United Kingdom

One Liberty Plaza, 20th Floor, New York, NY 10006, USA

477 Williamstown Road, Port Melbourne, VIC 3207, Australia

314–321, 3rd Floor, Plot 3, Splendor Forum, Jasola District Centre,
New Delhi – 110025, India

103 Penang Road, #05–06/07, Visioncrest Commercial, Singapore 238467

Cambridge University Press is part of the University of Cambridge.

It furthers the University's mission by disseminating knowledge in the pursuit of
education, learning, and research at the highest international levels of excellence.

www.cambridge.org
Information on this title: www.cambridge.org/9781009123341
DOI: 10.1017/9781009127684

First published 2010
Second edition 2022

A catalogue record for this publication is available from the British Library.

ISBN 978-1-009-12334-1 Hardback

Contents

Preface

This book is an outgrowth of a course taught by the author at MIT during fall 2007, on the subject of p-adic ordinary differential equations. The target audience was graduate students with some prior background in algebraic number theory, including exposure to p-adic numbers, but not necessarily any background in p-adic analytic geometry (of either the Tate or Berkovich flavors). The second edition was prepared during the 2020–2021 academic year.

Custom would dictate that this preface would continue with an explanation of what p-adic differential equations are and why they matter. Since we have included a whole chapter on this topic (Chapter 0), we instead devote this preface to a discussion of the origin of the book, its general structure, and what makes it different from previous books on the topic (including the first edition).

What was new in the first edition

The topic of p-adic differential equations has been treated in several previous books. Two that we used in preparing the MIT course, and to which we make frequent reference in the text, are those of Dwork, Gerotto, and Sullivan [149] and of Christol [89]. Another existing book is that of Dwork [145], but it is not a general treatise; rather, it focuses in detail on hypergeometric functions.

However, this book develops the theory of p-adic differential equations in a manner that differs significantly from most prior literature. The key differences include the following.

- We limit our use of cyclic vectors. This requires an initial investment in the study of matrix inequalities (Chapter 4) and lattice approximation arguments (especially Lemma 8.6.1), but pays off in the form of significantly stronger results.

- We introduce the notion of a Frobenius descendant (Chapter 10). This provides a complement to the older construction of Frobenius antecedents, particularly in dealing with certain boundary cases where the antecedent method does not apply.

As a result, we end up with some improvements of existing results, including the following (some of which can also be found in Christol's unpublished manuscript [93]):

- We refine the Frobenius antecedent theorem of Christol and Dwork (Theorem 10.4.2).
- We extend some results of Christol and Dwork, on the variation of the generic radius of convergence, to subsidiary radii (Theorem 11.3.2).
- We upgrade Young's geometric interpretation of subsidiary generic radii of convergence beyond the range of applicability of Newton polygons (Theorem 11.9.2).
- We quantify the Christol–Mebkhout decomposition theorem for differential modules on an annulus in ways which are applicable even when the modules are not solvable at a boundary (Theorems 12.2.2 and 12.3.1).
- We simplify the treatment of the theory of p-adic exponents (Theorems 13.5.5, 13.5.6, and 13.6.1).
- We sharpen the bound in the Christol transfer theorem to a disc containing a regular singularity with exponents in $\mathbb{Z}_p$ (Theorem 13.7.1).
- We generalize the Dieudonné–Manin classification theorem to difference modules over a complete nonarchimedean field (Theorem 14.6.3).
- We improve upon the Christol–Dwork–Robba effective bounds for solutions of p-adic differential equations (Theorem 18.2.1, Theorem 18.5.2) and some related bounds that apply in the presence of a Frobenius structure (Theorem 18.3.3). The latter can be used to recover a theorem of Chiarellotto and Tsuzuki concerning logarithmic growth of solutions of differential equations with Frobenius structure (Theorem 18.4.7).
- We state a relative version of the p-adic local monodromy theorem, formerly Crew's conjecture (Theorem 20.1.4). We describe in detail how it may be derived either from the p-adic index theory of Christol–Mebkhout, which we treat in detail in Chapter 13, or from the slope theory for Frobenius modules of Kedlaya, which we only sketch in Chapter 16.

Some of the new results are relevant in theory (in the study of higher-dimensional p-adic differential equations, largely in the context of the *semistable reduction problem* for overconvergent F-isocrystals, for which see [249] and [248]) or in practice (in the explicit computation of solutions of p-adic differential equations, e.g., for machine computation of zeta functions of particular varieties, for which see [244]). There is also some relevance entirely outside of number theory, to the study of flat connections on complex analytic varieties (see [250]).

Although some of the intended applications involve higher-dimensional p-adic analytic spaces, this book treats exclusively p-adic *ordinary* differential equations. In joint work with Liang Xiao [271], we have developed some extensions to higher-dimensional spaces.

What's new in the second edition

The second edition incorporates corrections to various errors in the original text. It also includes some new material, largely concerning developments that emerged after the publication of the first edition. We single out a few highlights:

- Many old and new references have been added to the chapter notes.
- In Chapter 7, we have added a discussion of the index of a meromorphic differential module (Section 7.4) and an example of the Stokes phenomenon (Section 7.6).
- In Chapter 12, we have expanded the discussion of clean modules and spectral decompositions (Section 12.8).
- In Chapter 13, we have eliminated the forward reference from Lemma 13.5.4 to Corollary 18.2.5 by providing an alternate proof of the former, corrected the proof of Theorem 13.6.1, and added a discussion of Liouville partitions (Section 13.8).
- In Chapter 18, we have expanded the discussion of logarithmic growth to include results of André and Ohkubo (Section 18.4), and corrected the statement and proof of Theorem 18.5.2.
- We have created a new Part VI consisting of Chapters 20 and 21 plus one new chapter (Chapter 22).
- In Chapter 20, we have added a theorem of Tsuzuki on minimal slope quotients (Theorem 20.6.7).
- We have added Chapter 22 to present a form of the p-adic local monodromy theorem that holds in the absence of a Frobenius structure. This includes the definition and analysis of modules of cyclic type (Section 22.2).
- Chapters 22, 23, and 24 from the first edition appear as Appendices A, B, and C in this edition.
- Part VII is entirely new to the second edition. See below.

Structure of the book

Each individual chapter of this book exhibits the following basic structure. Before the body of the chapter, we provide a brief introduction explaining what is to be discussed and often setting some running notations or hypotheses. After the body of the chapter, we include a section of afternotes, in which we provide detailed references for results in that chapter, fill in historical details, and add additional comments. (This practice is modeled on [172], although we cannot pretend to the level of detail achieved therein.) Note that we have a habit of attributing to various authors slightly stronger versions of their theorems than the ones they originally stated; to avoid complicating the discussion in the text, we resolve these misattributions in the afternotes instead. (See also the thematic bibliography of [259] for additional references, albeit without much context.) At the end of the chapter, we typically include a few exercises; a fair number of these request proofs of results which are stated and used in the text, but whose proofs pose no unusual difficulties.

The chapters themselves are grouped into several parts, which we now describe briefly. (Chapter 0, being introductory, does not fit into this grouping.)

Part I is preliminary, collecting some basic tools of p-adic analysis. However, it also includes some facts of matrix analysis (the variation of numerical invariants attached to matrices as a function of the matrix entries) which may not be familiar to the typical reader.

Part II introduces some formalism of differential algebra, such as differential rings and modules, twisted polynomials, and cyclic vectors, and applies these to fields equipped with a nonarchimedean norm.

Part III begins the study of p-adic differential equations in earnest, developing some basic theory for differential modules on rings and annuli, including the Christol–Dwork theory of variation of the generic radius of convergence, and the Christol–Mebkhout decomposition theory. We also include a treatment of p-adic exponents, culminating in the Christol–Mebkhout structure theorem for p-adic differential modules on an annulus satisfying the Robba condition (i.e., having intrinsic generic radius of convergence everywhere equal to 1).

Part IV introduces some formalism of difference algebra and presents (without full proofs) the theory of slope filtrations for Frobenius modules over the Robba ring.

Part V introduces the concept of a Frobenius structure on a p-adic differential module. We also discuss effective convergence bounds for solutions of p-adic differential equations.

Part VI presents the p-adic local monodromy theorem (formerly Crew's conjecture) and the proof techniques using either p-adic exponents or Frobenius slope filtrations. We also introduce a new approach that gives a version of the theorem that applies in the absence of a Frobenius structure.

Part VII revisits the theory of convergence of solutions of p-adic differential equations from a more global viewpoint. We introduce the Berkovich analytification of the projective line, define and study convergence polygons as functions on the projective line, and state a global index theorem in terms of the Laplacian of the convergence polygon.

The appendices consist of a series of brief discussions of several areas of application of the theory of p-adic differential equations. While they are formatted like chapters (without exercises), their textual style is somewhat more didactic and much less formal than in the main text (excluding Chapter 0); they are meant primarily as suggestions for further reading.

Prerequisites

As noted above, we have not assumed that the reader is familiar with rigid analytic geometry, and so have phrased all statements more concretely in terms of rings and modules. Although we expect that the typical reader has at least a passing familiarity with p-adic numbers (e.g., at the level of Gouvêa's text [178]), for completeness we begin with a rapid development of the algebra of complete rings and fields. This development, when read on its own, may appear somewhat idiosyncratic; its design is justified by the reuse of some material in later chapters.

We would ultimately like to think that the background needed is that of a two-semester undergraduate abstract algebra sequence. However, this may be a bit too optimistic; some basic notions from commutative algebra do occasionally intervene, including flat modules, exact sequences, and the snake lemma. It may be helpful to have a well-indexed text within arm's reach; we like Eisenbud's book [154], but the far slimmer book by Atiyah and Macdonald [24] should also suffice. (At the opposite extreme, we are also partial to the massive Stacks Project [378].)

Leitfaden

Figure 0.1 indicates logical dependencies among the chapters, with each part of the book represented in a single row. To keep the diagram manageable, we grouped together some chapters (1–3 and 9–12), and omitted Chapter 0 and the appendices.

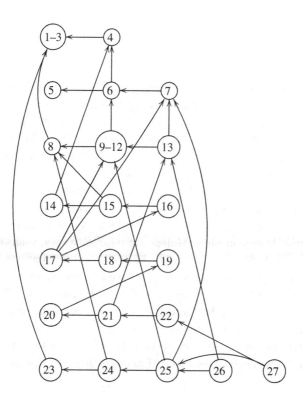

Figure 0.1 Diagram of logical dependencies among chapters

Acknowledgments

We thank the participants of the MIT course 18.787 (Topics in Number Theory, fall 2007) for numerous comments on the lecture notes which ultimately became this book. Particular thanks are due to Ben Brubaker and David Speyer for giving guest lectures, and to Chris Davis, Hansheng Diao, David Harvey, Raju Krishnamoorthy, Ruochuan Liu, Eric Rosen, and especially Liang Xiao for providing feedback. Additional feedback on drafts of the first edition was provided by Francesco Baldassarri, Laurent Berger, Bruno Chiarellotto, Gilles Christol, Ricardo García López, Tim Gowers, and Andrea Pulita. Feedback on the published first edition was provided by Francesco Baldassarri, Joshua Ciappara, Michel Matignon, Grant Molnar, Takahiro Nakagawa, Shun Ōhkubo, David Savitt, Atsushi Shiho, Junecue Suh, Peiduo Wang, and Shuyang Ye.

During the preparation of the course and of the first edition, the author was supported by a National Science Foundation CAREER grant (DMS-0545904), a Sloan Research Fellowship, MIT's NEC Research Support Fund, and the MIT Cecil and Ida Green Career Development Professorship. During the preparation of the second edition, the author was supported by NSF (DMS-1802161, DMS-2053473) and the Stefan E. Warschawski Professorship at UC San Diego.

0

Introductory remarks

The theory of ordinary differential equations is a fundamental instrument of *continuous* mathematics, in which the central objects of study are functions involving real numbers. It is not immediately apparent that this theory has anything useful to say about *discrete* mathematics in general, or number theory in particular.

In this book we consider ordinary differential equations in which the role of the real numbers is instead played by the field of p-adic numbers, for some prime number p. The p-adics form a number system with enough formal similarities to the real numbers to permit meaningful analogues of notions from calculus, such as continuity and differentiability. However, the p-adics incorporate data from arithmetic in a fundamental way; two numbers are p-adically close together if their difference is divisible by a large power of p.

In this chapter, we first survey some ways in which p-adic differential equations appear in number theory. We then focus on an example of Dwork, in which the p-adic behavior of Gauss's hypergeometric differential equation relates to the manifestly number-theoretic topic of the number of points on an elliptic curve over a finite field.

Since this chapter is meant only as an introduction, it is full of statements for which we give references instead of proofs. This practice is not typical of the rest of this book, except for the forward-looking discussions in the appendices. On a related note, the reader new to p-adic numbers should postpone this chapter's exercises until after reading Part I.

0.1 Why p-adic differential equations?

Although the very existence of a highly developed theory of p-adic ordinary differential equations is not entirely well known even within number theory, the subject is actually almost 50 years old. Here are some circumstances, past and present, in which it arises; some of these will be taken up again in the appendices.

Variation of zeta functions (see Appendix A). The original circumstance in which p-adic differential equations appeared in number theory was Dwork's work on the variation of zeta functions of algebraic varieties over finite fields. Roughly speaking, solving certain p-adic differential equations can give rise to explicit formulas for number of points on varieties over finite fields.

In contrast to methods involving étale cohomology, methods for studying zeta functions based on p-adic analysis (including p-adic cohomology) lend themselves well to numerical computation. Interest in computing zeta functions for varieties for which direct point-counting is not an option (e.g., curves over tremendously large finite fields) has been driven by applications in computer science, the principal example being cryptography based on elliptic or hyperelliptic curves.

p-adic cohomology (see Appendix B). Dwork's work suggested, but did not immediately lead to, a proper analogue of étale cohomology based on p-adic analytic techniques. Such an analogue was eventually developed by Berthelot by synthesizing work of Monsky and Washnitzer with ideas of Grothendieck); it is called *rigid cohomology* (see the chapter notes for the origin of the term "rigid"). The development of rigid cohomology has lagged somewhat behind that of étale cohomology, partly due to the emergence of some thorny problems related to the construction of a good category of coefficients. These problems, which have only recently been resolved, are rather closely related to questions concerning p-adic differential equations; in fact, some of the results presented in this book have been used to address these problems.

p-adic Hodge theory (see Appendix C). The subject of p-adic Hodge theory aims to do for the cohomology of varieties over p-adic fields what ordinary Hodge theory does for the cohomology of varieties over $\mathbb{C}$: namely, to provide a better understanding of the cohomology of a variety in its own right, independently of the geometry of the variety. In the p-adic case, the cohomology in question is often étale cohomology, which carries the structure of a Galois representation.

The study of such representations, as pioneered by Fontaine, involves a number of exotic auxiliary rings (rings of *p-adic periods*) which serve their intended purposes but are otherwise a bit mysterious. More recently, the work of Berger has connected much of the theory to the study of p-adic differential equations; notably, a key result that was originally intended for use in p-adic cohomology (the *p-adic local monodromy theorem*) turned out to imply an important conjecture of Fontaine on the potential semistability of Galois representations.

Ramification theory (see Chapter 19). There are some interesting analogies between properties of differential equations over $\mathbb{C}$ with meromorphic singularities and properties of wildly ramified Galois representations of p-adic fields. At some level, this is suggested by the parallel formulation of the Langlands conjectures in the number field and function field cases. One can use p-adic differential equations to interpolate between the two situations, by associating differential equations to Galois representations (as in the previous item) and then using differential invariants (such as irregularity) to recover Galois invariants (such as Artin and Swan conductors).

For representations of the étale fundamental group of a variety over a field of positive characteristic of dimension greater than 1, it is difficult to construct meaningful Galois-theoretic numerical invariants. Recent work of Abbes and Saito [1, 2] provides satisfactory definitions, but the resulting quantities are quite difficult to calculate. One can alternatively use p-adic differential equations to define invariants which can be somewhat easier to deal with; for instance, one can define a *differential Swan conductor* which is guaranteed to be an integer [238], whereas this is not clear for the Abbes–Saito conductor. One can then equate the two conductors, deducing integrality for the Abbes–Saito conductor; this has been carried out by Chiarellotto and Pulita [86] for one-dimensional representations, and by L. Xiao [411] in the general case.

0.2 Zeta functions of varieties

For the rest of this introduction, we return to Dwork's original example showing the role of p-adic differential equations and their solutions in number theory. This example refers to elliptic curves, for which see Silverman's book [373] for background.

Definition 0.2.1. For λ in some field K, let E_λ be the elliptic curve over K defined by the equation

$$E_\lambda : y^2 = x(x-1)(x-\lambda)$$

in the projective plane. Remember that there is one point $O = [0:1:0]$ at infinity. There is a natural commutative group law on $E_\lambda(K)$ with identity element O, characterized by the property that three points add to zero if and only if they are collinear. (It is better to say that three points add to zero if they are the three intersections of E_λ with some line, as this correctly permits degenerate cases. For instance, if two of the points coincide, the line must be the tangent to E_λ at that point.)

For elliptic curves over finite fields, one has the following result of Hasse, which generalizes some observations made by Gauss and others.

Theorem 0.2.2 (Hasse). *Suppose λ belongs to a finite field $\mathbb{F}_q$. If we write* $\#E_\lambda(\mathbb{F}_q) = q + 1 - a_q(\lambda)$, *then* $|a_q(\lambda)| \leq 2\sqrt{q}$.

Proof See [373, Theorem V.1.1]. □

Hasse's theorem was later vastly generalized as follows, originally as a set of conjectures by Weil. (Despite no longer being conjectural, these are still commonly referred to as the *Weil conjectures*.)

Definition 0.2.3. For X an algebraic variety over $\mathbb{F}_q$, the *zeta function* of X is defined as the formal power series

$$\zeta_X(T) = \exp\left(\sum_{n=1}^{\infty} \frac{T^n}{n} \#X(\mathbb{F}_{q^n})\right);$$

another way to write it, which makes it look like more familiar examples of zeta functions, is

$$\zeta_X(T) = \prod_x (1 - T^{\deg(x)})^{-1},$$

where x runs over Galois orbits of $X(\overline{\mathbb{F}}_q)$, and $\deg(x)$ denote the size of the orbit x. (If you prefer algebro-geometric terminology, you may run x over closed points of the scheme X, in which case $\deg(x)$ denotes the degree of the residue field of x over $\mathbb{F}_q$.)

Example 0.2.4. For $X = E_\lambda$, one can verify that

$$\zeta_X(T) = \frac{1 - a_q(\lambda)T + qT^2}{(1 - T)(1 - qT)}$$

using properties of the Tate module of E_λ; see [373, Theorem V.2.2].

The statement of the Weil conjectures is the following theorem.

Theorem 0.2.5 (Dwork, Grothendieck, Deligne, et al). *Let X be an algebraic variety over $\mathbb{F}_q$. Then $\zeta_X(T)$ represents a rational function of T. Moreover, if X is smooth and proper of dimension d, we can write*

$$\zeta_X(T) = \frac{P_1(T) \cdots P_{2d-1}(T)}{P_0(T) \cdots P_{2d}(T)},$$

where each $P_i(T)$ has integer coefficients, satisfies $P_i(0) = 1$, and has all roots in $\mathbb{C}$ on the circle $|T| = q^{-i/2}$.

Proof The proof of this theorem is a sufficiently massive undertaking that even a reference is not reasonable here; instead, we give [196, Appendix C] as a metareference. (Another useful exposition is [330]; see also the chapter notes.) □

Remark 0.2.6. It is worth pointing out that the first complete proof of Theorem 0.2.5 used the fact that for any prime $\ell \neq p$, one has

$$\#X(\mathbb{F}_{q^n}) = \sum_i (-1)^i \operatorname{Trace}(F^n, H^i_{\text{et}}(X, \mathbb{Q}_\ell)),$$

where $H^i_{\text{et}}(X, \mathbb{Q}_\ell)$ is the i-th étale cohomology group of X (or rather, the base change of X to $\overline{\mathbb{F}}_q$) with coefficients in $\mathbb{Q}_\ell$. This is an instance of the *Lefschetz trace formula* in étale cohomology.

0.3 Zeta functions and *p*-adic differential equations

Remark 0.3.1. The interpretation of Theorem 0.2.5 in terms of étale cohomology (Remark 0.2.6) is all well and good, but there are several downsides. One important one is that étale cohomology is not explicitly computable; for instance, it is not straightforward to describe étale cohomology to a computer well enough that the computer can make calculations. (The main problem is that while one can write down étale cocycles, it is very hard to tell whether or not any given cocycle is a coboundary.)

Another important downside is that étale cohomology does not yield good information about what happens to ζ_X when you vary X. This is where *p*-adic differential equations enter the picture. It was observed by Dwork that when you have a family of algebraic varieties defined over $\mathbb{Q}$, the same differential equations appear on one hand when you study variation of complex periods, and on the other hand when you study variation of zeta functions over $\mathbb{F}_p$.

Here is an explicit example due to Dwork.

Definition 0.3.2. Recall that the *hypergeometric series*

$$F(a, b; c; z) = \sum_{i=0}^{\infty} \frac{a(a+1)\cdots(a+i-1)b(b+1)\cdots(b+i-1)}{c(c+1)\cdots(c+i-1)i!} z^i$$

(0.3.2.1)

satisfies the *hypergeometric differential equation*

$$z(1-z)y'' + (c - (a+b+1)z)y' - aby = 0. \qquad (0.3.2.2)$$

Set

$$\alpha(z) = F(1/2, 1/2; 1; z).$$

Over $\mathbb{C}$, α is related to an elliptic integral, for instance, by the formula

$$\alpha(\lambda) = \frac{2}{\pi} \int_0^{\pi/2} \frac{d\theta}{\sqrt{1 - \lambda \sin^2 \theta}} \qquad (0 < \lambda < 1).$$

(One can extend this formula to complex λ, but this requires some care with branch cuts.) This elliptic integral can be viewed as a period integral for the curve E_λ, i.e., one is integrating some meromorphic differential form on E_λ around some loop (or more properly, around some homology class).

Let p be an odd prime. We now try to interpret $\alpha(z)$ as a function of a p-adic variable rather than a complex variable. Beware that this means that z can take *any* value in a field with a norm extending the p-adic norm on $\mathbb{Q}$, not just $\mathbb{Q}_p$ itself. (For the moment, you can imagine z running over a completed algebraic closure of $\mathbb{Q}_p$.)

Lemma 0.3.3. *The series $\alpha(z)$ converges p-adically for $|z| < 1$.*

Proof Exercise.   □

Dwork discovered that a closely related function admits a sort of analytic continuation.

Definition 0.3.4. Define the *Igusa polynomial*

$$H(z) = \sum_{i=0}^{(p-1)/2} \binom{(p-1)/2}{i}^2 z^i.$$

Modulo p, the roots of $H(z)$ are the values of $\lambda \in \overline{\mathbb{F}}_p$ for which E_λ is a *supersingular* elliptic curve, i.e., for which $a_q(\lambda) \equiv 0 \pmod{p}$. (In fact, the roots of $H(z)$ all belong to $\mathbb{F}_{p^2}$, by a theorem of Deuring; see [373, Theorem V.3.1].)

Dwork's analytic continuation result is the following.

Theorem 0.3.5 (Dwork). *There exists a series $\xi(z) = \sum_{i=0}^{\infty} P_i(z)/H(z)^i$, with each $P_i(z) \in \mathbb{Q}_p[z]$, converging uniformly for those z satisfying $|z| \leq 1$ and $|H(z)| = 1$ and such that*

$$\xi(z) = (-1)^{(p-1)/2} \frac{\alpha(z)}{\alpha(z^p)} \qquad (|z| < 1).$$

Proof See [402, §7]. □

Remark 0.3.6. Note that ξ itself satisfies a differential equation derived from the hypergeometric equation. We will see such equations again once we introduce the notion of a Frobenius structure on a differential equation, in Chapter 17.

In terms of the function ξ, we can compute zeta functions in the Legendre family as follows.

Definition 0.3.7. Let $\mathbb{Z}_q$ be the unique unramified extension of $\mathbb{Z}_p$ with residue field $\mathbb{F}_q$. For $\lambda \in \mathbb{F}_q$, let $[\lambda]$ be the unique q-th root of 1 in $\mathbb{Z}_q$ congruent to λ mod p. (See the notes for Chapter 14 for more discussion of this construction.)

Theorem 0.3.8 (Dwork). *If $q = p^a$ and $\lambda \in \mathbb{F}_q$ is not a root of $H(z)$, then*

$$T^2 - a_q(\lambda)T + q = (T - u)(T - q/u),$$

where

$$u = \xi([\lambda])\xi([\lambda]^p) \cdots \xi([\lambda]^{p^{a-1}}).$$

That is, the quantity u is the "unit root" (meaning the root of valuation 0) of the polynomial $T^2 - a_q(\lambda)T + q$ occurring (up to reversal) in the zeta function.

Proof See [402, §7]. □

0.4 A word of caution

Example 0.4.1. Before we embark on the study of p-adic ordinary differential equations, a cautionary note is in order, concerning the rather innocuous-looking differential equation $y' = y$. Over $\mathbb{R}$ or $\mathbb{C}$, this equation is nonsingular everywhere, and its solutions $y = ce^x$ are defined everywhere.

Over a p-adic field, things are quite different. As a power series around $x = 0$,

$$y = c \sum_{n=0}^{\infty} \frac{x^n}{n!}$$

and the denominators hurt us rather than helping. In fact, the series only converges for $|x| < p^{-1/(p-1)}$ (assuming that we are normalizing in such a way that $|p| = p^{-1}$). For comparison, note that the logarithm series

$$\log \frac{1}{1-x} = \sum_{n=1}^{\infty} \frac{x^n}{n}$$

converges for $|x| < 1$.

Remark 0.4.2. The conclusion to be taken away from the previous example is that there is no fundamental theorem of ordinary differential equations over the p-adics! In fact, the hypergeometric differential equation in the previous example was somewhat special; the fact that it had a solution in a disc where it had no singularities was not a foregone conclusion. One of Dwork's discoveries is that this typically happens for differential equations that "come from geometry", such as *Picard–Fuchs equations* which arise from integrals of algebraic functions (e.g., elliptic integrals). Another of Dwork's discoveries is that one can quantify the obstruction to solving a p-adic differential equation in a nonsingular disc, using similar techniques to those used to study obstructions to solving complex differential equations in singular discs. We will carry this out later in the book.

Notes

For detailed notes on the topics discussed in §0.1, see the notes for the chapters referenced.

We again mention [196, Appendix C] and [330] as starting points for further reading about the Weil conjectures. See also [261].

The notion of an analytic function in terms of a uniform limit of rational functions with poles prescribed to certain regions is the original such notion, introduced by Krasner. For this book, we will restrict our consideration of p-adic analysis to working with complete rings in this fashion, without attempting to introduce any notion of nonarchimedean analytic geometry. However, it must be noted that it is much better in the long run to work in terms of analytic geometry; for example, it is pretty hopeless to deal with partial differential equations without doing so.

That said, there are several ways to develop a theory of analytic spaces over a nonarchimedean field. The traditional method is Tate's theory of rigid analytic spaces, so-called because one develops everything "rigidly" by imitating the theory of schemes in algebraic geometry, but using rings of convergent power series instead of polynomials. The canonical foundational reference for rigid geometry is the book of Bosch, Güntzer, and Remmert [69], but novices may find the text of Fresnel and van der Put [171] or the lecture notes of Bosch [68] more approachable. Two more recent methods, which in some ways are more robust, is Berkovich's theory of nonarchimedean analytic spaces (commonly called *Berkovich spaces*), as introduced in [52] and further developed in [53]; and Huber's theory of *adic spaces*, as introduced in [211] and further developed in [212]. For all three points of view, see also the lecture notes of Conrad [111]. The Berkovich approach will manifest at a very superficial level in Part VII.

Dwork's original analysis of the Legendre family of elliptic curves via the associated hypergeometric equation (which expands upon earlier work of Tate) appears in [140, §8]. The treatment in [402] is more overtly related to p-adic cohomology.

The family of hypergeometric equations with $a, b, c \in \mathbb{Q} \cap \mathbb{Z}_p$ is rich enough that one could devote an entire book to the study of its p-adic properties. Indeed, Dwork did exactly this; the result is [145].

It is possible to resurrect partially the fundamental theorem of ordinary differential equations in the p-adic setting. The best possible results in that direction seem to be those of Priess-Crampe and Ribenboim [341]. One consequence of their work is that a differential equation over $\mathbb{Q}_p$ has a solution if and only if it has a sufficiently good approximate solution; this amounts to a differential version of Hensel's lemma. We too will need noncommutative forms of Hensel's lemma; see Theorem 2.2.2.

Christol [92] has given an interesting retrospective on some of the key ideas of Dwork, including generic points, the transfer principle, and Frobenius structures, which resonate throughout this book.

Exercises

0.1   Prove directly from the definition that the series $F(a, b; c; z)$ converges p-adically for $|z| < 1$ whenever a, b, c are rational numbers with denominators not divisible by p. This implies Lemma 0.3.3.

0.2   Using the fact that $\alpha(z)$ satisfies the hypergeometric equation, write down a nontrivial differential equation with coefficients in $\mathbb{Q}(z)$ satisfied by the function $\xi(z)$.

0.3   Check that the usual formula

$$\liminf_{n \to \infty} |a_n|^{-1/n}$$

for the radius of convergence of the power series $\sum_{n=0}^{\infty} a_n z^n$ still works over a nonarchimedean field. That is, the series converges when $|z|$ is less than this radius and diverges when $|z|$ is greater than this radius.

0.4   Show that in the previous exercise, just like in the archimedean case, a power series over a nonarchimedean field can either converge or diverge at a value of z for which $|z|$ equals the radius of convergence.

0.5   Check that (as claimed in Example 0.4.1) under the normalization

$|p| = p^{-1}$, the exponential series $\exp(z)$ over $\mathbb{Q}_p$ has radius of convergence $p^{-1/(p-1)}$, while the logarithm series $\log(1 - z)$ has radius of convergence 1.

0.6   Show that over $\mathbb{Q}_p$, a power series in z which converges for $|z| \leq 1$ may have an antiderivative which only converges for $|z| < 1$, but its derivative still converges for $|z| \leq 1$. This is the reverse of what happens over an archimedean field.

Part I

Tools of *p*-adic Analysis

1

Norms on algebraic structures

In this chapter, we recall some basic facts about norms (absolute values), primarily of the nonarchimedean sort, on groups, rings, fields, and modules. We also briefly discuss the phenomenon of spherical completeness, which is peculiar to the nonarchimedean setting. Our discussion is not particularly comprehensive; the reader new to nonarchimedean analysis is directed to [355] for a fuller treatment.

Several proofs in this chapter make forward references into Chapter 2. There should be no difficulty in verifying the absence of circular references.

Convention 1.0.1. In this book, a *ring* means a commutative ring unless this hypothesis is suppressed explicitly by describing the ring as "not necessarily commutative" or implicitly by usage in certain phrases, e.g., a *ring of twisted polynomials* (Definition 5.5.1).

Notation 1.0.2. For R a ring, we denote by $R^\times$ the multiplicative group of units of R.

1.1 Norms on abelian groups

Let us start by recalling some basic definitions from analysis, without yet specializing to the nonarchimedean case.

Definition 1.1.1. Let G be an abelian group. A *seminorm* (or *semiabsolute value*) on G is a function $|\cdot| : G \to [0, \infty)$ satisfying the following conditions.

(a) We have $|0| = 0$.

(b) For $f, g \in G$, $|f - g| \le |f| + |g|$. (Equivalently, $|g| = |-g|$ and $|f + g| \le |f| + |g|$. This condition is usually called the *triangle inequality*.)

We say the seminorm $|\cdot|$ is a *norm* (or *absolute value*) if the following additional condition holds.

(a') For $g \in G$, $|g| = 0$ if and only if $g = 0$.

We also express this by saying that G is *separated* under $|\cdot|$. A seminorm on an abelian group G induces a metric topology on G, in which the basic open subsets are the open balls, i.e., sets of the form $\{g \in G : |f - g| < r\}$ for some $f \in G$ and some $r > 0$.

Definition 1.1.2. Let G, G' be abelian groups equipped with seminorms $|\cdot|$, $|\cdot|'$, respectively, and let $\phi: G \to G'$ be a homomorphism. Note that ϕ is continuous for the metric topologies on G, G' if and only if there exists a function $h: (0, \infty) \to (0, \infty)$ such that for all $r > 0$,

$$\{g \in G : |g| < h(r)\} \subseteq \{g \in G : |\phi(g)|' < r\}.$$

We say that ϕ is *bounded* if there exists $c \geq 0$ such that $|\phi(g)|' \leq c|g|$ for all $g \in G$. We say that ϕ is *isometric* if $|\phi(g)|' = |g|$ for all $g \in G$. We say two seminorms $|\cdot|_1, |\cdot|_2$ on G are *topologically equivalent* if they induce the same metric topology, i.e., the identity morphism on G is continuous in both directions. We say that $|\cdot|_1, |\cdot|_2$ are *metrically equivalent* if there exist $c_1, c_2 > 0$ such that for all $g \in G$,

$$|g|_1 \leq c_1|g|_2, \qquad |g|_2 \leq c_2|g|_1;$$

this implies topological equivalence, but the reverse implication does not always hold.

Definition 1.1.3. Let G be an abelian group equipped with a seminorm. A *Cauchy sequence* in G under $|\cdot|$ is a sequence $\{x_n\}_{n=0}^{\infty}$ in G such that for any $\epsilon > 0$, there exists an integer N such that for all integers $m, n \geq N$, $|x_m - x_n| < \epsilon$. We say the sequence $\{x_n\}_{n=0}^{\infty}$ is *convergent* if there exists $x \in G$ such that for any $\epsilon > 0$, there exists an integer N such that for all integers $n \geq N$, $|x - x_n| < \epsilon$; in this case, the sequence is automatically Cauchy, and we say that x is a *limit* of the sequence. If G is separated under $|\cdot|$, then limits are unique when they exist. We say G is *complete* under $|\cdot|$ if every Cauchy sequence has a unique limit.

Theorem 1.1.4. *Let G be an abelian group equipped with a norm $|\cdot|$. Then there exists an abelian group G' equipped with a norm $|\cdot|'$ under which it is complete, and an isometric homomorphism $\phi: G \to G'$ with dense image.*

This is standard, so we only sketch the proof.

Proof We take the set of Cauchy sequences in G and declare two sequences $\{x_n\}_{n=0}^{\infty}, \{y_n\}_{n=0}^{\infty}$ to be equivalent if the sequence $x_0, y_0, x_1, y_1, \ldots$ is also Cauchy. This is indeed an equivalence relation; let G' be the set of equivalence classes. It is then straightforward to construct the group operation (termwise addition) and the norm on G' (the limit of the norms of the terms of the sequence). The map ϕ takes $g \in G$ to the constant sequence $g, g, \ldots$. $\qquad \square$

Definition 1.1.5. With notation as in Theorem 1.1.4, we call G' the *completion* of G; the group G', equipped with the norm $|\cdot|'$ and the homomorphism ϕ, is functorial in G. That is, any continuous homomorphism $G \to H$ extends uniquely to a continuous homomorphism $G' \to H'$ between the completions; in particular, G' is unique up to unique isomorphism. Note that one can also define the completion even if G is only equipped with a seminorm, but only by first quotienting by the kernel of the seminorm; in that case, the map from G to its completion need not be injective.

Definition 1.1.6. If R is a not necessarily commutative ring and $|\cdot|$ is a seminorm on its additive group, we say that $|\cdot|$ is *submultiplicative* if the following additional condition holds.

(c) For $f, g \in R$, $|fg| \leq |f||g|$.

We say that $|\cdot|$ is *multiplicative* if the following additional condition holds.

(c′) For $f, g \in R$, $|fg| = |f||g|$.

The completion of a ring R equipped with a submultiplicative seminorm admits a natural ring structure, because the termwise product of two Cauchy sequences is again Cauchy.

Lemma 1.1.7. *Let F be a field equipped with a multiplicative norm. Then the completion of F is also a field.*

Proof Note that if $\{f_n\}_{n=0}^{\infty}$ is a Cauchy sequence in F, then $\{|f_n|\}_{n=0}^{\infty}$ is a Cauchy sequence in $\mathbb{R}$ by the triangle inequality, and so has a limit since $\mathbb{R}$ is complete. Since F is equipped with a true norm, if $\{f_n\}_{n=0}^{\infty}$ does not converge to 0, then $\{|f_n|\}_{n=0}^{\infty}$ must also not converge to 0. In particular, $|f_n|_{n=0}^{\infty}$ is bounded below, from which it follows that $\{f_n^{-1}\}_{n=0}^{\infty}$ is also a Cauchy sequence. This proves that every nonzero element of the completion of F has a multiplicative inverse, as desired. $\qquad \square$

Proposition 1.1.8. *Two multiplicative norms $|\cdot|, |\cdot|'$ on a field F are topologically equivalent if and only if there exists $c > 0$ such that $|x|' = |x|^c$ for all $x \in F$.*

Proof Exercise, or see [149, Lemma I.1.2].   □

Definition 1.1.9. Let G be an abelian group equipped with a seminorm $|\cdot|_G$, and let G' be a subgroup of G. The *quotient seminorm* on the quotient G/G' is defined by the formula

$$|g + G'|_{G/G'} = \inf_{g' \in G'} \{|g + g'|_G\}. \qquad (1.1.9.1)$$

If $|\cdot|_G$ is a norm, then $|\cdot|_{G/G'}$ is a norm if and only if G' is closed in G.

Remark 1.1.10. The completion of a field with respect to a submultiplicative norm need not be a field. See the exercises.

It is also possible for a field to be complete with respect to a norm which is not topologically equivalent to any multiplicative norm, but this requires a more subtle construction. See [256] for discussion of this phenomenon.

1.2 Valuations and nonarchimedean norms

We now restrict attention to nonarchimedean absolute values, which can be described additively (using valuations) as well as multiplicatively (using norms). It will be convenient to switch back and forth between these points of view throughout the book.

Definition 1.2.1. A *real semivaluation* on an abelian group G is a function $v: G \to \mathbb{R} \cup \{\infty\}$ with the following properties.

(a) We have $v(0) = \infty$.
(b) For $f, g \in G$, $v(f - g) \geq \min\{v(f), v(g)\}$.

We say v is a *real valuation* if the following additional condition holds.

(a′) For $g \in G$, $v(g) = \infty$ if and only if $g = 0$.

If v is a real (semi)valuation on G, then the function $|\cdot| = \exp(-v(\cdot))$ is a (semi)norm on G which is *nonarchimedean* (or *ultrametric*), i.e., it satisfies the *strong triangle inequality*

(b′) For $f, g \in G$, $|f - g| \leq \max\{|f|, |g|\}$.

Conversely, for any nonarchimedean (semi)norm $|\cdot|$, $v(\cdot) = -\log|\cdot|$ is a real (semi)valuation. We will apply various definitions made for seminorms to semivaluations in this manner; for instance, if R is a ring and v is a real (semi)valuation on its additive group, we say that v is (sub)multiplicative if the corresponding nonarchimedean (semi)norm is.

Definition 1.2.2. We say a group is *nonarchimedean* if it is equipped with a nonarchimedean norm; we say a ring or field is *nonarchimedean* if it is equipped with a multiplicative nonarchimedean norm. Note that any nonarchimedean ring is an integral domain.

Definition 1.2.3. Let F be a nonarchimedean field. The *multiplicative value group* of a nonarchimedean field F is the image of $F^\times$ under $|\cdot|$, viewed as a subgroup of $\mathbb{R}^+$; we will often denote it simply as $|F^\times|$. The *additive value group* of F is the set of negative logarithms of the multiplicative value group. If these groups are discrete and nonzero (i.e., isomorphic to $\mathbb{Z}$), we say F is *discretely valued*. Define also

$$\mathfrak{o}_F = \{f \in F : v(f) \geq 0\}$$
$$\mathfrak{m}_F = \{f \in F : v(f) > 0\}$$
$$\kappa_F = \mathfrak{o}_F / \mathfrak{m}_F.$$

Note that $\mathfrak{o}_F$ is a local ring (the *valuation ring* of F), $\mathfrak{m}_F$ is the maximal ideal of $\mathfrak{o}_F$, and κ_F is a field (the *residue field* of F).

It is worth noting that there are comparatively few archimedean (i.e., not nonarchimedean) absolute values on fields.

Theorem 1.2.4 (Ostrowski). *Let F be a field equipped with a norm $|\cdot|$. Then $|\cdot|$ fails to be nonarchimedean if and only if the sequence $|1|, |2|, |3|, \ldots$ is unbounded. In that case, F is isomorphic to a subfield of $\mathbb{C}$ equipped with the restriction of the usual absolute value.*

Proof Exercise, or see [355, §2.1.6] and [82, Theorem 3.1.1], respectively. □

1.3 Norms on modules

When considering norms on modules, we usually require a compatibility with the underlying ring.

Definition 1.3.1. Let R be a ring equipped with a multiplicative seminorm $|\cdot|$, and let M be an R-module equipped with a seminorm $|\cdot|_M$. We say that $|\cdot|_M$ is *compatible with* $|\cdot|$ (or *with* R) if the following conditions hold.

(a) For $f \in R, x \in M$, $|fx|_M = |f||x|_M$.
(b) If $|\cdot|$ is nonarchimedean, then so is $|\cdot|_M$.

Note that (b) is not superfluous; see the exercises. If R is a field, then two norms

$|\cdot|_M, |\cdot|'_M$ on M compatible with R are metrically equivalent if and only if they are topologically equivalent (exercise).

One thing to be aware of is that if M' is a quotient of M and $|\cdot|_{M'}$ is the quotient norm on M' induced by $|\cdot|_M$, then in general we cannot say that $|\cdot|_{M'}$ is compatible with R. Rather, we only have the inequality

$$|fx'|_{M'} \le |f||x'|_{M'} \qquad (f \in R, x' \in M');$$

this implies compatibility only if R is a field.

We generate a rich supply of norms on modules via the following construction.

Definition 1.3.2. Let R be a ring equipped with a multiplicative (semi)norm $|\cdot|$, and let M be a finite free R-module. For B a basis of M, define the *supremum (semi)norm* of M with respect to B by setting

$$\left| \sum_{b \in B} c_b b \right| = \sup_{b \in B} \{|c_b|\} \qquad (c_b \in R).$$

This (semi)norm extends canonically to $M \otimes_R S$ for any isometric inclusion $R \hookrightarrow S$. (The situation is more complicated for arbitrary (semi)norms; see Definition 1.3.10 below.)

We say that a seminorm on M is *supremum-equivalent* if it is metrically equivalent to the supremum seminorm with respect to some basis; the same is then true of any basis by Lemma 1.3.3 below. In particular, if $|\cdot|$ is a norm, then any supremum-equivalent seminorm is a norm.

Lemma 1.3.3. *Let R be a ring equipped with a multiplicative seminorm $|\cdot|$, and let M be a finite free R-module. Then for any two bases B_1, B_2 of M, the supremum seminorms of M defined by B_1 and B_2 are metrically equivalent.*

Proof   Put $B_1 = \{m_{1,1}, \ldots, m_{1,n}\}$ and $B_2 = \{m_{2,1}, \ldots, m_{2,n}\}$. Define the $n \times n$ matrix A over R by the formula

$$m_{2,j} = \sum_{i=1}^{n} A_{ij} m_{1,i};$$

then A is invertible. In particular, we cannot have $|A_{ij}| = 0$ for all i, j.

For $x \in M$, we can uniquely write $x = a_{1,1}m_{1,1} + \cdots + a_{1,n}m_{1,n} = a_{2,1}m_{2,1} + \cdots + a_{2,n}m_{2,n}$ with $a_{i,j} \in R$. We then have

$$a_{1,i} = \sum_{j=1}^{n} A_{ij} a_{2,j} \qquad (i = 1, \ldots, n)$$

and so

$$\max_i\{|a_{1,i}|\} \leq \left(\sum_{i=1}^{n}\sum_{j=1}^{n}|A_{ij}|\right)\max_j\{|a_{2,j}|\}.$$

This inequality, together with the corresponding one with the bases reversed (involving the matrix A^{-1}), imply the claim. □

Corollary 1.3.4. *Let $R \hookrightarrow S$ be an isometric inclusion of rings equipped with multiplicative seminorms. Let M be a finite free R-module. Let $|\cdot|_M$ be a seminorm on M compatible with R, which is the restriction of a supremum-equivalent seminorm on $M \otimes_R S$ compatible with S. Then $|\cdot|_M$ is supremum-equivalent.*

Proof Put $N = M \otimes_R S$, and let $|\cdot|_N$ be a supremum-equivalent norm on N compatible with S, whose restriction to M equals $|\cdot|_M$. Pick any basis B of M; then B is also a basis of N. Since $|\cdot|_N$ is equivalent to the supremum norm on N defined by some basis, by Lemma 1.3.3 it is also equivalent to the supremum norm defined by B. By restriction, we see that $|\cdot|_M$ is equivalent to the supremum norm on M defined by B. □

The notion of supremum-equivalence behaves well under quotients.

Lemma 1.3.5. *Let R be a ring equipped with a multiplicative seminorm $|\cdot|$, let M be a finite free R-module, and let M_1 be a finite free R-submodule of M such that M/M_1 is also free. Let $|\cdot|_M$ be a supremum-equivalent norm on M compatible with R. Then the quotient norm $|\cdot|_{M_1}$ on M_1 induced by $|\cdot|_M$ is also supremum-equivalent.*

Proof Let $m_1, \ldots, m_k$ be a basis of M_1, and choose $m_{k+1}, \ldots, m_n \in M$ lifting a basis of M/M_1. Then $m_1, \ldots, m_n$ is a basis of M; by Lemma 1.3.3, $|\cdot|_M$ is equivalent to the supremum norm defined by $m_1, \ldots, m_n$. That is, there exist $c_1, c_2 > 0$ such that for any $x = a_1 m_1 + \cdots + a_n m_n \in M$,

$$c_1 \max\{|a_1|, \ldots, |a_n|\} \leq |x|_M \leq c_2 \max\{|a_1|, \ldots, |a_n|\}.$$

Then for any $a_{k+1}, \ldots, a_n \in R$,

$$
\begin{aligned}
c_1 \max\{|a_{k+1}|, \ldots, |a_n|\} &= c_1 \inf_{a_1, \ldots, a_k \in R}\{\max\{|a_1|, \ldots, |a_n|\}\} \\
&\leq \inf_{a_1, \ldots, a_k \in R}\{|a_1 m_1 + \cdots + a_n m_n|_M\} \\
&= |a_{k+1} m_{k+1} + \cdots + a_n m_n|_{M_1} \\
&\leq |a_{k+1} m_{k+1} + \cdots + a_n m_n|_M \\
&\leq c_2 \max\{|a_{k+1}|, \ldots, |a_n|\}.
\end{aligned}
$$

Thus $|\cdot|_{M_1}$ is equivalent to the supremum norm defined by the images of $m_{k+1}, \ldots, m_n$ in M_1, proving the desired result.   □

In general, even over a field, not every compatible norm on a vector space need be supremum-equivalent; see the exercises. However, one has such a statement for complete fields.

Theorem 1.3.6. *Let F be a field complete for a norm $|\cdot|$, and let V be a finite-dimensional vector space over F. Then any two norms on V compatible with F are metrically equivalent.*

Proof In the archimedean case, apply Theorem 1.2.4 to deduce that $F = \mathbb{R}$ or $F = \mathbb{C}$, then use compactness of the unit ball. In the nonarchimedean case, we proceed as follows. (See [149, Theorem I.3.2] for a variant proof.)

We proceed by induction on n, the case $n = 1$ being vacuous. Let $m_1, \ldots, m_n$ be any basis of V. It suffices to show that any given norm $|\cdot|$ on V compatible with F is equivalent to the supremum norm defined by $m_1, \ldots, m_n$. One inequality is evident: for any $a_1, \ldots, a_n \in F$,

$$|a_1 m_1 + \cdots + a_n m_n|_V \leq \max_i\{|m_i|\} \max_i\{|a_i|\}.$$

Put $V' = V/Fm_1$. Let $|\cdot|_{V'}$ denote the quotient seminorm on V' induced by $|\cdot|_V$. This seminorm is compatible with F, but we must check that it is indeed a norm. Suppose on the contrary that $a_2, \ldots, a_n \in F$ are such that $|a_2 m_2 + \cdots + a_n m_n|_{V'} = 0$. Then we can choose a sequence $a_{1,1}, a_{1,2}, \ldots$ of elements of F such that $|a_{1,i} m_1 + a_2 m_2 + \cdots + a_n m_n|_V \to 0$ as $i \to \infty$. But then $|a_{1,i} - a_{1,j}||m_1|_V = |(a_{1,i} - a_{1,j})m_1|_V \to 0$ as $i, j \to \infty$, so the $a_{1,i}$ form a Cauchy sequence. Since F is complete, this Cauchy sequence has a limit a_1, and $|a_1 m_1 + \cdots + a_n m_n|_V = 0$ contrary to the hypothesis that $|\cdot|_V$ is a norm.

Hence $|\cdot|_{V'}$ is indeed a norm. By the induction hypothesis, there exists $c_2 > 0$ such that

$$|a_2 m_2 + \cdots + a_n m_n|_{V'} \geq c_2 \max_i\{|a_i|\}.$$

By the definition of the quotient norm,

$$\begin{aligned}
|a_1 m_1 + \cdots + a_n m_n|_V &\geq |a_2 m_2 + \cdots + a_n m_n|_{V'} \\
&\geq c_2 \max\{|a_2|, \ldots, |a_n|\} \\
&\geq \frac{c_2}{\max\{|m_2|_V, \ldots, |m_n|_V\}} |a_2 m_2 + \cdots + a_n m_n|_V.
\end{aligned}$$

Put $c_1 = \min\{1, c_2/\max\{|m_2|_V, \ldots, |m_n|_V\}\}$; we then have

$$c_1^{-1}|a_1m_1 + \cdots + a_nm_n|_V = \max\{|a_1m_1 + \cdots + a_nm_n|_V,$$
$$|a_2m_2 + \cdots + a_nm_n|_V\}$$
$$\geq |a_1m_1|_V,$$

from which it follows that $|a_1m_1 + \cdots + a_nm_n|_V \geq c_1|a_1m_1|_V$. This proves that $|\cdot|_V$ is equivalent to the supremum norm defined by $m_1, \ldots, m_n$. □

Even if a norm is supremum-equivalent, it need not be equal to the supremum norm defined by any basis. However, one can approximate supremum-equivalent norms using supremum norms as follows. For a stronger result in the spherically complete case, see Lemma 1.5.5.

Lemma 1.3.7 (Approximation lemma). *Let F be a nonarchimedean field, let V be a finite-dimensional vector space over F, and let $|\cdot|_V$ be a supremum-equivalent norm on V compatible with F. Assume that either:*

(a) $c > 1$, and the value group of F is not discrete; or
(b) $c \geq 1$, and the value groups of F and V coincide and are discrete.

Then there exists a basis of V defining a supremum norm $|\cdot|'_V$ for which

$$c^{-1}|x|_V \leq |x|'_V \leq c|x|_V \qquad (x \in V).$$

Proof We induct on n, with vacuous base case $n = 0$. For $n > 0$, pick any nonzero $m_1 \in V$, and put $V_1 = V/Fm_1$. Using (a) or (b), we can rescale m_1 by an element of F to force $1 \leq |m_1|_V \leq c^{2/3}$.

Equip V_1 with the quotient seminorm $|\cdot|_{V_1}$ induced by $|\cdot|_V$. By Lemma 1.3.5, $|\cdot|_{V_1}$ is again supremum-equivalent. Moreover, in case (b), the infimum in (1.1.9.1) is always achieved, i.e., for every $x_1 \in V_1$, there exists $x \in V$ lifting x_1 with $|x|_V = |x_1|_{V_1}$. Hence V_1 again satisfies (b).

We may now apply the induction hypothesis to V_1, to produce a basis $m_{2,1}, \ldots, m_{n,1}$ of V_1 defining a supremum norm $|\cdot|'_{V_1}$ for which

$$c^{-1/3}|x_1|_{V_1} \leq |x_1|'_{V_1} \leq c^{1/3}|x_1|_{V_1} \qquad (x_1 \in V_1).$$

For $i = 2, \ldots, n$, choose $m_i \in V$ lifting $m_{i,1}$ such that $|m_i|_V \leq c^{1/3}|m_{i,1}|_{V_1}$; then

$$|m_i|_V \leq c^{1/3}|m_{i,1}|_{V_1} \leq c^{2/3}|m_{i,1}|'_{V_1} = c^{2/3}.$$

Let $|\cdot|'_V$ be the supremum norm defined by $m_1, \ldots, m_n$. For $x \in V$, write $x = a_1m_1 + \cdots + a_nm_n$ with $a_i \in F$. On one hand,

$$|x|_V \leq \max_{1 \leq i \leq n}\{|a_i||m_i|_V\} \leq c^{2/3}|x|'_V.$$

On the other hand, if x_1 is the image of x in V_1, then

$$|x_1|'_{V_1} \leq c^{1/3}|x_1|_{V_1} \leq c^{1/3}|x|_V,$$

so $|a_2|, \ldots, |a_n| \leq c^{1/3}|x|_V$. Moreover,

$$\begin{aligned}
|a_1m_1|_V &\leq \max\{|x|_V, |x - a_1m_1|_V\} \\
&\leq \max\{|x|_V, c^{2/3}|x - a_1m_1|'_V\} \\
&= \max\{|x|_V, c^{2/3}\max\{|a_2|, \ldots, |a_n|\}\} \\
&\leq c|x|_V.
\end{aligned}$$

Since $|m_1|_V \geq 1$, we deduce $|a_1| \leq c|x|_V$ and so $|x|'_V \leq c|x|_V$. This proves the desired inequalities. □

We need the following infinite-dimensional analogue of Theorem 1.3.6, taken from [363, Proposition 10.4]. Beware that the situation in the archimedean case is much subtler; see the chapter notes.

Lemma 1.3.8. *Let F be a complete nonarchimedean field. Let V be an F-vector space equipped with a norm $|\cdot|_V$ compatible with F. Suppose that V contains a dense F-subspace of countable infinite dimension over F. Then there exists a sequence $m_1, m_2, \ldots$ of elements of V with the following properties.*

(a) For each $m \in V$, there is a unique sequence $a_1, a_2, \ldots$ of elements of F such that the series $\sum_{i=1}^{\infty} a_i m_i$ converges to m.

(b) With notation as in (a), the function $|\cdot|'_V$ defined by

$$|m|'_V = \sup_i\{|a_im_i|_V\}$$

is a norm on V compatible with F and metrically equivalent to $|\cdot|_V$.

Proof Choose an ascending sequence of F-subspaces $0 = V_0 \subset V_1 \subset \cdots$, with $\dim_F V_n = n$, whose union is dense in V. For each $n > 0$, pick some $m_{n,0} \in V_n \setminus V_{n-1}$.

Let $|\cdot|_n$ be the quotient seminorm on V/V_n induced by $|\cdot|_V$. As in the proof of Theorem 1.3.6, we may show by induction on n that $|\cdot|_n$ is a norm, as follows. The claim for $n = 0$ is given. Supposing that $|\cdot|_{n-1}$ is a norm, let $m \in V$ be an element with $|m|_n = 0$. There must exist a sequence a_i of elements of F with $|a_im_{n,0}+m|_{n-1} \to 0$ as $i \to \infty$. Since $|m_{n,0}|_{n-1} \neq 0$ by the induction hypothesis, the a_i must form a Cauchy sequence in F whose limit a satisfies $|am_{n,0}+m|_{n-1} = 0$. Again by the induction hypothesis, m and $-am_{n,0}$ represent the same class in V/V_{n-1}, so m represents the zero class in V/V_n. Hence $|\cdot|_n$ is a norm.

Choose an increasing sequence of real numbers $0 < r_1 < r_2 < \cdots < 1$. Since $|\cdot|_{n-1}$ is a norm, $|m_{n,0}|_{n-1} \neq 0$. We can thus choose $m_n \in m_{n,0} + V_{n-1}$ with $m_1 = m_{1,0}$ and

$$|m_n|_{n-1} = |m_{n,0}|_{n-1} \geq \frac{r_n}{r_{n+1}} |m_n|_V \qquad (n > 1).$$

For $m \in V_{n-1}$ and $a \in F$, $|am_n + m|_V \geq |am_n|_{n-1} \geq (r_n/r_{n+1})|am_n|_V$. If $|am_n|_V = |m|_V$, this yields

$$|am_n + m|_V \geq \frac{r_n}{r_{n+1}} \max\{|am_n|_V, |m|_V\};$$

the same holds if $|am_n|_V \neq |m|_V$ since in that case we have $|am_n + m|_V = \max\{|am_n|_V, |m|_V\}$.

By induction on n, we deduce that for $a_1, \ldots, a_n \in F$,

$$|a_1 m_1 + \cdots + a_n m_n|_V \geq \frac{r_1}{r_{n+1}} \max\{|a_1 m_1|_V, \ldots, |a_n m_n|_V\}$$

$$\geq r_1 \max\{|a_1 m_1|_V, \ldots, |a_n m_n|_V\}.$$

Combining this with the evident inequality

$$|a_1 m_1 + \cdots + a_n m_n|_V \leq \max\{|a_1 m_1|_V, \ldots, |a_n m_n|_V\},$$

we conclude that on $\cup_n V_n$, the seminorms $|\cdot|_V$ and $|\cdot|'_V$ are metrically equivalent. Consequently, $|\cdot|'_V$ extends by continuity to a function on V which is metrically equivalent to $|\cdot|_V$, and hence is a norm (which is evidently compatible with F). This assertion will imply both (a) and (b) as soon as we establish the existence aspect of (a), which we do now.

Given $m \in V$, by hypothesis there exists a sequence $x_1, x_2, \ldots$ of elements of $\cup_n V_n$ converging to m. For $j = 1, 2, \ldots$, write $x_j = \sum_{i=1}^{\infty} a_{i,j} m_i$ with only finitely many $a_{i,j}$ nonzero. Since $\{x_j\}_{j=1}^{\infty}$ is a Cauchy sequence, for each $\epsilon > 0$ there exists N such that for $j, j' \geq N$, $|x_j - x_{j'}|_V \leq \epsilon$. Since $|\cdot|'_V$ is metrically equivalent to $|\cdot|_V$, for each $\epsilon > 0$ there also exists N such that for $j, j' \geq N$, $|x_j - x_{j'}|'_V \leq \epsilon$.

On one hand, this implies that for each fixed i, the sequence $\{a_{i,j}\}_{j=1}^{\infty}$ is Cauchy. Since F is complete, this sequence has a limit a_i. On the other hand, for $j = N$ there exists some i_0 such that $a_{i,j} = 0$ for all $i \geq i_0$. If we write $x_{i,j} = \sum_{h=i+1}^{\infty} a_{h,j} m_h$, then for all $j \geq N$ and all $i \geq i_0$, $|x_{i,j}|'_V \leq \epsilon$ and hence $|x_{i,j}|_V \leq \epsilon$. For fixed i, $x_{i,j}$ converges to $m - a_1 m_1 - \cdots - a_i m_i$ as $j \to \infty$; hence for all $i \geq i_0$, $|m - a_1 m_1 - \cdots - a_i m_i|_V \leq \epsilon$, and so the series $\sum_{i=1}^{\infty} a_i m_i$ converges to m. As noted earlier, both (a) and (b) now follow. □

Definition 1.3.9. For F a field complete for a norm $|\cdot|$, a *Banach space* over F is a vector space V over F equipped with a norm compatible with $|\cdot|$, under which it is complete. For V a Banach space and W a closed subspace, the quotient V/W is again complete. See [403] or [363] for a full development of the theory of Banach spaces and other topological vector spaces over complete nonarchimedean fields.

Definition 1.3.10. Let R be a nonarchimedean ring and let M, N be modules over R equipped with seminorms $|\cdot|_M, |\cdot|_N$ compatible with R. The *product seminorm* on $M \otimes_R N$ is defined by the formula

$$|x|_{M \otimes_R N} = \inf \left\{ \max_{1 \leq i \leq s} \{|m_i|_M |n_i|_N\} : x = \sum_{i=1}^{s} m_i \otimes n_i \right\}.$$

As in the case of the quotient seminorm, it is clear that the product seminorm is a seminorm, but it is not clear whether it is compatible with R unless R happens to be a field. Moreover, given norms $|\cdot|_M$ and $|\cdot|_N$, it is not clear whether the product seminorm is a norm. However, if M and N are finite free R-modules equipped with supremum-equivalent norms, then the product seminorm will be supremum-equivalent, which forces it to be a norm. See also the following lemma.

Lemma 1.3.11. *Let F be a complete nonarchimedean field. Let V and W be (possibly infinite-dimensional) vector spaces over F equipped with norms $|\cdot|_V$ and $|\cdot|_W$ compatible with F. Then the product seminorm on $V \otimes_F W$ is a norm.*

Proof Suppose first that V admits a dense F-subspace of at most countably infinite dimension. Then by Theorem 1.3.6 and/or Lemma 1.3.8, we can find a (finite or infinite) sequence $m_1, m_2, \ldots$ of elements of V such that every element of V can be uniquely written as a convergent series $\sum_{i=1}^{\infty} a_i m_i$ with $a_i \in F$, and $|\cdot|_V$ is equivalent to the norm $|\cdot|_V'$ defined by

$$\left| \sum_{i=1}^{\infty} a_i m_i \right|_V' = \sup_i \{|a_i m_i|_V\} \qquad (a_i \in F).$$

More precisely, $c|\cdot|_V' \leq |\cdot|_V \leq |\cdot|_V'$ for some $c > 0$. Let $\pi_j : V \to F$ be the projection carrying $\sum_{i=1}^{\infty} a_i m_i$ to a_j. By tensoring with W, we obtain a projection $\pi_{j,W} : V \otimes_F W \to W$. For $x \in V \otimes_F W$, we define

$$|x|_{V \otimes_F W}' = \sup_j \{|m_j|_V |\pi_{j,W}(x)|_W\}.$$

This gives a norm by the following argument. Let $\sum_{k=1}^{s} y_k \otimes z_k$ be any presentation of $x \in V \otimes_F W$ with $y_k \in V$ and $z_k \in W$, so that

$$\pi_{j,W}(x) = \sum_{k=1}^{s} \pi_j(y_k) z_k.$$

Suppose $|x|'_{V \otimes_F W} = 0$, then choose the presentation of x to minimize s. If $s > 0$, then $y_k \neq 0$ for all k and the z_k must be linearly independent over F. We can then choose j and k such that $\pi_j(y_k) \neq 0$, but then

$$0 = \pi_{j,W}(x) = \sum_{k=1}^{s} \pi_j(y_k) z_k,$$

a contradiction. Hence $s = 0$ and so $x = 0$.

For $x = \sum_{k=1}^{s} y_k \otimes z_k \in V \otimes_F W$ and any positive integer N, we can also express x as

$$m_1 \otimes \pi_{1,W}(x) + \cdots + m_N \otimes \pi_{N,W}(x)$$

$$+ \sum_{k=1}^{s} (y_k - \pi_1(y_k)m_1 - \cdots - \pi_N(y_k)m_N) \otimes z_k;$$

as $N \to \infty$, the product seminorm of the sum over k tends to zero. We thus conclude that

$$|x|_{V \otimes_F W} \leq |x|'_{V \otimes_F W}.$$

On the other hand,

$$\max_k \{ |y_k|_V |z_k|_W \} \geq c \sup_j \max_k \{ |\pi_j(y_k) m_j|_V |z_k|_W \}$$

$$\geq c \sup_j \left\{ |m_j|_V \left| \sum_{k=1}^{s} \pi_j(y_k) z_k \right|_W \right\}$$

$$= c \sup_j \{ |m_j|_V |\pi_{j,W}(x)|_W \},$$

and so

$$|x|_{V \otimes_F W} \geq c |x|'_{V \otimes_F W}.$$

That is, the product seminorm is equivalent to $| \cdot |'_{V \otimes_F W}$, and so is a norm.

In the general case, suppose to the contrary that $x = \sum_{j=1}^{s} m_j \otimes n_j \in V \otimes_F W$ has product seminorm 0. We can then find a sequence $x_i \in V \otimes_F W$ in which each x_i can be represented as $\sum_{j=1}^{s_i} m_{i,j} \otimes n_{i,j}$, and

$$\lim_{i \to \infty} \max_j \{ |m_{i,j}|_V |n_{i,j}|_W \} = 0.$$

Then the same data are available if we replace V by the closure of the F-subspace spanned by the m_j and the $m_{i,j}$, and similarly for W. We may thus apply the previous case to obtain a contradiction. $\qquad\square$

1.4 Examples of nonarchimedean norms

Example 1.4.1. For any field F, there is a *trivial norm* of F defined by

$$|f|_{\text{triv}} = \begin{cases} 1 & f \neq 0 \\ 0 & f = 0. \end{cases}$$

This norm is nonarchimedean, and F is complete under it. The case of the trivial norm will always be allowed unless explicitly excluded; it is often a useful input into a highly nontrivial construction, as in the next few examples.

Example 1.4.2. Let F be any field, and let $F((t))$ denote the field of formal Laurent series. The *t-adic valuation* v_t on F is defined as follows: for $f = \sum_i c_i t^i \in F((t))$, $v_t(f)$ is the least i for which $c_i \neq 0$. This exponentiates to give a *t-adic norm*, under which $F((t))$ is complete and discretely valued. (See Example 1.5.8 for a variation on this construction.)

Before introducing our next example, we make a more general definition for later use.

Definition 1.4.3. Let R be a ring equipped with a nonarchimedean submultiplicative (semi)norm $| \cdot |$. For $\rho \geq 0$, define the *ρ-Gauss (semi)norm* $| \cdot |_\rho$ on the polynomial ring $R[T]$ by

$$\left| \sum_i P_i T^i \right|_\rho = \max_i \{ |P_i| \rho^i \};$$

it is clearly submultiplicative. Moreover, it is also multiplicative if $| \cdot |$ is; however, we will postpone the verification of this to the next chapter (see Proposition 2.1.2). For $r \in \mathbb{R}$, we define the *r-Gauss (semi)valuation* v_r as the (semi)valuation associated to the e^{-r}-Gauss (semi)norm.

Remark 1.4.4. The definition of the ρ-Gauss norm depends on the choice of the indeterminate T; that is, it is not equivariant for arbitrary endomorphisms of the ring $R[T]$. For clarity, we will sometimes need to specify that the Gauss norm is being defined with respect to a particular indeterminate.

Example 1.4.5. For F a nonarchimedean field and $\rho > 0$, the *ρ-Gauss norm* on $F[t]$ (with respect to t) is a multiplicative norm, so it extends to the rational function field $F(t)$. Note that $F(t)$ is discretely valued under the ρ-Gauss norm if and only if either:

(a) F carries the trivial norm; or
(b) F is discretely valued, and ρ belongs to the divisible closure of the multiplicative value group of F.

In case (a), the ρ-Gauss norm is equivalent to the t-adic norm if $0 < \rho < 1$, the trivial norm if $\rho = 1$, and the t^{-1}-adic norm if $\rho > 1$.

So far we have not mentioned the principal examples from number theory; let us do so now.

Example 1.4.6. For p a prime number, the *p-adic norm* $|\cdot|_p$ on $\mathbb{Q}$ is defined as follows. Given $f = r/s$ with $r, s \in \mathbb{Z}$, write $r = p^a m$ and $s = p^b n$ with m, n not divisible by p, then put

$$|f|_p = p^{-a+b}.$$

In particular, we have normalized in such a way that $|p|_p = p^{-1}$; this convention has the effect of making the *product formula* hold. Namely, for any $f \in \mathbb{Q}$, if $|\cdot|_\infty$ denotes the usual archimedean absolute value, then

$$|f|_\infty \prod_p |f|_p = 1.$$

Completing $\mathbb{Q}$ under $|\cdot|_p$ gives the *field of p-adic numbers* $\mathbb{Q}_p$; it is discretely valued. Its valuation ring is denoted $\mathbb{Z}_p$ and called the *ring of p-adic integers*.

Remark 1.4.7. When converting the p-adic norm into a valuation, it is common to use base-p logarithms. We have instead opted to keep the factor of $\log p$ visible when we take logarithms. One may liken this practice to using metric units rather than normalizing some dimensioned constants to 1 (e.g., the speed of light).

Just as the only archimedean norm on $\mathbb{Q}$ is the usual one, every nontrivial nonarchimedean norm on $\mathbb{Q}$ is essentially a p-adic norm, again by a theorem of Ostrowski.

Theorem 1.4.8 (Ostrowski). *Any nontrivial nonarchimedean norm on $\mathbb{Q}$ is equivalent to the p-adic norm for some prime p.*

Proof See [355, §2.2.4]. □

To equip extensions of $\mathbb{Q}_p$ with norms, we use the following result. As usual, when E is a finite extension of a field F, we write $[E : F]$ for the degree of the field extension, i.e., the dimension of E as an F-vector space.

Theorem 1.4.9. *Let F be a complete nonarchimedean field. Then any finite extension E of F admits a unique extension of $|\cdot|$ to a norm on E (under which E is also complete).*

Proof We only prove uniqueness now; existence will be established in Section 2.3. Let $|\cdot|_1$ and $|\cdot|_2$ be two extensions of $|\cdot|$ to norms on E. Then these in particular give norms on E viewed as an F-vector space; by Theorem 1.3.6, these norms are metrically equivalent. That is, there exist $c_1, c_2 > 0$ such that

$$|x|_1 \leq c_1 |x|_2, \qquad |x|_2 \leq c_2 |x|_1 \qquad (x \in E).$$

They are also both metrically equivalent to the supremum norm for some basis of E over F, under which E is evidently complete.

We now use the extra information that $|\cdot|_1$ and $|\cdot|_2$ are multiplicative (because they are norms on E as a field in its own right). For any positive integer n, we may substitute x^n in place of x in the previous inequalities and then take n-th roots, to obtain

$$|x|_1 \leq c_1^{1/n} |x|_2, \qquad |x|_2 \leq c_2^{1/n} |x|_1 \qquad (x \in E).$$

Taking limits as $n \to \infty$ gives $|x|_1 = |x|_2$, as desired. $\qquad\square$

Remark 1.4.10. The completeness of F is crucial in Theorem 1.4.9. For instance, the 5-adic norm on $\mathbb{Q}$ extends in two different ways to the Gaussian rational numbers $\mathbb{Q}(i)$, depending on whether $|2 + i| = 5^{-1}$ and $|2 - i| = 1$, or vice versa.

Because of the uniqueness in Theorem 1.4.9, it also follows that any algebraic extension E of F, finite or not, inherits a unique extension of $|\cdot|$. However, if $[E : F] = \infty$, then E is not complete, so we may prefer to use its completion instead. For instance, if $F = \mathbb{Q}_p$, we define $\mathbb{C}_p$ to be the completion of an algebraic closure of $\mathbb{Q}_p$. One might worry that this may launch us into an endless cycle of completion and algebraic closure, but fortunately this does not occur.

Theorem 1.4.11. *Let F be an algebraically closed nonarchimedean field. Then the completion of F is also algebraically closed.*

For the proof, see Section 2.3.

1.5 Spherical completeness

For nonarchimedean fields, there is an important distinction between two different notions of completeness, which does not appear in the archimedean case.

Definition 1.5.1. A metric space is *complete* if any decreasing sequence of closed balls with radii tending to 0 has nonempty intersection. (For an abelian group equipped with a norm, this reproduces our earlier definition.) A metric space is *spherically complete* if any decreasing sequence of closed balls, regardless of radii, has nonempty intersection.

Example 1.5.2. The fields $\mathbb{R}$ and $\mathbb{C}$ with their usual absolute value are spherically complete. Any complete nonarchimedean field which is discretely valued, e.g., $\mathbb{Q}_p$ or $\mathbb{C}((t))$, is spherically complete. Any finite-dimensional vector space over a spherically complete nonarchimedean field equipped with a compatible norm is spherically complete (exercise); in particular, any finite extension of a spherically complete nonarchimedean field is again spherically complete. However, the completion of an infinite algebraic extension of $\mathbb{Q}_p$ is not spherically complete unless it is discretely valued; see the exercises.

Theorem 1.5.3 (Kaplansky–Krull). *Any nonarchimedean field embeds isometrically into a spherically complete, algebraically closed nonarchimedean field. (However, the construction is not functorial; see notes.)*

Proof Since completion is functorial, we may assume (by Theorem 1.4.11) we are starting with a complete, algebraically closed nonarchimedean field F. It was originally shown by Krull [283, Theorem 24] that F admits an extension which is algebraically closed and *maximally complete*, in the sense of not admitting any extensions preserving both the value group and the residue field. (In fact, this is not difficult to prove using Zorn's lemma.) The equivalence of the latter condition with spherical completeness was proved by Kaplansky [221, Theorem 4]. See also [403, p. 151].

One can also prove the result more directly; for instance, the case $F = \mathbb{Q}_p$ is explained in detail in [355, §3]. For the case $F = K((t))$, see Example 1.5.8. □

One benefit of the hypothesis of spherical completeness is that it can simplify the construction of quotient norms.

Lemma 1.5.4. *Let F be a spherically complete nonarchimedean field, let V be a finite-dimensional vector space over F, and let $|\cdot|_V$ be a norm on V compatible with F (which must be supremum-equivalent by Theorem 1.3.6). Let V' be a quotient of V, and let $|\cdot|_{V'}$ be the quotient norm on V' induced by $|\cdot|_V$. Then for every $x' \in V'$, there exists $x \in V$ lifting x' with $|x|_V = |x'|_{V'}$.*

Proof We first treat the case where $\dim_F(V') = \dim_F(V) - 1$. In this case, we can choose $m_1 \in V$ so that $V' = V/Fm_1$. Given $x' \in V'$, start with any lift x_0 of x' to V. Any other lift of x' to V can be written uniquely as $x_0 + am_1$ for some $a \in F$.

For $\epsilon > 0$, let B_ϵ be the set of $a \in F$ such that $|x_0 + am_1|_V \leq |x'|_{V'} + \epsilon$. By the definition of $|x'|_{V'}$, B_ϵ is nonempty. Pick any $a \in B_\epsilon$ and define

$$r(a, \epsilon) = \sup_{b \in B_\epsilon} \{|b - a|\}.$$

Then on one hand, B_ϵ is contained in the closed ball of radius $r(a, \epsilon)$ centered at a. On the other hand, for any $r < r(a, \epsilon)$ there exists $b \in B_\epsilon$ with $r \le |b - a|$, so

$$
\begin{aligned}
r|m_1|_V &\le |b - a||m_1|_V \\
&\le \max\{|x_0 + am_1|_V, |x_0 + bm_1|_V\} \\
&\le |x'|_{V'} + \epsilon.
\end{aligned}
$$

By taking limits, we deduce $r(a, \epsilon)|m_1|_V \le |x'|_{V'} + \epsilon$. Hence for any $c \in F$ with $|c - a| \le r(a, \epsilon)$,

$$
\begin{aligned}
|x_0 + cm_1|_V &\le \max\{|x_0 + am_1|_V, |(c - a)m_1|_V\} \\
&\le |x'|_{V'} + \epsilon
\end{aligned}
$$

and so $c \in B_\epsilon$.

We conclude that B_ϵ must equal the closed ball of radius $r(a, \epsilon)$ centered at a. As ϵ decreases, the B_ϵ form a decreasing family of closed balls in F. Since F is spherically complete, the intersection of the B_ϵ is nonempty. For any a in this intersection, $x = x_0 + am_1$ is a lift of x' to V satisfying $|x|_V = |x'|_{V_1}$.

Having completed the proof in case $\dim_F(V') = \dim_F(V) - 1$, we treat the general case by induction on $\dim_F(V) - \dim_F(V')$. There is nothing to check if the difference is 0, and the above argument applies if the difference is 1. Otherwise, we can choose a nontrivial quotient V'' of V which in turn has V' as a nontrivial quotient. Define the quotient norm $|\cdot|_{V''}$ on V'' induced by $|\cdot|_V$; then $|\cdot|_V$ and $|\cdot|_{V''}$ induce the same quotient norm $|\cdot|_{V'}$ on V'. To lift $x' \in V'$ to V while preserving the norm, it thus suffices to apply the induction hypothesis to first lift x' to $x'' \in V''$ while preserving the norm, then lift x'' to $x \in V$ while preserving the norm. $\square$

In the case of a spherically complete field, we obtain the following refinement of the approximation lemma (Lemma 1.3.7).

Lemma 1.5.5. *Let F be a spherically complete nonarchimedean field, let V be a finite-dimensional vector space over F, and let $|\cdot|_V$ be a norm on V compatible with F and having the same value group as F. Then $|\cdot|_V$ is the supremum norm defined by some basis of V.*

Proof Note that $|\cdot|_V$ is supremum-equivalent by Theorem 1.3.6. Define V_1 as in the proof of Lemma 1.3.7. By Lemma 1.5.4, any $x' \in V_1$ lifts to some $x \in V$

with $|x|_V = |x'|_{V_1}$. We may thus imitate the proof of case (b) of Lemma 1.3.7 to prove the desired result. □

We leave as an exercise the following alternate characterizations of spherical completeness.

Definition 1.5.6. Let X be a nonarchimedean metric space with distance function d. A sequence $\{x_n\}_{n=0}^{\infty}$ is *pseudoconvergent* if for some n, either $x_n = x_{n+1} = \cdots$ or $d(x_n, x_{n+1}) > d(x_{n+1}, x_{n+2}) > \cdots$. An element $x \in X$ is a *pseudolimit* if for all sufficiently large n, $d(x, x_n) \geq d(x_n, x_{n+1})$.

Proposition 1.5.7. *Let X be a nonarchimedean metric space. Then the following conditions are equivalent.*

(a) *The space X is spherically complete.*
(b) *For any sequence $B_1, B_2, \ldots$ of balls in X, if any two of the balls have nonempty intersection, then the intersection of all of the balls is nonempty.*
(c) *Every pseudoconvergent sequence in X has a pseudolimit.*

Proof Exercise. □

To conclude, we give one explicit example of a field which is spherically complete without being discretely valued. This example is used in the proof of the slope filtration theorem (Theorem 16.4.1).

Example 1.5.8. Let K be an arbitrary field. A *generalized power series*, or *Mal'cev–Neumann series*, over K is a formal sum $\sum_{i \in \mathbb{Q}} x_i t^i$ for which the set of $i \in \mathbb{Q}$ with $x_i \neq 0$ is a *well-ordered* subset of $\mathbb{Q}$ (i.e., it contains no infinite decreasing subsequence). Let $K((t^{\mathbb{Q}}))$ be the set of such series; we may equip $K((t^{\mathbb{Q}}))$ with a t-adic valuation by sending $\sum_{i \in \mathbb{Q}} x_i t^i$ to the least index i for which $x_i \neq 0$.

We see that $K((t^{\mathbb{Q}}))$ is a spherically complete abelian group as follows. Let $B_1, B_2, \ldots$ be a decreasing sequence of balls of radii $r_1, r_2, \ldots$. Put $r = \lim_{j \to \infty} r_j$; we may assume that $r \neq r_j$ for all j (otherwise the balls are all equal after some point and their intersection is clearly nonempty). For each $i < -\log r$, choose j with $i \leq -\log r_j$; then there is a single $x_i \in K$ which occurs as the coefficient of t^i for every element of B_j. The formal sum

$$x = \sum_{i < -\log r} x_i t^i$$

is then an element of $K((t^{\mathbb{Q}}))$ belonging to the intersection of all of the B_j.

It is somewhat less obvious that $K((t^{\mathbb{Q}}))$ is a field. Let $x = \sum_{i \in \mathbb{Q}} x_i t^i$ and

$y = \sum_{j \in \mathbb{Q}} y_j t^j$ be two elements of $K((t^{\mathbb{Q}}))$. We would like to define their product to be

$$xy = \sum_{k \in \mathbb{Q}} \left(\sum_{i,j \in \mathbb{Q}:\, i+j=k} x_i y_j \right) t^k.$$

To make this definition sensible, we must check two things:

(a) The set of $k \in \mathbb{Q}$ admitting at least one representation as $i + j$ where $x_i y_j \neq 0$ is well-ordered.
(b) For each $k \in \mathbb{Q}$, the number of representations of k as $i + j$ where $x_i y_j \neq 0$ is finite.

We leave both of these, plus the fact that the resulting ring $K((t^{\mathbb{Q}}))$ has multiplicative inverses, as an exercise. (Once the multiplication is known to be well-defined, such properties as associativity and distributivity over addition are fairly routine to check.)

Notes

The concept of a real valuation is a special case of Krull's notion of a *valuation* (sometimes called a *Krull valuation* for emphasis), in which the role of the real numbers is replaced by an arbitrary totally ordered group. For instance, on the polynomial ring $k[x, y]$, one can define a degree function taking monomials to elements of $\mathbb{Z} \times \mathbb{Z}$. If we then equip the latter with the lexicographic ordering (i.e., we compare pairs in their first component, only using the second component to break a tie in the first component), we may then define a valuation taking each polynomial to the lowest degree of any of its monomials.

The use of the terms *archimedean* and *nonarchimedean* to distinguish norms on abelian groups arises from the common attribution to Archimedes of the assertion (as an axiomatic property of real numbers) that given two positive real numbers x, y, there exists a positive integer n such that $nx > y$. While the oldest surviving reference stating this axiom is by Archimedes, the axiom is attributed therein to Eudoxus. See, e.g., [356] for further discussion.

While not all concepts from archimedean analysis generalize as nicely to Krull valuations as to real valuations, the Krull valuations are important in algebraic geometry; indeed, they were originally advocated by Zariski as a key component of rigorous foundations for the study of algebraic varieties. They were later ousted from this role by the more flexible theory of schemes, but they continue to play an important role in the study of birational properties of varieties and schemes, particularly in questions about resolution of singularities by blowups. See [347] for a full account of the theory of valuations, and [205] for discussion of Zariski's use of valuations in algebraic geometry. In addition, note that Krull valuations play an important role in nonarchimedean analytic geometry via Huber's theory of adic spaces; see the notes for Chapter 0.

Some prior references for Lemma 1.3.7 are [69, Proposition 2.6.2/3] and [305, §3, Lemme 2]. See also [170, Lemme 8].

A sequence $m_1, m_2, \ldots$ as in the statement of Lemma 1.3.8(a) is called a *Schauder basis* for V. For more discussion, see [69, §2.7.2]. Over a complete archimedean field (i.e., $\mathbb{R}$ or $\mathbb{C}$), the statement of Lemma 1.3.8 becomes false, as there are Banach spaces admitting countable dense subsets (i.e., which are *separable*) but not admitting Schauder bases. The first example is due to Enflo [156]. For a general discussion of the problem of trying to construct various sorts of bases of Banach spaces in the archimedean setting, see [292, Part I] or [79].

For a direct proof of Theorem 1.4.11 in the case of the completed algebraic closure of $\mathbb{Q}_p$, see [355, §3.3.3].

The fact that the completion of an infinite extension of $\mathbb{Q}_p$ fails to be spherically complete if it is not discretely valued (see Example 1.5.2 and the exercises) is a special case of a more general fact: any nonarchimedean metric space which admits a countable dense subset and whose metric takes values which are dense in $\mathbb{R}^+$ fails to be spherically complete [361, Theorem 20.5].

The condition of spherical completeness is quite important in nonarchimedean functional analysis, as it is needed for the Hahn–Banach theorem to hold. (By contrast, the nonarchimedean version of the open mapping theorem requires only completeness of the field.) For expansion of this remark, we recommend [363]; an older reference is [403].

Condition (c) of Proposition 1.5.7 is due to Ostrowski [331]; it is the definition used by Kaplansky in [221].

Example 1.5.8 is originally due to Hahn [189]. It was later generalized independently by Mal'cev and Neumann to the case where the ordered abelian group $\mathbb{Q}$ is replaced by a possibly nonabelian group. It was then used by Poonen [340] to describe the spherical completion of an arbitrary complete nonarchimedean field, even in the mixed-characteristic case. For instance, one obtains a description of a spherical completion of $\mathbb{C}_p$ in terms of what one might call "generalized power series in p".

Exercises

1.1 Prove Proposition 1.1.8. (Hint: first note that $|x| < 1$ if and only if the sequence $x, x^2, \ldots$ converges to 0.)

1.2 Prove Ostrowski's theorem (Theorem 1.2.4).

1.3 Give an example to show that, even for a finite-dimensional vector space V over a complete nonarchimedean field F, the requirement that a norm on $|\cdot|_V$ compatible with F must satisfy the strong triangle inequality is not superfluous; that is, condition (b) in Definition 1.3.1 is not a consequence of the rest of the definition. (Hint: use a modification of a supremum norm.)

1.4 Let M be a module over a nonarchimedean field F. Prove that any two norms on M compatible with F are topologically equivalent if and only if they are metrically equivalent.

1.5 Use Remark 1.4.10 to give an example showing that the statement of Theorem 1.3.6 may fail if F is not complete.

1.6 Prove that the valuation ring $\mathfrak{o}_F$ of a nonarchimedean field is noetherian if and only if F is trivially or discretely valued.

1.7 Use Theorem 1.4.9 to prove that for any field F, any nonarchimedean norm $|\cdot|$ on F, and any extension of E, there exists *at least one* extension of $|\cdot|$ to a norm on E. (Hint: reduce to the cases where E is a finite extension or a purely transcendental extension.)

1.8 Here is a more exotic variation of the t-adic valuation. Let F be a field, and choose $\alpha_1, \ldots, \alpha_n \in \mathbb{R}$.

    (a) Prove that on the rational function field $F(t_1, \ldots, t_n)$, there is a valuation v_α such that $v(f) = 0$ for all $f \in F^\times$ and $v(t_i) = \alpha_i$ for $i = 1, \ldots, n$. (Hint: you may construct it by iterating the definition of Gauss valuations.)

    (b) Prove that if $\alpha_1, \ldots, \alpha_n$ are linearly independent over $\mathbb{Q}$, the valuation v_α is uniquely determined by (a).

    (c) Prove that if $\alpha_1, \ldots, \alpha_n$ are not linearly independent over $\mathbb{Q}$, the valuation v_α is not uniquely determined by (a). (Hint: try the case $n = 2, \alpha_1 = \alpha_2 = 1$.)

1.9 Let E be the completion of an infinite algebraic extension of $\mathbb{Q}_p$ which is not discretely valued. Let $\alpha_1, \alpha_2, \ldots \in E$ be any sequence of elements such that $|\alpha_1|, |\alpha_2|, \ldots$ form a strictly decreasing sequence with positive limit not contained in $p^{\mathbb{Q}}$. (Since E is not discretely valued, such a sequence must exist.) Prove that the sequence of discs

$$\{z \in E : |z - \alpha_1 - \cdots - \alpha_i| \leq |\alpha_i|\}$$

is decreasing, but its intersection is empty. (Hint: note that if the intersection were nonempty, it would contain an element algebraic over $\mathbb{Q}_p$, since such elements are dense in E.) Deduce as a corollary that E is not spherically complete.

1.10 Prove that any finite-dimensional vector space over a spherically complete nonarchimedean field equipped with a compatible norm is again spherically complete. (Hint: use Lemma 1.5.5.)

1.11 Prove Proposition 1.5.7.

1.12 Prove that a subset S of $\mathbb{R}$ is well-ordered (contains no infinite decreasing sequence) if and only if every nonempty subset of S has a least element.

1.13 Check the unproved assertions at the end of Example 1.5.8. (Hint: reduce both (a) and (b) to the fact that, given any sequence of pairs $(i, j) \in \mathbb{Q}^2$ for which $i + j$ is nonincreasing, one can pass to a subsequence in which one of the components is nonincreasing. Reduce (c) to checking that $x \in K((t^{\mathbb{Q}}))$ has an inverse whenever $x - 1$ has positive t-adic valuation, then use a geometric series.)

2

Newton polygons

In this chapter, we recall the traditional theory of Newton polygons for polynomials over a nonarchimedean field. In the process, we introduce a general framework which will allow us to consider Newton polygons in a wider range of circumstances; this framework is based on a version of Hensel's lemma that applies in not necessarily commutative rings. As a first application, we fill in a few missing proofs from Chapter 1.

2.1 Newton polygons

We start with the possibly familiar notion of a Newton polygon associated to a polynomial over a nonarchimedean ring.

Definition 2.1.1. Let R be a ring equipped with a nonarchimedean submultiplicative (semi)norm $|\cdot|$. For $\rho > 0$ and $P = \sum_i P_i T^i \in R[T]$, define the *width* of P under the ρ-Gauss norm $|\cdot|_\rho$ as the difference between the maximum and minimum values of i achieving $\max_i\{|P_i|\rho^i\}$.

Proposition 2.1.2. *Let R be a ring equipped with a nonarchimedean multiplicative seminorm $|\cdot|$. For $\rho > 0$ and $P, Q \in R[T]$, the following results hold.*

(a) We have $|PQ|_\rho = |P|_\rho|Q|_\rho$. That is, $|\cdot|_\rho$ is multiplicative.

(b) The width of PQ under $|\cdot|_\rho$ is the sum of the widths of P and Q under $|\cdot|_\rho$.

Proof For $* \in \{P, Q\}$, let j_*, k_* be the minimum and maximum values of i for which $\max_i\{|*_i| \rho^i\}$ is achieved. Write

$$PQ = \sum_i (PQ)_i T^i = \sum_i \left(\sum_{g+h=i} P_g Q_h\right) T^i.$$

In the sum $(PQ)_i = \sum_{g+h=i} P_g Q_h$, each summand has norm at most $|P|_\rho |Q|_\rho \rho^{-i}$, with equality if and only if $|P_g| = |P|_\rho \rho^{-g}$ and $|Q_h| = |Q|_\rho \rho^{-h}$. This cannot occur for $i < j_P + j_Q$, and for $i = j_P + j_Q$ it can only occur for $g = j_P, h = j_Q$. Hence

$$|(PQ)_i| < |P|_\rho |Q|_\rho \rho^{-i} \qquad (i < j_P + j_Q)$$
$$|(PQ)_i| = |P|_\rho |Q|_\rho \rho^{-i} \qquad (i = j_P + j_Q)$$
$$|(PQ)_i| \le |P|_\rho |Q|_\rho \rho^{-i} \qquad (i > j_P + j_Q).$$

Similarly, we also have

$$|(PQ)_i| \le |P|_\rho |Q|_\rho \rho^{-i} \qquad (i < k_P + k_Q)$$
$$|(PQ)_i| = |P|_\rho |Q|_\rho \rho^{-i} \qquad (i = k_P + k_Q)$$
$$|(PQ)_i| < |P|_\rho |Q|_\rho \rho^{-i} \qquad (i > k_P + k_Q).$$

This proves both claims. $\qquad\qquad\qquad\qquad\qquad\qquad\qquad\qquad\qquad\qquad$ □

Definition 2.1.3. Let R be a ring equipped with a nonarchimedean submultiplicative seminorm $|\cdot|$. Let $v = -\log |\cdot|$ denote the corresponding valuation. Given a polynomial $P(T) = \sum_{i=0}^{n} P_i T^i \in R[T]$, draw the set of points

$$\{(-i, v(P_i)) : i = 0, \ldots, n; \ v(P_i) < \infty\} \subset \mathbb{R}^2,$$

then form the lower convex hull of these points. (That is, take the intersection of every closed half-plane which contains all of the points and lies above some nonvertical line.) The boundary of this region is called the *Newton polygon* of P. The *slopes* of P are the slopes of this open polygon, viewed as a multiset in which each slope r counts with multiplicity equal to the horizontal width of the segment of the Newton polygon of slope r (or equal to 0 if there is no such segment); the latter can also be interpreted as the width of P under $|\cdot|_{e^{-r}}$. (In case this multiset has cardinality less than $\deg(P)$, we include ∞ with sufficient multiplicity to make up the shortfall.)

Example 2.1.4. For $R = \mathbb{Q}_p$ equipped with the p-adic norm $|\cdot|_p$, the Newton polygon of the polynomial $T^3 + pT^2 + pT + p^3$ has vertices

$$(-3, 0), (-1, \log p), (0, 3 \log p).$$

Its slopes are $\frac{1}{2} \log p$ with multiplicity 2 and $2 \log p$ with multiplicity 1.

Proposition 2.1.5. *Let R be a nonarchimedean ring, and suppose that $P(T) = (T - \lambda_1) \cdots (T - \lambda_n)$. Then the slope multiset of P consists of*

$$-\log |\lambda_1|, \ldots, -\log |\lambda_n|.$$

Proof   This is immediate from the multiplicativity of $|\cdot|_{e^{-r}}$.   □

Here is an explicit example of Proposition 2.1.5 which we will use repeatedly.

Example 2.1.6. For p a prime and m a positive integer, let ζ_m be a primitive m-th root of unity in an algebraic closure of $\mathbb{Q}_p$. According to Theorem 1.4.9 (which we will finish proving later in this chapter), $\mathbb{Q}_p(\zeta_m)$ admits a unique extension of the p-adic norm $|\cdot|_p$ on $\mathbb{Q}_p$. Assuming this, let us calculate the norm of the element $1 - \zeta_m$.

The easiest case is when m is not divisible by p. In this case, $\zeta_m - 1$ is a root of the polynomial $((T+1)^m - 1)/T$, which has integer coefficients and constant term $m \not\equiv 0 \pmod{p}$. Hence the Newton polygon has all slopes equal to 0, so $|1 - \zeta_m|_p = 1$.

Suppose next that $m = p^h$ for some positive integer h. In this case ζ_{p^h} is a root of the polynomial

$$P(T) = \frac{T^{p^h} - 1}{T^{p^{h-1}} - 1} = \sum_{i=0}^{p-1} T^{i p^{h-1}},$$

so $\zeta_{p^h} - 1$ is a root of $Q(T) = P(T + 1)$. The image $\overline{Q}(T)$ of $Q(T)$ in $\mathbb{F}_p[T]$ satisfies

$$\overline{Q}(T) = \frac{(T+1)^{p^h} - 1}{(T+1)^{p^{h-1}} - 1} = \frac{T^{p^h} + 1^{p^h} - 1}{T^{p^{h-1}} + 1^{p^{h-1}} - 1} - T^{(p-1)p^{h-1}},$$

so all of the coefficients of Q except for its leading coefficient are divisible by p. Moreover $Q(0) = p$, so the Newton polygon of Q is a straight line segment with endpoints $(-(p-1)p^{h-1}, 0)$ and $(0, \log p)$. We conclude that

$$|1 - \zeta_{p^h}|_p = p^{-p^{-h+1}/(p-1)}. \tag{2.1.6.1}$$

(See the exercises for an alternate derivation of (2.1.6.1).)

Finally, suppose that $m = p^h j$ where j is not divisible by p. If $j = 1$, then $|1 - \zeta_m|_p = p^{-p^{-h+1}/(p-1)}$ by (2.1.6.1). Otherwise, we can write $\zeta_m = \zeta_{p^h} \zeta_j$ and

$$1 - \zeta_m = (1 - \zeta_{p^h}) + \zeta_{p^h}(1 - \zeta_j).$$

On the right side, the first term has norm less than 1 whereas the second term has norm equal to 1. We thus have $|1 - \zeta_m|_p = 1$.

Remark 2.1.7. The data of the Newton polygon is equivalent to the data of the ρ-Gauss norms for all ρ. This amounts to the statement that the graph of a convex continuous function is determined by the positions of its supporting lines of all slopes; this holds because the graph is the lower boundary of the intersection of the closed half-spaces bounded below by the supporting lines. For an example of the use of this observation, see the proof of Theorem 11.2.1.

2.2 Slope factorizations and a master factorization theorem

A key property of the operation of completing a nonarchimedean ring, first noticed by Hensel, is that it vastly enlarges the collection of polynomials over the ring which can be factored. Here is a sample statement of this form, which gives us a factorization that separates slopes in the Newton polygon.

Theorem 2.2.1. *Let F be a complete nonarchimedean ring. Suppose $S \in F[T]$, $r \in \mathbb{R}$, and $m \in \mathbb{Z}_{\geq 0}$ satisfy*

$$v_r(S - T^m) > v_r(T^m).$$

Then there exists a unique factorization $S = PQ$ satisfying the following conditions.

(a) *The polynomial $P \in F[T]$ has degree $\deg(S) - m$, and its slopes are all less than r.*

(b) *The polynomial $Q \in F[T]$ is monic of degree m, and its slopes are all greater than r.*

(c) *We have $v_r(P - 1) > 0$ and $v_r(Q - T^m) > v_r(T^m)$.*

Moreover, for this factorization,

$$\min\{v_r(P - 1), v_r(Q - T^m) - v_r(T^m)\} \geq v_r(S - T^m) - v_r(T^m). \quad (2.2.1.1)$$

(In fact, this inequality turns out to be an equality; see exercises.)

Let us translate this statement into plainer language. The hypothesis $v_r(S - T^m) > v_r(T^m)$ is equivalent to the following two conditions.

(a) The coefficient S_m of T^m in S satisfies $|S_m - 1| < 1$; in particular, S_m is a unit.

(b) The supporting line of slope r to the Newton polygon of S touches the polygon at the point $(-m, 0)$ and nowhere else.

In particular, r does not occur as a slope of the Newton polygon of S. Note that if (b) holds and S_m is a unit, we can apply the theorem to $S_m^{-1}S$ instead.

It is not so difficult to prove Theorem 2.2.1 directly. However, we will be stating a number of similar results as we go along, with similar proofs. To save some effort, we state a master factorization theorem applicable to not necessarily commutative rings, from which we can deduce Theorem 2.2.1 and all of the variants we will use later. The theorem, and the proof given here, are due to Christol [89, Proposition 1.5.1].

Theorem 2.2.2 (Master factorization theorem). *Let R be a not necessarily commutative ring equipped with a submultiplicative norm $|\cdot|$. Suppose the nonzero elements $a, b, c \in R$ and the additive subgroups $U, V, W \subseteq R$ satisfy the following conditions.*

(a) The spaces U, V are complete under the norm, and $UV \subseteq W$.
(b) The map $f(u, v) = av + ub$ is a surjection of $U \times V$ onto W.
(c) There exists $\lambda > 0$ such that

$$|f(u, v)| \geq \lambda \max\{|a||v|, |b||u|\} \qquad (u \in U, v \in V).$$

(d) We have $ab - c \in W$ and

$$|ab - c| \leq \lambda^2 |a||b|.$$

Then there exists a unique pair $(x, y) \in U \times V$ such that

$$c = (a + x)(b + y), \qquad |x| < \lambda|a|, \qquad |y| < \lambda|b|.$$

For this x, y, we also have

$$|x| \leq \lambda^{-1}|ab - c||b|^{-1}, \qquad |y| \leq \lambda^{-1}|ab - c||a|^{-1}.$$

Before proving Theorem 2.2.2, let us see how it implies Theorem 2.2.1.

Proof of Theorem 2.2.1   We apply Theorem 2.2.2 with the following parameters:

$$R = F[T]$$
$$|\cdot| = |\cdot|_{e^{-r}}$$
$$U = \{P \in F[T] : \deg(P) \leq \deg(S) - m\}$$
$$V = \{P \in F[T] : \deg(P) \leq m - 1\}$$
$$W = \{P \in F[T] : \deg(P) \leq \deg(S)\}$$
$$a = 1$$
$$b = T^m$$
$$c = S$$
$$\lambda = 1,$$

then put $P = a + x$ and $Q = b + y$.

To see that this works, let us verify explicitly that condition (c) is satisfied in this setup (the other conditions are more obvious). If $u \in U, v \in V$ are such that $\max\{|a||v|, |b||u|\}$ is achieved by some term of av, then that term appears unchanged in $av + ub$ because ub is divisible by T^m. Hence $|f(u, v)| \geq |av|$ in that case. Otherwise, $|av| < |bu|$ and so $|f(u, v)| = |bu|$. □

With this motivation in mind, we now prove Theorem 2.2.2.

Proof of Theorem 2.2.2 We define a norm on $U \times V$ by setting

$$|(u, v)| = \max\{|a||v|, |b||u|\}.$$

so that (c) implies

$$\lambda|(u, v)| \leq |f(u, v)| \leq |(u, v)|.$$

In particular, $\lambda \leq 1$, so $|ab - c| < |ab| = |c|$.

Since a, b are nonzero, (c) implies that f is injective. By (b), f is a bijective group homomorphism between $U \times V$ and W. It follows that for all $w \in W$,

$$|f^{-1}(w)| \leq \lambda^{-1}|w|.$$

By (d), we may choose $\mu \in (0, \lambda)$ with $|ab - c| \leq \lambda\mu|c|$. Define

$$B_\mu = \{(u, v) \in U \times V : |(u, v)| \leq \mu|a||b|\}.$$

For $(u, v) \in B_\mu$,

$$|a||v| \leq |(u, v)| \leq \mu|a||b|,$$

so $|v| \leq \mu|b|$. Similarly $|u| \leq \mu|a|$. As a result,

$$
\begin{aligned}
|f^{-1}(c - ab - uv)| &\leq \lambda^{-1}|c - ab - uv| \\
&\leq \lambda^{-1}\max\{|c - ab|, |uv|\} \\
&\leq \lambda^{-1}\max\{\lambda\mu|a||b|, \mu^2|a||b|\} \\
&= \mu|a||b|.
\end{aligned}
$$

Consequently, the map $g(u, v) = f^{-1}(c - ab - uv)$ carries B_μ into itself. We next show that g is contractive. For $(u, v), (t, s) \in B_\mu$,

$$
\begin{aligned}
|g(u, v) - g(t, s)| &\leq |f^{-1}(ts - uv)| \\
&\leq \lambda^{-1}|ts - uv| \\
&= \lambda^{-1}|t(s - v) + (t - u)v| \\
&\leq \lambda^{-1}\max\{\mu|a||s - v|, \mu|t - u||b|\} \\
&= \lambda^{-1}\mu|(u - t, v - s)| \\
&= \lambda^{-1}\mu|(u, v) - (t, s)|,
\end{aligned}
$$

which has the desired effect because $\lambda^{-1}\mu < 1$.

Since g is contractive on B_μ, and $U \times V$ is complete, by the Banach contraction mapping theorem (exercise) there is a unique $(x, y) \in U \times V$ fixed by g. That is,

$$ay + xb = f(x, y) = f(g(x, y)) = c - ab - xy$$

and so

$$c = (a + x)(b + y).$$

Moreover, there is a unique such (x, y) in the union of all of the B_μ, and that element belongs to the intersection of all of the B_μ. $\square$

Remark 2.2.3. One can also use Theorem 2.2.2 to recover other instances of Hensel's lemma. For instance, if F is a complete nonarchimedean field, $P(x) \in \mathfrak{o}_F[x]$, and the reduction of $P(x)$ into $\kappa_F[x]$ factors as $\overline{Q}\,\overline{R}$ with $\overline{Q}, \overline{R}$ coprime, then there exists a unique factorization $P = QR$ in $\mathfrak{o}_F[x]$ with Q, R lifting $\overline{Q}, \overline{R}$ (exercise).

2.3 Applications to nonarchimedean field theory

We now go back and apply Theorem 2.2.1 to prove some facts about extensions of nonarchimedean fields which were omitted from Chapter 1. We first complete the proof of Theorem 1.4.9; for this, we need the following lemma.

Lemma 2.3.1. *Let F be a complete nonarchimedean field. Let $P(T) \in F[T]$ be a polynomial whose slopes are all greater than or equal to r. Let $S(T) \in F[T]$ be any polynomial, and write $S = PQ + R$ with $\deg(R) < \deg(P)$ using the division algorithm. Then*

$$v_r(S) = \min\{v_r(P) + v_r(Q), v_r(R)\}.$$

Proof Exercise. $\square$

Proof of Theorem 1.4.9 (continued) It remains to show that if F is a complete nonarchimedean field, then any finite extension E of F admits an extension of $|\cdot|$ to a norm on E. See the exercises for Chapter 23 for an alternate approach.

For $\alpha \in E$, let $P(T)$ be the minimal polynomial of α over F, put $d = \deg(P)$, and define $|\alpha| = |P(0)|^{1/d}$. To check that this gives a multiplicative norm, choose an arbitrary $\beta \in E$ with minimal polynomial $Q(T)$ of degree f. The polynomials P and Q are irreducible, so by Theorem 2.2.1 their Newton polygons consist of single segments of some slopes r and s, respectively. Write $P(T) = \sum_i P_i T^i$ and $Q(T) = \sum_j Q_j T^j$; then $|P_i| \leq \exp(-r(d - i))$ and $|Q_j| \leq \exp(-s(f - j))$, with equality for $i = j = 0$.

Factor $P(T) = (T - \alpha_1) \cdots (T - \alpha_d)$ and $Q(T) = (T - \beta_1) \cdots (T - \beta_e)$ over some algebraic extension of F, and define

$$A(T) = \sum_k A_k T^k = \prod_{i=1}^{d} \prod_{j=1}^{e} (T - \alpha_i - \beta_j),$$

$$M(T) = \sum_k M_k T^k = \prod_{i=1}^{d} \prod_{j=1}^{e} (T - \alpha_i \beta_j).$$

Then A_k is an integer polynomial in the P_i and Q_j which is homogeneous of degree $df - k$ for the weighting giving degree $d - i$ to P_i and degree $f - j$ to Q_j. This implies that $|A_k| \leq \exp(-\min\{r, s\}(df - k))$, so the Newton polygon of A has no slopes less than $\min\{r, s\}$. By the multiplicativity of Newton polygons, the same holds for the minimal polynomial of $\alpha + \beta$, so $|\alpha + \beta| \leq \exp(-\min\{r, s\}) = \max\{|\alpha|, |\beta|\}$. Meanwhile, M_k is an integer polynomial in the P_i and Q_j which is homogeneous of bidegree $(df - k, df - k)$ for the weighting giving bidegree $(d - i, 0)$ to P_i and bidegree $(0, f - j)$ to Q_j. This implies that $|M_k| \leq \exp(-(r + s)(df - k))$ with equality for $k = 0$, so the Newton polygon of M has all slopes equal to $r + s$. By the multiplicativity of Newton polygons, the same holds for the minimal polynomial of $\alpha\beta$, so $|\alpha\beta| = \exp(-r - s) = |\alpha||\beta|$. $\square$

We next give the proof of Theorem 1.4.11. For this, we need a crude version of the principle that the roots of a polynomial over a complete algebraically closed nonarchimedean field should vary continuously in the coefficients.

Lemma 2.3.2. *Let F be an algebraically closed nonarchimedean field with completion E, and suppose $P \in E[T]$ is monic of degree d. Then for any $\epsilon > 0$, there exists $z \in F$ such that $|z| \leq |P(0)|^{1/d}$ and $|P(z)| < \epsilon$.*

Proof If $P(0) = 0$ we may pick $z = 0$, so assume $P(0) \neq 0$. Put $P = T^d + \sum_{i=0}^{d-1} P_i T^i$. For any $\delta > 0$, we can pick a polynomial $Q = T^d + \sum_{i=0}^{d-1} Q_i d^i \in F[T]$ with $|Q_i - P_i| < \delta$ for $i = 0, \ldots, d - 1$.

Now assume that $\delta < \min\{|P_0|, \epsilon, \epsilon/|P_0|\}$, so that $|Q_0| = |P_0|$. By Proposition 2.1.5, there exists $z \in F$ with $Q(z) = 0$ and $|z| \leq |Q_0|^{1/d} = |P_0|^{1/d}$. We now have

$$|P(z)| = |(P - Q)(z)| \leq \delta \max\{1, |z|\}^d \leq \delta \max\{1, |P(0)|\} < \epsilon,$$

as desired. $\square$

Proof of Theorem 1.4.11 We must check that the completion E of an algebraically closed nonarchimedean field F is itself algebraically closed. Let

$P(T) \in E[T]$ be a monic polynomial of degree d. Define a sequence of polynomials $P_0, P_1, \ldots$ as follows. Put $P_0 = P$. Given P_i, apply Lemma 2.3.2 to construct z_i with $|z_i| \leq |P_i(0)|^{1/d}$ and $|P_i(z_i)| < 2^{-i}$, then set $P_{i+1}(T) = P_i(T+z_i)$ so that $P_{i+1}(0) = P_i(z_i)$. If some P_i satisfies $P_i(0) = 0$, then $z_0 + \cdots + z_{i-1}$ is a root of P. Otherwise, we get an infinite sequence $z_0, z_1, \ldots$ such that $z_0 + z_1 + \cdots$ converges to a root of P. □

Notes

There is a good reason for Newton's name to be attached to the polygons considered in this chapter. He considered them in the context of finding a series expansion at the origin for a function $y = f(x)$ implicitly defined by a polynomial relation $P(x, y) = 0$ over $\mathbb{C}$. This was later reinterpreted by Puiseux, in terms of computing roots of polynomials over the subfield of $\mathbb{C}((x))$ consisting of series which represent meromorphic functions in a neighborhood of $x = 0$.

Remark 2.1.7 is perhaps best viewed within the broader context of duality for convex functions, which is a central theme in the study of linear optimization problems. One reference for the rigorous mathematical theory of convex functions is [357].

Although the formulation of Theorem 2.2.2 is due to Christol, the basic observation that Hensel's lemma does not rely on commutativity is rather old. One early instance is due to Zassenhaus [416].

The Banach contraction mapping theorem (also known as the Banach fixed point theorem), used in the proof of Theorem 2.2.2, is probably familiar to most readers except possibly in name. For instance, it appears in the most common proofs of the implicit function theorem, the inverse function theorem, and the fundamental theorem of ordinary differential equations.

The notion of a nonarchimedean metric space can be generalized to the case where the metric takes values in an arbitrary partially ordered set. One can then give a number of extensions of the Banach contraction mapping theorem, at least in the case of spherically complete metric spaces. For instance, Priess-Crampe has proved that if X is a spherically complete metric space with distance function d, and $f : X \to X$ is a function such that $d(f(x), f(y)) < d(x, y)$ for all $x, y \in X$ with $x \neq y$, then f has a fixed point. See [367] for an exposition of a number of results of this type.

Exercises

2.1 With notation as in Example 2.1.6, rederive (2.1.6.1) as follows. First note that for j coprime to p, $1 - \zeta_{p^h}$ and $1 - \zeta_{p^h}^j$ are multiples of each other in the ring $\mathbb{Z}[\zeta_{p^h}]$, so $|1 - \zeta_{p^h}|_p \geq |1 - \zeta_{p^h}^j|_p$ and vice versa. Then note that the product of $1 - \zeta_{p^h}^j$ over those $j \in \{0, \ldots, p^h - 1\}$ not divisible by p equals $Q(0) = p$.

2.2 Prove that equality holds in (2.2.1.1). (Hint: split $S - T^m = (P - 1)Q + (Q - T^m)$.)

2.3 Prove the claim in Remark 2.2.3. (Hint: since $\overline{Q}$ and $\overline{R}$ are coprime, we can choose $\overline{S}, \overline{T} \in \kappa_F[x]$ such that $\overline{QS} + \overline{RT} = 1$. We can also ensure that $\deg(\overline{S}) < \deg(\overline{R})$ and $\deg(\overline{T}) < \deg(\overline{Q})$. Use lifts of these to set up the conditions of Theorem 2.2.2.)

2.4 Prove Lemma 2.3.1. (Hint: separate P as a sum $P_1 + P_2$ in which $v_r(P)$ is achieved by the leading coefficient of P_1, while $v_r(P_2) > v_r(P)$.)

2.5 Prove the Banach contraction mapping theorem (used in the proof of Theorem 2.2.2). That is, suppose X is a nonempty complete metric space (not necessarily nonarchimedean) with distance function d, and let $g: X \to X$ be a map for which there exists $\mu \in (0, 1)$ such that $d(g(x), g(y)) \leq \mu d(x, y)$ for all $x, y \in X$. Prove that g has a unique fixed point. (Hint: show that for any $x \in X$, the sequence $x, g(x), g(g(x)), \ldots$ is Cauchy and its limit is the desired fixed point.)

3

Ramification theory

Recall (Theorem 1.4.9) that any finite extension of a complete nonarchimedean field carries a unique extension of the norm and is again complete. In this chapter, we study the relationship between a complete nonarchimedean field and its finite extensions; this relationship involves the residue fields, value groups, and Galois groups of the fields in question. We distinguish some important types of extensions, the *unramified* and *tamely ramified* extensions. See [347, Chapter 6] for a more thorough treatment.

We also briefly discuss the special case of discretely valued fields with perfect residue field, in which one can say much more. We introduce the standard ramification filtrations on the Galois groups of extensions of local fields; these will not reappear again until Part IV, at which point they will relate to the study of convergence of solutions of p-adic differential equations made in Part III. We make no attempt to be thorough in our treatment; instead, we direct the reader to the standard reference [369] for more details.

Notation 3.0.1. For E/F a Galois extension of fields, write $G_{E/F}$ for $\mathrm{Gal}(E/F)$. If E equals the separable closure F^{sep} of F, write G_F for the absolute Galois group $G_{F^{\mathrm{sep}}/F}$. Of course, if F has characteristic 0 (or is perfect, e.g., if F is a finite field), then F^{sep} coincides with the algebraic closure F^{alg}. (We avoid the usual notation $\overline{F}$ for the latter because we prefer to reserve the overbar to denote reduction modulo the maximal ideal of a local ring, e.g., from $\mathbb{Z}_p$ to $\mathbb{F}_p$.)

Remark 3.0.2. Throughout this chapter, when a nonarchimedean field F is assumed to be complete, it will be sufficient to assume it is only *henselian* instead. One of many equivalent formulations of this condition (for which see [319, 43.2]) is that for any monic polynomial $P(x) \in \mathfrak{o}_F[x]$ and any simple root $\overline{r} \in \kappa_F$ of $\overline{P} \in \kappa_F[x]$, there exists a unique root $r \in \mathfrak{o}_F$ of P lifting $\overline{r}$. (One may similarly define the henselian property for any local ring, not necessarily a valuation ring.)

3.1 Defect

Let F be a complete nonarchimedean field, and let E be a finite extension of F. Then the value group $|E^\times|$ contains $|F^\times|$, while the residue field κ_E may naturally be viewed as extension of κ_F. The first fundamental fact about the extension E/F is that both of these containments are finite in appropriate senses closely related to the degree of E over F.

Lemma 3.1.1. *Let F be a complete nonarchimedean field, and let E be a finite extension of F. Then*

$$[E : F] \geq [\kappa_E : \kappa_F][|E^\times| : |F^\times|], \tag{3.1.1.1}$$

with equality at least when F is discretely valued.

Proof Choose $\alpha_1, \ldots, \alpha_m \in \mathfrak{o}_E$ lifting a basis of κ_E over κ_F, and choose $\beta_1, \ldots, \beta_n \in E$ so that $|\beta_1|, \ldots, |\beta_n|$ form a set of coset representatives of $|F^\times|$ in $|E^\times|$. Then the $\alpha_i \beta_j$ are linearly independent over F, proving (3.1.1.1).

If F is discretely valued, then there exists a unique $\rho \in (0, 1)$ for which $|F^\times| = \rho^{\mathbb{Z}}$ and $|E^\times| = (\rho^{1/n})^{\mathbb{Z}}$. If we choose β_j to have norm ρ^{j-1}, then it is not hard to show that the $\alpha_i \beta_j$ form a basis of $\mathfrak{o}_E$ over $\mathfrak{o}_F$ (exercise). This proves the desired equality. □

If F is not discretely valued, the situation can be more complicated. One does however have the following refinement of (3.1.1.1), which we will not prove here.

Theorem 3.1.2 (Ostrowski). *Let F be a complete nonarchimedean field, and let E be a finite extension of F. Then the quantity*

$$\mathrm{defect}(E/F) = \frac{[E : F]}{[\kappa_E : \kappa_F][|E^\times| : |F^\times|]}$$

is a positive integer. Moreover, if F has characteristic 0 then $\mathrm{defect}(E/F) = 1$; otherwise $\mathrm{defect}(E/F)$ is a power of the characteristic of F.

Proof See [347, Theorem 6.2]. □

Definition 3.1.3. The quantity $\mathrm{defect}(E/F)$ in Theorem 3.1.2 is called the *defect* of E over F. (Note that some sources instead define the defect as $\log_p \mathrm{defect}(E/F)$, where p is the characteristic of F.) An extension for which defect(E/F) = 1 is said to be *defectless*. For example, any finite extension of a spherically complete field is defectless (see Remark 3.3.11). For an example of an extension with nontrivial defect, see [347, §6.3].

3.2 Unramified extensions

The easiest finite extensions of a complete nonarchimedean field to describe are the unramified extensions, which have a particularly simple effect on the residue field and the value group.

Definition 3.2.1. Let F be a complete nonarchimedean field. A finite extension E of F is *unramified* if κ_E is separable over κ_F and $[E : F] = [\kappa_E : \kappa_F]$. This forces E itself to be separable over F; see Proposition 3.2.3.

Lemma 3.2.2. *Let F be a complete nonarchimedean field, and let U be a finite extension of F. Then for any subextension E of U over F, U is unramified over F if and only if U is unramified over E and E is unramified over F.*

Proof Since κ_E sits between κ_U and κ_F, having κ_U separable over κ_F is equivalent to having both κ_U separable over κ_E and κ_E separable over κ_F. By Lemma 3.1.1,

$$[U : F] \geq [\kappa_U : \kappa_F], \qquad [U : E] \geq [\kappa_U : \kappa_E], \qquad [E : F] \geq [\kappa_E : \kappa_F];$$

since $[U : F] = [U : E][E : F]$ and $[\kappa_U : \kappa_F] = [\kappa_U : \kappa_E][\kappa_E : \kappa_F]$, the first of the three inequalities is an equality if and only if the other two are. □

What makes unramified extensions so simple to describe is that they are uniquely determined by their residue field extensions. Here is a precise statement to this effect.

Proposition 3.2.3. *Let F be a complete nonarchimedean field, and let E be a finite extension of F. Then for any separable subextension λ of κ_F over κ_E, there exists a unique unramified extension U of F contained in E with $\kappa_U \cong \lambda$; moreover, U is separable over F.*

Proof By the primitive element theorem, one can always write λ in the form $\kappa_F[x]/(\overline{P}(x))$ for some monic irreducible separable polynomial $\overline{P} \in \kappa_F[x]$. Lift $\overline{P}$ to a monic polynomial $P \in \mathfrak{o}_F[x]$. Choose $t \in \mathfrak{o}_E$ whose image in κ_E corresponds to x in $\kappa_F[x]/(\overline{P}(x))$; then the reduction of $P(x+t)$ into $\kappa_E[x]$ is divisible by x but not by x^2. We may thus apply the slope factorization theorem (Theorem 2.2.1) to deduce that $P(x+t)$ has a root in $\mathfrak{o}_E$. This proves existence and separability of U over F.

To prove uniqueness, let U' be another such extension. Then the previous argument applied to U' in place of E shows that $\mathfrak{o}_{U'}$ contains a root of $P(x+t)$ congruent to 0 modulo $\mathfrak{m}_{U'}$. However, there can only be one such root in U' because the Newton polygon of $P(x+t)$ has only one positive slope (since $\overline{P}$ is a separable polynomial), so in fact $U \subseteq U'$. Again by comparing degrees, $U = U'$. $\square$

Corollary 3.2.4. *For each finite separable extension λ of κ_F, there exists a unique unramified extension E of F with $\kappa_E \cong \lambda$.*

Proof Choose $P(x)$ as in the proof of Proposition 3.2.3. Then the field $E = F[x]/(P(x))$ is an unramified extension of F with residue field λ. The proof of Proposition 3.2.3 shows that any other unramified extension with residue field λ must contain E; by comparing degrees, we see that this containment must be an equality. $\square$

Lemma 3.2.5. *Let F be a complete nonarchimedean field, and let E be a finite extension of F. Let U_1, U_2 be unramified subextensions of E over F. Then the compositum $U = U_1 U_2$ is also unramified over F, and $\kappa_U = \kappa_{U_1} \kappa_{U_2}$ inside κ_E.*

Proof Put $U_3 = U_1 \cap U_2$ inside E; by Lemma 3.2.2, U_3 is unramified over F, and U_1 is unramified over U_3. By Proposition 3.2.3, inside κ_E, $\kappa_{U_1} \cap \kappa_{U_2} = \kappa_{U_3}$. Consequently,

$$
\begin{aligned}
[\kappa_U : \kappa_{U_2}] &\leq [U : U_2] \qquad \text{(by Lemma 3.1.1)} \\
&= [U_1 : U_3] \\
&= [\kappa_{U_1} : \kappa_{U_3}] \qquad \text{(because U_1 is unramified over U_3)} \\
&= [\kappa_{U_1} \kappa_{U_2} : \kappa_{U_2}] \\
&\leq [\kappa_U : \kappa_{U_2}] \qquad \text{(because $\kappa_{U_1} \kappa_{U_2} \subseteq \kappa_U$).}
\end{aligned}
$$

We deduce first that $\kappa_U = \kappa_{U_1} \kappa_{U_2}$, and second that $[U : U_2] = [\kappa_U : \kappa_{U_2}]$. Hence U is unramified over U_2, hence also over F by Lemma 3.2.2 again. $\square$

Definition 3.2.6. Let F be a complete nonarchimedean field, and let E be a finite extension of F. By Lemma 3.2.5, there is a maximal unramified subextension U of E over F; by Proposition 3.2.3, κ_U is the maximal separable subextension of κ_E over κ_F. (We will also say that $\mathfrak{o}_U$ is the "maximal unramified subextension" of $\mathfrak{o}_E$ over $\mathfrak{o}_F$.) We say E is *totally ramified* over F if $U = F$.

3.3 Tamely ramified extensions

For Galois extensions of complete nonarchimedean fields, we can refine the ramification theory introduced in the previous section. In the case of a discretely valued base field, this material may be familiar from [369]; we will continue the review of that case in Section 3.4.

Hypothesis 3.3.1. In this section, let F be a complete nonarchimedean field, and let E be a finite Galois extension of F.

Definition 3.3.2. The *inertia subgroup* $I_{E/F}$ of $G_{E/F}$ is the kernel of the map $G_{E/F} \to \mathrm{Aut}(\kappa_E)$; it may be interpreted as $G_{E/U}$ where U is the maximal unramified subextension of E over F. In particular, E is unramified over F if and only if $I_{E/F}$ is trivial, whereas E is totally ramified over F if and only if $I_{E/F} = G_{E/F}$.

Definition 3.3.3. Given $g \in I_{E/F}$ and $x \in E^\times$, let $\langle g, x \rangle$ denote the image of $g(x)/x$ in $\kappa_E^\times$. For fixed g, this is a homomorphism from $E^\times \to \kappa_E^\times$; moreover, it is trivial on $\mathfrak{o}_E^\times$ because $g \in I_{E/F}$. We thus obtain a homomorphism $I_{E/F} \to \mathrm{Hom}(|E^\times|, \kappa_E^\times)$; let $W_{E/F}$ denote the kernel of this map, called the *wild inertia subgroup* of $G_{E/F}$. Note that $\kappa_E^\times$ has no p-torsion, so neither does $\mathrm{Hom}(|E^\times|, \kappa_E^\times)$; hence $I_{E/F}/W_{E/F}$ is abelian of order not divisible by p.

Remark 3.3.4. Some readers may recall that if F is discretely valued, then $I_{E/F}/W_{E/F}$ is *cyclic* of order not divisible by p. This can be seen as follows. First, let μ denote the group of roots of unity in $\kappa_E^\times$ of order dividing $[E:F]$, and note that the image of $I_{E/F}/W_{E/F}$ in $\mathrm{Hom}(|E^\times|, \kappa_E^\times)$ includes only maps from $|E^\times|$ into μ. Second, note that μ is a finite cyclic group. Finally, note that $|F^\times| \cong \mathbb{Z}$ implies $|E^\times| \cong \mathbb{Z}$. (This last step does not apply if F is not discretely valued, and indeed $I_{E/F}/W_{E/F}$ need not be cyclic in general.)

Definition 3.3.5. We say E is *tamely ramified* over F if $W_{E/F}$ is trivial; in this case, the degree of E over its maximal unramified subextension is not divisible by p, and we call this the *tame degree* of E over F. Otherwise, we say E is *wildly ramified* over F; if $W_{E/F} = G_{E/F}$, we say E is *totally wildly ramified* over F. Note that if $p = 0$, then *every* finite extension of F is tamely ramified.

   The structure of tamely ramified extensions is almost as simple as that of unramified extensions, by an observation due at this level of generality to Abhyankar (although the special case of F discrete was known long previously).

Proposition 3.3.6 (Abhyankar). *Suppose E/F is tamely ramified (so that its inertia group $I_{E/F}$ is abelian of order not divisible by p). Let m be an integer not divisible by p annihilating $I_{E/F}$ (e.g., its order). Suppose $t_1, \ldots, t_h \in F^\times$*

have images in $|F^\times|$ which generate $|F^\times|/|F^\times|^m$. Then $E(t_1^{1/m}, \ldots, t_h^{1/m})$ is unramified over $F(t_1^{1/m}, \ldots, t_h^{1/m})$.

Proof Let m be a primitive m-th root of unity in an algebraic closure of F. Since m is not divisible by p, $F(\zeta_m)$ is unramified over F. It is thus harmless to assume $\zeta_m \in F$ (by Lemma 3.2.2). We may also assume E is totally ramified over F.

In this case, by Kummer theory, E is contained in an extension of F of the form $F(x_1^{1/m}, \ldots, x_n^{1/m})$ for some $x_1, \ldots, x_n \in F^\times$. To prove the claim, it suffices to check that $F(x_1^{1/m}, t_1^{1/m}, \ldots, t_h^{1/m})$ is unramified over $F(t_1^{1/m}, \ldots, t_h^{1/m})$. Since $|t_1|, \ldots, |t_h|$ generate $|F^\times|/|F^\times|^m$, we can choose integers $\ell_1, \ldots, \ell_h$ and an element $z \in F^\times$ such that $x_1 t_1^{\ell_1} \cdots t_h^{\ell_h} z^m \in \mathfrak{o}_F^\times$, and we can write

$$F(x_1^{1/m}, t_1^{1/m}, \ldots, t_h^{1/m}) = F(t_1^{1/m}, \ldots, t_h^{1/m})((x_1 t_1^{\ell_1} \cdots t_h^{\ell_h} z^m)^{1/m}).$$

It now suffices to check that if m is an integer not divisible by p and $y \in \mathfrak{o}_F^\times$, then $F(y^{1/m})$ is unramified over F. Again, it is safe to replace F by an unramified extension before checking this, so we may assume y reduces to an m-th power in κ_F. In this case, we will show that y already has an m-th root in F. Namely, we may now assume $y \equiv 1 \pmod{\mathfrak{m}_F}$; in this case, the binomial series

$$(1 + (y-1))^{1/m} = \sum_{i=0}^{\infty} \binom{1/m}{i} (y-1)^i$$

converges (since its coefficients are p-adically integral, given that m is not divisible by p) to an m-th root of y in F. □

Our next argument may be viewed as a preview of the filtration construction of the next section.

Proposition 3.3.7. *The group $W_{E/F}$ is a p-group.*

Proof We proceed by induction on $[E : F]$; we may assume that E is totally wildly ramified over F (so $W_{E/F} = G_{E/F}$), and that $E \neq F$. Let v_E denote the valuation on E. Pick any $x \in \mathfrak{o}_E \setminus \mathfrak{o}_F$, and set

$$j = \min\{v_E(1 - g(x)/x) : g \in G_{E/F}\}.$$

Note that $j < \infty$ because $x \notin F$, and $j > 0$ because E is totally wildly ramified over F.

The map $g \mapsto g(x)/x$ from $G_{E/F}$ to $\mathfrak{o}_E^\times$ is not a homomorphism; however, it does induce a nonzero homomorphism

$$G_{E/F} \to \frac{\{y \in \mathfrak{o}_E^\times : v_E(y-1) \geq j\}}{\{y \in \mathfrak{o}_E^\times : v_E(y-1) > j\}}.$$

Since $j > 0$, the group on the right is isomorphic to the additive group of κ_E, and so is a p-torsion group. Hence $G_{E/F}$ surjects onto a nontrivial p-group; let E' be the fixed field of the kernel of this surjection. It is clear that E is again totally wildly ramified over E', so $G_{E/E'}$ is also a p-group; hence $G_{E/F}$ is an extension of two p-groups and is thus a p-group itself.   □

The following are direct consequences of Proposition 3.3.6.

Lemma 3.3.8. *Let E' be a subextension of E over F. Then E is tamely ramified over F if and only if E is tamely ramified over E' and E' is tamely ramified over F.*

Lemma 3.3.9. *Let F be a complete nonarchimedean field, let E be a finite extension of F, and let T_1, T_2 be tamely ramified subextensions of E over F. Then $T = T_1 T_2$ is also tamely ramified over F.*

Remark 3.3.10. Let T be the maximal tamely ramified subextension of E over F, so that $G_{E/T} = W_{E/F}$ is a p-group by Proposition 3.3.7. A fact from elementary group theory (exercise) allows us to construct a tower $E_0 = T \subset E_1 \subset \cdots \subset E_m = E$ such that each E_i is Galois over T, and each group $G_{E_i/E_{i-1}}$ is isomorphic to $\mathbb{Z}/p\mathbb{Z}$. This is particularly important if F is of characteristic p, because every $\mathbb{Z}/p\mathbb{Z}$-extension of a field L of characteristic $p > 0$ is isomorphic to $L[z]/(z^p - z - x)$ for some $x \in L$. (Such an extension is called an *Artin–Schreier extension*.)

Remark 3.3.11. Using the result of Kaplansky mentioned in the proof of Theorem 1.5.3 (i.e., the fact that maximal completeness is equivalent to spherical completeness), it is possible to show that any finite extension E of a spherically complete field F is defectless. It suffices to check this first for E separable over F, and then for E purely inseparable over F (since any E can be obtained by first making a separable extension and then a purely inseparable extension on top of that). As we go along, keep in mind that any finite extension of F is also spherically complete (Example 1.5.2).

The key initial observation is that if $[E : F] = p$, then by Theorem 3.1.2, either E is defectless over F, or $[\kappa_E : \kappa_F] = [|E^\times| : |F^\times|] = 1$. If F is spherically complete, the latter case is ruled out by Kaplansky's theorem.

We now consider the separable case. Since it suffices to check defectlessness for an extension containing E (by Theorem 3.1.2), we may replace E with its Galois closure. Let T be the maximal tamely ramified subextension, so that T is defectless over F. By Remark 3.3.10, we can form E from T by a sequence of $\mathbb{Z}/p\mathbb{Z}$-extensions; by the previous paragraph, each of these is defectless. Hence E is defectless over F.

We next consider the purely inseparable case (which of course only applies if F has characteristic p). This case is even easier: we can obtain E from F by a sequence of extensions, each of which simply involves adjoining a p-th root. As above, each such extension is defectless.

3.4 The case of local fields

We now specialize the discussion of ramification theory to local fields, following [369, Chapter IV]. This material will not be used until Chapter 19, where we will relate ramification theory to convergence of solutions of p-adic differential equations.

Hypothesis 3.4.1. In this section, let F be a complete discretely valued nonarchimedean field whose residue field κ_F is perfect. (For more on what happens when the perfectness hypothesis is lifted, see the notes.) Let E be a finite Galois extension of F.

Definition 3.4.2. The *lower numbering filtration* of $G_{E/F}$ is defined as follows. For $i \geq -1$ an integer, put

$$G_{E/F,i} = \ker(G_{E/F} \to \mathrm{Aut}(\mathfrak{o}_E/\mathfrak{m}_E^{i+1})).$$

In particular,

$$G_{E/F,-1} = G_{E/F}$$
$$G_{E/F,0} = I_{E/F}$$
$$G_{E/F,1} = W_{E/F}.$$

For $i \geq -1$ real, we define $G_{E/F,i} = G_{E/F,\lceil i \rceil}$. The lower numbering filtration behaves nicely with respect to subgroups of $G_{E/F}$, but not with respect to quotients; it thus cannot be defined on the absolute Galois group G_F.

Definition 3.4.3. The *upper numbering filtration* of $G_{E/F}$ is defined by the relation $G_{E/F}^{\phi_{E/F}(i)} = G_{E/F,i}$, where

$$\phi_{E/F}(i) = \int_0^i \frac{1}{[G_{E/F,0} : G_{E/F,t}]}\, dt.$$

Note that the indices where the filtration jumps are now rational numbers, but not necessarily integers. In any case, Proposition 3.4.4 below implies that there is a unique filtration G_F^i on G_F which induces the upper numbering filtration on each $G_{E/F}$; that is, $G_{E/F}^i$ is the image of G_F^i under the surjection

$G_F \to G_{E/F}$. It is this filtration which plays the more important role, e.g., in class field theory.

Proposition 3.4.4 (Herbrand). *Let E' be a Galois subextension of E over F and put $H = G_{E/E'}$, so that H is normal in $G_{E/F}$ and $G_{E/F}/H = G_{E'/F}$. Then $G^i_{E'/F} = (G^i_{E/F}H)/H$; that is, the upper numbering filtration is compatible with forming quotients of $G_{E/F}$.*

Proof   The proof is elementary but slightly involved, so we will not give it here. See [369, § IV.3].   □

Notes

Ramification theory originally emerged in the study of algebraic number fields; the formalism we see nowadays is due largely to Hilbert. The upper and lower numbering filtrations were originally introduced by Hilbert as part of class field theory (the study of abelian extensions of number fields and their Galois groups), but not in the form we see today; the modern definitions were introduced slightly later by Herbrand. See [369] for more discussion.

Ramification theory for a complete discrete nonarchimedean field becomes substantially more complicated when one drops the requirement of a perfect residue field. However, the case of an imperfect residue field is of great interest in the study of finite covers of schemes of dimension greater than 1. A satisfactory theory for abelian extensions was introduced by Kato [222]. A generalization to nonabelian extensions was later introduced by Abbes and Saito [1, 2]. However, a number of alternate approaches exist, the relationships among which are not fully understood. These include Borger's theory of residual perfection [67], Kedlaya's differential Swan conductor [238], and methods from higher local class field theory [417, 418]. See also the notes for Chapter 19.

A henselian valued field is called *stable* if every finite extension of it is defectless. A deep theorem of Kuhlmann [282, Theorem 1] (generalizing earlier theorems of Grauert–Remmert and Gruson) states that if F is a stable henselian field, and E is a henselian extension of F of finite transcendence degree with transcendence defect equal to 0, then E is again stable. In this statement, the *transcendence defect* of E/F is defined as

$$\mathrm{trdeg}(E/F) - \dim_{\mathbb{Q}}(|E^\times|^{\mathbb{Q}}/|F^\times|^{\mathbb{Q}}) - \mathrm{trdeg}(\kappa_E/\kappa_F),$$

where trdeg denotes transcendence degree. A theorem of Abhyankar [404, Théorème 9.2] implies that the transcendence defect is always nonnegative; moreover, when the transcendence defect is 0, the group $|E^\times|/|F^\times|$ and the extension κ_E/κ_F are both finite. By contrast, both of these can fail when the transcendence degree is positive.

Exercises

3.1 Complete the proof of Lemma 3.1.1. (Hint: construct a sequence of elements in the span of the purported basis, converging to a given element of $\mathfrak{o}_E$.)

3.2 Let G be a group of order p^n. Prove that there exists a chain of subgroups $G_0 = \{e\} \subset G_1 \subset \cdots \subset G_n = G$ such that each inclusion is proper and each G_i is normal in G. Deduce that for any finite Galois extension L/K of fields of characteristic p of p-power degree, there exists a sequence $K = K_0 \subset K_1 \subset \cdots \subset K_n = L$ of subextensions such that each inclusion is an extension of degree p and each K_i is Galois over K. (A somewhat easier exercise is to produce such a sequence in which each K_i is only Galois over K_{i-1}.)

3.3 Within an algebraic closure of $\mathbb{Q}_p$, let π be a root of the polynomial $\pi^{p-1} = -p$ and let ζ be a primitive p-th root of unity. Prove that the fields $\mathbb{Q}_p(\pi)$ and $\mathbb{Q}_p(\zeta)$ are isomorphic. (Hint: both fields are tamely ramified over $\mathbb{Q}_p$, so Proposition 3.3.6 applies.)

3.4 Let k be a perfect field of characteristic p and let F be the local field $k((t))$. Let $z \in k((t))$ be an element of t-adic valuation $-m$ where m is positive and not divisible by p. Let E be the Artin–Schreier extension $F[u]/(u^p - u - z)$. Prove that

$$G_{E/F,i} = \begin{cases} \mathbb{Z}/p\mathbb{Z} & i \le m \\ \{e\} & i > m. \end{cases}$$

(Hint: recall that $G_{E/F}$ is generated by the map $u \mapsto u + 1$.)

4

Matrix analysis

We come now to the subject of metric properties of matrices over a field complete for a specified norm. While this topic is central to our study of differential modules over nonarchimedean fields, it is based on ideas which have their origins largely outside of number theory. We have thus opted to first present the main points in the archimedean setting, then repeat the presentation for nonarchimedean fields.

The main theme is the relationship between the norms of the eigenvalues of a matrix, which are core invariants but depend on the entries of the matrix in a somewhat complicated fashion, and some less structured but more readily visible invariants. The latter are the *singular values* of a matrix, which play a key role in numerical linear algebra in controlling numerical stability of certain matrix operations (including the extraction of eigenvalues). Their role in our work is similar.

Before proceeding, we set some basic notation and terminology for matrices.

Notation 4.0.1. Let $\mathrm{Diag}(\sigma_1, \ldots, \sigma_n)$ denote the $n \times n$ diagonal matrix D with $D_{ii} = \sigma_i$ for $i = 1, \ldots, n$.

Notation 4.0.2. For A a matrix, let A^T denote the transpose of A. For A an invertible square matrix, let A^{-T} denote the inverse transpose of A.

Definition 4.0.3. An $n \times n$ *elementary matrix* over a ring R is an $n \times n$ matrix obtained from the identity matrix by performing one of the following operations:

(a) exchanging two rows;
(b) adding c times one row to another row, for some $c \in R$;
(c) multiplying one row by some $c \in R^\times$.

If B is an elementary matrix, multiplying another $n \times n$ matrix C on the left by B effects the corresponding operation on C; such an operation is called an *elementary row operation*. Multiplying on the right by B instead effects an *elementary column operation*. (It is possible to omit type (a); see the exercises.)

4.1 Singular values and eigenvalues (archimedean case)

Hypothesis 4.1.1. In this section and the next, let A be an $n \times n$ matrix over $\mathbb{C}$. Identify $\mathbb{C}^n$ with the space of column vectors equipped with the L^2 norm, i.e.,

$$|(z_1, \ldots, z_n)| = (|z_1|^2 + \cdots + |z_n|^2)^{1/2}.$$

View A as a linear transformation on $\mathbb{C}^n$, and write

$$|A| = \sup_{v \in \mathbb{C}^n \setminus \{0\}} \{|Av|/|v|\};$$

that is, $|A|$ is the *operator norm* of A as will be defined in Definition 6.1.2.

We are interested in two sets of numerical invariants of A. One of these is the familiar set of eigenvalues.

Definition 4.1.2. Let $\lambda_1, \ldots, \lambda_n$ be the list of eigenvalues of A, ordered so that $|\lambda_1| \geq \cdots \geq |\lambda_n|$.

A second set of numerical invariants of A, which is better behaved from the point of view of numerical analysis, is the set of singular values.

Definition 4.1.3. Let A^* denote the conjugate transpose (or *Hermitian transpose*) of A. The matrix A^*A is Hermitian and nonnegative definite, and so has nonnegative real eigenvalues. The (nonnegative) square roots of these eigenvalues comprise the *singular values* of A; we denote them by $\sigma_1, \ldots, \sigma_n$ with $\sigma_1 \geq \cdots \geq \sigma_n$. These are not invariant under conjugation, but they are invariant under multiplying A on either side by a unitary matrix.

Theorem 4.1.4 (Singular value decomposition). *There exist unitary $n \times n$ matrices U, V such that $UAV = \mathrm{Diag}(\sigma_1, \ldots, \sigma_n)$.*

Proof This is equivalent to showing that there is an orthonormal basis of $\mathbb{C}^n$ which remains orthogonal upon applying A. To construct it, start with a vector $v \in \mathbb{C}^n$ maximizing $|Av|/|v|$, then show that for any $w \in \mathbb{C}^n$ orthogonal to v, Aw is also orthogonal to Av. For further details, see the chapter notes for references. □

Corollary 4.1.5. *The singular values of A^{-1} are $\sigma_n^{-1}, \ldots, \sigma_1^{-1}$.*

From the singular value decomposition, we may infer a convenient interpretation of σ_i.

Corollary 4.1.6. *The number σ_i is the smallest value of λ for which the following holds: for any i-dimensional subspace V of $\mathbb{C}^n$, there exists a nonzero element $v \in V$ such that $|Av| \leq \lambda|v|$.*

Proof Theorem 4.1.4 provides an orthonormal basis $v_1, \ldots, v_n$ of V such that $Av_1, \ldots, Av_n$ is again orthogonal, and $|Av_i| = \sigma_i|v_i|$ for $i = 1, \ldots, n$. Let W be the span of $v_i, \ldots, v_n$. On one hand, for any i-dimensional subspace V of $\mathbb{C}^n$, $V \cap W$ is nonzero and any nonzero $v \in V \cap W$ satisfies $|Av| \leq \sigma_i|v|$. On the other hand, if we take V to be the span of $v_1, \ldots, v_i$, then $|Av| \geq \sigma_i|v|$ for all $v \in V$. This proves the claim. $\qquad\square$

The relationship between the singular values and the eigenvalues is controlled by the following inequality of Weyl [409].

Theorem 4.1.7 (Weyl). *We have*

$$\sigma_1 \cdots \sigma_i \geq |\lambda_1 \cdots \lambda_i| \qquad (i = 1, \ldots, n),$$

with equality for $i = n$.

For the proof, we will need a construction that will recur frequently in what follows.

Definition 4.1.8. Let M be a module over a ring R. The i-th *exterior power* or *wedge power* $\wedge_R^i M$ (or $\wedge^i M$ if there is no ambiguity about R) of M is the R-module generated by the symbols $m_1 \wedge \cdots \wedge m_i$ for $m_1, \ldots, m_i \in M$, modulo those relations that force the map $(m_1, \ldots, m_i) \mapsto m_1 \wedge \cdots \wedge m_i$ to be R-linear in each variable (while the others are held fixed) and alternating. The latter means that $m_1 \wedge \cdots \wedge m_i = 0$ if $m_1, \ldots, m_i$ are not all distinct.

Any element of $\wedge^i M$ of the form $m_1 \wedge \cdots \wedge m_i$ is said to be *decomposable*; it is also called the *exterior product* of $m_1, \ldots, m_i$. The set of decomposable elements is not closed under addition, and consequently may not fill out $\wedge^i M$ (see exercises). However, if M is freely generated by $e_1, \ldots, e_n$, then the decomposable elements of the form $e_{j_1} \wedge \cdots \wedge e_{j_i}$ with $1 \leq j_1 < \cdots < j_i \leq n$ do form a basis of $\wedge^i M$.

Note also that the exterior power is a functor on the category of R-modules; in particular, any linear transformation $T \colon M \to N$ induces a linear transformation $\wedge^i T \colon \wedge^i M \to \wedge^i N$. In the special case where $M = N$ is free of rank i, $\wedge^i M$ is a one-dimensional space, and $\wedge^i T$ turns out to be multiplication by $\det(T)$.

One checks that the formula

$$(m_1 \wedge \cdots \wedge m_i, m_1' \wedge \cdots \wedge m_j') \mapsto m_1 \wedge \cdots \wedge m_i \wedge m_1' \wedge \cdots \wedge m_j'$$

gives a well-defined bilinear map $\wedge^i M \times \wedge^j M \to \wedge^{i+j} M$. This map is usually denoted by $\wedge$, since it corresponds to adding an extra wedge between an i-fold and a j-fold wedge product.

Lemma 4.1.9. *The singular values (resp. eigenvalues) of $\wedge^i A$ are the i-fold products of the singular values (resp. eigenvalues) of A.*

Proof Starting with any basis of $\mathbb{C}^n$, we can obtain a basis of $\wedge^i A$ by taking i-fold exterior products of basis elements. In particular, if we choose $U \in \mathrm{GL}_n(\mathbb{C})$ so that $U^{-1}AU$ is upper triangular with the eigenvalues of A on the diagonal, then

$$\wedge^i(U^{-1}AU) = (\wedge^i U)^{-1}(\wedge^i A)(\wedge^i U)$$

will be upper triangular with the i-fold products of the eigenvalues of A on the diagonal. Similarly, if we apply Theorem 4.1.4 to construct unitary matrices U, V such that UAV is diagonal with the singular values of A on the diagonal, then

$$\wedge^i(UAV) = (\wedge^i U)(\wedge^i A)(\wedge^i V)$$

will be diagonal with the i-fold products of the singular values of A on the diagonal.   □

We now return to Theorem 4.1.7.

Proof of Theorem 4.1.7 The equality for $i = n$ holds because $\det(A^*A) = |\det(A)|^2$. We check the inequality first for $i = 1$. Note that $\sigma_1 = |A|$ is the operator norm. Since there exists $v \in \mathbb{C}^n \setminus \{0\}$ with $Av = \lambda_1 v$, we deduce that $\sigma_1 \geq |\lambda_1|$.

To handle the general case, we consider the exterior power $\wedge^i \mathbb{C}^n$ with the action of $\wedge^i A$. By Lemma 4.1.9, the largest singular value (resp. eigenvalue) of $\wedge^i A$ is equal to the product of the i largest singular values (resp. eigenvalues) of A. Consequently, the previous inequality applied to $\wedge^i A$ gives exactly the desired result.   □

We mention in passing the following converse of Theorem 4.1.7, due to Horn [207, Theorem 4].

Theorem 4.1.10. *For $\lambda_1, \ldots, \lambda_n \in \mathbb{C}$ and $\sigma_1, \ldots, \sigma_n \in \mathbb{R}_{\geq 0}$ satisfying*

$$\sigma_1 \cdots \sigma_i \geq |\lambda_1 \cdots \lambda_i| \qquad (i = 1, \ldots, n)$$

with equality for $i = n$, there exists an $n \times n$ matrix A over $\mathbb{C}$ with singular values $\sigma_1, \ldots, \sigma_n$ and eigenvalues $\lambda_1, \ldots, \lambda_n$.

Equality in Weyl's theorem at an intermediate stage has a structural meaning.

Lemma 4.1.11. *Let F be any field. If $v_1, \ldots, v_i, w_1, \ldots, w_i \in F^n$ are such that $v_1 \wedge \cdots \wedge v_i$ and $w_1 \wedge \cdots \wedge w_i$ are nonzero and equal, then the F-span of $v_1, \ldots, v_i$ equals the F-span of $w_1, \ldots, w_i$.*

Proof If $v_1 \wedge \cdots \wedge v_i$ is nonzero, then $v_1, \ldots, v_i$ must be linearly independent. We may thus extend $v_1, \ldots, v_i$ to a basis $v_1, \ldots, v_n$ of F^n.

Consider the bilinear map

$$\wedge : \wedge^i F^n \times F^n \to \wedge^{i+1} F^n.$$

Suppose $w \in F^n$ pairs to zero with $v_1 \wedge \cdots \wedge v_i$. If we write $w = c_1 v_1 + \cdots + c_n v_n$, we must then have $c_{i+1} = 0$ or else the coefficient of $v_1 \wedge \cdots \wedge v_{i+1}$ in $v_1 \wedge \cdots \wedge v_i \wedge w$ will be nonzero. Similarly, $c_{i+2} = \cdots = c_n = 0$, so w belongs to the F-span of $v_1, \ldots, v_i$.

Since $(v_1 \wedge \cdots \wedge v_i) \wedge w_1 = (w_1 \wedge \cdots \wedge w_i) \wedge w_1 = 0$, w_1 must belong to the F-span of $v_1, \ldots, v_i$, and likewise for $w_2, \ldots, w_i$. Consequently, one of the two spans is contained in the other, and vice versa by the same argument. $\square$

Theorem 4.1.12. *Suppose that for some $i \in \{1, \ldots, n-1\}$,*

$$\sigma_i > \sigma_{i+1}, \qquad |\lambda_i| > |\lambda_{i+1}|,$$
$$\sigma_1 \cdots \sigma_i = |\lambda_1 \cdots \lambda_i|.$$

Then there exists a unitary matrix U such that $U^{-1} A U$ is block diagonal, with the first block accounting for the first i singular values and eigenvalues and the second block accounting for the others.

Proof Let $v_1, \ldots, v_n$ be a basis of $\mathbb{C}^n$ such that $v_1, \ldots, v_i$ span the generalized eigenspaces with eigenvalues $\lambda_1, \ldots, \lambda_i$, and $v_{i+1}, \ldots, v_n$ span the generalized eigenspaces with eigenvalues $\lambda_{i+1}, \ldots, \lambda_n$. Apply the singular value decomposition (Theorem 4.1.4) to construct an orthonormal basis $w_1, \ldots, w_n$ such that $Aw_1, \ldots, Aw_n$ are also orthogonal and $|Aw_i| = \sigma_i |w_i|$.

Since $\sigma_i > \sigma_{i+1}$, the only nonzero vectors $v \in \wedge^i \mathbb{C}^n$ for which $|Av|/|v|$ achieves its maximum value $\sigma_1 \cdots \sigma_i$ are the nonzero multiples of $w_1 \wedge \cdots \wedge w_i$. However, this is also true for $v_1 \wedge \cdots \wedge v_i$. By Lemma 4.1.11, $w_1, \ldots, w_i$ span V; this implies that the orthogonal complement of V is spanned by $w_{i+1}, \ldots, w_n$, and so it is also preserved by A. This yields the desired result. $\square$

Theorem 4.1.13. *The following are equivalent.*

(a) There exists a unitary matrix U such that $U^{-1} A U$ is diagonal.

(b) The matrix A is normal, i.e., $A^* A = A A^*$.

(c) The eigenvalues $\lambda_1, \ldots, \lambda_n$ and singular values $\sigma_1, \ldots, \sigma_n$ of A satisfy $|\lambda_i| = \sigma_i$ for $i = 1, \ldots, n$.

Proof To deduce that (a) implies (b), note that the condition of U being unitary is equivalent to the equality $U^* = U^{-1}$. Given (b), we can perform a joint eigenspace decomposition for the commuting matrices A and A^*. On any common generalized eigenspace, A has some eigenvalue λ, A^* has eigenvalue $\overline{\lambda}$, and so A^*A has eigenvalue $|\lambda|^2$. This implies (c).

Given (c), Theorem 4.1.12 implies that A can be conjugated by a unitary matrix into a block diagonal matrix in which each block has a single eigenvalue λ and a single singular value σ, such that $|\lambda| = \sigma$. Let B be such a block, corresponding to a subspace V of $\mathbb{C}^n$. If $\sigma = 0$, then $B = 0$. Otherwise, $\lambda \neq 0$ and $\lambda^{-1}B$ is unitary. Hence given orthogonal eigenvectors $v_1, \ldots, v_i \in V$ of B, the orthogonal complement in V of their span is preserved by B, so it either is zero or contains another eigenvector v_{i+1}. This shows that B is diagonalizable with a single eigenvalue, and thus is itself a scalar matrix. (One can also argue this last step using compactness of the unitary group.) $\square$

In general, we can conjugate any matrix into an almost normal matrix; the qualifier "almost" is only needed when the matrix is not semisimple.

Lemma 4.1.14. *For any $\eta > 1$, we can choose $U \in \mathrm{GL}_n(\mathbb{C})$ such that for $i = 1, \ldots, n$, the i-th singular value of $U^{-1}AU$ is at most $\eta|\lambda_i|$. If A is semisimple (i.e., diagonalizable), we can also take $\eta = 1$.*

Proof Put A in Jordan normal form, then rescale so that for each eigenvalue λ, the superdiagonal terms have absolute value at most $(\eta^2 - 1)^{1/2}|\lambda|$, and all other terms are zero. $\square$

4.2 Perturbations (archimedean case)

Another inequality of Weyl [408] shows that the singular values do not change much under a small (additive) perturbation.

Theorem 4.2.1 (Weyl). *Let B be an $n \times n$ matrix over $\mathbb{C}$, and let $\sigma'_1, \ldots, \sigma'_n$ be the singular values of $A + B$. Then*

$$|\sigma'_i - \sigma_i| \leq |B| \qquad (i = 1, \ldots, n).$$

It is more complicated to describe what happens to the eigenvalues under a small additive perturbation. The best we will do here is to quantify the effect of an additive perturbation on the characteristic polynomial.

Theorem 4.2.2. *Let B be an $n \times n$ matrix over $\mathbb{C}$. Let $P(T) = T^n + \sum_{i=0}^{n-1} P_i T^i$*

and $Q(T) = T^n + \sum_{i=0}^{n-1} Q_i T^i$ be the characteristic polynomials of A and $A + B$.
Then

$$|P_{n-i} - Q_{n-i}| \leq \left|2^i \binom{n}{i}\right| |B| \prod_{j=1}^{i-1} \max\{\sigma_j, |B|\} \qquad (i = 1, \ldots, n).$$

The superfluous enclosure of the integer $2^i \binom{n}{i}$ in absolute value signs is quite deliberate; it will be relevant in the nonarchimedean setting.

Proof Note that Q_{n-i} is the sum of the $\binom{n}{i}$ principal $i \times i$ minors of $A + B$. (A *minor* is the determinant of the $i \times i$ submatrix obtained by choosing a set of i rows and i columns. A *principal minor* is a minor in which the rows and columns correspond; for instance, if the first row is included, then the first column must also be included.) By multilinearity of the determinant, each principal minor can be written as a sum of 2^i terms, each of which is the product of a sign, a $k \times k$ minor of A, and an $(i - k) \times (i - k)$ minor of B. The terms with $k = i$ sum to P_{n-i} itself; the others all have $k < i$ and so have norm bounded by $\sigma_1 \cdots \sigma_k |B|^{i-k}$. This proves the claim. $\square$

We also need to consider multiplicative perturbations. For a vast generalization of the following inequality, see Theorem 4.5.2.

Proposition 4.2.3. *Let $B \in \mathrm{GL}_n(\mathbb{C})$ satisfy $|B| \leq \eta$. Let $\sigma_1', \ldots, \sigma_n'$ be the singular values of BA. Then*

$$\sigma_i' \leq \eta \sigma_i \qquad (i = 1, \ldots, n).$$

(The analogous result with BA replaced by AB follows from this, since transposal does not change singular values.)

Proof We use the interpretation of singular values given by Corollary 4.1.6. Choose an i-dimensional subspace V of $\mathbb{C}^n$ such that $|BAv| \geq \sigma_i'|v|$ for all $v \in V$. Then choose a nonzero element $v \in V$ such that $|Av| \leq \sigma_i|v|$. We have

$$\sigma_i'|v| \leq |BAv| \leq |B||Av| \leq \sigma_i|B||v|,$$

proving the claim. $\square$

This implies that the norms of the eigenvalues can be recovered from the singular values, provided that we consider not just the matrix A but also its powers.

Proposition 4.2.4. *Let $\sigma_{k,1}, \ldots, \sigma_{k,n}$ be the singular values of A^k. Then*

$$\lim_{k \to \infty} \sigma_{k,i}^{1/k} = |\lambda_i| \qquad (i = 1, \ldots, n).$$

Proof Pick $\eta > 1$ and choose U as in Lemma 4.1.14; then $U^{-1}AU$ is upper triangular and each block of eigenvalue λ differs from the scalar matrix $\mathrm{Diag}(\lambda, \ldots, \lambda)$ by a matrix of operator norm at most $(\eta^2 - 1)^{1/2}|\lambda|$. In a block with eigenvalue λ, the singular values of the k-th power are bounded below by $|\lambda|^k$ and above by $\eta^k|\lambda|^k$. Consequently, we may apply Proposition 4.2.3 to deduce that

$$|\lambda_i|^k|U|^{-1}|U^{-1}|^{-1} \leq \sigma_{k,i} \leq \eta^k|\lambda_i|^k|U||U^{-1}|.$$

Taking k-th roots and then taking $k \to \infty$, we deduce

$$|\lambda_i| \leq \liminf_{k\to\infty} \sigma_{k,i}^{1/k}, \qquad \limsup_{k\to\infty} \sigma_{k,i}^{1/k} \leq \eta|\lambda_i|.$$

Since $\eta > 1$ was arbitrary, we deduce the desired result. □

Remark 4.2.5. The case $i = 1$ of Proposition 4.2.4, and the corresponding case of its nonarchimedean analogue (Proposition 4.4.11), are instances of a general fact about spectra of bounded operators on Banach spaces. See Remark 6.1.7.

4.3 Singular values and eigenvalues (nonarchimedean case)

We now pass to nonarchimedean analogues.

Hypothesis 4.3.1. Throughout this section and the next, let F be a complete nonarchimedean field and let A be an $n \times n$ matrix over F. View A as a linear transformation on F^n, equip F^n with the supremum norm

$$|(z_1, \ldots, z_n)| = \max\{|z_1|, \ldots, |z_n|\},$$

and again define the operator norm

$$|A| = \sup_{v \in \mathbb{C}^n \setminus \{0\}} \{|Av|/|v|\}.$$

Note, however, that we also have the simpler expression

$$|A| = \max_{i,j}\{|A_{ij}|\}.$$

Definition 4.3.2. Given a sequence $s_1, \ldots, s_n$, we define the *associated polygon* for this sequence to be the polygonal line joining the points

$$(-n + i, s_1 + \cdots + s_i) \qquad (i = 0, \ldots, n).$$

This polygon is the graph of a convex function on $[-n, 0]$ if and only if $s_1 \leq \cdots \leq s_n$.

Definition 4.3.3. Let $s_1, \ldots, s_n$ be the sequence with the property that for $i = 1, \ldots, n$, $s_1 + \cdots + s_i$ is the minimum valuation of an $i \times i$ minor of A; that is, s_i are the *elementary divisors* (or *invariant factors*) of A. The associated polygon is called the *Hodge polygon* of A (see the chapter notes for an explanation of the terminology). Define the *singular values* of A as $\sigma_1, \ldots, \sigma_n = e^{-s_1}, \ldots, e^{-s_n}$; these are invariant under multiplication on either side by a matrix in $\mathrm{GL}_n(\mathfrak{o}_F)$. Note that as in the archimedean case,

$$\sigma_1 = |A|.$$

We also have an analogue of the singular value decomposition.

Theorem 4.3.4 (Smith normal form). *There exist $U, V \in \mathrm{GL}_n(\mathfrak{o}_F)$ such that UAV is a diagonal matrix whose entries have norms $\sigma_1, \ldots, \sigma_n$.*

Proof It is equivalent to prove that starting with A, one can perform elementary row and column operations defined over $\mathfrak{o}_F$ so as to produce a diagonal matrix (this amounts to a limited Gauss–Jordan elimination over A). To do this, find the largest entry of A, permute rows and columns to put this entry at the top left, then use it to clear the remainder of the first row and column. Repeat with the matrix obtained by removing the first row and column, and so on. $\quad\square$

Corollary 4.3.5. *The slopes $s_1, \ldots, s_n$ of the Hodge polygon satisfy $s_1 \leq \cdots \leq s_n$.*

Proof The i-th slope s_i is evidently the i-th smallest valuation of a diagonal entry of the Smith normal form. $\quad\square$

We again have a characterization as in Corollary 4.1.6.

Corollary 4.3.6. *The number σ_i is the smallest value of λ for which the following holds: for any i-dimensional subspace V of F^n, there exists a nonzero element $v \in V$ such that $|Av| \leq \lambda|v|$.*

Definition 4.3.7. Let $\lambda_1, \ldots, \lambda_n$ be the eigenvalues of A in some algebraic extension of F equipped with the unique extension of $|\cdot|$, sorted so that $|\lambda_1| \geq \cdots \geq |\lambda_n|$. The associated polygon is the *Newton polygon* of A; this is invariant under conjugation by any element of $\mathrm{GL}_n(F)$.

The nonarchimedean analogue of Weyl's inequality is the following.

Theorem 4.3.8 (Newton above Hodge). *We have*

$$\sigma_1 \cdots \sigma_i \geq |\lambda_1 \cdots \lambda_i| \qquad (i = 1, \ldots, n),$$

with equality for $i = n$. In other words, the Hodge and Newton polygons have *the same endpoints, and the Newton polygon is everywhere on or above the Hodge polygon.*

Proof Again, the case $i = 1$ is clear because σ_1 is the operator norm of A, and the general case follows by considering exterior powers (using the obvious analogue of Lemma 4.1.9).   □

Like its archimedean analogue, Theorem 4.3.8 also has a converse, but in this case we can write the construction down quite explicitly.

Definition 4.3.9. For $P = T^n + \sum_{i=0}^{n-1} P_i T^i$ a monic polynomial of degree n over a ring R, the *companion matrix* of P is defined as the matrix

$$
\begin{pmatrix}
0 & \cdots & 0 & -P_0 \\
1 & \cdots & 0 & -P_1 \\
\vdots & \ddots & & \vdots \\
0 & \cdots & 1 & -P_{n-1}
\end{pmatrix}
$$

with 1's on the subdiagonal, the negated coefficients of P in the rightmost column, and 0's elsewhere. The companion matrix is constructed to have characteristic polynomial equal to P (exercise).

Proposition 4.3.10. *Choose* $\lambda_1, \ldots, \lambda_n \in F^{\mathrm{alg}}$ *such that* $|\lambda_1| \geq \cdots \geq |\lambda_n|$, *and such that the polynomial* $P(T) = (T - \lambda_1) \cdots (T - \lambda_n) = T^n + \sum_{i=0}^{n-1} P_i T^i$ *has coefficients in* F. *Choose* $c_1, \ldots, c_n \in F$ *with* $\sigma_i = |c_i|$, *such that* $\sigma_1 \geq \cdots \geq \sigma_n$, *and*

$$
\sigma_1 \cdots \sigma_i \geq |\lambda_1 \cdots \lambda_i| \qquad (i = 1, \ldots, n),
$$

with equality for $i = n$. *Then the matrix*

$$
\begin{pmatrix}
0 & \cdots & 0 & -c_1^{-1} \cdots c_{n-1}^{-1} P_0 \\
c_{n-1} & \cdots & 0 & -c_1^{-1} \cdots c_{n-2}^{-1} P_1 \\
\vdots & \ddots & & \vdots \\
0 & \cdots & c_1 & -P_{n-1}
\end{pmatrix}
$$

has singular values $\sigma_1, \ldots, \sigma_n$ *and eigenvalues* $\lambda_1, \ldots, \lambda_n$.

Proof The given matrix is conjugate to the companion matrix of P, so its eigenvalues are also $\lambda_1, \ldots, \lambda_n$. To compute the singular values, we note that for $i = 1, \ldots, n - 1$,

$$
\begin{aligned}
\left| -c_1^{-1} \cdots c_{n-i-1}^{-1} P_i \right| &= \sigma_1^{-1} \cdots \sigma_{n-i-1}^{-1} |P_i| \\
&\leq \sigma_1^{-1} \cdots \sigma_{n-i-1}^{-1} |\lambda_1 \cdots \lambda_{n-i}| \\
&\leq \sigma_{n-i}.
\end{aligned}
$$

Thus we can perform column operations over $\mathfrak{o}_F$ to clear everything in the rightmost column except $-c_1^{-1} \cdots c_{n-1}^{-1} P_0$, which has norm $\sigma_1^{-1} \cdots \sigma_{n-1}^{-1} |\lambda_1 \cdots \lambda_n| = \sigma_n$. By permuting the rows and columns, we obtain a diagonal matrix whose diagonal entries have norms $\sigma_1, \ldots, \sigma_n$. This proves the claim. $\quad\square$

Equality again has a structural meaning, but the proof requires a bit more work than in the archimedean case since we no longer have access to orthogonality. However, this extra work is rewarded by a slightly stronger result.

Theorem 4.3.11 (Hodge–Newton decomposition). *Suppose that for some $i \in \{1, \ldots, n-1\}$,*

$$|\lambda_i| > |\lambda_{i+1}|, \qquad \sigma_1 \cdots \sigma_i = |\lambda_1 \cdots \lambda_i|.$$

(That is, the Newton polygon has a vertex with x-coordinate $-n + i$ and this vertex also lies on the Hodge polygon.) Then there exists $U \in \mathrm{GL}_n(\mathfrak{o}_F)$ such that $U^{-1}AU$ is block upper triangular, with the top left block accounting for the first i singular values and eigenvalues and the bottom right block accounting for the others. Moreover, if $\sigma_i > \sigma_{i+1}$, we can ensure that $U^{-1}AU$ is block diagonal.

Proof We first note that by Theorem 2.2.1 applied to the characteristic polynomial of A, $P(T) = (T - \lambda_1) \cdots (T - \lambda_i)$ and $Q(T) = (T - \lambda_{i+1}) \cdots (T - \lambda_n)$ have coefficients in F. Since P and Q have no common roots, we can write $1 = PB + QC$ for some $B, C \in F[T]$, and then $P(A)B(A)$ and $Q(A)C(A)$ give projectors for a direct sum decomposition separating the first i generalized eigenspaces from the others.

In other words, we can find a basis $v_1, \ldots, v_n$ of F^n such that $v_1, \ldots, v_i$ span the generalized eigenspaces with eigenvalues $\lambda_1, \ldots, \lambda_i$, and $v_{i+1}, \ldots, v_n$ span the generalized eigenspaces with eigenvalues $\lambda_{i+1}, \ldots, \lambda_n$. Choose a basis $w_1, \ldots, w_n$ of $\mathfrak{o}_F^n$ such that $w_1, \ldots, w_i$ is a basis of $\mathfrak{o}_F^n \cap (Fv_1 + \cdots + Fv_i)$. Let $e_1, \ldots, e_n$ be the standard basis of F^n, and define $U \in \mathrm{GL}_n(\mathfrak{o}_F)$ by $w_j = \sum_i U_{ij} e_i$. Then

$$U^{-1}AU = \begin{pmatrix} B & C \\ 0 & D \end{pmatrix}$$

is block upper triangular. By Cramer's rule, each entry of $B^{-1}C$ is an $i \times i$ minor of A divided by the determinant of B. Since $|\det(B)| = \sigma_1 \cdots \sigma_i$, $B^{-1}C$ must thus have entries in $\mathfrak{o}_F$. Writing

$$U^{-1}AU = \begin{pmatrix} B & 0 \\ 0 & D \end{pmatrix} \begin{pmatrix} I_i & B^{-1}C \\ 0 & I_{n-i} \end{pmatrix},$$

we see that the singular values of B and D together must comprise $\sigma_1, \ldots, \sigma_n$.

The only way for this to happen, given the constraint that the product of the singular values of B equals $\sigma_1 \cdots \sigma_i$, is for B to account for $\sigma_1, \ldots, \sigma_i$ and for D to account for $\sigma_{i+1}, \ldots, \sigma_n$.

This proves the first claim; we may thus assume now that $\sigma_i > \sigma_{i+1}$. In that case, conjugating by the matrix

$$\begin{pmatrix} I_i & -B^{-1}C \\ 0 & I_{n-i} \end{pmatrix}$$

gives a new matrix

$$\begin{pmatrix} B & C_1 \\ 0 & D \end{pmatrix}$$

with $C_1 = B^{-1}CD$. Since

$$|C_1| \le |B^{-1}||C||D| = \sigma_i^{-1}|C|\sigma_{i+1} < |C|,$$

this process converges. More explicitly, we obtain a sequence of matrices $U_i \in \mathrm{GL}_n(\mathfrak{o}_F)$ converging to the identity, such that the convergent product $U = U_1 U_2 \cdots$ satisfies

$$U^{-1}AU = \begin{pmatrix} B & 0 \\ 0 & D \end{pmatrix},$$

as desired.   □

Note that the slopes of the Hodge polygon are forced to be in the additive value group of F, whereas the slopes of the Newton polygon need only lie in the divisible closure of the additive value group. Consequently, it is possible for a matrix to have no conjugates over $\mathrm{GL}_n(F)$ for which the Hodge and Newton polygons coincide. However, the following is true; see also Corollary 4.4.8 below.

Lemma 4.3.12. *Suppose that one of the following holds.*

(a) The value group of $|F^\times|$ is dense in $\mathbb{R}_{>0}$, and $\eta > 1$.
(b) We have $|\lambda_i| \in |F^\times|$ for $i = 1, \ldots, n$ (so in particular $\lambda_i \ne 0$), and $\eta \ge 1$.

Then there exists $U \in \mathrm{GL}_n(F)$ such that the i-th singular value of $U^{-1}AU$ is at most $\eta|\lambda_i|$.

Proof Case (a) will follow from Corollary 4.4.8 below. Case (b) is directly analogous to Lemma 4.1.14.   □

One also has the following variant.

Lemma 4.3.13. *Suppose that $|F^\times|$ is discrete. Then there exists $U \in GL_n(F)$ such that for each positive integer m, $|U^{-1}A^mU|$ is the least element of $|F^\times|$ greater than or equal to $|\lambda_1^m|$.*

Proof We may normalize the valuation on F for convenience so that $\log |F^\times| = \mathbb{Z}$. As in the proof of Theorem 4.3.11, we may also reduce to the case where all of the eigenvalues of A have the same norm.

Let E be a finite extension of F containing an element λ with $|\lambda| = |\lambda_1|$. (For instance, we could take $E = F(\lambda_1)$ and $\lambda = \lambda_1$, but any other choice would also work.) By Lemma 4.3.12, there exists $U_0 \in GL_n(E)$ such that $\lambda^{-1}U_0^{-1}AU_0 \in GL_n(\mathfrak{o}_E)$; in other words, there exists a supremum norm $|\cdot|_0$ on E^n such that $|\lambda^{-1}Av|_0 = |v|_0$ for all $v \in E^n$.

Put $V = \{v \in F^n : |v|_0 \le 1\}$. Let $v_1, \ldots, v_n$ be a basis of V over $\mathfrak{o}_F$, and let $|\cdot|_1$ be the supremum norm on F^n defined by $v_1, \ldots, v_n$. Given $v \in F^n$ nonzero, choose $\mu \in F^\times$ with $-\log|\mu v|_0 \in [0, 1)$; then μv is an element of V which is not divisible by $\mathfrak{m}_F$, so $|\mu v|_1 = 1$. We conclude that

$$e^{-1}|v|_1 < |v|_0 \le |v|_1 \qquad (v \in F^n). \tag{4.3.13.1}$$

Let $e_1, \ldots, e_n$ be the standard basis of F^n, and define $U \in GL_n(F)$ by $v_j = \sum_i U_{ij}e_i$. Then for each positive integer m, $-\log|U^{-1}A^mU| \in \mathbb{Z}$ is at most $-m\log|\lambda_1|$ by Theorem 4.3.8, but is strictly greater than $-m\log|\lambda_1| - 1$ by (4.3.13.1). We thus obtain the desired equality. □

4.4 Perturbations (nonarchimedean case)

Again, we can ask about the effect of perturbations. We start with the analogue of Weyl's second inequality.

Proposition 4.4.1. *If B is a matrix with $|B| < \sigma_i$, then the first i singular values of $A + B$ are $\sigma_1, \ldots, \sigma_i$.*

Proof Exercise. □

We next consider the effect on the characteristic polynomial.

Theorem 4.4.2. *Let B be an $n \times n$ matrix. Let $P(T) = T^n + \sum_{i=0}^{n-1} P_iT^i$ and $Q(T) = T^n + \sum_{i=0}^{n-1} Q_iT^i$ be the characteristic polynomials of A and $A + B$. Then*

$$|P_{n-i} - Q_{n-i}| \le |B| \prod_{j=1}^{i-1} \max\{\sigma_j, |B|\} \qquad (i = 1, \ldots, n).$$

Proof The proof is as for Theorem 4.2.2, except now the factor $|2^i \binom{n}{i}|$ is dominated by 1. □

Question 4.4.3. *Is the inequality in Theorem 4.4.2 best possible?*

We may also consider multiplicative perturbations.

Proposition 4.4.4. *Let $B \in \mathrm{GL}_n(F)$ satisfy $|B| \leq \eta$. Let $\sigma'_1, \ldots, \sigma'_n$ be the singular values of AB. Then*

$$\sigma'_i \leq \eta \sigma_i \qquad (i = 1, \ldots, n).$$

Proof As for Proposition 4.2.3, but using the Smith normal form (Theorem 4.3.4) instead of the singular value decomposition. □

Corollary 4.4.5. *Suppose that the Newton and Hodge slopes of A coincide, and that $U \in \mathrm{GL}_n(F)$ satisfies $|U| \cdot |U^{-1}| \leq \eta$. Then each Newton slope of $U^{-1}AU$ differs by at most $\log \eta$ from the corresponding Hodge slope.*

Here is a weak converse to Corollary 4.4.5. (We leave the archimedean analogue to the reader's imagination.)

Proposition 4.4.6. *Suppose that the Newton slopes of A are nonnegative and that $\sigma_1 \geq 1$. Then there exists $U \in \mathrm{GL}_n(F)$ such that*

$$|U^{-1}AU| \leq 1, \quad |U^{-1}| \leq 1, \quad |U| \leq \sigma_1^{n-1}.$$

Proof Let $e_1, \ldots, e_n$ denote the standard basis vectors of F^n. Let M be the smallest $\mathfrak{o}_F$-submodule of F^n containing $e_1, \ldots, e_n$ and stable under A. For each i, if $j = j(i)$ is the least integer such that $e_i, Ae_i, \ldots, A^j e_i$ are linearly dependent, then $A^j e_i = \sum_{h=0}^{j-1} c_h A^h e_i$ for some $c_h \in F$. The polynomial $T^j - \sum_{h=0}^{j-1} c_h T^h$ has roots which are eigenvalues of A, so the nonnegativity of the Newton slopes forces $|c_h| \leq 1$. Hence M is finitely generated, and thus free, over $\mathfrak{o}_F$.

Let $v_1, \ldots, v_n$ be a basis of M, and let U be the change-of-basis matrix $v_j = \sum_i U_{ij} e_i$; then $|U^{-1}AU| \leq 1$ because M is stable under A, and $|U^{-1}| \leq 1$ because M contains $e_1, \ldots, e_n$. The desired bound on $|U|$ will follow from the fact that for any $x = c_1 e_1 + \cdots + c_n e_n \in M$,

$$\max_i \{|c_i|\} \leq \sigma_1^{n-1}. \tag{4.4.6.1}$$

It suffices to check (4.4.6.1) for $x = A^h e_i$ for $i = 1, \ldots, n$ and $h = 0, \ldots, j(i)-1$, as these generate M over $\mathfrak{o}_F$. But it is evident that $|A^h e_i| \leq \sigma_1^h |e_i| = \sigma_1^h$; since $j(i) \leq n$, we are done. □

Example 4.4.7. The example

$$A = \begin{pmatrix} 1 & c & 0 \\ 0 & 1 & c \\ 0 & 0 & 1 \end{pmatrix}$$

with $|c| > 1$ shows that this bound of Proposition 4.4.6 is sharp; in particular, the bound $|U| \le \sigma_1^{n-1}$ cannot be improved to $|U| \le \sigma_1$, as one might initially expect. However, one should be able to get a more precise bound (which agrees with the given bound in this example) by accounting for the other singular values; see the exercises.

Corollary 4.4.8. *There exists a continuous function*

$$f_n(\sigma_1, \ldots, \sigma_n, \sigma_1', \ldots, \sigma_n', \delta) : (0, \infty)^{2n} \times [0, \infty) \to (0, \infty)$$

(independent of F) with the following properties.

(a) Suppose for each $i = 1, \ldots, n$, either $\sigma_i = \sigma_i'$ or $\delta \ge \max\{\sigma_i, \sigma_i'\}$. Then

$$f_n(\sigma_1, \quad , \sigma_n, \sigma_1', \ldots, \sigma_n', \delta) = 1.$$

(b) If A has singular values $\sigma_1, \ldots, \sigma_n$ and eigenvalues $\lambda_1, \ldots, \lambda_n$, $\sigma_i' = |\lambda_i|$
for $i = 1, \ldots, n$, and $\sigma_i' \in |F^\times|$ whenever $\sigma_i' > \delta$, then there exists
$U \in \mathrm{GL}_n(F)$ such that

$$|U^{-1}| \le 1, \quad |U| \le f_n(\sigma_1, \ldots, \sigma_n, \sigma_1', \ldots, \sigma_n', \delta)$$

for which the multiset of singular values of $U^{-1}AU$ matches $\sigma_1', \ldots, \sigma_n'$ in
its values greater than δ.

Proof This follows by induction on n as follows. If $\sigma_1' \le \delta$, we deduce the whole claim by Proposition 4.4.6 (after rescaling in case $\delta \ne 1$). Otherwise, again by Proposition 4.4.6 (again after rescaling), we can find $U_1 \in \mathrm{GL}_n(F)$ such that

$$|U_1^{-1}A_1U_1| \le \sigma_1', \quad |U_1^{-1}| \le 1, \quad |U_1| \le (\sigma_1/\sigma_1')^{n-1}.$$

Let i be the largest index such that $\sigma_i' = \sigma_1'$. Then the first i singular values of $U_1^{-1}AU_1$ are all at most σ_1', but at least σ_i'. Hence they are all equal, and $A_1 = U_1^{-1}AU_1$ satisfies the hypothesis of Theorem 4.3.11. We may thus choose $U_2 \in \mathrm{GL}_n(\mathfrak{o}_F)$ such that $A_2 = U_2^{-1}A_1U_2$ is block upper triangular, with the top left block accounting for the first i singular values and eigenvalues and the bottom right block accounting for the others.

If $i = n$, then we may take $U = U_1U_2$ and be done. Otherwise, note that by applying Proposition 4.4.4, we may bound the singular values of A_2 by a continuous function of $\sigma_1, \ldots, \sigma_n, \sigma'_1, \ldots, \sigma'_n, \delta$. We may then apply the induction hypothesis to construct a block diagonal matrix U_3, where the top left block of U_3 is the identity, $|U_3^{-1}| \leq 1$, $|U_3|$ is bounded by a continuous function of $\sigma_1, \ldots, \sigma_n, \sigma'_1, \ldots, \sigma'_n, \delta$, and the bottom right block of $A_3 = U_3^{-1} A_2 U_3$ has a multiset of singular values which agrees with $\sigma'_{i+1}, \ldots, \sigma'_n$ in its values greater than δ.

We may bound the norm of the top right block of A_3 by a continuous function of $\sigma_1, \ldots, \sigma_n, \sigma'_1, \ldots, \sigma'_n, \delta$. We can then conjugate by a suitable block diagonal matrix U_4, with scalar matrices in the diagonal blocks, to ensure that $A_4 = U_4^{-1} A_3 U_4$ has a multiset of singular values which agrees with $\sigma'_1, \ldots, \sigma'_n$ in its values greater than δ. We then take $U = U_1 \cdots U_4$. $\qquad\square$

For the purposes of this book, it is immaterial what the function f_n is, as long as it is continuous. However, for numerical applications, it may be quite helpful to identify a good function f_n; here is a conjectural best possible result in the case $\delta = 0$, phrased in a somewhat stronger form. (One can also formulate an archimedean analogue. It should also be possible to prove using the Horn inequalities that this conjecture cannot be improved; see next section.)

Conjecture 4.4.9. *Suppose that A has singular values $\sigma_1, \ldots, \sigma_n$ and eigenvalues $\lambda_1, \ldots, \lambda_n$, none equal to 0, and that $\sigma'_1, \ldots, \sigma'_n \in |F^\times|$ satisfy*

$$\sigma_1 \cdots \sigma_i \geq \sigma'_1 \cdots \sigma'_i \geq |\lambda_1 \cdots \lambda_i| \qquad (i = 1, \ldots, n).$$

Then there exists $U \in \mathrm{GL}_n(F)$ such that

$$|U^{-1}| \leq 1, \quad |U| \leq \max_i \{(\sigma_1 \cdots \sigma_i)/(\sigma'_1 \cdots \sigma'_i)\},$$

for which $U^{-1} A U$ has singular values $\sigma'_1, \ldots, \sigma'_n$.

In another direction, we give a variant of Corollary 4.4.8 where we make the bound explicit, but give up on condition (a).

Corollary 4.4.10. *Suppose that A has singular values $\sigma_1, \ldots, \sigma_n$ and eigenvalues $\lambda_1, \ldots, \lambda_n$; $\sigma'_i = |\lambda_i|$ for $i = 1, \ldots, n$; and $\sigma'_i \in |F^\times|$ whenever $\sigma'_i > \delta$. Then there exists $U \in \mathrm{GL}_n(F)$ such that*

$$|U^{-1}| \leq 1, \quad |U| \leq \max\{1, \sigma_1/\delta\}^{2^n n! - 1}$$

for which the multiset of singular values of $U^{-1} A U$ matches $\sigma'_1, \ldots, \sigma'_n$ in its values greater than δ.

Proof Following the proof of Corollary 4.4.8, we obtain matrices U_1, U_2, U_3, U_4 for which

$$|U_1| \leq (\sigma_1/\delta)^{n-1}, \quad |U_2| \leq 1, \quad |U_4| \leq (\sigma_1/\delta)|U_1|^2|U_3|^2,$$

and $|U_3|$ is bounded by the induction hypothesis with n replaced by $n-i \le n-1$ and σ_1 replaced by $|A_2| \le \sigma_1|U_1|$.   □

By imitating the proof of Proposition 4.2.4 (after enlarging F to contain the eigenvalues of A), we obtain the following.

Proposition 4.4.11. *Let* $\sigma_{k,1}, \ldots, \sigma_{k,n}$ *be the singular values of* A^k. *Then*

$$\lim_{k \to \infty} \sigma_{k,i}^{1/k} = |\lambda_i| \qquad (i = 1, \ldots, n).$$

4.5 Horn's inequalities

Although they will not be needed in this book, it is quite natural to mention here some stronger versions of the perturbation inequalities in the archimedean and nonarchimedean cases, introduced conjecturally by Horn [208] in the archimedean case and resolved by work of Klyachko, Knutson, Speyer, Tao, Woodward, and others. See the beautiful survey article of Fulton [173] for more information.

Definition 4.5.1. To introduce the stronger inequalities, we must set up some notation. Put

$$U_r^n = \{(I, J, K): I, J, K \subseteq \{1, \ldots, n\}, \#I = \#J = \#K = r,$$
$$\sum_{i \in I} i + \sum_{j \in J} j = \sum_{k \in K} k + \tfrac{r(r+1)}{2}\}.$$

For $(I, J, K) \in U_r^n$, write $I = \{i_1 < \cdots < i_r\}$ and similarly for J, K. For $r = 1$, put $T_1^n = U_1^n$. For $r > 1$, put

$$T_r^n = \{(I, J, K) \in U_r^n : \text{for all } p < r \text{ and } (F, G, H) \in T_p^r,$$
$$\sum_{f \in F} i_f + \sum_{g \in G} j_g \le \sum_{h \in H} k_h + \tfrac{p(p+1)}{2}\}.$$

For multiplicative perturbations, we obtain the following results, which include Propositions 4.2.3 and 4.4.4. It is important for the proofs that one can rephrase the Horn inequalities in terms of Littlewood–Richardson numbers; see [173, §3].

Theorem 4.5.2. *For* $* \in \{A, B, C\}$, *let* $\sigma_{*,1}, \ldots, \sigma_{*,n}$ *be a nonincreasing sequence of nonnegative real numbers. Then the following are equivalent.*

(a) There exist $n \times n$ *matrices* A, B, C *over* $\mathbb{C}$ *with* $AB = C$ *such that for* $* \in \{A, B, C\}$, $*$ *has singular values* $\sigma_{*,1}, \ldots, \sigma_{*,n}$.

(b) *We have* $\prod_{i=1}^{n} \sigma_{A,i} \prod_{j=1}^{n} \sigma_{B,j} = \prod_{k=1}^{n} \sigma_{C,k}$, *and for all* $r < n$ *and* $(I, J, K) \in T_r^n$,

$$\prod_{k \in K} \sigma_{C,k} \leq \prod_{i \in I} \sigma_{A,i} \prod_{j \in J} \sigma_{B,j}.$$

Proof See [173, Theorem 16]. Beware that the first condition in (b) is omitted in the statement given in [173], but this was a typographical error. □

Theorem 4.5.3. *Let F be a complete nonarchimedean field. For* $* \in \{A, B, C\}$, *let* $\sigma_{*,1}, \ldots, \sigma_{*,n}$ *be a nonincreasing sequence of elements of* $|F|$. *Then the following are equivalent.*

(a) *There exist* $n \times n$ *matrices* A, B, C *over* F *with* $AB = C$ *such that for* $* \in \{A, B, C\}$, $*$ *has singular values* $\sigma_{*,1}, \ldots, \sigma_{*,n}$.
(b) *We have* $\prod_{i=1}^{n} \sigma_{A,i} \prod_{j=1}^{n} \sigma_{B,j} = \prod_{k=1}^{n} \sigma_{C,k}$, *and for all* $r < n$ *and* $(I, J, K) \in T_r^n$,

$$\prod_{k \in K} \sigma_{C,k} \leq \prod_{i \in I} \sigma_{A,i} \prod_{j \in J} \sigma_{B,j}.$$

Proof See [173, Theorem 7]. □

Example 4.5.4. Let us see explicitly how Theorem 4.5.2 implies Proposition 4.2.3. Since $T_1^n = U_1^n$, condition (b) includes all cases with $(I, J, K) \in U_1^n$. In particular, we may take

$$I = \{i\}, \qquad J = \{1\}, \qquad K = \{i\}$$

to obtain the inequality $\sigma_{C,i} \leq \sigma_{A,i} \sigma_{B,1}$; this is precisely Proposition 4.2.3.

Remark 4.5.5. For additive perturbations, one has an analogous result in the archimedean case; see [173, Theorem 15]. We am not aware of an additive result in the nonarchimedean case. Also, in the archimedean case one has analogous results (with slightly different statements) in which one restricts to Hermitian matrices.

Notes

The subject of archimedean matrix inequalities is an old one, with many important applications. A good reference for this is [63]; for instance, see [63, §I.2] for the singular value decomposition, [63, Theorem II.3.6] for the Weyl inequalities in a much stronger form known as *Weyl's majorant theorem*, [63, Theorem III.4.5] for a strong form of Proposition 4.2.3 (also a consequence of the Horn inequalities), and so on. (A variant of our Theorem 4.2.2 appears as

[63, Problem I.6.11].) Another standard reference for this topic is [209]; see especially [209, Chapter 3] for the development of the Weyl inequalities.

The strong analogy between archimedean and p-adic matrix inequalities appears to be a little-known piece of folklore. As a result, we have been unable to locate a suitable reference; the closest match we have found is the work of Caruso–Roe–Vaccon on numerical stability for linear algebra over a nonarchimedean field [80, 81].

It should be pointed out that most of what we have done here is the special case for GL_n of a more general theory encompassing the other reductive algebraic groups. This point of view can be seen in [173], where GL_n makes some explicit appearances for which other groups can be substituted.

In Theorem 4.1.13, the equivalence of (a) and (b) is standard. We do not have a reference for the equivalence with (c), although it is implicit in most proofs of the equivalence of (a) and (b).

The reader familiar with the notions of elementary divisors or invariant factors may be wondering why the terminology "Hodge polygon" is necessary or reasonable. The answer is that the Hodge numbers of a variety over a p-adic field are reflected by the elementary divisors of the action of Frobenius on crystalline cohomology. The fact that the Newton polygon lies above the Hodge polygon then implies a relation between the characteristic polynomial of Frobenius and the Hodge numbers of the original variety; this relationship was originally conjectured by Katz and proved by Mazur. See [62] for further discussion of this point, and of crystalline cohomology as a whole.

Much of the work in this chapter can be carried over to the case of a transformation which is only semilinear for some isometric endomorphism of F. We will adopt that point of view in Chapter 14; for instance, this will lead to a generalization (in Theorem 14.5.5) of the Hodge–Newton decomposition theorem (Theorem 4.3.11). In this case, the carrying over is really in the other direction: it is the latter result (due to Katz; see the notes for Chapter 14) which inspired our presentation of Theorem 4.3.11 and its archimedean analogue (Theorem 4.1.12). Similarly, our treatment of Proposition 4.4.11 and its archimedean analogue (Proposition 4.2.4) are modeled on [224, Corollary 1.4.4].

The question of quantifying the sensitivity to perturbation of the characteristic polynomial of a square matrix arises in numerical applications. The question is familiar in the archimedean case, but perhaps less so in the nonarchimedean case; numerical applications of the latter include using p-adic cohomology to compute zeta functions of varieties over finite fields. See for instance [3, §1.6], [177, §3].

Exercises

4.1 Prove that any elementary matrix of type (a) (exchanging two rows) can be factored as a product of elementary matrices of types (b) and (c).

4.2 Let $e_1, \ldots, e_4$ be a basis of $\mathbb{C}^4$. Prove that in $\wedge^2 \mathbb{C}^4$, the element $e_1 \wedge e_2 + e_3 \wedge e_4$ is not decomposable.

4.3 Check that the characteristic polynomial of the companion matrix of a polynomial P (Definition 4.3.9) is equal to P.

4.4 Prove Proposition 4.4.1. (Hint: use Corollary 4.3.6.)

4.5 With notation as in Theorem 4.3.11, suppose $U, V \in GL_n(\mathfrak{o}_F)$ are congruent to the identity matrix modulo $\mathfrak{m}_F$. Prove that the product of the i largest eigenvalues of UAV again has norm $|\lambda_1 \cdots \lambda_i|$. (Hint: use exterior powers to reduce to the case $i = 1$.) This yields as a corollary [75, Lemma 5]: if $D \in GL_n(F)$ is diagonal and $U, V \in GL_n(\mathfrak{o}_F)$ are congruent to the identity matrix modulo $\mathfrak{m}_F$, then the Newton polygons of D and UDV coincide.

4.6 State and prove an archimedean analogue of the previous problem.

4.7 Prove the following improved version of Proposition 4.4.6. Suppose that the Newton slopes of A are nonnegative. Then there exists $U \in GL_n(F)$ such that

$$|U^{-1}AU| \leq 1, \quad |U^{-1}| \leq 1, \quad |U| \leq \prod_{i=1}^{n-1} \max\{1, \sigma_i\}.$$

We do not know of an appropriate archimedean analogue.

Part II

Differential Algebra

5

Formalism of differential algebra

In this chapter, we introduce some basic formalism of differential algebra. This may viewed as a mild perturbation of commutative algebra, in which we consider commutative rings equipped with the additional noncommutative structure of a derivation. This allows us to manipulate differential equations and differential systems in a manner that keeps a bit more useful structure visible, though we will need to convert back and forth from this point of view. (One thing we will not do is generalize to the realm of differential schemes and sheaves; we leave this as a thought exercise for the curious reader.)

A particularly important result we introduce is the *cyclic vector theorem*, which gives a compact but highly noncanonical way to represent a finite differential module over a field. While the cyclic vector theorem will prove indispensable at a few key points in our treatment of p-adic differential equations, we will ultimately make more progress by limiting its use. See Remark 5.7.1 for further discussion.

5.1 Differential rings and differential modules

Definition 5.1.1. A *differential ring* is a commutative ring R equipped with a derivation $d: R \to R$, the latter being an additive map satisfying the Leibniz rule

$$d(ab) = ad(b) + bd(a) \qquad (a, b \in R).$$

We expressly allow $d = 0$ unless otherwise specified; this will come in handy in some situations. A differential ring which is also a domain, field, etc., will be called a *differential domain, field,* etc. Note that there is a unique extension of d to any localization of R, using the quotient rule; in particular, if R is a domain, then there is a unique extension of d to $\text{Frac}(R)$.

Definition 5.1.2. A *differential module* over a differential ring (R, d) is a module M equipped with an additive map $D: M \to M$ satisfying

$$D(am) = aD(m) + d(a)m;$$

such a D will also be called a *differential operator* on M relative to d. For example, (R, d) is a differential module over itself; any differential module isomorphic to a direct sum of copies of (R, d) is said to be *trivial*. (If we refer to "the trivial differential module", though, we mean (R, d) itself.) A differential module which is a successive extension of trivial modules is said to be *unipotent* (see Proposition 7.2.5 for the reason why). A *differential ideal* of R is a differential submodule of R itself, i.e., an ideal stable under d.

Definition 5.1.3. For (M, D) a differential module, define

$$H^0(M) = \ker(D), \qquad H^1(M) = \operatorname{coker}(D) = M/D(M).$$

The latter computes Yoneda extensions; see Lemma 5.3.3 below. Elements of $H^0(M)$ are said to be *horizontal*, and are often referred to as *horizontal sections*; these terms have a geometric significance described in the chapter notes. Note that $H^0(R) = \ker(d)$ is a subring of R; if R is a field, then $\ker(d)$ is a subfield. We call this the *constant subring* (or *constant subfield*) of R.

We make an observation about base change of the constant subring. For the definition of a more general base change, see Definition 5.3.2. (See also Proposition 6.9.1.)

Lemma 5.1.4. *Let R_0 be the constant subring of the differential ring (R, d), and let R_0' be an R_0-algebra. Let (M, D) be a differential module over (R, d). View $R' = R \otimes_{R_0} R_0'$ as a differential ring by defining the derivation d' by*

$$d'\left(\sum_i a_i \otimes r_i\right) = \sum_i d(a_i) \otimes r_i \qquad (a_i \in R, r_i \in R_0').$$

Similarly, view $M' = M \otimes_{R_0} R_0'$ as a differential module over (R', d') by defining the differential operator

$$D'\left(\sum_i m_i \otimes r_i\right) = \sum_i D(m_i) \otimes r_i \qquad (m_i \in M, r_i \in R_0').$$

(a) *There are natural maps $H^i(M) \otimes_{R_0} R_0' \to H^i(M')$ for $i = 0, 1$.*

(b) *The map in (a) is always an isomorphism for $i = 1$.*

(c) *If R_0' is flat over R_0, then the map in (a) is an isomorphism for $i = 0$.*

Proof (a) Tensoring the structure morphism $R_0 \to R_0'$ with M induces a map $M \to M'$. This in turn induces maps $H^i(M) \to H^i(M')$ of R_0-modules; using the R'-module structure on $H^i(M')$, we also obtain maps $H^i(M) \otimes_{R_0} R_0' \to H^i(M')$.

(b) Tensoring with R_0' is always a right exact functor on R_0-modules. Since

$$M \xrightarrow{D} M \to H^1(M) \to 0$$

is an exact sequence in the category of R_0-modules,

$$M' \xrightarrow{D'} M' \to H^1(M) \otimes_{R_0} R_0' \to 0$$

is also exact. Hence the induced map $H^1(M) \otimes_{R_0} R_0' \to H^1(M')$ is an isomorphism.

(c) Recall that by definition, R_0' is flat over R_0 if and only if tensoring with R_0' is an exact functor on R_0-modules. Since

$$0 \to H^0(M) \to M \xrightarrow{D} M$$

is an exact sequence in the category of R_0-modules,

$$0 \to H^0(M) \otimes_{R_0} R_0' \to M' \xrightarrow{D'} M'$$

is also exact. Hence the induced map $H^0(M) \otimes_{R_0} R_0' \to H^0(M')$ is an isomorphism. $\square$

Another frequently used observation is the following.

Lemma 5.1.5. *Let (R, d) be a differential field with constant subfield R_0. Then for any differential module (M, D) over (R, d), the natural map $H^0(M) \otimes_{R_0} R \to M$ is injective. In particular, $\dim_{R_0} H^0(M) \le \dim_R M$.*

Proof An equivalent statement is that if $m_1, \ldots, m_n \in H^0(M)$ are linearly dependent over R, then they are also linearly dependent over R_0. Suppose on the contrary that for some positive integer n, there exist $m_1, \ldots, m_n \in H^0(M)$ which are linearly dependent over R but linearly independent over R_0. Choose n as small as possible with this property; then there exist nonzero elements $c_1, \ldots, c_n \in R$ such that $c_1 m_1 + \cdots + c_n m_n = 0$. We may rescale the c_i so that $c_1 = 1$.

Since $m_1, \ldots, m_n \in H^0(M)$, we also have

$$d(c_1) m_1 + \cdots + d(c_n) m_n = 0.$$

That is, $d(c_1), \ldots, d(c_n)$ form another linear dependence relation among the elements $m_1, \ldots, m_n$. But $d(c_1) = d(1) = 0$, so to avoid contradicting the choice of n we must have $d(c_2) = \cdots = d(c_n) = 0$. That is, $c_1, \ldots, c_n \in R_0$, contradicting the hypothesis that $m_1, \ldots, m_n$ are linearly independent over R_0. This contradiction proves the claim. $\qquad \square$

Definition 5.1.6. Let M be a differential module over a differential ring R admitting a finite exhaustive filtration with irreducible successive quotients. (For instance, R could be a differential field and M could be a finitely generated differential module over R.) Then the multiset of these quotients is independent of the choice of the filtration; we call them the *(Jordan–Hölder) constituents* of M.

Remark 5.1.7. Let M be a differential module over R. Suppose that

$$0 \to M_1 \to M \to M_2 \to 0$$

is a short exact sequence of R-modules. One can then define a function $f: M_1 \to M_2$ by taking $v \in M_1$ to the class of $D(v)$ in M_2. This function is R-linear: it is additive because D is, and

$$D(rv) = rD(v) + d(r)v \equiv rD(v) \pmod{M_2}.$$

The function f is zero if and only if M_1 is a differential submodule of M. It is an algebraic analogue of the *Kodaira–Spencer map* in differential geometry.

The following matrix form of the Leibniz rule is useful for some calculations

Lemma 5.1.8. *Let* $u_1, \ldots, u_n, v$ *be elements of a differential field* (F, d) *of characteristic* 0. *Define the* $\infty \times n$ *matrix* U *by*

$$U_{ij} = \frac{1}{i!} d^i(u_j) \qquad (i = 0, 1, \ldots; j = 1, \ldots, n)$$

and the lower triangular $\infty \times \infty$ *matrix* P *by*

$$P_{ij} = \frac{1}{(i-j)!} d^{(i-j)}(v) \qquad (i = 0, 1, \ldots; j = 0, 1, \ldots).$$

Then

$$(PU)_{ij} = \frac{1}{i!} d^i(u_j v) \qquad (i = 0, 1, \ldots; j = 1, \ldots, n).$$

Proof Exercise. $\qquad \square$

5.2 Differential modules and differential systems

We now describe the link between differential modules and linear differential systems.

Definition 5.2.1. Let R be a differential ring, and let M be a finite free differential module of rank n over R. Let $e_1, \ldots, e_n$ be a basis of M. Then for any $v \in M$, we can write $v = v_1 e_1 + \cdots + v_n e_n$ for some $v_1, \ldots, v_n \in R$ and compute

$$D(v) = v_1 D(e_1) + \cdots + v_n D(e_n) + d(v_1) e_1 + \cdots + d(v_n) e_n.$$

Define the *matrix of action of D* on the basis $e_1, \ldots, e_n$ to be the $n \times n$ matrix N over R given by the formula

$$D(e_j) = \sum_{i=1}^{n} N_{ij} e_i.$$

We then have

$$D(v) = \sum_{i=1}^{n} \left(d(v_i) + \sum_{j=1}^{n} N_{ij} v_j \right) e_i.$$

That is, if we identify v with the column vector $[v_1 \cdots v_n]$, then

$$D(v) = Nv + d(v).$$

Conversely, it is clear that given the underlying finite free R-module, any differential module structure is given by such an equation.

Definition 5.2.2. With notation as in Definition 5.2.1, let $v_1, \ldots, v_n$ be a second basis of M. The *change-of-basis matrix* from $e_1, \ldots, e_n$ to $v_1, \ldots, v_n$ is the $n \times n$ matrix U defined by

$$v_j = \sum_i U_{ij} e_i.$$

The effect of changing basis is that the matrix of action of D on $v_1, \ldots, v_n$ is

$$U^{-1} N U + U^{-1} d(U).$$

Remark 5.2.3. In other words, differential modules are a coordinate-free version of differential systems. If you are a geometer, you may wish to go further and think of *differential bundles*, i.e., vector bundles equipped with a differential operator. A differential operator on a vector bundle is usually called a *connection*.

5.3 Operations on differential modules

We now describe the basic operations in the category of differential modules over a differential ring.

Definition 5.3.1. For R a differential ring, we regard the differential modules over R as a category in which the morphisms (or *homomorphisms*) from M_1 to M_2 are R-module homomorphisms $f : M_1 \to M_2$ satisfying $D(f(m)) = f(D(m))$ (we sometimes say these maps are *horizontal*).

Definition 5.3.2. The category of differential modules over a differential ring admits certain functors corresponding to familiar functors on the category of modules over an ordinary ring, such as the following. (Beware that in the following notations, the subscripted R on such symbols as the tensor product will often be suppressed when it is unambiguous. Our habit tends to be to drop the subscript when tensoring modules over a single ring, but not when performing a base change.)

Given two differential modules M_1, M_2, the tensor product $M_1 \otimes_R M_2$ in the category of rings may be viewed as a differential module via the formula

$$D(m_1 \otimes m_2) = D(m_1) \otimes m_2 + m_1 \otimes D(m_2).$$

This in particular gives meaning to the *base change* $M \otimes_R R'$ of a differential R-module M to a differential R-algebra R'; we also denote this by $f^* M$ if f is the name of the map from R to R'. Similarly, the exterior power $\wedge_R^n M$ may be viewed as a differential module via the formula

$$D(m_1 \wedge \cdots \wedge m_n) = \sum_{i=1}^{n} m_1 \wedge \cdots \wedge m_{i-1} \wedge D(m_i) \wedge m_{i+1} \wedge \quad \wedge m_n.$$

(A similar fact is true for the symmetric power $\mathrm{Sym}_R^n M$, but we will have no need for this.) The module of R-homomorphisms $\mathrm{Hom}_R(M_1, M_2)$ may be viewed as a differential module via the formula

$$D(f)(m) = D(f(m)) - f(D(m));$$

the homomorphisms from M_1 to M_2 as differential modules are precisely the horizontal elements of $\mathrm{Hom}_R(M_1, M_2)$. (In this case, the subscript is quite crucial: we have $\mathrm{Hom}(M_1, M_2) = H^0(\mathrm{Hom}_R(M_1, M_2))$.)

We write $M_1^\vee$ for $\mathrm{Hom}_R(M_1, R)$ and call it the *dual* of M_1; if M_1 is finite free, then $\mathrm{Hom}_R(M_1, M_2) \cong M_1^\vee \otimes M_2$ and the natural map $M_1 \to (M_1^\vee)^\vee$ is an isomorphism. In particular, $M_1^\vee \otimes M_1$ contains a horizontal element corresponding to the identity map $M_1 \to M_1$; we call this the *trace (element)* of $M_1^\vee \otimes M_1$, and we call the trivial submodule generated by the trace element the *trace component* of $M_1^\vee \otimes M_1$. If R is a $\mathbb{Q}$-algebra, then $M_1^\vee \otimes M_1$ splits as the direct sum of the trace component with the set of elements of $\mathrm{Hom}_R(M_1, M_2)$ of trace zero; we call the latter the *trace-zero component* of $M_1^\vee \otimes M_1$. Even if R is not a $\mathbb{Q}$-algebra, we can still view the trace component as a quotient of

$M_1^\vee \otimes M_1$ by duality, but the map to itself given by embedding into $M_1^\vee \otimes M_1$ and then projecting need not be an isomorphism.

Lemma 5.3.3. *Let M, N be differential modules with M finite free. Then the group $H^1(M^\vee \otimes N)$ is canonically isomorphic to the Yoneda extension group $\mathrm{Ext}(M, N)$.*

Proof The group $\mathrm{Ext}(M, N)$ consists of equivalence classes of exact sequences $0 \to N \to P \to M \to 0$ under the relation that this sequence is equivalent to a second sequence $0 \to N \to P' \to M \to 0$ if there is an isomorphism $P \cong P'$ that induces the identity maps on M and N. Addition of two sequences produces the *Baer sum* $0 \to N \to Q/\Delta \to M \to 0$, where Q is the set of elements in $P \oplus P'$ whose images in M coincide and $\Delta = \{(n, -n) \in Q : n \in N\}$. The identity element is the split sequence $0 \to N \to M \oplus N \to M \to 0$. The inverse of a sequence $0 \to N \to P \to M \to 0$ is the same sequence with the map $N \to P$ negated. (See [407, §3.4] for the proof that this indeed gives a group.)

We first construct a canonical isomorphism $\mathrm{Ext}(M, N) \to \mathrm{Ext}(R, M^\vee \otimes N)$. Given an extension $0 \to N \to P \to M \to 0$, tensor with $M^\vee$ to get

$$0 \to M^\vee \otimes N \to M^\vee \otimes P \to M^\vee \otimes M \to 0.$$

Let Q be the inverse image of the trace component of $M^\vee \otimes M$; we then get an extension

$$0 \to M^\vee \otimes N \to Q \to R \to 0,$$

yielding a map $\mathrm{Ext}(M, N) \to \mathrm{Ext}(R, M^\vee \otimes N)$, which turns out to be a homomorphism. In the other direction, given an extension

$$0 \to M^\vee \otimes N \to Q \to R \to 0,$$

tensor with M to get

$$0 \to M \otimes M^\vee \otimes N \to M \otimes Q \to M \to 0,$$

then quotient $M \otimes Q$ by the kernel of the projection $M \otimes M^\vee \to R$ tensored with N. These are seen to be inverses by a diagram-chasing argument, which we omit.

By the previous paragraph, we may reduce the statement of the lemma to the case $M = R$. (One can also describe the construction without first making this reduction, but it is a bit harder to follow.) Given an extension $0 \to N \to P \to R \to 0$, compute H^0 and H^1 and apply the snake lemma to obtain a connecting homomorphism $H^0(R) \to H^1(N)$. The image of $1 \in R$ under this homomorphism determines an element of $H^1(N)$, thus giving a map $\text{Ext}(R, N) \rightarrow H^1(N)$. This map can be shown to be a homomorphism (exercise).

It remains to construct an inverse map. Given an element of $H^1(N)$ represented by $x \in N$, we equip $N \oplus R$ with the structure of a differential module by setting

$$D(n, r) = (D(n) + rx, d(r)).$$

This module is indeed an extension of the desired form, and the image of $1 \in R$ under the resulting connecting homomorphism is precisely the class of x in $H^1(N)$. In the other direction, any extension splits at the level of modules, and so must have this form for some $x \in N$. This yields the claim. □

Remark 5.3.4. The isomorphisms $\text{Hom}_R(M_1, M_2) \cong M_1^\vee \otimes M_2$ and $M_1 \rightarrow (M_1^\vee)^\vee$ and the assertion of Lemma 5.3.3 carry over to the case where M_1 is a finite *projective* R-module, i.e., a direct summand of a finite free R-module. Such a module is always flat. A finitely generated R-module M is projective if and only if it is finitely presented and *locally free*, i.e., if there exists a finite subset $f_1, \ldots, f_m$ of R generating the unit ideal, such that $M[f_i^{-1}]$ is free over $R[f_i^{-1}]$ for each i [154, Exercise 4.12].

Remark 5.3.5. Let M be a finite projective module over a $\mathbb{Q}$-algebra R. The identity map on $M^\vee \otimes M$ can be thought of as an element of

$$(M^\vee \otimes M)^\vee \otimes (M^\vee \otimes M) \cong M \otimes M^\vee \otimes M^\vee \otimes M.$$

If we now swap the two factors of $M^\vee$ in the middle and reverse the construction, we get back a different map on $M^\vee \otimes M$. That map turns out to be the projection onto the trace component: when M is free this may be checked by computing in terms of a basis of M; this implies the general case by working locally on R (see Remark 5.3.4).

Now let N be a second projective R-module, let $f : M \rightarrow N$ be an isomorphism, and let $f^\vee : N^\vee \rightarrow M^\vee$ be the transpose map, which is also an isomorphism. We can then combine f and $f^\vee$ to obtain an isomorphism $M^\vee \otimes M \rightarrow N^\vee \otimes N$, which we view as an element of

$$(M^\vee \otimes M)^\vee \otimes (N^\vee \otimes N) \cong M \otimes M^\vee \otimes N^\vee \otimes N \cong (M^\vee \otimes N)^\vee \otimes (M^\vee \otimes N).$$

This, in turn, corresponds to a map from $M^\vee \otimes N$ to itself which is a projector of rank 1.

Finally, let $f : M^\vee \otimes M \rightarrow N^\vee \otimes N$ be an isomorphism; then f defines an element of

$$(M^\vee \otimes M)^\vee \otimes N^\vee \otimes N \cong M \otimes M^\vee \otimes N^\vee \otimes N \cong (M^\vee \otimes N)^\vee \otimes (M^\vee \otimes N),$$

but it is not automatic that the resulting map on $M^\vee \otimes N$ is a projector of rank 1. However, this is guaranteed if the diagram

$$
\begin{array}{ccc}
(M^\vee \otimes M) \otimes (M^\vee \otimes M) & \xrightarrow{\;-\circ-\;} & M^\vee \otimes M \\
\Big\downarrow{\scriptstyle f \otimes f} & & \Big\downarrow{\scriptstyle f} \\
(N^\vee \otimes N) \otimes (N^\vee \otimes N) & \xrightarrow{\;-\circ-\;} & N^\vee \otimes N
\end{array}
$$

commutes. Although this can again be checked by a direct computation in terms of basis elements, there is also a more conceptual interpretation: the commutativity promotes f to an isomorphism of $M^\vee \otimes M$ and $N^\vee \otimes N$ in the category of *Azumaya algebras* over R in the sense of [26] (as generalized to schemes by Grothendieck [184]). Such an isomorphism implies the existence of a projective module Q of rank 1 and an isomorphism $M \cong N \otimes Q$ such that f is defined by the identifications

$$
M^\vee \otimes M \cong (N \otimes Q)^\vee \otimes (N \otimes Q) \cong N^\vee \otimes N \otimes Q^\vee \otimes Q \cong N^\vee \otimes N;
$$

this follows from [25, Proposition A.6] as applied in the proof of [26, Proposition 5.3]. Then the map on $M^\vee \otimes N$ is obtained by identifying the latter with $N^\vee \otimes N \otimes Q^\vee$ and projecting onto the trace component of $N^\vee \otimes N$.

5.4 Cyclic vectors

Definition 5.4.1. Let R be a differential ring, and let M be a finite free differential module of rank n over R. A *cyclic vector* for M is an element $m \in M$ such that $m, D(m), \ldots, D^{n-1}(m)$ form a basis of M.

Theorem 5.4.2 (Cyclic vector theorem). *Let R be a differential field of characteristic 0 with nonzero derivation. Then every finite differential module over R has a cyclic vector.*

Many proofs are possible; we give here the proof from [149, Theorem III.4.2]. For another proof that applies over some rings other than fields, see Theorem 5.7.3. See also the notes for further discussion. (For a comment on characteristic p, see the exercises.)

Proof We start by normalizing the derivation. For $u \in R^\times$, given one differential module (M, D) over (R, d), we get another differential module (M, uD) over (R, ud), and m is a cyclic vector for one if and only if it is a cyclic vector for the other (because the image of m under $(uD)^j$ is in the span of $m, D(m), \ldots, D^j(m)$). We may thus assume (thanks to the assumption that the derivation is nontrivial) that there exists a nonzero element $x \in R$ such that $d(x) = x$.

Let M be a differential module of dimension n, and choose $m \in M$ so that the dimension μ of the span of $m, D(m), \ldots$ is as large as possible. We derive a contradiction under the hypothesis $\mu < n$.

For $z \in M$ and $\lambda \in \mathbb{Q}$,

$$(m + \lambda z) \wedge D(m + \lambda z) \wedge \cdots \wedge D^{\mu}(m + \lambda z) = 0$$

in the exterior power $\wedge^{\mu+1} M$. If we write this expression as a polynomial in λ, it vanishes for infinitely many values, so it must be identically zero. Hence each coefficient must vanish separately, including the coefficient of λ^1, which is

$$\sum_{i=0}^{\mu} m \wedge \cdots \wedge D^{i-1}(m) \wedge D^i(z) \wedge D^{i+1}(m) \wedge \cdots \wedge D^{\mu}(m). \qquad (5.4.2.1)$$

Pick $s \in \mathbb{Z}$, substitute $x^s z$ for z in (5.4.2.1), divide by x^s, and set the result equal to zero. We get

$$\sum_{i=0}^{\mu} s^i \Lambda_i(m, z) = 0 \qquad (s \in \mathbb{Z}) \qquad (5.4.2.2)$$

for

$$\Lambda_i(m, z) = \sum_{j=0}^{\mu-i} \binom{i+j}{i} m \wedge \cdots \wedge D^{i+j-1}(m) \wedge D^j(z) \wedge D^{i+j+1}(m) \wedge \cdots \wedge D^{\mu}(m).$$

Again because we are in characteristic 0, we may conclude that (5.4.2.2), viewed as a polynomial in s, has all coefficients equal to zero; that is, $\Lambda_i(m, z) = 0$ for all $m, z \in M$.

We now take $i = \mu$ to obtain

$$(m \wedge \cdots \wedge D^{\mu-1}(m)) \wedge z = 0 \qquad (m, z \in M);$$

since $\mu < n$, we may use this to deduce

$$m \wedge \cdots \wedge D^{\mu-1}(m) = 0 \qquad (m \in M).$$

But that means that the dimension of the span of $m, D(m), \ldots$ is at most $\mu - 1$, contradicting the definition of μ. $\qquad\qquad\square$

5.5 Differential polynomials

We now give the interpretation of differential modules as modules over a mildly noncommutative ring.

Definition 5.5.1. Let (R, d) be a differential ring. The *ring of twisted polynomials* $R\{T\}$ over R in the variable T is the additive group

$$R \oplus (R \cdot T) \oplus (R \cdot T^2) \oplus \cdots,$$

with noncommuting multiplication given by the formula

$$\left(\sum_{i=0}^{\infty} a_i T^i\right)\left(\sum_{j=0}^{\infty} b_j T^j\right) = \sum_{i,j=0}^{\infty} \sum_{h=0}^{i} \binom{i}{h} a_i d^h(b_j) T^{i+j-h}.$$

In other words, we impose the relation

$$Ta = aT + d(a) \qquad (a \in R)$$

and check that the result is a not necessarily commutative ring (see the exercises). We define the *degree* of a twisted polynomial in the usual way, as the exponent of the largest power of T with a nonzero coefficient; the degree of the zero polynomial may be taken to be any particular negative value.

Proposition 5.5.2 (Ore). *For R a differential field, the ring $R\{T\}$ admits a left division algorithm. That is, if $f, g \in R\{T\}$ and $g \neq 0$, then there exist unique $q, r \in R\{T\}$ with $\deg(r) < \deg(g)$ and $f = gq + r$. (There is also a right division algorithm.)*

Proof Exercise. □

Using the Euclidean algorithm, this yields the following consequence as in the untwisted case.

Theorem 5.5.3 (Ore). *Let R be a differential field. Then $R\{T\}$ is both left principal and right principal; that is, any left ideal (resp. right ideal) has the form $R\{T\}f$ (resp. $fR\{T\}$) for some $f \in R\{T\}$.*

Definition 5.5.4. Note that the opposite ring of $R\{T\}$, i.e., the ring in which multiplication is performed by first switching the order of the factors, is a twisted polynomial ring for the derivation $-d$. Given $f \in R\{T\}$, we define the *formal adjoint* of f as the element f in the opposite ring. This operation looks a bit less formal if you also push the coefficients over to the other side, giving what we will call the *adjoint form* of f. For instance, the adjoint form of $T^3 + aT^2 + bT + c$ is

$$T^3 + T^2 a + T(b - 2d(a)) + d^2(a) - d(b) + c.$$

Remark 5.5.5. The twisted polynomial ring is engineered precisely so that for any differential module M over R, we get an action of $R\{T\}$ on M under which T acts like D. In particular, $R\{T\}$ acts on R itself with T acting like d. In fact, the category of differential modules over R is equivalent to the category of left $R\{T\}$-modules. Moreover, if M is a finite differential module over R, any cyclic vector $m \in M$ corresponds to an isomorphism $M \cong R\{T\}/R\{T\}P$ for some monic twisted polynomial P, where the isomorphism carries m to the class of 1. (You might want to think of f as a sort of "characteristic polynomial" for M, except that it depends strongly on the choice of the cyclic vector.) Under such an isomorphism, a factorization $P = P_1P_2$ corresponds to a short exact sequence $0 \rightarrow M_1 \rightarrow M \rightarrow M_2 \rightarrow 0$ with

$$M_1 \cong R\{T\}P_2/R\{T\}P \cong R\{T\}/R\{T\}P_1, \qquad M_2 \cong R\{T\}/R\{T\}P_2.$$

5.6 Differential equations

You may have been wondering when differential equations will appear, as these purport to be the objects of study of this book. If so, your wait is over. (Note that we only consider *linear* differential equations here.)

Definition 5.6.1. A *differential equation of order n* over the differential ring (R, d) is an equation of the form

$$(a_n d^n + \cdots + a_1 d + a_0)y = b,$$

with $a_0, \ldots, a_n, b \in R$ and y indeterminate. We say the equation is *homogeneous* if $b = 0$ and *inhomogeneous* otherwise.

Remark 5.6.2. Using our setup, we may write this equation as $f(d)y = b$ for some $f \in R\{T\}$. Similarly, we may view systems of differential equations as equations of the form $f(D)y = b$ where b lives in some differential module (M, D). By the usual method of introducing extra variables corresponding to derivatives of y, we can convert any differential system into a first-order system $Dy = b$. We can also convert an inhomogeneous system into a homogeneous one by adding an extra variable, with the understanding that we would like the value of that last variable to be 1 in order to get back a solution of the original equation.

Remark 5.6.3. Here is a more explicit relationship between adjoint polynomials and solving differential equations. Suppose that we start with the cyclic differential module $M \cong R\{T\}/R\{T\}f$ and want to find a horizontal element. That means that we want $g \in R\{T\}$ such that $Tg \in R\{T\}f$; we may as well assume that $\deg(g) < \deg(f)$. By comparing degrees, we see that $Tg = rf$ for some $r \in R$. Write f in adjoint form as $f_0 + Tf_1 + \cdots + T^n$; then

$$rf \equiv rf_0 - d(r)f_1 + d^2(r)f_2 - \cdots + (-1)^n d^n(r) \quad \bmod TR\{T\}.$$

In this manner, finding a horizontal element becomes equivalent to solving a differential equation.

Definition 5.6.4. We recall a classical terminology here. For a sequence $u_1, \ldots, u_n$ in a differential ring (R, d), the *Wronskian matrix* is the $n \times n$ matrix

$$W(u_1, \ldots, u_n) = \begin{pmatrix} u_1 & u_2 & \cdots & u_n \\ d(u_1) & d(u_2) & \cdots & d(u_n) \\ \vdots & \vdots & \ddots & \vdots \\ d^{n-1}(u_1) & d^{n-1}(u_2) & \cdots & d^{n-1}(u_n) \end{pmatrix}.$$

If R is a differential field, then this matrix is singular if and only if $u_1, \ldots, u_n$ satisfy some differential equation of order less than n. (The term *Wronskian* is often used to refer to the determinant of the Wronskian matrix, but we will not do this.)

5.7 Cyclic vectors: a mixed blessing

The reader may at this point be wondering why so many points of view are necessary, since the cyclic vector theorem can be used to transform any differential module into a differential equation, and ultimately differential equations are the things one writes down and wants to solve. Please permit me to interject here a countervailing opinion.

Remark 5.7.1. In ordinary linear algebra (or in other words, when considering differential modules for the trivial derivation), one can pass freely between linear transformations on a vector space and square matrices if one is willing to choose a basis. The merits of making such a choice depend on the situation, so it is valuable to have both the matricial and coordinate-free viewpoints well in hand. One can then pass to the characteristic polynomial, but not all information is retained (one loses information about nilpotency) and even information that in principle is retained is sometimes not so conveniently accessed. In short, no one would seriously argue that one can dispense with studying matrices because of the existence of the characteristic polynomial.

The situation is not so different in the differential case. The difference between a differential module and a differential system is merely the choice of a basis, and again it is valuable to have both points of view in mind. However, the cyclic vector theorem may seduce one into thinking that collapsing a differential system into a differential polynomial is an operation without drawbacks, whereas this is far from the case. For instance, determining whether two differential polynomials correspond to the same differential system is not straightforward.

More seriously for our purposes, the cyclic vector theorem only applies over a differential field. Many differential modules are more naturally defined over some ring which is not a field. For instance, differential modules arising from geometry (as Picard–Fuchs modules) are usually defined over a ring of functions on some geometric space. While there are forms of the cyclic vector theorem available over nonfields (for instance, see Theorem 5.7.3 below), these do not suffice for our purposes. We also find that working with differential modules instead of differential polynomials has a tremendously clarifying effect, partly because it improves the parallelism with difference algebra, where there is no good analogue of the cyclic vector theorem even over a field. (See Part IV.)

We find it unfortunate that much of the literature on complex ordinary differential equations, and nearly all of the literature on p-adic ordinary differential equations, is mired in the language of differential polynomials. By instead switching between differential modules and differential polynomials as appropriate, we will be able to demonstrate strategies that lead to a more systematic development of the p-adic theory.

As promised, we offer some results concerning cyclic vectors over rings, due to Katz [227].

Theorem 5.7.2. *Let R be a differential local $\mathbb{Q}$-algebra. Suppose that the maximal ideal of R contains an element t such that $d(t) = 1$. Let M be a finite free differential module over R and let $e_1, \ldots, e_n$ be a basis of M. Then*

$$v = \sum_{j=0}^{n-1} \frac{t^j}{j!} \sum_{k=0}^{j} (-1)^k \binom{j}{k} D^k(e_{j+1-k})$$

is a cyclic vector of M.

Proof For $i, j \geq 0$, put

$$c(i, j) = \sum_{k=0}^{j} (-1)^k \binom{j}{k} D^k(e_{i+j+1-k})$$

with the convention that $e_h = 0$ for $h > n$. From the definition,

$$c(i + 1, j) = D(c(i, j)) + c(i, j + 1).$$

By induction on i, we find that

$$D^i(v) = \sum_{j=0}^{n-1} \frac{t^j}{j!} c(i, j) \equiv e_{i+1} \pmod{t}.$$

Hence $v, D(v), \ldots, D^{n-1}(v)$ freely generate M modulo the maximal ideal of R. They thus freely generate M itself by Nakayama's lemma [154, Corollary 4.8].

□

Theorem 5.7.3. *Let R be a differential $\mathbb{Q}$-algebra containing an element t for which $d(t) = 1$. Let M be a finite free differential module over R, and let I be a prime ideal of R. Then there exists $f \in R \setminus I$ such that $M \otimes_R R[f^{-1}]$ contains a cyclic vector.*

Proof Choose a basis $e_1, \ldots, e_n$ of M, and define $c(i, j)$ as in the proof of Theorem 5.7.2. For $x \in R$, put

$$c_i(x) = \sum_{j=0}^{n-1} \frac{x^j}{j!} c(i, j).$$

Then there exists a polynomial $P(T) \in R[T]$ such that

$$c_0(x) \wedge \cdots \wedge c_{n-1}(x) = P(x)(e_1 \wedge \cdots \wedge e_n) \qquad (x \in R).$$

Since $c(i, 0) = e_{i+1}$, $P(0) = 1$. In particular, the image $Q(T) \in (R/I)[T]$ of $P(T)$ is not identically zero.

Since R contains $\mathbb{Q}$, the images of $t - a$ for $a \in \mathbb{Z}$ are all distinct, so only finitely many of them are roots of $Q(T)$. We can thus choose $a \in \mathbb{Z}$ so that $P(t - a) \notin I$. Since

$$D^i(c_0(t - a)) = c_i(t - a)$$

as in the previous proof, $c_0(t - a)$ is a cyclic vector of $M \otimes_R R[f^{-1}]$ for $f = P(t - a)$, as desired.

□

5.8 Taylor series

Definition 5.8.1. Let R be a *topological* differential ring, i.e., a ring equipped with a topology and a derivation such that all operations are continuous. Assume also that R is a $\mathbb{Q}$-algebra. Let M be a topological differential module over R, i.e., a differential module such that all operations are continuous. For $r \in R$ and $m \in M$, we define the *Taylor series* $T(r, m)$ as the infinite sum

$$\sum_{i=0}^{\infty} \frac{r^i}{i!} D^i(m)$$

whenever the sum converges absolutely (i.e., all rearrangements converge to the same value).

Remark 5.8.2. The expression $T(r, m)$ is *de facto* additive in m: if $m_1, m_2 \in M$, then

$$T(r, m_1) + T(r, m_2) = T(r, m_1 + m_2)$$

whenever at least two of the three terms are well-defined (as then is the third term). For $s \in R$, $T(r, s)$ is also *de facto* multiplicative: if $s_1, s_2 \in R$, then (by the Leibniz rule)

$$T(r, s_1)T(r, s_2) = T(r, s_1 s_2)$$

whenever the two terms on the left are well-defined (as then is the third term). More generally, for $m \in M$, $T(r, m)$ is *de facto* semilinear: if $s \in R$, $m \in M$, then

$$T(r, s)T(r, m) = T(r, sm)$$

whenever the two terms on the left are well-defined (as then is the third term).

Example 5.8.3. A key instance of the previous remark is the case where R is a completion of a rational function field $F(t)$ and $d = d/dt$. In this case, the ring homomorphism $T(r, \ )$ is the substitution $t \mapsto t + r$; note that this cannot make sense except possibly if $|r| \leq 1$.

Remark 5.8.4. Another use for Taylor series is to construct horizontal sections. Note that

$$D(T(r, m)) = \sum_{i=1}^{\infty} d(r) \frac{r^{i-1}}{(i-1)!} D^i(m) + \sum_{i=0}^{\infty} \frac{r^i}{i!} D^{i+1}(m)$$
$$= (1 + d(r))T(r, D(m))$$

if everything converges absolutely. In particular, if $d(r) = -1$, then $T(r, m)$ is horizontal.

Notes

The subject of differential algebra is rather well-developed; a classic treatment, though possibly too dry to be useful to the casual reader, is the book of Ritt [348]. As in abstract algebra in general, development of differential algebra was partly driven by *differential Galois theory*, i.e., the study of when solutions of differential equations can be expressed in terms of solutions to ostensibly simpler differential equations. A relatively lively introduction to the latter is [375].

Calling an element of a differential module *horizontal* when it is killed by the derivation makes sense if you consider *connections* in differential geometry. In that setting, the differential operator is measuring the extent to which a section of a vector bundle deviates from some prescribed "horizontal" direction identifying points on one fiber with points on nearby fibers.

Remark 5.3.5 is the point of departure for a study of the Brauer group of a differential ring. See [206].

The history of the cyclic vector theorem is rather complicated. It appears to have been first proved by Loewy [296] in the case of meromorphic functions, and (independently) by Cope [113] in the case of rational functions. For a detailed historical discussion, see [101]. An application of Katz's method to some differential modules over Banach rings can be found in [344].

Twisted polynomials were introduced by Ore [329]. They are actually somewhat more general than we have discussed; for instance, one can also twist by an endomorphism $\tau \colon R \to R$ by imposing the relation $Ta = \tau(a)T$. (This enters the realm of the analogue of differential algebra called *difference algebra*, which we will treat in Part IV.) Moreover, one can twist by both an endomorphism and a derivation if they are compatible in an appropriate way, and one can even study differential or difference Galois theory in this setting. A unifying framework for doing so, which is also suitable for considering multiple derivations and automorphisms, is given by André [15].

Differential algebra in positive characteristic has a rather different flavor than in characteristic 0; for instance, the p-th power of the derivation d/dt on $\mathbb{F}_p(t)$ is the zero map. A brief discussion of the characteristic-p situation is given in [149, §III.1].

Exercises

5.1 Prove that if M is a locally free differential module over R of rank 1, then $M^\vee \otimes M$ is trivial (as a differential module).

5.2 Check that the bijection $\mathrm{Ext}(R, N) \to H^1(N)$ constructed in the proof of Lemma 5.3.3 is indeed a homomorphism.

5.3 Verify the assertions of Remark 5.3.5.

5.4 Check that in characteristic $p > 0$, the cyclic vector theorem holds for modules of rank at most p, but may fail for modules of rank $p + 1$.

5.5 Give a counterexample to the cyclic vector theorem for a differential field of characteristic 0 with trivial derivation.

5.6 Verify that $R\{T\}$ is indeed a not necessarily commutative ring; the content in this is to check the associativity of multiplication.

5.7 Prove the left division algorithm for a twisted polynomial ring over a differential field (see Proposition 5.5.2).

5.8 Prove Lemma 5.1.8.

6

Metric properties of differential modules

In this chapter, we study the metric properties of differential modules over nonarchimedean differential rings. The principal invariant that we identify is a familiar quantity from functional analysis, the *spectral radius* of a bounded endomorphism. When the endomorphism is the derivation acting on a differential module, the spectral radius can be related to the least slope of the Newton polygon of the corresponding twisted polynomial.

We give meaning to the other slopes as well by proving that over a complete nonarchimedean differential field, any differential module decomposes into components whose spectral radii are computed by the various slopes of the Newton polygon. However, this theorem will provide somewhat incomplete results when we apply it to p-adic differential modules in Part III; we will have to remedy the situation using Frobenius descendants and antecedents.

This chapter provides important foundational material for much of what follows, but on its own it may prove undigestably abstract at first. The reader experiencing this is advised to read Chapter 7 in conjunction with this one, to see how the constructions of this chapter become explicit in a simple but important class of examples.

6.1 Spectral radii of bounded endomorphisms

Before considering differential operators, let us recall the difference between the operator norm and the spectral radius of a bounded endomorphism of an abelian group.

Hypothesis 6.1.1. Throughout this section, let G be a nonzero abelian group equipped with a norm $| \cdot |$, and let $T : G \to G$ be a bounded endomorphism of G. Recall that this means there exists $c \geq 0$ such that $|T(g)| \leq c|g|$ for all $g \in G$.

Definition 6.1.2. The *operator norm* $|T|_G$ of T is defined to be the least $c \geq 0$ for which $|T(g)| \leq c|g|$ for all $g \in G$, i.e.,

$$|T|_G = \sup_{g \in G, g \neq 0} \{|T(g)|/|g|\}.$$

Recall that if M is a finite free module over a nonarchimedean ring R, $|\cdot|_M$ is the supremum norm for some basis, and T is an R-linear transformation, then $|T| = |A|$ where A is the matrix of action of T on the chosen basis of M and $|A| = \sup_{ij}\{|A_{ij}|\}$ as in Chapter 4.

One can similarly define the operator norm for a map between two different abelian groups, each equipped with a norm. We may even allow seminorms as long as we take the supremum over elements of the source group which are not in the kernel of the seminorm.

Although the condition that T is bounded is preserved by replacing the norm by a metrically equivalent norm, the operator norm is not preserved. To obtain a less fragile numerical invariant, we introduce the spectral radius.

Definition 6.1.3. The *spectral radius* of T is defined as

$$|T|_{\mathrm{sp},G} = \lim_{s \to \infty} |T^s|_G^{1/s};$$

the existence of the limit follows from the fact $|T^{m+n}|_G \leq |T^m|_G |T^n|_G$ and the following lemma.

Lemma 6.1.4 (Fekete). *Let $\{a_n\}_{n=1}^{\infty}$ be a sequence of real numbers such that $a_{m+n} \geq a_m + a_n$ for all m, n. Then the sequence $\{a_n/n\}_{n=1}^{\infty}$ either converges to its supremum or diverges to ∞.*

Proof Exercise.   □

Proposition 6.1.5. *The spectral radius of T depends on the norm $|\cdot|$ only up to metric equivalence.*

Proof Suppose $|\cdot|'$ is a norm metrically equivalent to $|\cdot|$. We can then choose $c > 0$ such that $c^{-1}|g| \leq |g|' \leq c|g|$ for all $g \in G$. We then have $|T(g)|/|g| \leq c^2 |T(g)|'/|g|'$ for all $g \in G \setminus \{e\}$. Applying this with T replaced by T^s gives $|T^s|_G \leq c^2 |T^s|'_G$, so

$$|T^s|_{\mathrm{sp},G} \leq \lim_{s \to \infty} c^{2/s}(|T^s|'_{\mathrm{sp},G})^{1/s}.$$

Since $c^{2/s} \to 1$ as $s \to \infty$, this gives $|T^s|_{\mathrm{sp},G} \leq |T^s|'_{\mathrm{sp},G}$. The reverse inequality is seen to hold by reversing the roles of the norms. □

We have the following relationship with the concepts studied in Chapter 4.

Remark 6.1.6. Let F be a field equipped with a norm, and let V be a finite-dimensional vector space over F. Pick a basis for V and equip V with either the L^2 norm or the supremum norm defined by this basis, according as whether F is archimedean or nonarchimedean. Let A be the matrix of action of T on this basis. Then $|T|_V$ equals the largest singular value of A, whereas $|T|_{\mathrm{sp},V}$ equals the largest norm of an eigenvalue of A (by Proposition 4.2.4 in the archimedean case, and Proposition 4.4.11 in the nonarchimedean case).

Remark 6.1.7. The spectral radius derives its name from the fact that by a celebrated theorem of Gelfand, for a bounded endomorphism of a commutative Banach algebra over $\mathbb{R}$ or $\mathbb{C}$, the spectral radius computes the radius of the smallest disc containing the entire spectrum of the operator. (This includes the archimedean case of the previous remark.)

When dealing with the spectral radius, we will frequently use the following observation.

Lemma 6.1.8. *For any $\epsilon > 0$, there exists $c = c(\epsilon)$ such that for all $s \geq 0$,*

$$|T^s|_G \leq c(|T|_{\mathrm{sp},G} + \epsilon)^s.$$

Proof For $c = 1$, the claim already holds for all but finitely many s. It thus suffices to increase c to cover the finite set of exceptions. □

Remark 6.1.9. Some authors writing about p-adic differential equations (including the present author) tend to refer to the spectral radius as the *spectral norm*. This is somewhat dangerous, because the spectral radius is not in general a norm or even a seminorm, even for matrices over a complete field (exercise). On the other hand, we will be using the spectral radius as a measure of size, and so the normlike notation $|\cdot|_{\mathrm{sp},V}$ is useful. We will compromise by keeping this notation but referring to spectral radii rather than spectral norms.

6.2 Spectral radii of differential operators

We now specialize the previous discussion to the case of differential modules.

Definition 6.2.1. By a *nonarchimedean differential ring* or a *nonarchimedean differential field*, we mean a nonarchimedean ring or field equipped with a *bounded* derivation. For F a nonarchimedean differential field, we can define the operator norm $|d|_F$ and the spectral radius $|d|_{\mathrm{sp},F}$; by hypothesis the former is finite, so the latter is too.

Definition 6.2.2. Let F be a nonarchimedean differential field. By a *normed differential module* over F, we mean a vector space V over F equipped with a compatible norm $|\cdot|_V$ and a derivation D with respect to d which is bounded as a endomorphism of the additive group of V. For V nonzero, we may then consider the operator norm $|D|_V$ and the spectral radius $|D|_{\mathrm{sp},V}$.

Remark 6.2.3. If V is finite-dimensional over F and F is complete, then the spectral radius does not depend on the norm on V, since by Theorem 1.3.6 any two norms on V compatible with the norm on F are metrically equivalent.

In general, one cannot have differential modules with arbitrarily small spectral radius.

Lemma 6.2.4. *Let F be a nonarchimedean differential field and let V be a nonzero normed differential module over F. Then*

$$|D|_{\mathrm{sp},V} \geq |d|_{\mathrm{sp},F}.$$

Proof (This proof was suggested by Liang Xiao.) For $a \in F$ and $v \in V$ nonzero, the Leibniz rule gives

$$D^{s-i}(aD^i(v)) = d^{s-i}(a)D^i(v) + \sum_{j=1}^{s-i}\binom{s-i}{j}d^{s-i-j}(a)D^{i+j}(v) \qquad (0 \leq i \leq s).$$

Inverting this system of equations gives an identity of the form

$$d^s(a)v = \sum_{i=0}^{s} c_{s,i}D^{s-i}(aD^i(v))$$

for certain universal constants $c_{s,i} \in \mathbb{Z}$. Consequently,

$$|d^s(a)v|_V \leq \max_{0 \leq i \leq s}\{|D^{s-i}(aD^i(v))|_V\}. \qquad (6.2.4.1)$$

By Lemma 6.1.8, given $\epsilon > 0$, we can choose $c = c(\epsilon)$ such that for all $s \geq 0$,

$$|D^s|_V \leq c(|D|_{\mathrm{sp},V} + \epsilon)^s.$$

Using (6.2.4.1), we deduce

$$|d^s(a)v|_V \leq c^2(|D|_{\mathrm{sp},V} + \epsilon)^s|a||v|_V.$$

Dividing by $|a||v|_V$ and taking the supremum over $a \in F$, we obtain

$$|d^s|_F \leq c^2(|D|_{\mathrm{sp},V} + \epsilon)^s.$$

Taking s-th roots and then letting $s \to \infty$, we get

$$|d|_{\mathrm{sp},F} \leq |D|_{\mathrm{sp},V} + \epsilon.$$

Since $\epsilon > 0$ was arbitrary, this yields the claim. □

It is sometimes useful to compute in terms of a basis of V over F.

Lemma 6.2.5. *Let F be a complete nonarchimedean differential field and let V be a nonzero finite differential module over F. Fix a basis $e_1, \ldots, e_n$ of V and let D_s be the matrix of action of D^s on this basis (i.e., $D^s(e_j) = \sum_i (D_s)_{ij} e_i$). Then*

$$|D|_{\mathrm{sp},V} = \max\{|d|_{\mathrm{sp},F}, \limsup_{s \to \infty} |D_s|^{1/s}\}. \tag{6.2.5.1}$$

Proof (Compare [96, Proposition 1.3].) Equip V with the supremum norm defined by $e_1, \ldots, e_n$; then $|D^s|_V \geq \max_{i,j} |(D_s)_{i,j}|$. This plus Lemma 6.2.4 imply that the left side of (6.2.5.1) is greater than or equal to the the right side.

Conversely, for any $x \in V$, if we write $x = x_1 e_1 + \cdots + x_n e_n$ with $x_1, \ldots, x_n \in F$, then

$$D^s(x) = \sum_{i=1}^{n} \sum_{j=0}^{s} \binom{s}{j} d^j(x_i) D^{s-j}(e_i),$$

so

$$|D^s|_V^{1/s} \leq \max_{0 \leq j \leq s} \{|d^j|_F^{1/s} |D_{s-j}|^{1/s}\}. \tag{6.2.5.2}$$

Given $\epsilon > 0$, apply Lemma 6.1.8 to choose $c = c(\epsilon)$ such that for all $s \geq 0$,

$$|d^s|_F \leq c(|d|_{\mathrm{sp},F} + \epsilon)^s$$
$$|D_s|_V \leq c(\limsup_{s \to \infty} |D_s|^{1/s} + \epsilon)^s.$$

Then (6.2.5.2) implies

$$|D^s|_V^{1/s} \leq c^{2/s} \max\{|d|_{\mathrm{sp},F} + \epsilon, \limsup_{s \to \infty} |D_s|^{1/s} + \epsilon\}.$$

As in the previous proof, the factor $c^{2/s}$ tends to 1 as $s \to \infty$. From this it follows that the right side of (6.2.5.1) is greater than or equal to the left side minus ϵ; since $\epsilon > 0$ was arbitrary, we get the same inequality with $\epsilon = 0$. □

Remark 6.2.6. In Lemma 6.2.5, if $|D|_{\mathrm{sp},V} > |d|_{\mathrm{sp},F}$, then the limit superior can be replaced by a limit; see exercises.

Using Lemma 6.2.5, we may infer the following base-change property for the spectral radius. We will need to refine this result later; see Corollary 6.5.5 and Proposition 10.6.6.

Corollary 6.2.7. *Let $F \to F'$ be an isometric embedding of complete nonarchimedean differential fields. (In particular, the differential on F' must restrict to the differential on F.) Then for any nonzero finite differential module V over F,*

$$|D|_{\mathrm{sp},V'} = \max\{|d|_{\mathrm{sp},F'}, |D|_{\mathrm{sp},V}\}.$$

Consequently, if $|d|_{\mathrm{sp},F'} = |d|_{\mathrm{sp},F}$, then $|D|_{\mathrm{sp},V'} = |D|_{\mathrm{sp},V}$ by Lemma 6.2.4.

Proof Choose a basis of V and use Lemma 6.2.5 to compute both $|D|_{\mathrm{sp},V}$ and $|D|_{\mathrm{sp},V'}$ in terms of this basis. This yields the desired formula. □

Here is how spectral radius behaves with respect to basic operations on the category of differential modules.

Lemma 6.2.8. *Let F be a complete nonarchimedean differential field.*

(a) *For a short exact sequence $0 \to V_1 \to V \to V_2 \to 0$ of nonzero finite differential modules over F,*

$$|D|_{\mathrm{sp},V} = \max\{|D|_{\mathrm{sp},V_1}, |D|_{\mathrm{sp},V_2}\}.$$

(b) *For V a nonzero finite differential module over F,*

$$|D|_{\mathrm{sp},V^\vee} = |D|_{\mathrm{sp},V}.$$

(c) *For V_1, V_2 nonzero finite differential modules over F,*

$$|D|_{\mathrm{sp},V_1 \otimes V_2} \leq \max\{|D|_{\mathrm{sp},V_1}, |D|_{\mathrm{sp},V_2}\},$$

with equality when $|D|_{\mathrm{sp},V_1} \neq |D|_{\mathrm{sp},V_2}$.

Proof To prove (a), choose a splitting $V = V_1 \oplus V_2$ of the short exact sequence in the category of vector spaces over F. The action of D on V is then given by $D(v_1, v_2) = (D(v_1) + f(v_2), D(v_2))$ for some F-linear map $f : V_2 \to V_1$ (as in the proof of Lemma 5.3.3). From this, the rest of the proof of (a) is an exercise in the spirit of the proofs of Lemma 6.2.4 and Lemma 6.2.5.

We next recall from Definition 5.3.2 that for any finite differential modules V_1, V_2 over F, the action of D on $V = \mathrm{Hom}_F(V_1, V_2)$ is given by

$$D(f)(v_1) = D(f(v_1)) - f(D(v_1)).$$

The Leibniz rule in this setting is that for any nonnegative integer s,

$$D^s(f)(v_1) = \sum_{i=0}^{s} (-1)^i \binom{i}{s} D^i(f(D^{s-i}(v_1))).$$

We may again (as an exercise) deduce that $|D|_{\mathrm{sp},V} \leq \max\{|D|_{\mathrm{sp},V_1}, |D|_{\mathrm{sp},V_2}\}$.

This implies first (b) (since we get $|D|_{\mathrm{sp},V^\vee} \leq |D|_{\mathrm{sp},V}$ by Lemma 6.2.4, and similarly in reverse), then the first assertion of (c) (since $V_1 \otimes_F V_2 = \mathrm{Hom}_F(V_1^\vee, V_2)$).

It remains to prove the second assertion of (c). Suppose that $|D|_{\mathrm{sp},V_1} > |D|_{\mathrm{sp},V_2}$. Then by (b) and the first assertion of (c),

$$\begin{aligned}
|D|_{\mathrm{sp},V_1} &= \max\{|D|_{\mathrm{sp},V_1}, |D|_{\mathrm{sp},V_2}\} \\
&\geq \max\{|D|_{\mathrm{sp},V_1 \otimes V_2}, |D|_{\mathrm{sp},V_2^\vee}\} \\
&\geq |D|_{\mathrm{sp},V_1 \otimes V_2 \otimes V_2^\vee}.
\end{aligned}$$

Moreover, $V_2 \otimes V_2^\vee$ contains a trivial submodule (the trace), so $V_1 \otimes V_2 \otimes V_2^\vee$ contains a copy of V_1. Hence by (a), $|D|_{\mathrm{sp},V_1 \otimes V_2 \otimes V_2^\vee} \geq |D|_{\mathrm{sp},V_1}$. We thus obtain a chain of inequalities leading to $|D|_{\mathrm{sp},V_1} \geq |D|_{\mathrm{sp},V_1}$; this forces the intermediate equality $|D|_{\mathrm{sp},V_1} = \max\{|D|_{\mathrm{sp},V_1 \otimes V_2}, |D|_{\mathrm{sp},V_2^\vee}\}$. Since $|D|_{\mathrm{sp},V_1} \neq |D|_{\mathrm{sp},V_2} = |D|_{\mathrm{sp},V_2^\vee}$, we can only have $|D|_{\mathrm{sp},V_1} = |D|_{\mathrm{sp},V_1 \otimes V_2}$, as desired. □

Corollary 6.2.9. *If V_1, V_2 are irreducible finite differential modules over a complete nonarchimedean differential field and $|D|_{\mathrm{sp},V_1} \neq |D|_{\mathrm{sp},V_2}$, then every irreducible submodule W of $V_1 \otimes V_2$ satisfies $|D|_{\mathrm{sp},W} = \max\{|D|_{\mathrm{sp},V_1}, |D|_{\mathrm{sp},V_2}\}$.*

Proof Suppose the contrary; we may assume that $|D|_{\mathrm{sp},V_1} > |D|_{\mathrm{sp},V_2}$. The inclusion $W \hookrightarrow V_1 \otimes V_2$ corresponds to a nonzero horizontal section of $W^\vee \otimes V_1 \otimes V_2 \cong (W \otimes V_2^\vee)^\vee \otimes V_1$, which in turn corresponds to a nonzero map $W \otimes V_2^\vee \to V_1$. Since V_1 is irreducible, the map has image V_1; that is, $W \otimes V_2^\vee$ has a quotient isomorphic to V_1.

However, we can contradict this using Lemma 6.2.8. Namely,

$$|D|_{\mathrm{sp},W \otimes V_2^\vee} \leq \max\{|D|_{\mathrm{sp},W}, |D|_{\mathrm{sp},V_2}\} < |D|_{\mathrm{sp},V_1},$$

so each nonzero subquotient of $W \otimes V_2^\vee$ has spectral radius strictly less than $|D|_{\mathrm{sp},V_1}$. □

Remark 6.2.10. By contrast, when $|D|_{\mathrm{sp},V_1} = |D|_{\mathrm{sp},V_2}$, it is possible for an irreducible submodule W of $V_1 \otimes V_2$ to satisfy $|D|_{\mathrm{sp},W} \neq \max\{|D|_{\mathrm{sp},V_1}, |D|_{\mathrm{sp},V_2}\}$. For instance, take V_1 with $|D|_{\mathrm{sp},V_1} > |d|_{\mathrm{sp},F}$, put $V_2 = V_1^\vee$, and let W be the trace component of $V_1 \otimes V_2$.

Remark 6.2.11. It would be convenient to have the analogue of Corollary 6.2.9 also for irreducible subquotients of $V_1 \otimes V_2$. We will prove something slightly weaker later (Corollary 6.6.3).

We now refine the notion of the spectral radius to give a working notion of the spectrum of a differential operator.

Definition 6.2.12. For V a finite differential module over a nonarchimedean differential field F, let $V_1, \ldots, V_l$ be the Jordan–Hölder constituents of V (listed with multiplicity). Define the *full spectrum* of V to be the multiset consisting of $|D|_{\mathrm{sp},V_i}$ with multiplicity $\dim_F V_i$, for $i = 1, \ldots, l$. We say that V is *pure* if its full spectrum consists of a single element (with multiplicity).

We say that V is *refined* if $|D|_{\mathrm{sp},V^\vee \otimes V} < |D|_{\mathrm{sp},V}$. This implies that V is pure, by the following reasoning. If V is not pure, then it admits a subquotient V_1 with $|D|_{\mathrm{sp},V_1} = |D|_{\mathrm{sp},V}$ and another subquotient V_2 with $|D|_{\mathrm{sp},V_2} < |D|_{\mathrm{sp},V}$. Since $V_2^\vee \otimes V_1$ occurs as a subquotient of $V^\otimes \otimes V$, $|D|_{\mathrm{sp},V^\vee \otimes V} \geq |D|_{\mathrm{sp},V_2^\vee \otimes V_1} = |D|_{\mathrm{sp},V}$ and so V is not refined.

If V and W are two refined finite differential modules over F, we write $V \sim W$ if

$$|D|_{\mathrm{sp},V} = |D|_{\mathrm{sp},W} > |D|_{\mathrm{sp},V^\vee \otimes W}.$$

We will see that this is an equivalence relation (Lemma 6.2.14).

Remark 6.2.13. It may be helpful to keep in mind how the above notions behave when the derivation on F is zero. In this case, a finite differential module over F is simply a finite-dimensional vector space V equipped with a linear transformation. The full spectrum consists of the norms of the eigenvalues in F^{alg} of this linear transformation. The module V is pure if these eigenvalues all have the same norm. The module V is refined if it is pure and additionally the ratio of any two eigenvalues is congruent to 1 modulo $\mathfrak{m}_F$. In particular, we can always decompose a finite differential module into pure summands, but not necessarily into refined summands; for instance, one cannot separate a linear transformation over $\mathbb{Q}_p$ with characteristic polynomial $T^2 - p$.

On the other hand, we can achieve a refined decomposition after a finite *tamely ramified* extension. We may see this from Definition 3.3.3 as follows. Let E be a finite Galois extension of F containing the eigenvalues of the linear transformation. Then for any g in the wild inertia subgroup $W_{E/F}$ of $G_{E/F}$ and any eigenvalue λ, $|g(\lambda)/\lambda - 1| < 1$. We thus obtain a refined decomposition over the fixed field of $W_{E/F}$.

We can also describe a suitable tamely refined extension F' of F explicitly, by ensuring that it satisfies the following conditions in terms of the characteristic polynomial Q of the linear transformation.

(a) Let p be the characteristic of κ_F. For each root λ of Q,

$$|\lambda| \in \begin{cases} |(F')^\times| & p = 0 \\ \bigcup_{h \geq 0} |(F')^\times|^{1/p^h} & p > 0. \end{cases}$$

(b) For any two roots λ, μ of Q,

$$\overline{\lambda/\mu} \in \kappa_{F'}.$$

Lemma 6.2.14. *Let F be a nonarchimedean differential field. Then the relation $\sim$ on refined finite differential modules over F is an equivalence relation.*

Proof The reflexivity of $\sim$ holds because we only consider refined modules. The symmetry of $\sim$ holds because we have $|D|_{\mathrm{sp},V^\vee \otimes W} = |D|_{\mathrm{sp},W^\vee \otimes V}$ by Lemma 6.2.8(b). To check the transitivity of $\sim$, suppose $V \sim W$ and $W \sim X$. Since $V^\vee \otimes X$ occurs as a direct summand of $(V^\vee \otimes W) \otimes (W^\vee \otimes X)$, by Lemma 6.2.8(a),(c),

$$
\begin{aligned}
|D|_{\mathrm{sp},V^\vee \otimes X} &\le |D|_{\mathrm{sp},V^\vee \otimes W \otimes W^\vee \otimes X} \\
&\le \max\{|D|_{\mathrm{sp},V^\vee \otimes W}, |D|_{\mathrm{sp},W^\vee \otimes X}\} \\
&< |D|_{\mathrm{sp},V} = |D|_{\mathrm{sp},W} = |D|_{\mathrm{sp},X}.
\end{aligned}
$$

Hence $V \sim X$. $\qquad\qquad\square$

Lemma 6.2.15. *Let F be a complete nonarchimedean differential field and let F′ be a finite tamely ramified extension of F. Then d extends uniquely to F′ and $|d|_{F'} = |d|_F$.*

Proof Suppose first that F' is unramified over F. For $\alpha \in F'$, let $P(T) = \sum_i P_i T^i \in F[T]$ be the minimal polynomial of α. Then the unique extension of d to F' is characterized by

$$0 = d(\alpha)P'(\alpha) + \sum_i d(P_i)\alpha^i. \qquad (6.2.15.1)$$

If $\alpha \in \mathfrak{o}_{F'}^\times$, then the unramified condition implies that $|P'(\alpha)| = 1$, so

$$|d(\alpha)| \le \max_i\{|d(P_i)||\alpha|^i\} \le |d|_F.$$

In general, we can write any $\alpha \in F'$ as $\beta\gamma$ with $\beta \in \mathfrak{o}_{F'}^\times$ and $\gamma \in F$. We then have $d(\alpha) = \gamma d(\beta) + \beta d(\gamma)$, so

$$|d(\alpha)| \le \max\{|\gamma||d(\beta)|, |\beta||d(\gamma)|\} \le |d|_F |\beta||\gamma| = |d|_F |\alpha|.$$

This proves the claim.

We next verify the claim in case $F' = F(t^{1/m})$ for some $t \in F$ and some integer m not divisible by the characteristic p of κ_F, such that $|t|^{1/e} \notin |F^\times|$ for any divisor $e > 1$ of m. Again, the unique extension of d to F' is given by (6.2.15.1). Also, each element of F' has a unique expression of the form $\sum_{i=0}^{m-1} c_i t^{i/m}$ with $c_i \in F$, and the norm on F' is given by

$$\left| \sum_{i=0}^{m-1} c_i t^{i/m} \right| = \max_i \{|c_i||t|^{i/m}\}.$$

It thus suffices to check that $|d(c_i t^{i/m})| \leq |d|_F |c_i t^{i/m}|$ for $i = 0, \ldots, m-1$. By the Leibniz rule, this reduces to checking $|d(t^{i/m})| \leq |d|_F |t^{i/m}|$. But

$$|d(t^{i/m})| = |t^{i/m-1} d(t)| \leq |t^{i/m}||t|^{-1}|d|_F |t| = |d|_F |t^{i/m}|,$$

so the claim follows.

To treat the general case, we may choose an integer m not divisible by p which annihilates $I_{E/F}$ and some $t_1, \ldots, t_h \in F^\times$ such that $|t_1|, \ldots, |t_h|$ generate $|F^\times|/|F^\times|^m$. Put $F'' = F(t_1^{1/m}, \ldots, t_h^{1/m})$. By the two previous paragraphs, $|d|_{F''} = |d|_F$. (More precisely, when adjoining $t_i^{1/m}$, we split into two cases. For e the largest divisor of m such that $|t_i|^{1/e}$ is already present, adjoining $t_i^{1/e}$ gives an unramified extension as in the proof of Proposition 3.3.6, so the operator norm of d does not change. After that, adjoining $t_i^{1/m}$ proceeds as in the second paragraph.) By Proposition 3.3.6, F' is contained in an unramified extension of F'', so the claim follows by applying the first paragraph again. □

6.3 A coordinate-free approach

We note in passing that our definition of the spectral radius is rather sensitive to the choice of the derivation d. (See the exercises for Chapter 9 for an explicit example.) A coordinate-free approach is suggested by the work of Baldassarri and Di Vizio.

Proposition 6.3.1. *Let F be a nonarchimedean differential field. Let $F\{T\}^{(s)}$ be the set of twisted polynomials of degree at most s, equipped with the seminorm $|P| = |P(d)|_F$ compatible with F (that is, consider $P(d)$ as an operator on F). Let V be a nonzero finite differential module over F and fix a norm on V compatible with F. Let $L(V)$ be the space of bounded endomorphisms of the additive group of V, equipped with the operator norm. Let $D_s : F\{T\}^{(s)} \to L(V)$ be the map $P \mapsto P(D)$. Then*

$$|D|_{\text{sp},V} \leq |d|_{\text{sp},F} \liminf_{s \to \infty} |D_s|^{1/s}. \qquad (6.3.1.1)$$

Conversely, suppose that for any nonnegative integer s and any $c_0, \ldots, c_s \in F$,

$$|c_0 + c_1 d + \cdots + c_s d^s|_F = \max\{|c_0|_F, |c_1 d|_F, \ldots, |c_s d^s|_F\}. \qquad (6.3.1.2)$$

Then equality holds in (6.3.1.1); moreover, if $|d|_{\mathrm{sp},F} > 0$, then the limit inferior on the right is also a limit.

The condition (6.3.1.2) will be satisfied in the situation of greatest interest to us; see Proposition 9.10.2.

Proof By taking $T^s \in F\{T\}^{(s)}$, we obtain the inequality $|D^s|_V \le |d^s|_F |D_s|$. Taking s-th roots of both sides, then taking limits as $s \to \infty$ yields (6.3.1.1).

Conversely, suppose that (6.3.1.2) holds. Given $\epsilon > 0$, apply Lemma 6.1.8 to choose $c > 0$ such that for all $s \ge 0$,

$$|D^s|_V \le c(|D|_{\mathrm{sp},V} + \epsilon)^s.$$

Given a nonnegative integer s such that $|D_s| > \epsilon$, choose a nonzero twisted polynomial $P = \sum_{i=0}^s P_i T^i \in F\{T\}^{(s)}$ such that $|P(D)|_V \ge |P(d)|_F(|D_s| - \epsilon)$. (This could only fail to be possible if we had $|P(d)|_F = 0$ for all $P \in F\{T\}^{(s)}$, but the presence of constant polynomials eliminates this possibility.) Then

$$\max_{i \le s}\{|P_i||d|_{\mathrm{sp},F}^i(|D_s| - \epsilon)\} \le \max_{i \le s}\{|P_i d^i|_F(|D_s| - \epsilon)\}$$
$$= |P(d)|_F(|D_s| - \epsilon)$$
$$\le |P(D)|_V$$
$$\le \max_{i \le s}\{|P_i D^i|_V\}$$
$$\le \max_{i \le s}\{|P_i|c(|D|_{\mathrm{sp},V} + \epsilon)^i\}.$$

For the index i which maximizes the right side,

$$|d|_{\mathrm{sp},F}^i(|D_s| - \epsilon) \le c(|D|_{\mathrm{sp},V} + \epsilon)^i.$$

Since $|d|_{\mathrm{sp},F} < |D|_{\mathrm{sp},V} + \epsilon$ by Lemma 6.2.4, we can increase the exponent from i to s on both sides while maintaining the truth of the inequality; we may then also include values of s for which $|D_s| < \epsilon$. Taking s-th roots, then taking the limit as $s \to \infty$, yields

$$|d|_{\mathrm{sp},F} \limsup_{s \to \infty}(|D_s| - \epsilon)^{1/s} \le |D|_{\mathrm{sp},V} + \epsilon.$$

Since this holds for any $\epsilon > 0$, we obtain the inequality

$$|D|_{\mathrm{sp},V} \ge |d|_{\mathrm{sp},F} \limsup_{s \to \infty}|D_s|^{1/s}.$$

This forces equality in (6.3.1.1) and forces the limit inferior therein to be a limit as long as $|d|_{\mathrm{sp},F} > 0$. $\qquad \square$

6.4 Newton polygons for twisted polynomials

Twisted polynomials admit a partial analogue of the theory of Newton polygons; we will use these polygons in the next section to compute spectral radii of differential operators.

Definition 6.4.1. Let R be a nonarchimedean differential domain. For $\rho \geq |d|_R$, define the ρ-*Gauss norm* on the twisted polynomial ring $R\{T\}$ by

$$\left| \sum_i P_i T^i \right| = \max_i \{|P_i|\rho^i\};$$

it is the same as the ρ-Gauss norm of the untwisted polynomial $\sum_i P_i T^i \in R[T]$. For $r \leq -\log |d|_R$, we obtain a corresponding r-*Gauss valuation* $v_r(P) = -\log |P|_{e^{-r}}$.

Lemma 6.4.2. *For $\rho \geq |d|_R$, the ρ-Gauss norm is multiplicative. Moreover, any polynomial and its formal adjoint have the same ρ-Gauss norm.*

Proof It suffices to check for $\rho > |d|_R$, as the boundary case may be inferred from continuity of the map $\rho \mapsto |P|_\rho$ for fixed P. The key observation (and the source of the restriction on ρ) is that for $P, Q \in R\{T\}$ and $\rho > |d|_R$,

$$|PQ - QP|_\rho \leq \rho^{-1}|d|_R |P|_\rho |Q|_\rho < |P|_\rho |Q|_\rho.$$

(The first inequality is evident in case $P = T$ and $Q \in R$; the reduction of the general case to this one is an exercise.) This allows us to deduce multiplicativity on $R\{T\}$ from multiplicativity on $R[T]$ (Proposition 2.1.2). The claim about the adjoint follows by a similar argument. □

Definition 6.4.3. We define the *Newton polygon* of $P = \sum_i P_i T^i \in R\{T\}$ by taking the Newton polygon of the corresponding untwisted polynomial $\sum P_i T^i \in R[T]$, then omitting all slopes greater than or equal to $-\log |d|_R$. We similarly define multiplicity for slopes less than $-\log |d|_R$. By Lemma 6.4.2 and an argument as in Remark 2.1.7, the multiplicity of a slope $r < -\log |d|_R$ in a product PQ is equal to the sum of the multiplicities of r as a slope of P and as a slope of Q. However, this fails for $r = -\log |d|_R$ because we only control the *left* endpoint of the segment of the Newton polygon with this slope (see the exercises); hence we omit $-\log |d|_R$ and all larger slopes from the Newton polygon.

By another application of the master factorization theorem (Theorem 2.2.2), we obtain the following.

Theorem 6.4.4. *Let R be a complete nonarchimedean differential domain. Suppose $S \in R\{T\}$, $r < -\log |d|_R$, and $m \in \mathbb{Z}_{\geq 0}$ satisfy*

$$v_r(S - T^m) > v_r(T^m).$$

Then there exists a unique factorization $S = PQ$ satisfying the following conditions.

(a) The polynomial $P \in R\{T\}$ has degree $\deg(S) - m$, and its slopes are all less than r.

(b) The polynomial $Q \in R\{T\}$ is monic of degree m, and its slopes are all greater than r.

(c) We have $v_r(P - 1) > 0$ and $v_r(Q - T^m) > v_r(T^m)$.

Moreover, for this factorization,

$$\min\{v_r(P-1), v_r(Q - T^m) - v_r(T^m)\} \geq v_r(S - T^m) - v_r(T^m).$$

Similarly, we can factor $S = QP$ with the same properties (but the factors themselves may differ).

Proof Use the same construction as in Theorem 2.2.1. □

Corollary 6.4.5. *Let F be a complete nonarchimedean differential field. If $P \in F\{T\}$ is irreducible, then either it has no slopes, or it has all slopes equal to some value less than $-\log |d|_F$.*

Remark 6.4.6. Instead of defining the Newton polygon as above and then truncating, one may prefer to declare all of the missing slopes to be equal to $-\log |d|_R$. One can achieve this by modifying the set whose lower convex hull is used to define the Newton polygon, to build in the right truncation behavior. The correct modified set is

$$\bigcup_{i=0}^{n} \{(x, y) \in \mathbb{R}^2 : x \geq -i, y \geq v(P_i) - (x + i) \log |d|_F\}$$

provided that d is nonzero. (If $d = 0$, then $-(x + i) \log |d|_F$ must be interpreted to mean 0 if $x + i = 0$ and ∞ if $x + i > 0$.)

6.5 Twisted polynomials and spectral radii

One can use twisted polynomials over nonarchimedean differential fields to detect part of the full spectrum of a normed differential module.

Definition 6.5.1. For V a finite differential module over a nonarchimedean differential field F, define the *visible spectrum* of V to be the submultiset of the full spectrum of V consisting of those values greater than $|d|_F$.

Remark 6.5.2. In the application to regular singularities (Chapter 7), we will consider a case where $|d|_F = |d|_{sp,F}$. In such a case, there is no real loss in restricting to the visible spectrum: the only missing norm is $|d|_F$ itself, and one can infer its multiplicity from the dimension of the module. However, in the applications to p-adic differential equations in Part III, we will have $|d|_F > |d|_{sp,F}$, so the restriction to the visible spectrum will cause real problems; these will have to be remedied using pullback and pushforward along a Frobenius map.

The key theorem relating spectral radii to Newton polygons is the following.

Theorem 6.5.3 (Christol–Dwork)**.** *Let F be a complete nonarchimedean differential field. For $P \in F\{T\}$ nonconstant, put $V = F\{T\}/F\{T\}P$. Let r be the least slope of the Newton polygon of P, or $-\log|d|_F$ if no such slope exists. Then*

$$\max\{|d|_F, |D|_{sp,V}\} = e^{-r}.$$

Proof Let $r_1 \leq \cdots \leq r_k$ be the slopes of P counted with multiplicity, and define $r_{k+1} = \cdots = r_n = -\log|d|_F$. Equip V with the norm

$$\left|\sum_{i=0}^{n-1} a_i T^i\right|_V = \max_i\{|a_i|\exp(-r_{n-1} - \cdots - r_{n-i})\}.$$

As in the proof of Proposition 4.3.10, we then have $|D|_V = e^{-r_1}$ and so $|D|_{sp,V} \leq e^{-r_1}$.

To finish, we must check that if $r_1 < -\log|d|_F$, then $|D|_{sp,V} = e^{-r_1}$. Let δ be the operation

$$\delta\left(\sum_{i=0}^{n-1} a_i T^i\right) = \sum_{i=0}^{n-1} d(a_i)T^i;$$

then $|\delta|_V = |d|_F$, $D - \delta$ is F-linear, and $|D - \delta|_V = e^{-r_1}$. Then for all positive integers s,

$$|(D - \delta)^s|_V = e^{-r_1 s}, \qquad |D^s - (D - \delta)^s|_V \leq e^{-r_1(s-1)}|d|_F < e^{-r_1 s},$$

so $|D^s|_V = e^{-r_1 s}$ and $|D|_{sp,V} = e^{-r_1}$ as desired. $\square$

Corollary 6.5.4. *Let F be a complete nonarchimedean differential field. For any $P \in F\{T\}$, the visible spectrum of the differential module $F\{T\}/F\{T\}P$*

consists of e^{-r} *for r running over the slope multiset of the Newton polygon of* P.

Proof Write down a maximal factorization of P; it corresponds to a maximal filtration of $F\{T\}/F\{T\}P$. By Corollary 6.4.5, each factor in the factorization has only a single slope, so Theorem 6.5.3 gives what we want. □

We obtain a partial refinement of Corollary 6.5.5. For a further refinement, see Proposition 10.6.6.

Corollary 6.5.5. *Let $F \to F'$ be an isometric embedding of complete nonarchimedean differential fields. For any finite differential module V over F, the visible spectrum of $V \otimes_F F'$ is the submultiset of the visible spectrum of V consisting of those values greater than $|d|_{F'}$.*

Proof We may reduce to the case where V is irreducible (but $V \otimes_F F'$ need not be). In this case, any nonzero element of V is a cyclic vector, so we may apply Corollary 6.5.4 to deduce the claim. □

Using Theorem 6.5.3, we can give a differential version of Proposition 4.4.6.

Proposition 6.5.6. *Let F be a complete nonarchimedean differential field with $|d|_F \leq 1$. Let V be a finite differential module of rank $n > 0$ over F with $|D|_{\mathrm{sp},V} \leq 1$. Let $|\cdot|_V$ be the supremum norm on V defined by a basis $e_1, \ldots, e_n$ and suppose that $|D|_V = c \geq 1$. Then there exists a basis of V defining a second supremum norm $|\cdot|'_V$ for which $|D|'_V \leq 1$ and $|x|'_V \leq |x|_V \leq c^{n-1}|x|'_V$ for all $x \in V$.*

Proof We proceed as in Proposition 4.4.6. Let M be the smallest $\mathfrak{o}_F$-submodule of V containing $e_1, \ldots, e_n$ and stable under D. For each i, if $j = j(i)$ is the least integer such that $e_i, D(e_i), \ldots, D^j(e_i)$ are linearly dependent, then $D^j(e_i) = \sum_{h=0}^{j-1} c_h D^h(e_i)$ for some $c_h \in F$. Since $|D|_{\mathrm{sp},V} \leq 1$ and $|d|_F \leq 1$, Theorem 6.5.3 implies that $|c_h| \leq 1$ for each h. Hence M is finitely generated, and thus free, over $\mathfrak{o}_F$.

Let $|\cdot|'_V$ be the supremum norm on V defined by a basis of M. Then $|x|'_V \leq |x|_V$ because $e_1, \ldots, e_n \in M$. Conversely, for $i = 1, \ldots, n$ and $h = 0, \ldots, j(i) - 1$, $|D^h(e_i)|_V \leq |D|_V^h \leq c^h \leq c^{n-1}$. Since the $D^h(e_i)$ generate M, this implies that $|x|_V \leq c^{n-1}|x|'_V$ for all $x \in V$. □

6.6 The visible decomposition theorem

Using twisted polynomials, we can split V into components corresponding to the elements of the visible spectrum (see Definition 6.5.1).

Theorem 6.6.1 (Visible decomposition theorem). *Let F be a complete nonar-chimedean differential field of characteristic 0 and let V be a finite differential module over F. Then there exists a unique decomposition*

$$V = V_0 \oplus \bigoplus_{s > |d|_F} V_s$$

of differential modules such that every subquotient of V_s has spectral radius s and every subquotient of V_0 has spectral radius at most $|d|_F$.

Proof If the derivation on F is zero, then we are just considering a vector space equipped with an endomorphism, and the claim becomes an elementary exercise in linear algebra (as noted in Remark 6.2.13). We may thus assume hereafter that the derivation is nonzero.

We induct on $\dim(V)$. By Theorem 5.4.2 and our hypothesis that $d \neq 0$, there exists a cyclic vector for V. We thus obtain an isomorphism $V \cong F\{T\}/F\{T\}P$ for some $P \in F\{T\}$. If the Newton polygon of P is empty, we may put $V = V_0$ and be done. Otherwise, let $r < -\log|d|_F$ be the least slope. By applying Theorem 6.4.4 once to P, we obtain a short exact sequence $0 \rightarrow V_1 \rightarrow V \rightarrow V_2 \rightarrow 0$ in which (by Theorem 6.5.3) every subquotient of V_1 has spectral radius e^{-r} and every subquotient of V_2 has spectral radius less than e^{-r}. Applying Theorem 6.4.4 again to P but with the factors in the opposite order, we get a short exact sequence $0 \rightarrow V_2' \rightarrow V \rightarrow V_1' \rightarrow 0$ where every subquotient of V_1' has spectral radius e^{-r} and every subquotient of V_2' has spectral radius less than e^{-r}. This yields $V_1 \cap V_2' = 0$, so $V_1 \oplus V_2'$ injects into V. Moreover, $\dim V_1 = \dim V_1'$ and $\dim V_2 = \dim V_2'$ because P and its formal adjoint have the same Newton polygon (Lemma 6.4.2). By a dimension count, $V_1 \oplus V_2'$ must equal V. Applying the induction hypothesis to V_2' gives the claim. $\square$

Corollary 6.6.2. *Let F be a complete nonarchimedean differential field of characteristic 0. Let V be a finite differential module over F such that every subquotient of V has spectral radius greater than $|d|_F$. Then $H^0(V) = H^1(V) = 0$.*

Proof The claim about H^0 is clear: a nonzero element of $H^0(V)$ would generate a differential submodule of V which would be trivial, and thus would have spectral radius $|d|_{\mathrm{sp},F} \leq |d|_F$. As for H^1, let $0 \rightarrow V \rightarrow W \rightarrow F \rightarrow 0$ be a short exact sequence of differential modules. Decompose $W = W_0 \oplus W_1$ according to Theorem 6.6.1, so that every subquotient of W_0 has spectral radius at most $|d|_F$ and every subquotient of W_1 has spectral radius greater than $|d|_F$. The map $V \rightarrow W_0$ must vanish (its image is a subquotient of both V and W_0), so $V \subseteq W_1$. But $W_1 \neq W$ as otherwise W could not surject onto a trivial module, so for dimensional reasons we must have $V = W_1$. Hence the sequence splits, proving that $H^1(V) = 0$ by Lemma 5.3.3. ◻

We also obtain a refinement of Corollary 6.2.9 in this setting, in which we allow subquotients rather than just submodules.

Corollary 6.6.3. *Let F be a complete nonarchimedean differential field of characteristic 0. If V_1, V_2 are finite irreducible differential modules over F, $|D|_{\mathrm{sp},V_1} > |d|_F$, and $|D|_{\mathrm{sp},V_1} > |D|_{\mathrm{sp},V_2}$, then every irreducible subquotient W of $V_1 \otimes V_2$ satisfies $|D|_{\mathrm{sp},W} = |D|_{\mathrm{sp},V_1}$.*

Proof Decompose $V_1 \otimes V_2 = V_0 \oplus \bigoplus_{s > |d|_F} V_s$ according to Theorem 6.6.1; then $V_s = 0$ whenever $s > |D|_{\mathrm{sp},V_1}$. If either V_0 or some V_s with $s < |D|_{\mathrm{sp},V_1}$ were nonzero, then $V_1 \otimes V_2$ would have an irreducible submodule of spectral radius less than $|D|_{\mathrm{sp},V_1}$, in violation of Corollary 6.2.9. ◻

These results are quite sufficient for applications to the study of singularities of complex meromorphic differential equations, at which we hint in Chapter 7. However, in the p-adic situation, we will have to further decompose V_0; we will do this using Frobenius antecedents in Chapter 10.

6.7 Matrices and the visible spectrum

The proof of Theorem 6.5.3 relies on the fact that one can detect the spectral radius of a differential module admitting a cyclic vector, using the characteristic polynomial of the matrix of the action of D on the cyclic basis. For some applications, we need to extend this to some bases not necessarily generated by cyclic vectors; for this, the relationship between singular values and eigenvalues considered in Chapter 4 will be crucial.

Lemma 6.7.1. *Let R be a complete nonarchimedean differential domain. Let N be a 2×2 block matrix over R with the following properties.*

(a) The matrix N_{11} has an inverse A over R.

(b) We have $|A| \max\{|d|_F, |N_{12}|, |N_{21}|, |N_{22}|\} < 1$.

Then there exists a block upper triangular unipotent matrix U over R such that

$$|U_{12}| \le |A| \max\{|N_{12}|, |N_{21}|, |N_{22}|\}$$

and $U^{-1}NU + U^{-1}d(U)$ is block lower triangular.

Proof Put

$$\delta = |A| \max\{|N_{12}|, |N_{21}|, |N_{22}|\} < 1, \qquad \epsilon = |A||N_{12}| \le \delta.$$

Let X be the block upper triangular nilpotent matrix with $X_{12} = AN_{12}$, and put $U = I - X$ and

$$N' = U^{-1}NU + U^{-1}d(U).$$

Since $U^{-1} = I + X$, $N' = N + XN - NX - XNX - d(X)$. In block form,

$$N' = \begin{pmatrix} N_{11} + X_{12}N_{21} & N_{12} - N_{11}X_{12} + X_{12}N_{22} - X_{12}N_{21}X_{12} - d(X_{12}) \\ N_{21} & N_{22} - N_{21}X_{12} \end{pmatrix}.$$

We claim that

$$|N'_{12}| \le \epsilon \max\{\delta, |d|_F|A|\}|A|^{-1}$$
$$|N'_{21}| \le \delta|A|^{-1}$$
$$|N'_{22}| \le \delta|A|^{-1}.$$

The second and third lines hold because

$$|(U^{-1}NU - N)_{2j}| = |(-NX)_{2j}| \le \epsilon\delta|A|^{-1} \qquad (j = 1, 2).$$

The first line holds because $N_{12} - N_{11}X_{12} = 0$, so we can write

$$N'_{12} = X_{12}N_{22} - X_{12}N_{21}X_{12} - d(X_{12}),$$

in which the first two terms have norm at most $\epsilon\delta|A|^{-1}$ and the third has norm at most $|d|_F\epsilon$.

To analyze N'_{11}, we write it as $(I + X_{12}N_{21}A)N_{11}$. Because $|X_{12}N_{21}A| \le |X_{12}|(|N_{21}||A|) < \epsilon < 1$, the first factor is invertible, and it and its inverse both have norm 1. Hence N'_{11} is invertible, $|N'_{11}| = |N_{11}|$, and $|(N'_{11})^{-1}| = |A|$.

Since $\max\{\delta, |d|_F|A|\} \le \mu$ for some fixed $\mu < 1$, iterating the construction of N' from N yields a convergent sequence of conjugations whose limit has the desired property. $\square$

We need a refinement of the argument used in Theorem 6.5.3 which is no longer restricted to cyclic vectors.

Lemma 6.7.2. *Let F be a complete nonarchimedean differential field. Let V be a nonzero finite differential module over F. Let $e_1, \dots, e_n$ be a basis of V and let N be the matrix of action of D on $e_1, \dots, e_n$. Suppose that $|N| = \sigma > |d|_F$ and $|N^{-1}| = \sigma^{-1}$. Then the full spectrum of V consists entirely of σ.*

Proof As in the proof of Theorem 6.5.3, we find that for the supremum norm for $e_1, \dots, e_n$, $|D^s(v)|_V = \sigma^s|v|_V$ for all nonnegative integers s and all $v \in V$.

Consequently, for any nonzero differential submodule W of V, $|D|_{\mathrm{sp},W} = \sigma$. By Theorem 6.6.1, it follows that every irreducible subquotient of V also has spectral radius σ, as desired. □

By combining what we have so far, we get a further refinement of the previous lemma, in which we can detect elements of the visible spectrum which need not all be equal.

Lemma 6.7.3. *Let F be a complete nonarchimedean differential field. Let V be a finite differential module over F. Let $e_1, \ldots, e_n$ be a basis of V, and let N be the matrix of action of D on $e_1, \ldots, e_n$. Let $\sigma_1 \geq \cdots \geq \sigma_n$ be the singular values of N and let $\lambda_1, \ldots, \lambda_n$ be the eigenvalues of N, sorted so that $|\lambda_1| \geq \cdots \geq |\lambda_n|$. Suppose that the following conditions hold for some $i = 1, \ldots, n$ and some $\delta \geq |d|_F$.*

(a) We have $\sigma_i > \delta$.

(b) Either $i = n$ or $\sigma_{i+1} \leq \delta$.

(c) We have $\sigma_j = |\lambda_j|$ for $j = 1, \ldots, i$.

Then the elements of the full spectrum greater than δ are precisely $\sigma_1, \ldots, \sigma_i$.

Proof We first check that conditions (a)–(c) are invariant under the change of basis

$$N \mapsto U^{-1}NU + U^{-1}d(U) = U^{-1}(N + d(U)U^{-1})U$$

for $U \in \mathrm{GL}_n(\mathfrak{o}_F)$. Note that by Proposition 4.4.1, N and $N + d(U)U^{-1}$ have the same singular values greater than δ. Thanks to conditions (a)–(c), we may apply Theorem 4.4.2 to deduce that N and $N + d(U)U^{-1}$ also have the same norms of eigenvalues greater than δ. Since neither the singular values nor the eigenvalues are altered by conjugating by U, we draw the same conclusion about N and $U^{-1}NU + U^{-1}d(U)$. In particular, the truth of (a)–(c) is not affected by this change of basis.

If $\sigma_1 \leq |d|_F$, then we have nothing to check. If $\sigma_1 = \cdots = \sigma_n > |d|_F$, then Lemma 6.7.2 implies the claim. If neither of these applies, we may induct on n: choose i with $\sigma_1 = \cdots = \sigma_i > \sigma_{i+1}$, so that necessarily $\sigma_1 > |d|_F$. View N as a 2×2 block matrix with block sizes $i, n - i$. At this point there is no harm in extending the constant subfield of F, as by Corollary 6.5.5 this preserves the visible spectrum; consequently, we may reduce to the case $|\sigma_1| = 1$. We may then apply Lemma 6.7.1 to obtain an upper triangular unipotent block matrix U over $\mathfrak{o}_F$ such that $N' = U^{-1}NU + U^{-1}d(U)$ is lower triangular. We may then reduce to checking the claim with N replaced by the two diagonal blocks of N'. □

To put everything together, we relax the condition on the singular values.

Theorem 6.7.4. *Let F be a complete nonarchimedean differential field. Let V be a finite differential module over F. Let $e_1, \ldots, e_n$ be a basis of V, and let N be the matrix of action of D on $e_1, \ldots, e_n$. Let $\sigma_1 \geq \cdots \geq \sigma_n$ be the singular values of N and let $\lambda_1, \ldots, \lambda_n$ be the eigenvalues of N, sorted so that $|\lambda_1| \geq \cdots \geq |\lambda_n|$. Suppose either that*

(i) $\theta = f_n(\sigma_1, \ldots, \sigma_n, |\lambda_1|, \ldots, |\lambda_n|, |d|_F)$ for f_n as in Corollary 4.4.8, or
(ii) $\theta = \max\{1, \sigma_1/|d|_F\}^{2^n n! - 1}$.

Suppose that the following conditions hold for some $i = 1, \ldots, n$ and some $\delta \geq |d|_F \theta$.

(a) We have $|\lambda_i| > \delta$.
(b) Either $i = n$ or $|\lambda_{i+1}| \leq \delta$.

Then the elements of the full spectrum greater than δ are precisely $|\lambda_1|, \ldots, |\lambda_i|$.

Proof By Corollary 6.5.5, there is no harm in replacing F by a larger complete nonarchimedean differential field F', provided that $|d|_{F'} = |d|_F$. In particular, we may take F' to be the completion of $F(t)$ for the ρ-Gauss norm for any $\rho > 0$, extending d so that $d(t) = 0$ (exercise). After enlarging F suitably in this manner, by Corollary 4.4.8 in case (i) or Corollary 4.4.10 in case (ii), we can choose a matrix $U \in \mathrm{GL}_n(F)$ such that the following conditions hold.

(a) We have $|U^{-1}| \leq 1$ and $|U| \leq \theta$.
(b) The first i singular values of $U^{-1}NU$ are $|\lambda_1|, \ldots, |\lambda_i|$.
(c) Either $i = n$, or the $(i + 1)$-st singular value of $U^{-1}NU$ is at most δ.

Since $|U^{-1}d(U)| \leq \theta|d|_F \leq \delta$, by Proposition 4.4.1, the new conditions (b) and (c) hold when $U^{-1}NU$ is replaced by $U^{-1}NU + U^{-1}d(U)$. We may thus apply Lemma 6.7.3 to obtain the desired result. □

6.8 A refined visible decomposition theorem

We give a refinement of the visible decomposition theorem (Theorem 6.6.1). This can be used to obtain the Turrittin–Hukuhara–Levelt decomposition theorem; see Theorem 7.5.1. We first give an extension of Remark 6.2.13.

Lemma 6.8.1. *Let F be a complete nonarchimedean differential field of characteristic 0. Suppose $P = \sum_i P_i T^i \in F\{T\}$ and $\tilde{P} = \sum_i \tilde{P}_i T^i \in F\{T\}$ are nonconstant twisted polynomials such that for some $s < -\log|d|_F$, the slopes* *of P and $\tilde{P}$ are all equal to s. Put $V = F\{T\}/F\{T\}P$ and $W = F\{T\}/F\{T\}\tilde{P}$. Let $Q = \sum_i Q_i U^i \in F[T]$ and $\tilde{Q} = \sum_i \tilde{Q}_i U^i \in F[T]$ be the untwisted polynomials with the same coefficients as P and $\tilde{P}$, respectively. Then $|D|_{\mathrm{sp},V^\vee \otimes W} < |D|_{\mathrm{sp},V}$ if and only if for any root $\lambda \in F^{\mathrm{alg}}$ of Q and any root $\mu \in F^{\mathrm{alg}}$ of $\tilde{Q}$, $|\lambda/\mu - 1| < 1$.*

Proof As in the proof of Theorem 6.7.4, we may enlarge F to reduce to the case where $|D|_{\mathrm{sp},V} = |\eta|$ for some $\eta \in F^\times$. Equip V and W with the bases given by $1, \eta^{-1}T, \eta^{-2}T^2, \ldots$, equip $V^\vee$ with the dual basis, and use these bases to define supremum norms. As in the proof of Theorem 6.5.3, $|D|_V = |D|_W = |\eta|$.

The matrix of action of D on the resulting basis of $V^\vee \otimes W$ has eigenvalues of the form $\mu - \lambda$ where λ is a root of Q and μ is a root of $\tilde{Q}$. By Theorem 6.7.4, $|D|_{\mathrm{sp},V^\vee \otimes W} < |\eta|$ if and only if each of these eigenvalues has norm strictly less than $|\eta|$, which occurs if and only if $|\lambda/\mu - 1| < 1$ for all λ, μ. (Note that in this application, we must apply case (ii) of Theorem 6.7.4 so that the bound holds for δ in some neighborhood of σ_1 depending only on n and $|d|_F$.) $\square$

Theorem 6.8.2 (Refined visible decomposition theorem). *Let F be a complete nonarchimedean differential field of characteristic 0, and let V be a finite differential module over F such that no subquotient of V_s has norm less than or equal to $|d|_F$. Then for some finite tamely ramified extension F' of F, there exists a (unique) decomposition*

$$V \otimes_F F' = \bigoplus_i V_i$$

of $V \otimes_F F'$ into refined differential modules, such that for $i \neq j$, $V_i \nsim V_j$ in the sense of Definition 6.2.12.

Proof The uniqueness is a direct consequence of $\sim$ being an equivalence relation (by Lemma 6.2.14). To check existence, by Theorem 6.6.1 we may reduce to the case where V is pure. We may also assume $d \neq 0$, as otherwise the result follows from Remark 6.2.13. Note that for any finite tamely ramified extension F' of F, $|d|_{F'} = |d|_F$ by Lemma 6.2.15.

Apply Theorem 5.4.2 to choose an isomorphism $V \cong F\{T\}/F\{T\}P$. By Corollary 6.5.4, P has all slopes equal to a single value. Let $Q \in F[U]$ be the untwisted polynomial with the same coefficients as P. By Remark 6.2.13, we can choose a finite tamely ramified extension F' of F such that Q factors in $F'[U]$ into factors Q_i such that for any two roots λ, μ of any Q_i, $|\lambda/\mu - 1| < 1$. By applying Theorem 2.2.2 (as in Remark 2.2.3), we may correspondingly factor P in $F'\{T\}$ so that each factor corresponds to a refined differential module. By performing this factorization again with the residual roots in the opposite order, then arguing as in Theorem 6.6.1, we obtain the desired splitting. $\square$

Remark 6.8.3. From the proof of Theorem 6.8.2, one may extract a map from the classes of refined finite differential modules V with $|D|_{\mathrm{sp},V} = s > |d|_F$ to the quotient

$$\frac{\{x \in \mathfrak{o}_{F^{\mathrm{alg}}} : |x| \leq s\}}{\{x \in \mathfrak{o}_{F^{\mathrm{alg}}} : |x| < s\}},$$

namely, given an isomorphism $V \cong F\{T\}/F\{T\}P$, one associates to V the class of the roots of the untwisted polynomial with the same coefficients as P. However, one must check with some care that this map does not depend on the initial choice of a cyclic vector. A more robust approach is to identify certain "test objects" for which one can read off the map; this has been done by Xiao [410] (see also [413, §1] and Lemma 12.8.5).

From the proof of Theorem 6.8.2, we may also read off the following simple but important observations. These will be used in the proofs of both the Turrittin–Levelt–Hukuhara decomposition (Theorem 7.5.1) and the p-adic local monodromy theorem (Theorem 20.1.4).

Proposition 6.8.4. *Let F be a complete nonarchimedean differential field of characteristic 0. Let V be a finite differential module over F of rank $n > 0$ which is refined with spectral radius $s > |d|_F$.*

(a) *For any positive integer m which is nonzero in κ_F, $V^{\otimes m}$ is again refined with spectral radius s.*

(b) *The spectral radius of $(\wedge^n V^\vee) \otimes V^{\otimes n}$ is strictly less than s.*

(c) *If $p > 0$, then the spectral radius of $V^{\otimes p}$ is strictly less than s.*

Proof The result is elementary (as in Remark 6.2.13) if $d = 0$, so we may assume $d \neq 0$. By Theorem 5.4.2, we may choose an isomorphism $V \cong F\{T\}/F\{T\}P$. Let Q be the untwisted polynomial with the same coefficients as P and let μ be a root of Q.

As in the proof of Theorem 6.7.4, we may enlarge F to reduce to the case where $|D|_{\mathrm{sp},V} = |\eta|$ for some $\eta \in F^\times$. Equip V with the basis given by $1, \eta^{-1}T, \ldots, \eta^{-n+1}T^{n-1}$ and the corresponding supremum norm. Equip $V^{\otimes m}$ and $(\wedge^n V^\vee) \otimes V^{\otimes n}$ with the induced bases and norms.

On $V^{\otimes m}$, the matrix of action of D has eigenvalues which are m-fold sums of roots of Q. In particular, each eigenvalue λ satisfies $|\lambda/(m\mu) - 1| < 1$. We deduce (a) from Theorem 6.7.4.

On $(\wedge^n V^\vee) \otimes V^{\otimes n}$, the matrix of action of D has eigenvalues which are n-fold sums of roots of Q, minus the sum of the roots of Q. In particular, each eigenvalue λ satisfies $|\lambda/\mu| < 1$. We deduce (b) from Theorem 6.7.4. A similar argument implies (c). $\qquad \square$

6.9 Changing the constant field

We now check that in many cases, the operation of computing the horizontal sections of a differential module commutes with the formation of *completed* tensor products.

Proposition 6.9.1. *Let R be a complete nonarchimedean differential domain such that R and* $\mathrm{Frac}(R)$ *have the same constant ring* R_0 *(which is necessarily a complete nonarchimedean field). Let* R'_0 *be a complete field extension of* R_0*. Let* R' *be the completed tensor product of* $R \otimes_{R_0} R'_0$ *for the product norm (which is a norm by Lemma 1.3.11). View* R' *as a differential ring by equipping it with the unique continuous extension of d with* R'_0 *in its kernel. Then for any differential module M over R, the natural map*

$$H^0(M) \otimes_{R_0} R'_0 \to H^0(M \otimes_R R')$$

is an isomorphism of R'_0*-modules.*

Proof We first check injectivity. Since R'_0 is flat over R_0, the map $H^0(M) \otimes_{R_0} R'_0 \to H^0(M \otimes_{R_0} R'_0)$ is bijective by Lemma 5.1.4. Since M is finite over R, we may identify $M \otimes_R R'$ with the completed tensor product of $M \otimes_R (R \otimes_{R_0} R'_0) = M \otimes_{R_0} R'_0$, on which the product seminorm is a norm by Lemma 1.3.11. In particular, $M \otimes_{R_0} R'_0$ injects into $M \otimes_R R'$, so the map $H^0(M \otimes_{R_0} R'_0) \to H^0(M \otimes_R R')$ is also injective.

We next check surjectivity in case R'_0 is the completion of a finitely generated field extension S_0 of R_0. Then S_0 is of countable dimension over R_0, so the hypothesis of Lemma 1.3.8 is satisfied by $F = R_0, V = R'_0$. Let $m_1, m_2, \ldots$ be the sequence of elements of R'_0 given by Lemma 1.3.8. As in the proof of Lemma 1.3.11, these define projection maps $\lambda_{j,R} : R' \to R$ and $\lambda_{j,M} : M \otimes_R R' \to M$; these maps are both horizontal.

Given an element $x \in H^0(M \otimes_R R')$, choose a presentation $x = \sum_{k=1}^s y_k \otimes z_k$ with $y_k \in M$ and $z_k \in R'$. Then

$$\lambda_{j,M}(x) = \sum_{k=1}^{s} \lambda_{j,R}(z_k) y_k$$

is an element of $H^0(M)$ for each j. Moreover, we can write x as a convergent sum

$$x = \sum_{j=1}^{\infty} \lambda_{j,M}(x) \otimes m_j.$$

Hence x lies in the closure of $H^0(M) \otimes_{R_0} R'_0$ under the product norm. However, since $H^0(M)$ is finite-dimensional over R_0 by Lemma 5.1.5, $H^0(M) \otimes_{R_0} R'_0$ is complete under the product norm, hence closed. Thus $x \in H^0(M) \otimes_{R_0} R_0'$ as desired.

We next check surjectivity in general. Pick any $x \in H^0(M \otimes_R R')$. As noted earlier, we may identify $M \otimes_R R'$ with the completion of $M \otimes_{R_0} R_0'$ under the product norm. We can thus choose a convergent series

$$x = \sum_{k=1}^{\infty} y_k \otimes z_k \qquad (y_k \in M, z_k \in R_0').$$

Let R_0'' be the completion of the subfield of R_0' generated over R_0 by $z_1, z_2, \ldots$. Then x is a horizontal element of the completion of $M \otimes_{R_0} R_0''$, so the previous paragraph shows that $x \in H^0(M) \otimes_{R_0} R_0''$. This proves the claim. □

Notes

Lemma 6.2.5 is tacitly assumed at various places in the literature (including by the present author), but we were unable to locate even an explicit statement, let alone a proof. We thank Liang Xiao for contributing the proof given here.

Newton polygons for differential operators were considered by Dwork and Robba [150, §6.2.3]; the first systematic treatment seems to have been made by Robba [350]. Our treatment using Theorem 2.2.2 follows [89].

The proof of Theorem 6.5.3 given here is close to the original proof of Christol and Dwork [96, Théorème 1.5], save that we avoid a small logical gap in the latter. The gap is in the implication $1 \implies 2$; there one makes a finite extension of the differential field, without accounting for the possibility that this might increase $|d|_F$ to the point where Corollary 6.2.7 may fail to show that $|D|_V$ is preserved. (It would be obvious that this possibility would not occur if the finite extension were being made in the constant subfield, but that is not the case here.) Compare also [149, Lemma VI.2.1].

Proposition 6.5.6 answers a conjecture of Christol and Dwork [95, Introduction, Conjecture A]. This conjecture was posed in the context of giving effective convergence bounds, and that is one use to which we will put it here; see Theorem 18.2.1 and its proof.

The use in Lemma 6.7.3 of a well-chosen norm (meaning one in which at least some of the singular values and eigenvalues of the matrix of action of d match) is an extension of the notion of the *canonical lattice* (or *Deligne–Malgrange lattice*) introduced by Malgrange in the study of irregular meromorphic connections. See [301].

The refined visible decomposition theorem is due to Liang Xiao [410]. It is motivated by applications to *refined Swan conductors* of étale sheaves and overconvergent isocrystals, as suggested by the work of Kato [222] in the rank-1 case.

We thank Andrea Pulita for the suggestion to include Proposition 6.9.1.

Exercises

6.1 Prove Fekete's lemma (Lemma 6.1.4).

6.2 (a) Let A, B be *commuting* bounded endomorphisms on an abelian group G equipped with a norm. Prove that

$$|A + B|_{\mathrm{sp},G} \leq |A|_{\mathrm{sp},G} + |B|_{\mathrm{sp},G}.$$

(b) Prove that if the norm on G is nonarchimedean, then the inequality in (a) can be improved to

$$|A + B|_{\mathrm{sp},G} \leq \max\{|A|_{\mathrm{sp},G}, |B|_{\mathrm{sp},G}\}$$

and equality occurs (at least) when the maximum is achieved only once.

(c) Prove that both of these assertions may fail in the case where A and B do not commute. (Hint: write an identity matrix as a sum of nilpotent matrices.)

6.3 Prove that in Lemma 6.2.5, if $|D|_{\mathrm{sp},V} > |d|_{\mathrm{sp},F}$, then $|D_s|^{1/s}$ converges to a limit as $s \to \infty$. (Hint: again reduce to Lemma 6.1.4.)

6.4 Fill in the missing arguments in the proofs of Lemma 6.2.8(a),(b), using Lemma 6.1.8.

6.5 Prove the claim from the proof of Lemma 6.4.2: for R a nonarchimedean differential domain, $\rho > |d|_R$, and $P, Q \in R\{T\}$, we have $|PQ - QP|_\rho \leq \rho^{-1}|d|_R|P|_\rho|Q|_\rho$. (Hint: reduce to the case where P and Q are monomials, then to the case where $P = T$ and $Q \in R$.)

6.6 Exhibit an example to show that in Definition 6.4.3, if we had not omitted slopes greater than or equal to $-\log|d|_R$, then it would no longer be the case that the multiplicity of a slope r in a product PQ is the sum of the multiplicities of r as a slope of P and as a slope of Q. (Hint: this can already be seen using the t-adic valuation on $\mathbb{Q}(t)$ with $d = \frac{d}{dt}$.)

6.7 Let (F, d) be a complete nonarchimedean differential field. Let F_ρ be the completion of $F(t)$ for the ρ-Gauss norm. Prove that there is a unique continuous extension of d to F_ρ with $d(t) = 0$, and that

$$|d|_{F_\rho} = |d|_F, \qquad |d|_{\mathrm{sp},F_\rho} = |d|_{\mathrm{sp},F}.$$

(Hint: first check everything on $F[t]$.)

7

Regular and irregular singularities

In the next part (starting with Chapter 8), we will use the results from the previous chapters to make a detailed analysis of ordinary differential equations over nonarchimedean fields of characteristic 0, the motivating case being that of positive residual characteristic. However, before doing so it may be helpful to demonstrate how they apply in a somewhat simpler setting.

In this chapter, we reconstruct some of the traditional Fuchsian theory of regular singular points of meromorphic differential equations. (The treatment is heavily modeled on [149, §3].) We first introduce a quantitative measure of the *irregularity* of a singular point. We then recall how in the case of a *regular singularity* (i.e., a singularity with irregularity equal to zero), one has an algebraic interpretation of the eigenvalues of the monodromy operator around the singular point, using the notion of *exponents*. We then describe how to compute formal solutions of meromorphic differential equations and sketch the proof of Fuchs's theorem: the formal solutions of a regular meromorphic differential equation all converge in some disc. We finally establish the Turrittin–Levelt–Hukuhara decomposition theorem, which gives a decomposition of an arbitrary formal differential module analogous to the eigenspace decomposition of a complex linear transformation. The search for an appropriate p-adic analogue of this result will lead us in Part V to the p-adic local monodromy theorem.

Although this chapter focuses on different issues than much of the rest of the book, it should not be considered optional. In particular, the discussion of formal solution matrices in Section 7.3 will recur in several places.

Hypothesis 7.0.1. Throughout this chapter, we view $\mathbb{C}((z))$ as a complete nonarchimedean differential field with the valuation given by the z-adic valuation v_z and the derivation given by $d = z\frac{d}{dz}$; note that $|d|_{\mathbb{C}((z))} = 1$. Let K be a field of characteristic 0; although we do not need to think of K as a subfield of $\mathbb{C}$, it is harmless to do so. (The reason for this is the *Lefschetz principle*: every statement we make about K will refer to at most countably many elements of K, so each individual instance of the statement can be realized over a subfield of K which is countably generated over $\mathbb{Q}$ and hence embeds into $\mathbb{C}$.)

7.1 Irregularity

Definition 7.1.1. Let V be a finite differential module over $\mathbb{C}((z))$, and decompose V according to Theorem 6.6.1. Define the *irregularity* of V as

$$\mathrm{irr}(V) = \sum_{s>1} (\log s) \dim(V_s).$$

For F a subfield of $\mathbb{C}((z))$ stable under d and V a finite differential module over F, we define the irregularity of V to be the irregularity of $V \otimes_F \mathbb{C}((z))$. We say that V is *regular* if $\mathrm{irr}(V) = 0$.

Theorem 7.1.2. *For any isomorphism $V \cong F\{T\}/F\{T\}P$, the irregularity of V is equal to minus the sum of the slopes of P; consequently, it is always an integer. More explicitly, if $P = T^d + \sum_{i=0}^{d-1} P_i T^i$, then*

$$\mathrm{irr}(V) = \max \left\{0, \max_i \{-v_z(P_i)\} \right\}.$$

Proof The second assertion follows from Corollary 6.5.4, since in this case $|d|_F = |d|_{\mathrm{sp},F} = 1$. The first assertion follows from the second because by Theorem 5.4.2, V always admits a cyclic vector. □

This theorem gives rise to several criteria for regularity.

Corollary 7.1.3. *Let F be any subfield of $\mathbb{C}((z))$ containing z and stable under d, and let V be a finite differential module over F. Then the following conditions are equivalent.*

(a) *The module V is regular, i.e., $\mathrm{irr}(V) = 0$.*

(b) *For some isomorphism $V \cong F\{T\}/F\{T\}P$ with P monic, P has coefficients in $\mathfrak{o}_F$.*

(c) *For any isomorphism $V \cong F\{T\}/F\{T\}P$ with P monic, P has coefficients in $\mathfrak{o}_F$.*

(d) *There exists a basis of V on which D acts via a matrix over $\mathfrak{o}_F$.*

(e) *For any basis B of V, the $\mathfrak{o}_F$-span of the set $\{D^i(v) : i \in \{0,\ldots,\dim(V) - 1\}, v \in B\}$ is stable under D.*

Proof By Theorem 7.1.2, (a) implies (c). It is obvious that (c) implies (b) and that (b) implies (d). Given (d), let $| \cdot |_V$ be the supremum norm defined by the chosen basis of V; then $|D|_V \leq 1$, which implies (a).

This proves that (a), (b), (c), (d) are all equivalent. To add (e) to the circle, note that on one hand, (e) implies (d). On the other hand, given (a), pick any $v \in B$; then $v, D(v), D^2(v), \ldots$ generate a differential submodule W of V for which v is a cyclic vector. Since V is regular, its submodule W is also regular by Lemma 6.2.8; since (a) implies (c), the $\mathfrak{o}_F$-span of $\{D^i(v) : i \in \{0, \ldots, \dim(W) - 1\}\}$ is stable under D. This implies (e). □

Remark 7.1.4. One can also view $\mathbb{C}((z))$ as a differential field with the derivation $\frac{d}{dz}$ instead of $z\frac{d}{dz}$. The categories of differential modules for these two choices of derivation are equivalent in the obvious fashion: given an action of $z\frac{d}{dz}$, we obtain an action of $\frac{d}{dz}$ by dividing by z. If V is a differential module for $z\frac{d}{dz}$ with spectral radius s, then the spectral radius of V for $\frac{d}{dz}$ is $s|z|^{-1}$ (exercise). The notion of irregularity translates naturally: for instance, if V is a differential module for $\frac{d}{dz}$ which is isomorphic to $F\{T\}/F\{T\}P$ for some $P = T^n + \sum_{i=0}^{n-1} P_i T^i$, then V is regular if and only if $v_z(P_i) \geq -n + i$ for $i = 1, \ldots, n$. For example, for $a, b, c \in \mathbb{C}$, the differential system corresponding to the hypergeometric differential equation

$$y'' + \frac{(c - (a + b + 1)z)}{z(1 - z)} y' - \frac{ab}{z(1 - z)} y = 0$$

is regular.

Example 7.1.5. For another example from the classical theory of ordinary differential equations, consider the Bessel equation

$$y'' + \frac{1}{z} y' + \frac{z^2 - n^2}{z^2} y = 0$$

for some parameter n. This gives rise to a regular differential system if we expand around $z = 0$. However, if we expand around ∞, i.e., substitute $1/z$ for z and then expand around $z = 0$, we get an irregular differential system (exercise).

7.2 Exponents in the complex analytic setting

To see why regular singularities are so important in the complex analytic setting (by way of motivation for our p-adic studies), let us consider the monodromy transformation. First, we recall a familiar fact.

Theorem 7.2.1 (Cauchy). *Fix $\rho > 0$ and let $R \subset \mathbb{C}[\![z]\!]$ be the ring of power* *series convergent for $|z| < \rho$. Let N be an $n \times n$ matrix over R. Then the differential system $D(v) = Nv + \frac{d}{dz}(v)$ has a basis of horizontal sections.*

Proof This can be deduced from the fundamental theorem of ordinary differential equations; however, it will be useful for future reference to give a slightly more detailed explanation. (Specifically, we are anticipating Definition 7.3.1.)

Note that there exists a unique $n \times n$ matrix U over $\mathbb{C}[\![z]\!]$ such that $U \equiv I_n$ (mod z) and $NU + \frac{d}{dz}(U) = 0$; this follows by writing $N = \sum_{i=0}^{\infty} N_i z^i$ and $U = \sum_{i=0}^{\infty} U_i z^i$, then rewriting the equation $NU + \frac{d}{dz}(U) = 0$ as a recurrence

$$(i+1)U_{i+1} = -\sum_{j=0}^{i} N_j U_{i-j} \qquad (i = 0, 1, \dots). \qquad (7.2.1.1)$$

Following an argument of Cauchy [149, Appendix III], we may deduce that the entries of U converge on a disc of positive radius, as follows. (In this argument, use the L^2 operator norm on matrices over $\mathbb{C}$.) Pick any $\eta \in (0, \rho)$. Since N converges in the open disc of radius ρ, $|N_i|\eta^i \to 0$ as $i \to \infty$; in particular, we may choose $c > 1$ with $|N_i|\eta^{i+1} \le c$ for all i. We then have by induction on i that

$$|U_i|\eta^i \le c^i \qquad (i \ge 0) :$$

namely, this holds for $j = 0$, and (7.2.1.1) implies that

$$|U_{i+1}|\eta^{i+1} \le \frac{1}{i+1} \sum_{j=0}^{i} |N_j|\eta^{j+1}|U_{i-j}|\eta^{i-j} \le c^{i+1}.$$

Consequently, the entries of U converge on the open disc of radius η/c.

The previous argument applied to the system $(-N)U^{-T} + \frac{d}{dz}(U^{-T}) = 0$ shows that the entries of U^{-1} also converge on an open disc. Consequently, we can find an open neighborhood of $z = 0$ on which the differential system admits a basis of horizontal sections. By translating, we may derive the same conclusion around any point of the original disc; since that disc is simply connected, we obtain a basis of horizontal sections over the entire disc. $\qquad \square$

Remark 7.2.2. In the p-adic setting, we will see that the first step of the proof of Theorem 7.2.1 remains valid, but there is no analogue of the second step (analytic continuation), and indeed the whole conclusion becomes false; see Example 0.4.1.

Let us now consider a punctured disc and look at monodromy.

Notation 7.2.3. In this chapter only, for K a subfield of $\mathbb{C}$, let $K\{z\}$ be the subfield of $K((z))$ consisting of formal Laurent series which represent meromorphic functions on some neighborhood of $z = 0$. The exact choice of that neighborhood may vary with the series.

Definition 7.2.4. Let V be a finite differential module over $\mathbb{C}\{z\}$; choose a basis of V and let N be the matrix of action of D on this basis. On some disc centered at $z = 0$, the entries of N are meromorphic with no poles away from $z = 0$. On any subdisc not containing 0, by Theorem 7.2.1 we obtain a basis of horizontal sections. If we start with a basis of horizontal sections in a neighborhood of some point away from 0, then analytically continue around a circle proceeding once counterclockwise around the origin, we end up with a new basis of local horizontal sections. The linear transformation taking the old basis to the new one is called the *monodromy transformation* of V (or of its associated differential system). The *(topological) exponents* of V are defined (modulo translation by $\mathbb{Z}$) to be the multiset of numbers $\alpha_1, \ldots, \alpha_n$ for which $e^{-2\pi i \alpha_1}, \ldots, e^{-2\pi i \alpha_n}$ are the eigenvalues of the monodromy transformation.

The monodromy transformation controls our ability to construct global horizontal sections, by the following statement whose proof is evident.

Proposition 7.2.5. *In Definition 7.2.4, any fixed vector under the monodromy transformation corresponds to a horizontal section defined on some punctured disc, rather than on the universal covering space of a punctured disc. As a result, the monodromy transformation is unipotent (i.e., the exponents are all zero) if and only if there exists a basis on which D acts via a nilpotent matrix.*

Definition 7.2.6. In Definition 7.2.4, we say that V is *quasiunipotent* if its exponents are rational; equivalently, the monodromy transformation of V becomes unipotent after pulling back along $z \mapsto z^m$ for some positive integer m. This situation arises for Picard–Fuchs modules; see Appendix A.

Remark 7.2.7. The relationship between the properties of the monodromy transformation and the existence of horizontal sections of the differential module begs the question: is it possible to extract the monodromy transformation for a differential module, whose definition is purely analytic, from the algebraic data that defines the differential system? The only case in which this is straightforward is that of a regular module, which we consider next.

7.3 Formal solutions of regular differential equations

One can formally imitate the proof of Cauchy's theorem (Theorem 7.2.1) in the regular case, as follows.

Definition 7.3.1. Let $N = \sum_{i=0}^{\infty} N_i t^i$ be an $n \times n$ matrix with entries in $K[\![z]\!]$. A *fundamental solution matrix* for N is an $n \times n$ matrix U with $U \equiv I_n \pmod{z}$ such that $U^{-1} N U + U^{-1} z \frac{d}{dz}(U) = N_0$.

Remark 7.3.2. Note that if U is a fundamental solution matrix for N, then

$$-U^T N^T U^{-T} + U^T z \frac{d}{dz}(U^{-T}) = -U^T N^T U^{-T} - U^T U^{-T}(z \frac{d}{dz}(U^T))U^{-T}$$

$$= -U^T N^T U^{-T} - (z \frac{d}{dz}(U^T))U^{-T}$$

$$= -N_0^T.$$

That is, U^{-T} is a fundamental solution matrix for $-N^T$. Consequently, by proving a general result about U, we also obtain a corresponding result for U^{-T} and hence for U^{-1}.

To specify when a fundamental solution matrix exists, we need the following definition.

Definition 7.3.3. We say that a square matrix N with entries in a field of characteristic zero has *prepared eigenvalues* if the eigenvalues $\lambda_1, \ldots, \lambda_n$ of N satisfy the following conditions:

$$\lambda_i \in \mathbb{Z} \Leftrightarrow \lambda_i = 0$$
$$\lambda_i - \lambda_j \in \mathbb{Z} \Leftrightarrow \lambda_i = \lambda_j.$$

If only the second condition holds, we say that N has *weakly prepared eigenvalues*.

We will also need the following lemma, which will come up again several times later.

Definition 7.3.4. Let N be a nilpotent $n \times n$ matrix over K. The *nilpotency index* of N is the smallest positive integer e such that $N^e = 0$.

Lemma 7.3.5. *Let N_1, N_2 be matrices of respective sizes $m \times m$, $n \times n$ over K. Let $\lambda_{1,1}, \ldots, \lambda_{1,m}$ and $\lambda_{2,1}, \ldots, \lambda_{2,n}$ be the eigenvalues of N_1 and N_2, respectively. Then the eigenvalues of the K-linear endomorphism $X \mapsto N_1 X + X N_2$ on the space of $m \times n$ matrices over K are equal to $\lambda_{1,i} + \lambda_{2,j}$ for $i = 1, \ldots, m$, $j = 1, \ldots, n$. Moreover, if N_1, N_2 are themselves nilpotent with respective nilpotency indices e_1, e_2, then the nilpotency index of $X \mapsto N_1 X + X N_2$ equals $e_1 + e_2 - 1$.*

Proof There is no harm in enlarging K, so we may assume it is algebraically closed. Moreover, if $U_1 \in \mathrm{GL}_m(K)$ and $U_2 \in \mathrm{GL}_n(K)$, then it is equivalent to calculate the eigenvalues of the conjugated endomorphism

$$X \mapsto U_1^{-1}(N_1(U_1 X U_2^{-1}) + (U_1 X U_2^{-1})N_2)U_2 = (U_1^{-1}N_1 U_1)X + X(U_2^{-1}N_2 U_2).$$

Consequently, we may conjugate N_1 and N_2 into Jordan normal form. By separating X into blocks, we may reduce to the case where N_1 and N_2 consist of single Jordan normal blocks with the eigenvalues λ_1 and λ_2. By subtracting λ_1 and λ_2 from the overall endomorphisms, we may reduce to the case $\lambda_1 = \lambda_2 = 0$, where we want to show that the map $X \mapsto N_1 X + X N_2$ is nilpotent with nilpotency index at most $e_1 + e_2 - 1$. This can be shown by hand, but can also be seen as follows: put $f(X) = N_1 X + X N_2$, and write the i-th composition of f as

$$f^i(X) = \sum_{j=0}^{i} \binom{i}{j} N_1^j X N_2^{i-j}.$$

If $i \geq e_1 + e_2 - 1$, then in each term either $j \geq e_1$ or $i - j \geq e_2$, so the whole sum vanishes. If $i = e_1 + e_2 - 2$, then similarly every term vanishes except possibly $\binom{e_1+e_2-2}{e_1-1} N_1^{e_1-1} X N_2^{e_2-1}$. Since we are assuming K has characteristic 0, the binomial coefficient $\binom{e_1+e_2-2}{e_1-1}$ is nonzero and the matrices $N_1^{e_1-1}$ and $N_2^{e_2-1}$ are nonzero, so we can make this product nonzero by choosing suitable X. □

Proposition 7.3.6. *Let $N = \sum_{i=0}^{\infty} N_i t^i$ be an $n \times n$ matrix with entries in $K[\![z]\!]$ such that N_0 has weakly prepared eigenvalues. Then N admits a unique fundamental solution matrix.*

Proof Let $\lambda_1, \ldots, \lambda_n \in K^{\mathrm{alg}}$ be the eigenvalues of N_0. Rewrite the defining equation as $NU + z\frac{d}{dz}(U) = UN_0$, then expand $U = \sum_{i=0}^{\infty} U_i t^i$ and write the new defining equation as a recurrence:

$$iU_i = U_i N_0 - N_0 U_i - \sum_{j=1}^{i} N_j U_{i-j} \qquad (i > 0). \tag{7.3.6.1}$$

Viewing the map $X \mapsto X N_0 - N_0 X$ as a linear transformation on the space of $n \times n$ matrices over $K((t))$, we see by Lemma 7.3.5 that its eigenvalues are the differences $\lambda_j - \lambda_k$ for $j, k = 1, \ldots, n$. Likewise, the eigenvalues of $X \mapsto iX - XN_0 + N_0X$ are $i - \lambda_j + \lambda_k$; for i a positive integer, the condition that the λ's are weakly prepared ensures that $i - \lambda_j + \lambda_k$ cannot vanish (indeed, it cannot be an integer unless it equals i). Consequently, given N and $U_0, \ldots, U_{i-1}$, there is a unique choice of U_i satisfying (7.3.6.1); this proves the desired result. □

Remark 7.3.7. Suppose that N is an $n \times n$ matrix with entries in $\mathbb{C}[\![z]\!]$ whose constant term has prepared eigenvalues. If we convert the differential system defined by N into a differential module M, then the $\mathbb{C}$-span of the columns of the fundamental solution matrix forms a $\mathbb{C}$-submodule of M stable under D. In fact, the action of D on M is the linear transformation defined by N_0. Since this transformation may be nonzero, the elements of this span are not all horizontal; however, one can show (exercise) that every horizontal element of M does appear in this span. This justifies the name "fundamental solution matrix".

This formal argument becomes relevant in the complex analytic setting by virtue of the following fact, whose proof is not at all formal. For proof, see [149, §III.8, Appendix II] or the exercises.

Theorem 7.3.8 (Fuchs). *Let $N = \sum_{i=0}^{\infty} N_i t^i$ be an $n \times n$ matrix with entries in $\mathbb{C}\{z\}$ such that N_0 has weakly prepared eigenvalues. Then the fundamental solution matrix for N over $\mathbb{C}[[z]]$ also has entries in $\mathbb{C}\{z\}$ (as does its inverse).*

Corollary 7.3.9. *With notation as in Theorem 7.3.8, let $\lambda_1, \ldots, \lambda_n$ be the eigenvalues of N_0. Then the eigenvalues of the monodromy transformation of the system $D(v) = Nv + d(v)$ are $e^{-2\pi i \lambda_1}, \ldots, e^{-2\pi i \lambda_n}$.*

Proof In terms of a basis on which the matrix of action of D is N_0, the matrix $\exp(-N_0 \log(z))$ provides a basis of horizontal elements. (The case $N_0 = 0$ is Theorem 7.2.1.) $\square$

In order to enforce the condition on prepared eigenvalues, we use what are classically known as *shearing transformations*.

Proposition 7.3.10 (Shearing transformations). *Let N be an $n \times n$ matrix over $K[[z]]$, with constant term N_0. Let $\alpha_1, \ldots, \alpha_m \in K^{\mathrm{alg}}$ be eigenvalues of N forming a single Galois orbit over K. Then there exists $U \in \mathrm{GL}_n(K[z, z^{-1}])$ such that $U^{-1}NU + U^{-1}d(U)$ again has entries in $K[[z]]$, and its matrix of constant terms has the same eigenvalues as N_0 except that every instance of each α_i has been replaced by $\alpha_i + 1$. The same conclusion holds with $\alpha_i - 1$ in place of $\alpha_i + 1$.*

Proof Exercise. $\square$

Corollary 7.3.11 (Fuchs). *Let V be a regular finite differential module over $\mathbb{C}\{z\}$. Then any horizontal element of $V \otimes_{\mathbb{C}\{z\}} \mathbb{C}((z))$ belongs to V itself; that is, any formal horizontal section is convergent. (This is false in the irregular case; see the notes.)*

Proof Using Proposition 7.3.10, we may construct a basis of V on which D acts via a matrix in $\mathbb{C}[[z]] \cap \mathbb{C}\{z\}$ whose constant term has weakly prepared eigenvalues. By Remark 7.3.7, any horizontal element of $V \otimes_{\mathbb{C}\{z\}} \mathbb{C}((z))$ is a

$\mathbb{C}$-linear combination of column vectors of the fundamental solution matrix; by Theorem 7.3.8, any such linear combination converges in a disc. □

To put everything together, we state the following result.

Proposition 7.3.12. *Let V be a regular finite differential module over $K((z))$. Then there exists a basis of V on which the matrix of action of D has entries in K and prepared eigenvalues.*

Proof By Proposition 6.5.6, we can find a $K[\![z]\!]$-lattice L in V (i.e., a finitely generated $K[\![z]\!]$-submodule whose $K((z))$-span is the entirety of V) which is stable under D. By Proposition 7.3.10, we can modify L so that the constant term of the matrix of action of D on some basis of L has prepared eigenvalues. (Because we cannot separate Galois conjugates, the previous statement relies on the fact that no two distinct Galois conjugates α, β over K may differ by a nonzero integer. This fact holds because the trace from the Galois closure of $K(\alpha, \beta)$ to K vanishes on $\alpha - \beta$, but is injective on K because K is of characteristic 0.) We may then apply Proposition 7.3.6 to deduce the claim. □

7.4 Index and irregularity

We mention an alternate interpretation of irregularity considered by Malgrange and its relation to the index of a meromorphic differential module on $\mathbb{P}^1$.

Definition 7.4.1. For V a vector space over K and $f : V \to V$ a K-linear transformation, define the *index* of f as

$$\chi(f, V) = \dim_K \ker(f, V) - \dim_K \operatorname{coker}(f, V),$$

provided that both terms are finite. When this holds, we say that the index exists, or that f has an index, or that f is a *Fredholm map*).

Let F be any subfield of $K((z))$ stable under d and let (V, D) be a finite differential module over F. We then define the index of V as $\chi(V) = \chi(D, V)$.

Proposition 7.4.2. *For any finite differential module V over $K((z))$,*

$$\dim_K H^0(V) = \dim_K H^1(V) < \infty,$$

so $\chi(V) = 0$.

Proof Exercise. □

In the convergent case, the index carries more information.

Theorem 7.4.3. *Let V be a finite differential module over* $\mathbb{C}\{z\}$. *Then V has an index and* $\chi(V) = -\operatorname{irr}(V)$.

Proof See [299, Théorème 2.1]. □

Definition 7.4.4. Let Z be a finite subset of the Riemann sphere $\mathbb{P}^1(\mathbb{C}) = \mathbb{C} \cup \{\infty\}$. By a *a finite meromorphic differential module over* $\mathbb{P}^1_{\mathbb{C}} \setminus Z$, we will mean a finite differential module over the ring of rational functions within $\mathbb{C}(z)$ having poles only within Z.

For U an open domain of $\mathbb{P}^1(\mathbb{C})$ and Z a finite subset of U, we may also define a finite meromorphic differential module over $U \setminus Z$ viewing U as a complex-analytic space. For $U = \mathbb{P}^1(\mathbb{C})$, we will see in Theorem 7.4.5 that this agrees with the previous definition.

Let M be a finite meromorphic differential module over $U \setminus Z$. For $z \in Z$, define the *irregularity* $\operatorname{irr}_z(M)$ of M at z by making a linear fractional change of coordinates to move z to 0 and using Definition 7.1.1.

We study such modules using the *GAGA principle*; see the chapter notes.

Theorem 7.4.5 (GAGA principle). *The following statements hold.*

(a) *Every vector bundle on the Riemann sphere is the pullback of a unique vector bundle on the scheme* $\mathbb{P}^1_{\mathbb{C}}$.
(b) *The global sections of a vector bundle on* $\mathbb{P}^1_{\mathbb{C}}$ *coincide with those of its pullback to the Riemann sphere.*

Proof By a theorem of Grothendieck [181], any vector bundle on either $\mathbb{P}^1_{\mathbb{C}}$ or the Riemann sphere splits (nonuniquely) as a direct sum of line bundles, each of the form $O(n)$ for some $n \in \mathbb{Z}$; this implies the existence of the pullback in (a). Using Liouville's theorem that every bounded entire function on $\mathbb{C}$ is constant, one then deduces that the sections of $O(n)$ in either category are exactly the polynomials of degree at most n; this implies (b). Finally, using (b), we may deduce the uniqueness in (a) by interpreting maps of bundles $\mathcal{E}_1 \to \mathcal{E}_2$ as sections of $\mathcal{E}_1^\vee \otimes \mathcal{E}_2$. □

Corollary 7.4.6. *Let M be a finite meromorphic differential module over* $\mathbb{P}^1_{\mathbb{C}} \setminus Z$. *Let* M^{an} *be the associated differential module over the complex-analytic space* $\mathbb{P}^1(\mathbb{C}) \setminus Z$. *Then the natural maps* $H^i(M) \to H^i(M^{\mathrm{an}})$ *for* $i = 0, 1$ *are isomorphisms. In particular, M has an index if and only if* M^{an} *does, in which case* $\chi(M) = \chi(M^{\mathrm{an}})$.

Proof The equality of H^0 follows from Theorem 7.4.5(b). The equality of H^1 follows from Theorem 7.4.5(a) via the Yoneda interpretation (Lemma 5.3.3). □

Lemma 7.4.7. *Let U be an open domain of $\mathbb{P}^1(\mathbb{C})$ with finite Euler characteristic $\chi(U)$. Let Z be a finite subset of U. Let M be a finite meromorphic differential module of rank n over $U \setminus Z$. Then M has an index equal to*

$$\chi(M) = n\chi(U) - \sum_{z \in Z} \mathrm{irr}_z(M).$$

Proof If $U = U_1 \cup U_2$, then

$$\chi(M) = \chi(M|_{U_1}) + \chi(M|_{U_2}) - \chi(M_{U_1 \cap U_2})$$

and a similar identity holds for the right-hand side of the desired equality. By covering U with sufficiently small opens, we may reduce to the case where either U is simply connected and $Z = \emptyset$, or U is a disc and $Z = \{z\}$ is a single point. In the former case, M is trivial on U and so $\chi(M) = n = n\chi(U)$. In the latter case, we may invoke Theorem 7.4.3 to conclude. $\qquad\square$

Theorem 7.4.8. *Let Z be a finite subset of $\mathbb{P}^1(\mathbb{C})$. Let M be a finite meromorphic differential module of rank n over $\mathbb{P}^1_{\mathbb{C}} \setminus Z$. Then M has an index equal to*

$$\chi(M) = n(2 - \#Z) - \sum_{z \in Z} \mathrm{irr}_z(M).$$

Proof Using Corollary 7.4.6, we may reduce the claim to Lemma 7.4.7. $\qquad\square$

7.5 The Turrittin–Levelt–Hukuhara decomposition theorem

One can classify differential modules over $K((z))$ rather simply, provided that one is willing to admit finite field extensions. Note that the results do not descend from $\mathbb{C}((z))$ to $\mathbb{C}\{z\}$ on account of the failure of Corollary 7.3.11 in the irregular case. Nonetheless, even a formal classification is extremely useful; see the chapter notes.

Note the use of the refined visible decomposition theorem (Theorem 6.8.2) in the following proof.

Theorem 7.5.1. *Let V be a finite differential module over $K((z))$. Then there exist a positive integer h and a finite extension K' of K such that $V \otimes_{K((z))} K'((z^{1/h}))$ admits a direct sum decomposition $\bigoplus_i V_i$ in which each $V_i^{\vee} \otimes V_i$ is regular.*

Proof We induct on the dimension of V. Let v be a generator of $\wedge^n V$, write $D(v) = sv$, and let W be the differential module over $K((z))$ with one generator w satisfying $D(v) = n^{-1}sw$. Note that $W^{\vee} \otimes V$ cannot be refined, as otherwise

Proposition 6.8.4(a) would imply that $W^\vee \otimes V$ has the same spectral radius as $\wedge^n(W^\vee \otimes V) \cong (W^\vee)^{\otimes n} \otimes \wedge^n V$, which is trivial.

By Theorem 6.8.2, either $W^\vee \otimes V$ is regular or refined, or for suitable h, K', $(W^\vee \otimes V) \otimes_{K((z))} K'((z^{1/h}))$ splits as a nontrivial direct sum. Since we ruled out the refined case, either $W^\vee \otimes V$ is regular and so $V^\vee \otimes V$ is regular, or we may invoke the induction hypothesis. □

Remark 7.5.2. If we insist that the decomposition in Theorem 7.5.1 be minimal (i.e., that it uses as few summands as possible), then it is unique. Consequently, such a minimal decomposition must be respected by any extra structures on V which respect the action of D. For instance, if K carries one or more derivations, and we equip V with actions of these derivations, then the decomposition is preserved by these actions.

From the proof of Theorem 7.5.1, we immediately read off the following classification of the components appearing in the direct sum decomposition therein.

Theorem 7.5.3. *Let V be a finite differential module over $K((z))$ such that $V^\vee \otimes V$ is regular. Then there exists a differential module W of rank 1 such that $W^\vee \otimes V$ is regular.*

Corollary 7.5.4. *Suppose that K is algebraically closed. Let V be a finite differential module over $K((z))$ such that $V^\vee \otimes V$ is regular. Then there is a unique decomposition $V = \bigoplus_i V_i$ such that $V_i^\vee \otimes V_j$ has all exponents zero if $i = j$ and all exponents nonzero if $i \neq j$.*

Proof If V itself is regular, the assertion follows from Proposition 7.3.12 and the fact that a linear transformation over an algebraically closed field decomposes as a direct sum of generalized eigenspaces. In general, we may replace V with $W^\vee \otimes V$ for some V as in Theorem 7.5.3 and reduce to the regular case. □

The strongest form of the Turrittin–Levelt–Hukuhara decomposition is the following statement, which both eliminates the base extension in Theorem 7.5.1 and incorporates the statement of Theorem 7.5.3.

Definition 7.5.5. Let h be a positive integer, and suppose $P = \{P_1, \ldots, P_h\}$ is a Galois orbit over $K((z))$. Let F be a finite Galois extension of $K((z))$ containing $P_1, \ldots, P_h$. Then the differential module of rank h over F with generators $e_1, \ldots, e_h$ satisfying

$$D(e_i) = P_i e_i \qquad (i = 1, \ldots, h)$$

descends uniquely to a differential module over $K((z))$, which we denote $E(P)$. When the orbit is a singleton $\{P_1\}$, we also write $E(P_1)$ as shorthand for $E(\{P_1\})$. (See the proof of Theorem 7.5.6 for more on the Galois descent construction.)

Theorem 7.5.6. *Let V be a finite differential module over $K((z))$. Then V admits a direct sum decomposition*

$$V = \bigoplus_i E(P_i) \otimes X_i$$

for some Galois orbits P_i and some regular differential modules X_i.

Proof Suppose first that the conclusion of Theorem 7.5.1 applies by taking $K'((z^{1/h})) = K((z))$. Then Theorem 7.5.3 implies that $V = E(P) \otimes X$ for some $P \in K((z))$ and some regular differential module X. In fact, there is a unique choice of $P \in z^{-1}K[z^{-1}]$ for which such an X exists, and the decomposition is also unique.

In general, we get a decomposition of the desired form over $K'((z^{1/h}))$ for some finite extension K' of K and some positive integer h. We deduce the desired assertion from this using a Galois descent argument. Put $G = \mathrm{Gal}(K'((z^{1/h}))/K((z)))$. For $\tau \in G$, note that we can perform base change along τ (as in Definition 5.3.2) to define $\tau^* V$.

As above, we choose each P_i to consist entirely of elements of $z^{-1/h}K'[z^{-1/h}]$. For each i and each $\tau \in G$, we must then have $\tau(P_i) = P_j$ for some j. From the uniqueness of the decomposition, we get an isomorphism

$$\iota_\tau : \tau^*(E(P_i) \otimes X_i) \to E(P_j) \otimes X_j.$$

These isomorphisms satisfy a cocycle condition: for $\sigma, \tau \in G$,

$$\iota_{\sigma\tau} = \sigma^*(\iota_\tau) \circ \iota_\sigma.$$

We may canonically identify $\tau^* E(P_i)$ with $E(P_j)$ by matching up the generator $w \otimes 1$ with the generator w. We may then canonically identify ι_τ with a horizontal element of $E(P_j)^\vee \otimes E(P_j) \otimes \tau^* X_i^\vee \otimes X_j$. By projecting onto the trace component of $E(P_j)^\vee \otimes E(P_j)$, we get a horizontal element of $\tau^* X_i^\vee \otimes X_j$, which we in turn identify with a morphism $\iota'_\tau : \tau^* X_i \to X_j$. These maps again satisfy the cocycle condition, so Galois descent allows us to identify each X_i with the base extension of a differential module X_i' over $K((z))$. Under this identification, if P_i' denotes the Galois orbit of P_i, we get a canonical identification

$$(E(P_i') \otimes X_i') \otimes_{K((z))} K'((z^{1/h})) \cong \bigoplus_{j : P_j \in P_i'} E(P_j) \otimes X_j.$$

This proves the claim.   □

Remark 7.5.7. The proof of Theorem 7.5.6 shows that both the direct sum decomposition and the identification of the tensor factors in each summand can be made canonical, by insisting that each P_i consist of elements of $z^{-1/h} K'[z^{-1/h}]$ for some finite extension K' of K and some positive integer h. However, everything still depends implicitly on the choice of the series parameter z in the field $K((z))$; in many applications, such a choice is not at all natural.

7.6 Asymptotic behavior

Another manifestation of the difference between regular and irregular singularities, of great interest in the classical theory, is in the asymptotic behavior of horizontal sections near the singular point. This gives rise to an observation first made by Stokes [379] and subsequently known as the *Stokes phenomenon*.

Theorem 7.6.1. *Let U be the open unit disc in $\mathbb{C}$, and let Z be the subset $\{0\}$ of U. Let V be a finite meromorphic differential module over $U \setminus Z$ admitting a basis on which D acts via the matrix N. Then the following conditions are equivalent.*

(a) The module $V \otimes \mathbb{C}((z))$ is regular.

(b) For any open subspace W of U and any solution v of the equation $Nv + d(v) = 0$ in the ring of holomorphic functions on $W \setminus Z$, there exists an integer m such that on some neighborhood of 0, the entries of the vector v are bounded in absolute value by $|z|^{-m}$. (That is, these entries are functions of moderate growth.)

Proof If (a) holds, then so does (b) by Fuchs's theorem (Theorem 7.3.8). Conversely, if (a) fails, we may use the decomposition of $V \otimes \mathbb{C}((z))$ given by Theorem 7.5.1 to read off the asymptotic expansions of solutions near 0, and any basis of solutions must contain at least one whose asymptotics are dominated by an exponential term. See [126, Théorème 1.19] for a more detailed analysis.   □

Example 7.6.2. Consider the differential equation

$$d^2(y) + d(y) - z^{-3}y = 0;$$

it arises from the classical *Airy equation* $y'' - zy = 0$ via the substitution $z \mapsto 1/z$. This equation has irregularity 3. Making a branch cut along the negative real axis, we have the following asymptotic expansion for a solution [8, 10.4.59]:

$$y \sim \frac{1}{2\sqrt{\pi}} z^{1/4} \exp\left(-\frac{2}{3} z^{-3/2}\right) \sum_{n=0}^{\infty} \frac{(2n+1)(2n+3)\cdots(6n-1)}{(-144)^n n!} z^{3n/2}.$$

In particular, the primary asymptotic behavior is given by the term with $n = 0$, which on account of the factor of $\exp(-(2/3)z^{-3/2})$ is not a function of moderate growth.

We draw attention to another feature of this example which is entirely typical of the general case: the infinite series over n has radius of convergence 0! That is, the series does not correspond to the germ of a holomorphic function at $z = 0$. Instead, it must be interpreted as follows: for each positive integer m, y is asymptotic to

$$\frac{1}{2\sqrt{\pi}} z^{1/4} \exp\left(-\frac{2}{3} z^{-3/2}\right) \left(\sum_{n=0}^{m-1} \frac{(2n+1)(2n+3)\cdots(6n-1)}{(-144)^n n!} z^{3n/2} + O(z^{3m/2}) \right)$$

but the implied constants in the big-O notation do not behave well as m varies.

Another feature typical of the general case is that the growth of the dominant factor $\exp\left(-\frac{2}{3}z^{-3/2}\right)$ as z approaches 0 depends strongly on the direction of approach. For example, say we take $z = re^{i\theta}$ with θ fixed and $r \to 0^+$; then

$$\left| \exp\left(-\frac{2}{3} z^{-3/2}\right) \right| = \left| \exp\left(-\frac{2}{3} r^{-3/2} e^{-3i\theta/2}\right) \right| = \exp\left(-\frac{2}{3} r^{-3/2} \cos\frac{3\theta}{2}\right).$$

This decays rapidly for $\theta = 0$, but does not decay at all for $\theta = \pi/3$.

Notes

Much of this material can be found in [375]. In particular, see Chapter 3 for the formal classification, Chapter 7 for asymptotic expansions, and Chapters 8 and 9 for the Stokes phenomenon.

The notion of a regular singularity was introduced by Fuchs in the nineteenth century, as part of a classification of differential equations with everywhere-meromorphic singularities on the Riemann sphere which have algebraic solutions. Regular singularities are sometimes referred to as *Fuchsian singularities*. Much of our modern understanding of the regularity condition, especially in higher dimensions, comes from the book of Deligne [126].

As noted in the text, Corollary 7.3.11 is completely false for irregular modules. This was originally noticed in numerous examples of particular differential equations (e.g., the Bessel equation at infinity; see Example 20.2.1), and motivated the definition of irregularity in the first place.

The relationship between the interpretations of irregularity using the Newton polygon and indices is due to Malgrange [299]. (Other contemporary treatments are [176, 223].) Our treatment, in which the spectral radius plays a pivotal role, is based on [149, §3]; this point of view is ultimately due to Robba.

A complex-analytic interpretation of the Newton polygon, in the manner of the relation between irregularity and index, has been given by Ramis [346]. It involves considering subrings of $\mathbb{C}\{z\}$ composed of functions with certain extra convergence restrictions (*Gevrey functions*), and looking at the index of $z\frac{d}{dz}$ after tensoring the given differential module with one of these subrings.

The term "GAGA principle" used in reference to Theorem 7.4.5 is an acronym for the title of Serre's paper [368]. The main theorem of that paper asserts that for a complex projective variety, coherent sheaves on the variety are the same as coherent sheaves on the analytification, and the cohomology is the same whether computed in the algebraic or analytic category; this includes Theorem 7.4.5 as a very special case. A sample corollary is that every analytic hypersurface in projective space is algebraic; more directly relevant for our purposes, the de Rham cohomology of a meromorphic connection may be computed equally well in either the algebraic or analytic categories.

This type of result was considered further by Grothendieck in SGA 1 [183, Expose XII], where it is shown that a similar result holds for proper algebraic varieties (but see below). Moreover, the nature of analytification is clarified: for X a complex algebraic variety, its analytification X^{an} admits a map to X in the category of locally ringed spaces satisfying a certain universal property.

Beware that while one motivation for GAGA is Riemann's theorem that every compact Riemann surface is the analytification of an algebraic curve, the latter statement does not generalize as far as one might expect; for example, there are two-dimensional complex tori with no algebraic structure. These are examples of proper *analytic* varieties which are not themselves algebraic; this does not violate GAGA because these are not contained in the analytification of a proper algebraic variety. See [179, §2.6] for further discussion.

Our application of the GAGA principle to study the index of differential modules (Theorem 7.4.8) follows Deligne [126, (I.6.21.1)]. In connection with the use of Grothendieck's classification of vector bundles on $\mathbb{P}^1$, one has an analogue of (and ultimately a corollary of) the residue theorem in complex analysis: for a vector bundle carrying a connection with regular singularities, the degree is equal to the sum over all residues at all singular points. See [327], and also [23] for a precursor result.

Theorem 7.5.6 is a slight reformulation of a classification result due to

Turrittin [391] (building on earlier work of Hukuhara in the rank one case) for the existence aspect and Levelt [290] for the uniqueness aspect. See [19, §5.3.1] for further discussion. (Keep in mind that the name "Turrittin" is misspelled "Turritin" in some references.)

The literature on the Stokes phenomenon is too vast to catalog here. Some early algebraic descriptions were given by Balser–Jurkat–Lutz [43], Sibuya [372], and Wasow [406]. For more modern treatments, see [294], [300], [375], and [405].

One can describe a higher-dimensional analogue of the situation considered in this chapter by considering flat meromorphic connections on complex algebraic or analytic varieties. (Here *flat* means that the curvature tensor vanishes; to avoid confusion with the algebraic notion of flatness, the term *integrable* is also used.) In this context, the equivalence between regularity of singularities and moderate asymptotic growth of solutions is due to Deligne [126, Théorème II.4.1]. The comparison between algebraic and analytic cohomology again persists; see [19]. A higher-dimensional analogue of Turrittin's theorem is given by the theory of *good formal structures* on flat meromorphic connections; see [250, 251, 260] for an approach drawing on results from this book, and [310, 311] for an alternate point of view due to Mochizuki. This then leads to an irregular Riemann–Hilbert correspondence [122] and to a theory of Stokes torsors [312, 359, 386, 387, 388]; the latter includes a group filtration bearing a striking resemblance to the higher ramification filtration on the Galois group of a local field (Section 3.4).

Exercises

7.1 Let V be a differential module over $K((z))$ with spectral radius s for $z\frac{d}{dz}$. Prove that the spectral radius of V for $\frac{d}{dz}$ is $s|z|^{-1}$.

7.2 Verify the claims of Example 7.1.5.

7.3 Let N be an $n \times n$ matrix over $K[[z]]$ whose constant term has prepared eigenvalues. Prove that any vector $v \in K((z))^n$ with $Nv + z\frac{d}{dz}(v) = 0$ is a K-linear combination of columns of the fundamental solution matrix of N. (Hint: change basis first.)

7.4 Prove Fuchs's theorem (Theorem 7.3.8). (Hint: let $U = \sum_{i=0}^{\infty} U_i z^i$ be the fundamental solution matrix of N. Relate the norms of U_i and $iU_i - U_i N_0 + N_0 U_i$ by considering the singular values of the map $X \mapsto$

$X - XN_0 + N_0X$. Then use (7.3.6.1) to obtain

$$\left| U_i - \frac{U_i N_0 - N_0 U_i}{i} \right| \leq \frac{1}{i} \sum_{j=1}^{i} |N_j| |U_{i-j}| \leq \max_{1 \leq j \leq i} \{ |N_j| |U_{i-j}| \}$$

and proceed as in the proof of Theorem 7.2.1.)

7.5   Prove Proposition 7.3.10. (Hint: show that for any space V of constant vectors stable under multiplication by N_0, the set of vectors with entries in $K[\![z]\!]$ whose reductions modulo z belong to V are closed under the operation $v \mapsto Nv + d(v)$. Then take V to be a union of generalized eigenspaces.)

7.6   Prove Proposition 7.4.2. (Hint: change basis by preparing eigenvalues and then applying Theorem 7.3.8.)

7.7   Let $Q \in K((z))[U]$ be a polynomial whose roots $\lambda_1, \ldots, \lambda_n \in K((z))^{\mathrm{alg}}$ satisfy $|\lambda_i/\lambda_j - 1| < 1$ for all i, j. Prove that there exists $\lambda \in K((z))$ such that $|\lambda_i/\lambda - 1|$ for all i. Then show that this fails without the hypothesis that K has characteristic 0.

Part III

p-adic Differential Equations on Discs and Annuli

8

Rings of functions on discs and annuli

In Part III, we focus our attention specifically on p-adic ordinary differential equations (although most of our results apply also to compete nonarchimedean fields of residual characteristic 0). To do this at the ideal level of generality, one would first need to introduce a category of geometric spaces over which to work. This would require a fair bit of discussion of either rigid analytic geometry in the manner of Tate, or nonarchimedean analytic geometry in the manner of Berkovich, neither of which we want to assume or introduce (except at a very superficial level in Part VII). Fortunately, since we only need to consider one-dimensional spaces, we can manage by working completely algebraically, considering differential modules over appropriate rings.

In this chapter, we introduce those rings and collect their basic algebraic properties. This includes the fact that they carry Newton polygons analogous to those for polynomials. Another key fact is that one has a form of the approximation lemma (Lemma 1.3.7) valid over some of these rings.

Notation 8.0.1. Throughout this part, let K be a field of characteristic 0, complete for a nontrivial nonarchimedean norm $|\cdot|$. (The hypothesis of characteristic 0 is not used in this chapter; it will become crucial when we start discussing differential modules again.) Let p denote the characteristic of the residue field κ_K. We do not assume $p > 0$ (as the case $p = 0$ may be useful for some applications), but when $p > 0$ we do require the norm to be normalized so that $|p| = p^{-1}$.

Definition 8.0.2. By a *piecewise affine* function on an interval I, we will mean a continuous function $f \colon I \to \mathbb{R}$ such that I can be covered with intervals of positive length, on each of which f restricts to an affine function (a function of the form $ax + b$ for some $a, b \in \mathbb{R}$). By compactness of a closed interval (see Lemma 8.0.4), it is equivalent to require that each point of I admit a one-sided neighborhood on each side within I, on which f restricts to an affine function.

(If I is infinite, we allow the possibility of having infinitely many different slopes unless we say otherwise explicitly.)

Example 8.0.3. An example specifically excluded by Definition 8.0.2 is the function $f : [0, 1] \to \mathbb{R}$ defined by

$$f(x) = \begin{cases} \frac{1}{2} & x = 0 \\ \frac{1}{2} - \frac{1}{N+1} + \frac{N}{N+2}\left(\frac{1}{N} - x\right) & x \in \left[\frac{1}{N+1}, \frac{1}{N}\right], N = 1, 2, \ldots; \end{cases}$$

this function is piecewise affine on $(0, 1]$ and continuous on $[0, 1]$, but there is no one-sided neighborhood of 0 on which f is affine.

Lemma 8.0.4. *Let $I = [\alpha, \beta]$ be a (bounded) closed interval. Let S be a set of closed subintervals of I which cover every one-sided neighborhood of every point in I. (That is, for any $\gamma \in [\alpha, \beta)$, there exist $\delta \in (\gamma, \beta]$ and $J \in S$ such that $[\gamma, \delta] \subseteq J$, and similarly for any $\gamma \in (\alpha, \beta]$, there exist $\delta \in [\alpha, \gamma)$ and $J \in S$ such that $[\delta, \gamma] \subseteq J$.) Then there exists a finite subset of S with union I.*

Proof Exercise.   □

8.1 Power series on closed discs and annuli

We start by introducing some rings that should be thought of as the analytic functions on a closed disc $|t| \le \beta$ or a closed annulus $\alpha \le |t| \le \beta$. As noted above, this is more properly done in a framework of p-adic analytic geometry, but we have deliberately chosen to avoid this framework.

Definition 8.1.1. For $\alpha, \beta > 0$, put

$$K\langle \alpha/t, t/\beta \rangle = \left\{ \sum_{i \in \mathbb{Z}} c_i t^i \in K[[t, t^{-1}]] : \lim_{i \to \pm\infty} |c_i|\rho^i = 0 \quad (\rho \in [\alpha, \beta]) \right\}.$$

That is, consider the set of formal bidirectional power series which converge when t is specialized to any value with $|t| \in [\alpha, \beta]$, or in other words, when $\alpha/|t|$ and $|t|/\beta$ are both at most 1; it suffices to check for $\rho = \alpha$ and $\rho = \beta$. Although formal bidirectional power series do not form a ring, the subset $K\langle \alpha/t, t/\beta \rangle$ does form a ring under the expected operations. (In this and all similar notation, we will omit β when it is equal to 1.)

Definition 8.1.2. If $\alpha = 0$, the only reasonable interpretation of the previous definition is to require $c_i = 0$ for $i < 0$. When there are no negative powers of

t, it is redundant to require convergence for $\rho < \beta$. In other words, we shall define

$$K\langle 0/t, t/\beta\rangle = K\langle t/\beta\rangle = \left\{\sum_{i=0}^{\infty} c_i t^i \in K[\![t]\!] : \lim_{i\to\infty} |c_i|\beta^i = 0\right\}.$$

We shall also define the ring $K\langle\alpha/t\rangle$ by implicitly taking $\beta = \infty$, or by identifying $K\langle\alpha/t, t/\beta\rangle$ with $K\langle\beta^{-1}/t^{-1}, t^{-1}/\alpha^{-1}\rangle$ and invoking the previous convention.

Remark 8.1.3. Note that if $\alpha \leq \gamma \leq \beta \leq \delta$, then inside $K\langle\gamma/t, t/\beta\rangle$,

$$K\langle\alpha/t, t/\beta\rangle \cap K\langle\gamma/t, t/\delta\rangle = K\langle\alpha/t, t/\delta\rangle.$$

This will serve as the basis for some gluing arguments later.

Definition 8.1.4. We will also occasionally use the intermediate ring

$$K[\![t/\beta]\!]_0 = \left\{\sum_{i=0}^{\infty} c_i t^i \in K[\![t]\!] : \sup_i\{|c_i|\beta^i\} < \infty\right\};$$

these are the power series which converge and take bounded values on the open disc $|t| < \beta$. (The notation will make more sense once we also define $K[\![t/\beta]\!]_\delta$ for $\delta > 0$; see Definition 18.4.1.) Note that for any $\delta \in (0, \beta)$,

$$K\langle t/\beta\rangle \subset K[\![t/\beta]\!]_0 \subset K\langle t/\delta\rangle.$$

We will most often use this construction with $\beta = 1$, in which case we can also write

$$K[\![t]\!]_0 = \mathfrak{o}_K[\![t]\!] \otimes_{\mathfrak{o}_K} K.$$

Definition 8.1.5. An analogue of the previous construction for an annulus is

$$K\langle\alpha/t, t/\beta]\!]_0 = \left\{\sum_{i\in\mathbb{Z}} c_i t^i : c_i \in K, \lim_{i\to-\infty} |c_i|\alpha^i = 0, \sup_i\{|c_i|\beta^i\} < \infty\right\};$$

these are the Laurent series which converge and take bounded values on the half-open annulus $\alpha \leq |t| < \beta$. For any $\delta \in (\alpha, \beta)$, this ring satisfies

$$K\langle\alpha/t, t/\beta\rangle \subset K\langle\alpha/t, t/\beta]\!]_0 \subset K\langle\alpha/t, t/\delta\rangle.$$

One can also use the boundedness condition on both sides, defining

$$K[\![\alpha/t, t/\beta]\!]_0 = \left\{\sum_{i\in\mathbb{Z}} c_i t^i : c_i \in K, \sup_i\{|c_i|\alpha^i\} < \infty, \sup_i\{|c_i|\beta^i\} < \infty\right\}.$$

Remark 8.1.6. The rings of bounded series behave well only in case K is discretely valued; otherwise, they are not even noetherian (exercise). For general K, it is better to work with rings of analytic elements; see Section 8.5.

8.2 Gauss norms and Newton polygons

The rings $K\langle \alpha/t, t/\beta \rangle$ behave rather like polynomial rings (or Laurent polynomial rings, in case $\alpha \neq 0$) in one variable. The next few statements are all instances of this analogy.

Definition 8.2.1. From the definition of $K\langle \alpha/t, t/\beta \rangle$, we see that it carries a well-defined ρ-Gauss norm

$$\left| \sum_i c_i t^i \right|_\rho = \max_i \{ |c_i| \rho^i \}$$

for any $\rho \in [\alpha, \beta]$. For $\rho = \alpha = 0$, this reduces to simply $|c_0|$. The fact that this is a multiplicative norm follows as in Proposition 2.1.2. The additive version (Gauss valuation) is to take $r \in [-\log \beta, -\log \alpha]$ and put

$$v_r \left(\sum_i c_i t^i \right) = \min_i \{ v(c_i) + ri \},$$

where $v(c) = -\log |c|$. One also has a β-Gauss norm on $K\langle \alpha/t, t/\beta \rrbracket_0$, although it must be defined as a supremum that may fail to be achieved if K is not discretely valued.

Definition 8.2.2. One may define the *Newton polygon* for an element $x = \sum_i x_i t^i \in K\langle \alpha/t, t/\beta \rangle$ as the boundary of the lower convex hull of the set

$$\{ (-i, v(x_i)) : i \in \mathbb{Z}, x_i \neq 0 \},$$

but retaining only those slopes in $[-\log \beta, -\log \alpha]$.

Proposition 8.2.3. *Let $x = \sum_i x_i t^i \in K\langle \alpha/t, t/\beta \rangle$ be nonzero.*

(a) *The Newton polygon of x has finite width.*

(b) *The function $r \mapsto v_r(x)$ on $[-\log \beta, -\log \alpha]$ is continuous, piecewise affine, and concave. Moreover, even if $\alpha = 0$ there are only finitely many different slopes.*

(c) *The function $\rho \mapsto |x|_\rho$ on $[\alpha, \beta]$ is continuous and log-convex. The log-convexity means that if $\rho, \sigma \in [\alpha, \beta]$, $c \in [0, 1]$, and $\tau = \rho^c \sigma^{1-c}$, then*

$$|x|_\tau \leq |x|_\rho^c |x|_\sigma^{1-c}.$$

(d) If $\alpha = 0$, then $v_r(x)$ is increasing on $[-\log\beta, \infty)$; in other words, for all $\rho \in [0, \beta]$, $|x|_\rho \leq |x|_\beta$.

Part (c) should be thought of as a nonarchimedean analogue of the Hadamard three-circle theorem.

Proof We have (a) because there is a least i for which $|c_i|\alpha^i$ is maximal, and there is a greatest j for which $|c_j|\beta^j$ is maximal. This implies (b) because as in the polynomial case, we may interpret $v_r(x)$ as the y-intercept of the supporting line of the Newton polygon of slope r. This in turn implies (c), and (d) is a remark made earlier. □

Remark 8.2.4. The analogue of Proposition 8.2.3 for $x \in K\langle\alpha/t, t/\beta\rrbracket_0$ also holds *except* that if K is not discretely valued, then $v_r(x)$ need not be piecewise affine in a one-sided neighborhood of $r = -\log\beta$ (exercise).

When dealing with the ring $K\langle\alpha/t, t/\beta\rangle$, the following completeness property will be extremely useful.

Proposition 8.2.5. *The rings $K\langle\alpha/t, t/\beta\rangle$ and $K\langle\alpha/t, t/\beta\rrbracket_0$ are Fréchet complete for the norms $|\cdot|_\rho$ for all $\rho \in [\alpha, \beta]$. That is, if $\{x_n\}_{n=0}^\infty$ is a sequence which is simultaneously Cauchy under $|\cdot|_\rho$ for all $\rho \in [\alpha, \beta]$, then it is convergent. (By Proposition 8.2.3, it suffices to check the Cauchy property at $\rho = \alpha, \beta$, or just at $\rho = \beta$ in the case $\alpha = 0$.)*

Proof Exercise. □

The completeness property is used in the construction of multiplicative inverses.

Lemma 8.2.6. *The following statements hold.*

(a) A nonzero element $f \in K\langle t/\beta\rangle$ is a unit if and only if there exists $c \in K^\times$ such that $|f - c|_\beta < |f|_\beta$.

(b) A nonzero element $f \in K\langle\alpha/t, t/\beta\rangle$ is a unit if and only if there exist $c \in K^\times$ and $i \in \mathbb{Z}$ such that $|f - ct^i|_\rho < |f|_\rho$ for $\rho = \alpha, \beta$.

(c) If $\alpha < \beta$, then a nonzero element $f \in K\langle\alpha/t, t/\beta\rrbracket_0$ is a unit if and only if there exist $c \in K^\times$ and $i \in \mathbb{Z}$ such that $|f - ct^i|_\rho < |f|_\rho$ for all $\rho \in [\alpha, \beta)$.

Proof If f is a unit, then the Newton polygon of f has no slopes in $[\alpha, \beta]$; this implies the forward implication in all cases. For the reverse implication in cases (a) and (b), it suffices to check that if $|f - 1|_\rho < |f|_\rho$ for $\rho = \alpha$ in case (a) or for $\rho = \alpha, \beta$ in case (b), then f is a unit. This holds because the series

$$\sum_{j=0}^\infty (1 - f)^j \tag{8.2.6.1}$$

converges by Proposition 8.2.5, and its limit is an inverse of f.

For the reverse implication in case (c), we note that the series (8.2.6.1) converges in $K\langle \alpha/t, t/\delta \rangle$ for each $\delta \in [\alpha, \beta]$. Moreover, the terms of the sum have bounded β-norm, so the limit does also. $\square$

Remark 8.2.7. As a consequence of Lemma 8.2.6, note that if $\alpha > 0$ is not in the divisible closure of $|K^\times|$, then $K\langle \alpha/t, t/\alpha \rangle$ is a field. Namely, for $f = \sum_i f_i t^i \in K\langle \alpha/t, t/\alpha \rangle$ nonzero, there is always a unique index i for which $|f_i t_i|$ is maximized, so Lemma 8.2.6(b) always applies.

8.3 Factorization results

We need a number of results to the effect that elements of one of the rings we are considering can be factored into "positive" and "negative" parts. The basic result of this form may be viewed as a form of the Weierstrass preparation theorem.

Proposition 8.3.1 (Weierstrass preparation). *Assume one of the following sets of hypotheses.*

(a) *Put $R = K\langle \alpha/t, t/\beta \rangle$ or $R = K\langle \alpha/t, t/\beta \rrbracket_0$. Given $f = \sum_{i \in \mathbb{Z}} f_i t^i \in R$ and $\rho \in [\alpha, \beta]$, suppose that there is a unique $m \in \mathbb{Z}$ maximizing $|f_m| \rho^m$.*

(b) *Put $R = K\langle \alpha/t, t/\alpha \rrbracket_0$ and put $\rho = \alpha$. Given $f = \sum_{i \in \mathbb{Z}} f_i t^i \in R$, suppose that the supremum of $|f_i| \rho^i$ is achieved by at least one i (this is only guaranteed to hold if K is discretely valued), and let m be the least such i.*

Then there is a unique factorization $f = f_m t^m g h$ with

$$g \in R \cap K\llbracket t \rrbracket, \qquad h \in R \cap K\llbracket t^{-1} \rrbracket,$$

such that $|g|_\rho = |g_0| = 1$, and $|h - 1|_\rho < 1$.

Proof As in Theorem 2.2.1, we invoke the master factorization theorem (Theorem 2.2.2). This gives a factorization of the desired form in the completion of R with respect to $|\cdot|_\rho$.

However, within this completion, define the subring $R' = K\langle \rho/t, t/\beta \rangle$ if $R = K\langle \alpha/t, t/\beta \rangle$, or $R' = K\langle \rho/t, t/\beta \rrbracket_0$ if $R = K\langle \alpha/t, t/\beta \rrbracket_0$. Then h is a unit in $R' \cap K\llbracket t^{-1} \rrbracket = K\langle \rho/t \rangle$ and hence in R'. We thus have $g = fh^{-1} \in R' \cap K\llbracket t \rrbracket = R \cap K\llbracket t \rrbracket$. We similarly deduce that $h \in R \cap K\llbracket t^{-1} \rrbracket$. $\square$

In light of the finite-width property of the Newton polygon, the following should not be a surprise. (One can only replace $K\langle \alpha/t, t/\beta \rangle$ with $K\langle \alpha/t, t/\beta \rrbracket_0$ if K is discretely valued; see the exercises.)

Proposition 8.3.2 (More Weierstrass preparation). *For $f \in K\langle\alpha/t, t/\beta\rangle$ with $\beta < \infty$, there exists a polynomial $P \in K[t]$ and a unit $g \in K\langle\alpha/t, t/\beta\rangle^{\times}$ such that $f = Pg$. In particular, $K\langle\alpha/t, t/\beta\rangle$ is a principal ideal domain (even if $\beta = \infty$).*

Proof If $\alpha > 0$, then we may apply Proposition 8.3.1(b) to f since the Newton polygon has only finitely many slopes. We may thus factor f in $R = K\langle\alpha/t, t/\beta\rangle_0$ as $f_m t^m g h$ with $f_m \in K$, $m \in \mathbb{Z}$, $g \in R \cap K[\![t]\!] = K[\![t/\alpha]\!]_0$, $h \in R \cap K[\![t^{-1}]\!] = K\langle\alpha/t\rangle$, $|g|_\alpha = |g_0| = 1$, and $|h - 1|_\alpha < 1$. In particular, h is a unit in $K\langle\alpha/t\rangle$ by Lemma 8.2.6, so

$$h^{-1}f \in K\langle\alpha/t, t/\beta\rangle \cap t^m K[\![t/\alpha]\!]_0 \subseteq t^m K\langle t/\beta\rangle.$$

If $\alpha = 0$, the same conclusion holds with $h = 1$, $m = 0$.

Next, factor $h^{-1}f$ in $K\langle\beta^{-1}/t^{-1}, t^{-1}/\beta^{-1}\rangle_0$ and argue as above, to get an associate of f in $K\langle\alpha/t, t/\beta\rangle$ which belongs to $t^m K\langle t/\beta\rangle \cap t^n K\langle\alpha/t\rangle$ for some integer n. This element must now belong to $K[t]$ if $\alpha = 0$, or to $K[t, t^{-1}]$ if $\alpha > 0$; in either case, we may deduce the claim. $\square$

We next wish to generalize the previous considerations to matrices, but this will require a bit more care.

Lemma 8.3.3. *Let A be an invertible $n \times n$ matrix over $K\langle\alpha/t, t/\beta\rangle$. Then A can be factored as a product of elementary matrices.*

Proof We first note that we can perform a sequence of elementary row operations after which A_{11} becomes a unit and A_{i1} becomes zero for $i = 2, \ldots, n$. Namely, using Proposition 8.3.2, multiply each row by a suitable unit to replace each A_{i1} by an element of $K[t]$. Then perform the Euclidean algorithm in $K[t]$ on the A_{i1} to achieve the desired result.

To finish, it suffices to note that we can now perform the usual Gauss–Jordan elimination over A: repeatedly apply the previous paragraph to construct a sequence of row operations putting A into upper triangular form, then perform column operations to eliminate entries above the diagonal. $\square$

Lemma 8.3.4. *Let A be an invertible $n \times n$ matrix over $K\langle t/\rho\rangle$ (resp. over $K\langle\rho/t, t/\rho\rangle$.) Then there exists $U \in \mathrm{GL}_n(K[t])$ (resp. $U \in \mathrm{GL}_n(K[t, t^{-1}])$) such that $|UA - I_n|_\rho < 1$.*

Proof Note that in the notation of Definition 4.0.3, conjugating an elementary matrix of type (c) (scaling a row) by one of type (a) (exchanging rows) produces another one of type (c), whereas conjugating an elementary matrix of type (b) (adding a multiple of one row to another) by one of type (c) produces another one of type (b). Consequently, any sequence of elementary row operations has the same effect as another sequence in which all steps of type (a) and (b) happen before all steps of type (c).

By Lemma 8.3.3, we can perform a sequence of elementary row operations on A that produces the identity matrix. By this plus the previous paragraph, we can perform another sequence of elementary row operations of types (a) and (b) on A that produces a diagonal matrix D. By appending additional row operations of type (c) with $c \in K^\times$ (resp. $c \in K^\times t^{\mathbb{Z}}$), we can force $|D - I_n|_\rho < 1$.

Let $R_1, \ldots, R_m$ be elementary matrices as above with $R_1 \cdots R_m A = D$; note that those R_j of type (a) and (c) are already elementary matrices over $K[t]$ (resp. over $K[t, t^{-1}]$), the case of type (c) being handled by the previous paragraph. Put

$$\delta = \max\{1, |A|^{-1}\} \prod_{j=1}^{m} \max\{1, |R_j|^{-1}\}.$$

For R_j of type (b), let R'_j be an elementary matrix of type (b) over $K[t]$ (resp. over $K[t, t^{-1}]$) with $|R_j - R'_j|_\rho < \delta^{-1}$; for R_j not of type (b), put $R'_j = R_j$. Then $|R'_1 \cdots R'_m A - I_n|_\rho < 1$, proving the claim. $\qquad\square$

We can now deduce a Weierstrass preparation theorem for matrices.

Proposition 8.3.5. *Let A be an invertible $n \times n$ matrix over $K\langle \alpha/t, t/\beta \rangle$. Then there exist $U \in \mathrm{GL}_n(K\langle t/\beta \rangle[t^{-1}])$ and $V \in \mathrm{GL}_n(K\langle \alpha/t \rangle)$ such that $A = UV$.*

Proof Pick any $\rho \in [\alpha, \beta]$. By Lemma 8.3.4, we can find $U_1 \in \mathrm{GL}_n(K[t, t^{-1}])$ such that $|U_1 A - I_n|_\rho < 1$. By applying the master factorization theorem (Theorem 2.2.2) in the (noncommutative) ring of $n \times n$ matrices over $K\langle \rho/t, t/\rho \rangle$, then arguing as in Proposition 8.3.1, we can factor $U_1 A$ as $U_2 V_2$ with $U_2 \in \mathrm{GL}_n(K\langle t/\beta \rangle)$ and $V_2 \in \mathrm{GL}_n(K\langle \alpha/t \rangle)$. We may then set $U = U_1^{-1} U_2$ and $V = V_2$. $\qquad\square$

Our main application of Proposition 8.3.5 is the following gluing lemma, which we will invoke frequently and often implicitly.

Lemma 8.3.6 (Gluing lemma). *Suppose $\alpha \leq \gamma \leq \beta \leq \delta$. Let M_1 be a finite free module over $K\langle \alpha/t, t/\beta \rangle$, let M_2 be a finite free module over $K\langle \gamma/t, t/\delta \rangle$, and suppose we are given an isomorphism*

$$\psi : M_1 \otimes K\langle \gamma/t, t/\beta \rangle \cong M_2 \otimes K\langle \gamma/t, t/\beta \rangle.$$

Then we can find a finite free module M over $K\langle \alpha/t, t/\delta \rangle$ and isomorphisms $M_1 \cong M \otimes K\langle \alpha/t, t/\beta \rangle$, $M_2 \cong M \otimes K\langle \gamma/t, t/\delta \rangle$ inducing ψ. Moreover, M is determined by this requirement up to unique isomorphism.

Proof We will only explain the case $\alpha > 0$; the case $\alpha = 0$ is similar. Choose bases $v_{1,1}, \ldots, v_{1,n}$ and $v_{2,1}, \ldots, v_{2,n}$ of M_1 and M_2, respectively. Let A be the change-of-basis matrix from the $v_{1,i}$ to the $v_{2,i}$, viewing both bases as bases of $M_1 \otimes K\langle \gamma/t, t/\beta \rangle \cong M_2 \otimes K\langle \gamma/t, t/\beta \rangle$ via ψ. By Proposition 8.3.5, we can factor $A = UV$ with $U \in \mathrm{GL}_n(K\langle t/\beta \rangle[t^{-1}])$ and $V \in \mathrm{GL}_n(K\langle \gamma/t \rangle)$.

We can then construct a finite free module M over $K\langle \alpha/t, t/\delta \rangle$ equipped with a basis $v_1, \ldots, v_n$ such that the change-of-basis matrices from this basis to the $v_{1,i}$ and to the $v_{2,i}$ are U^{-1} and V, respectively. This is the desired module. $\square$

Remark 8.3.7. If $\alpha \leq \gamma \leq \beta < \delta$, one can similarly glue a finite free module over $K\langle \alpha/t, t/\beta \rangle$ and a finite free module over $K\langle \gamma/t, t/\delta \rrbracket_0$ whose base extensions to $K\langle \gamma/t, t/\beta \rangle$ are isomorphic; the result is a finite free module over $K\langle \alpha/t, t/\delta \rrbracket_0$. As we will not use this fact, we omit further details; see Lemma 8.5.8 for a similar result.

8.4 Open discs and annuli

Although we have been talking about closed discs so far, it is quite natural to also consider open discs. One important reason is that the antiderivative of an analytic function on the closed disc of radius β is only defined on the *open* disc of radius β (see the exercises for Chapter 9).

Definition 8.4.1. Define the ring

$$K\{t/\beta\} = \left\{ \sum_{i=0}^{\infty} c_i t^i : c_i \in K, \lim_{i \to \infty} |c_i| \rho^i = 0 \quad (\rho \in (0, \beta)) \right\};$$

these are the power series convergent on the open disc $|t| < \beta$, with no boundedness restriction. Note that we can write

$$K\{t/\beta\} = \bigcap_{\delta \in (0,\beta)} K\langle t/\delta \rangle;$$

in particular, for any $\delta \in (0, \beta)$,

$$K\llbracket t/\beta \rrbracket_0 \subset K\{t/\beta\} \subset K\langle t/\delta \rangle.$$

Definition 8.4.2. An analogue of the previous construction for an annulus is

$$K\langle\!\langle \alpha/t, t/\beta \rangle\!\rangle = \left\{ \sum_{i \in \mathbb{Z}} c_i t^i : c_i \in K, \lim_{i \to -\infty} |c_i| \alpha^i = 0, \lim_{i \to \infty} |c_i| \rho^i = 0 \quad (\rho \in (0, \beta)) \right\};$$

these are the Laurent series convergent on the half-open annulus $\alpha \leq |t| < \beta$.

These rings have a more complicated module theory than their closed counterparts (see exercises), so we will only make occasional use of them. More often, we will work with the following definition motivated by considerations from rigid analytic geometry.

Definition 8.4.3. Consider the region $|t| \in I$, for $I \subseteq [0, \infty)$ any interval; this could be an open or closed disc, or a closed, open, or half-open annulus. By a *coherent locally free module* M on this region, we will mean a sequence of finite free modules M_i over $K\langle \alpha_i/t, t/\beta_i \rangle$ with $[\alpha_1, \beta_1] \subseteq [\alpha_2, \beta_2] \subseteq \ldots$ an increasing sequence of closed intervals with union I, together with isomorphisms $M_{i+1} \otimes K\langle \alpha_i/t, t/\beta_i \rangle \cong M_i$. Using Lemma 8.3.6, we check that the construction is canonically independent of the choice of the sequence.

Using techniques from rigid analytic geometry (e.g., see [45, Corollary 2.1.8]), it can be shown that a coherent locally free module is indeed finitely generated; that is, with notation as above, the intersection of the modules M_i is a finite locally free module over the intersection of the rings $K\langle \alpha_i/t, t/\beta_i \rangle$. However, it is not necessarily a free module unless K is spherically complete; see Proposition 16.1.4 for a special case of this assertion. For further discussion, see the chapter notes.

8.5 Analytic elements

One intermediate construction between open and closed discs is the ring $K[\![t/\beta]\!]_0$ of bounded power series, but as noted previously, it behaves badly if K is not discretely valued. Another intermediate construction that behaves somewhat better is the following.

Definition 8.5.1. Define the ring $K[\![t/\beta]\!]_{\mathrm{an}}$ by starting with the subring of $K(t)$ consisting of rational functions with no poles in the disc $|t| < \beta$, then completing for the β-Gauss norm. This is the ring of *analytic elements* on the open disc $|t| < \beta$; it satisfies

$$K\langle t/\beta \rangle \subset K[\![t/\beta]\!]_{\mathrm{an}} \subset K[\![t/\beta]\!]_0.$$

Analogously, we define the ring $K\langle \alpha/t, t/\beta]\!]_{\mathrm{an}}$ of analytic elements on the half-open annulus $\alpha \le |t| < \beta$ as follows. Start with the subring of $K(t)$ consisting of rational functions with no poles in the annulus $\alpha \le |t| < \beta$. Then take the Fréchet completion for the ρ-Gauss norms with $\alpha \le \rho \le \beta$.

One may also define the ring $K[\![\alpha/t, t/\beta]\!]_{\mathrm{an}}$ of analytic elements on the open annulus $\alpha < |t| < \beta$. For this, we start with the subring of $K(t)$ consisting of rational functions with no poles in the annulus $\alpha < |t| < \beta$, then again take the Fréchet completion for the ρ-Gauss norms with $\alpha \leq \rho \leq \beta$.

Keep in mind that the construction of $K[\![\alpha/t, t/\beta]\!]_{\mathrm{an}}$ behaves slightly differently in the cases $\alpha < \beta$ and $\alpha = \beta$. In the latter case, the ring is a field and $K[t, t^{-1}]$ is not dense in it (one must use $K(t)$ instead); we will see this field again in Definition 9.4.1.

For analytic elements, we have analogues of many of the properties asserted for analytic functions.

Proposition 8.5.2. *Let R be one of $K[\![t/\beta]\!]_{\mathrm{an}}$, $K\langle\alpha/t, t/\beta]\!]_{\mathrm{an}}$, or $K[\![\alpha/t, t/\beta]\!]_{\mathrm{an}}$.*

(a) For any $x \in R$, the function $r \mapsto -\log|x|_{e^{-r}}$ is continuous and concave in r.

(b) For any $x \in R$, the Newton polygon of x has finite width.

(c) Any $x \in R$ can be written as a polynomial $P \in K[t]$ times a unit in R.

(d) The ring R is a principal ideal domain.

(e) Any invertible square matrix over R can be factored as a product of elementary matrices.

Proof Since we are taking the Fréchet completion for norms over a closed interval, the convergence of any Cauchy sequence must be uniform in the different norms. We may thus deduce (a) from the corresponding assertions in case $x \in K(t)$ with no poles in the appropriate disc/annulus.

We check (b) for $K[\![t/\beta]\!]_{\mathrm{an}}$; the other cases are analogous. Let $x = \sum_{i=0}^{\infty} x_i t^i \in K[\![t/\beta]\!]_{\mathrm{an}}$ be any nonzero ring element. We can then choose a rational function $f \in K(t)$ with no poles in the disc $|t| < \beta$, such that $|x - f|_\beta < |x|_\beta$. By (a), $|x|_\rho$ is continuous in ρ, so we also have $|x - f|_\rho < |x|_\rho$ for ρ in a neighborhood of β. Consequently, those slopes of x sufficiently close to β occur with the same multiplicities as they occur as slopes of f. But f is a rational function, so it has no slopes in some punctured neighborhood of β. This proves that x has no slopes in some punctured neighborhood of β either, so its Newton polygon has finite width.

To check (c) and (d), we may use the same proof as in Proposition 8.3.2. To check (e), we may apply (c) as in the proof of Lemma 8.3.3. □

Corollary 8.5.3. *For fixed β, the ring $\bigcup_{\alpha \in (0,\beta)} K\langle\alpha/t, t/\beta]\!]_{\mathrm{an}}$ is a field.*

Proof By Proposition 8.5.2(c), every element of $K\langle\alpha/t, t/\beta]\!]_{\mathrm{an}}$ can be written as a unit in $K\langle\alpha/t, t/\beta]\!]_{\mathrm{an}}$ times a polynomial. It thus suffices to observe that for any $P \in K[t]$, we can choose α so that none of the roots of P lie in the annulus $\alpha \leq |t| < \beta$; for such α, P is a unit in $K\langle\alpha/t, t/\beta]\!]_{\mathrm{an}}$. □

We will use the following optimal approximation property.

Lemma 8.5.4. *Let R be one of $K\langle t/\beta \rangle$, $K\langle \beta/t, t/\beta \rangle$, or $K[\![t/\beta]\!]_{an}$. Let F be the completion of* Frac(R) *under* $|\cdot|_\beta$ *(in other words, $F = K[\![\beta/t, t/\beta]\!]_{an}$). Then for any $f \in F$, there exists $g \in R$ minimizing $|f - g|_\beta$.*

Proof We may assume $f \notin R$ as otherwise $g = f$ has the desired property. Put $c = \inf\{|f - g|_\beta : g \in R\}$. Since R is complete under $|\cdot|_\beta$, $c > 0$. Since $K(t)$ is dense in F, we can choose $h \in K(t)$ such that $|f - h|_\beta < c$. Let $P \in K[t]$ be the (monic) denominator of h. By Theorem 2.2.1, we may factor P as $P_1 P_2$ so that no irreducible factor of P in $K[t]$ is a unit in R, whereas every irreducible factor of P in $K[t]$ is a unit in R. (The point is that each irreducible polynomial has only a single slope in its Newton polygon, and whether or not the polynomial is a unit is determined entirely by this slope.) Since the claims for f and fP_2 are equivalent, we may assume that $P_2 = 1$. Using the division algorithm, write $hP = gP + S$ with $g, S \in K[t]$ and $\deg(S) < \deg(P)$. We claim that this choice of g works; that is, for any $g' \in R$, $|f - g'|_\beta \geq |f - g|_\beta$.

Suppose first that $R = K\langle t/\beta \rangle$; in this case, $R \cap K[t]$ is dense in R. Given $g' \in R$, we may choose $g'' \in K[t]$ such that $|g' - g''|_\beta < c$. Applying Lemma 2.3.1 to the instance $(h - g'')P = (g - g'')P + S$ of the division algorithm yields

$$|(h - g'')P|_\beta = \max\{|(g - g'')P|_\beta, |S|_\beta\} \geq |S|_\beta$$

Therefore $|h - g''|_\beta \geq |S/P|_\beta = |h - g|_\beta$; since $|f - h|_\beta, |g' - g''|_\beta < c \leq |h - g|_\beta$, we also have $|f - g'|_\beta \geq |f - g|_\beta$ as claimed.

Suppose next that $R = K\langle \beta/t, t/\beta \rangle$; in this case, $R \cap K[t, t^{-1}]$ is dense in R. Given $g' \in R$, we may choose $g'' \in K[t]$ and $m \geq 0$ such that $|g' - g'' t^{-m}|_\beta < c$. Using the division algorithm to divide $(ht^m - g'')P$ by P now returns the same remainder as does dividing $t^m S$ by P. We may argue as in the previous case after checking that for $S \in K[t]$ with $\deg(S) < \deg(P)$, the remainder S' upon dividing tS by P satisfies $|S'|_\beta = \beta|S|_\beta$. To establish this, we emulate the proof of Lemma 2.3.1 as follows. Put $d = \deg(P)$ and let S_{d-1} be the coefficient of t^{d-1} in S. If $|S|_\beta = |S_{d-1} t^{d-1}|_\beta$, then the constant term of S' is equal to the remainder upon dividing $S_{d-1} t^d$ by P, which has norm $\beta|S|_\beta$. Otherwise, $|S' - t(S - S_{d-1} t^{d-1})|_\beta < \beta|S|_\beta$, so the claim follows again.

Suppose finally that $R = K[\![t/\beta]\!]_{an}$. Given $g' \in R$, we may choose $g'', Q \in K[t]$ such that Q is monic with all roots of norm β, and $|g' - g''/Q|_\beta < c$. Using the division algorithm to divide $(hQ - g'')P$ by P now returns the same remainder as does dividing QS by P. If we denote the latter by S', we may now argue as in the previous cases after checking that $|S'|_\beta = |S|_\beta$. To establish this, put $d = \deg(P)$; since P has all roots of norm less than β, $|P - t^d|_\beta < 1$.

Hence $|S'|_\beta = |S''|_\beta$ for S'' the remainder upon dividing QS by t^d. But it is clear that $|S''|_\beta = |S|_\beta$: namely, we may check this after dividing S and t^d by any common factors of t, at which point $|S|_\beta$ and $|S''|_\beta$ are achieved by the respective constant terms of S and S''. □

Remark 8.5.5. Note that the proof of Lemma 8.5.4 actually shows something a bit stronger: the element $g \in R$ constructed continues to minimize $|f - g|_\beta$ even if we replace K with a complete extension L.

Remark 8.5.6. If K is discretely valued, then the conclusion of Lemma 8.5.4 also holds for $R = K[\![t]\!]_0$ and $\beta = 1$, since we can then write F as the completion of $K[\![t]\!]_0[t^{-1}]$.

Proposition 8.5.7. *Assume that $\alpha < \beta$. Let A be an invertible $n \times n$ matrix over $K[\![\alpha/t, t/\beta]\!]_{\mathrm{an}}$. Then there exist $U \in \mathrm{GL}_n(K\langle t/\beta\rangle[t^{-1}])$ and $V \in \mathrm{GL}_n(K\langle \alpha/t\rangle)$ such that $A = UV$.*

Proof Since $\alpha < \beta$, we can choose γ, δ with $\alpha < \gamma < \delta < \beta$, so that $K[\![\alpha/t, t/\beta]\!]_{\mathrm{an}} \subseteq K\langle \gamma/t, t/\delta\rangle$. We may then directly apply Proposition 8.3.5 to conclude. □

We need the following analogues of the gluing lemma (Lemma 8.3.6).

Lemma 8.5.8 (Gluing lemma). *Suppose that $0 < \alpha \le \gamma \le \beta < \delta$.*

(a) *Let M_1 be a finite free module over $K\langle \alpha/t, t/\beta\rangle$, let M_2 be a finite free module over $K\langle \gamma/t, t/\delta\rangle_{\mathrm{an}}$, and suppose we are given an isomorphism*

$$\psi : M_1 \otimes K\langle \gamma/t, t/\beta\rangle \cong M_2 \otimes K\langle \gamma/t, t/\beta\rangle.$$

Then we can find a finite free module M over $K\langle \alpha/t, t/\delta\rangle_{\mathrm{an}}$ and isomorphisms $M_1 \cong M \otimes K\langle \alpha/t, t/\beta\rangle$, $M_2 \cong M \otimes K\langle \gamma/t, t/\delta\rangle_{\mathrm{an}}$ inducing ψ. Moreover, M is determined by this requirement up to unique isomorphism.

(b) *Let M_1 be a finite free module over $K\langle \gamma/t, t/\delta\rangle$, let M_2 be a finite free module over $K[\![t/\delta]\!]_{\mathrm{an}}$, and suppose we are given an isomorphism*

$$\psi : M_1 \otimes K\langle \gamma/t, t/\delta\rangle_{\mathrm{an}} \cong M_2 \otimes K\langle \gamma/t, t/\delta\rangle_{\mathrm{an}}.$$

Then we can find a finite free module M over $K\langle t/\delta\rangle$ and isomorphisms $M_1 \cong M \otimes K\langle \gamma/t, t/\delta\rangle$, $M_2 \cong M \otimes K[\![t/\delta]\!]_{\mathrm{an}}$ inducing ψ. Moreover, M is determined by this requirement up to unique isomorphism.

Proof To prove (a), we may emulate the proof of Lemma 8.3.6. The same holds for (b), except that we must replace Proposition 8.3.5 with Proposition 8.5.7. □

8.6 More approximation arguments

We now give some variants of the approximation lemmas (Lemma 1.3.7 and Lemma 1.5.5) involving rings of power series.

Lemma 8.6.1. *Let R be one of $K\langle t\rangle$, $K\langle 1/t, t\rangle$, $K[\![t]\!]_{an}$, or (if K is discretely valued) $K[\![t]\!]_0$ equipped with the 1-Gauss norm. Let F be the completion of $\mathrm{Frac}(R)$ under $|\cdot|_1$. Let M be a finite free R-module. Let $|\cdot|_M$ be a norm on M compatible with R, obtained by restriction from the supremum norm defined by some basis of $M \otimes_R F$. Then $|\cdot|_M$ is the supremum norm defined by some basis of M.*

Proof (Thanks to Liang Xiao for his help with this proof.) Let N be the $\mathfrak{o}_F$-span of a basis of $M \otimes_R F$ whose supremum norm restricts to $|\cdot|_M$. Put $R_1 = \{r \in R : |r|_1 \le 1\}$ and $M_1 = \{x \in M : |x|_M \le 1\}$. Note that $\mathfrak{m}_K R_1$ coincides with the subring of R consisting of series whose coefficients all belong to $\mathfrak{m}_K$. (If $R = K[\![t]\!]_0$, this holds only if K is discretely valued.) Hence the ring $R_1/\mathfrak{m}_K R_1$ is either $\kappa_K[\![t]\!]$ or a localization of $\kappa_K[t]$, so in any case it is a principal ideal domain.

Note that $M_1/\mathfrak{m}_K M_1$ embeds into $N/\mathfrak{m}_K N$, so in particular it is torsion-free as a module over $R_1/\mathfrak{m}_K R_1$. Since $\mathrm{Frac}(R)$ is dense in F, we can choose a basis $y_1, \ldots, y_n$ of N which is also a basis of $M \otimes_R \mathrm{Frac}(R)$. We can then find a nonzero element $f \in R$ such that M contains $fy_1, \ldots, fy_n$; by multiplying by a unit in K, we may normalize f so that $|f|_1 = 1$. Then $fy_1, \ldots, fy_n$ project to elements of $M_1/\mathfrak{m}_K M_1$ which form a basis of $N/\mathfrak{m}_K N$ over κ_F.

We can also find a nonzero element $g \in R$ such that M is contained in the R-span of $g^{-1}y_1, \ldots, g^{-1}y_n$, which we again normalize so that $|g|_1 = 1$. This means that $M_1/\mathfrak{m}_K M_1$ is contained in the $(R_1/\mathfrak{m}_K R_1)$-submodule of $N/\mathfrak{m}_K N$ generated by $g^{-1}y_1, \ldots, g^{-1}y_n$. Since $R_1/\mathfrak{m}_K R_1$ is a principal ideal domain, it follows that $M_1/\mathfrak{m}_K M_1$ is finitely generated and torsion-free, and hence finite free, as a module over $R_1/\mathfrak{m}_K R_1$. By the previous paragraph, any basis of $M_1/\mathfrak{m}_K M_1$ over $R_1/\mathfrak{m}_K R_1$ freely generates $N/\mathfrak{m}_K N$ over κ_F, and hence freely generates N over $\mathfrak{o}_F$.

Let $x_1, \ldots, x_n \in M_1$ lift a basis of $M_1/\mathfrak{m}_K M_1$ over $R_1/\mathfrak{m}_K R_1$. For any $x \in M_1$, we have a unique representation $x = r_1 x_1 + \cdots + r_n x_n$ with $r_1, \ldots, r_n \in \mathfrak{o}_F$. By Lemma 8.5.4 and Remark 8.5.6, we may choose $r_i' \in R$ to maximize $|r_i - r_i'|_1$. Choose $\lambda \in K$ so that $|\lambda| = \max_i\{|r_i - r_i'|_1\}$; if $\lambda \ne 0$, then the image of $\lambda^{-1}(x - r_1'x_1 - \cdots - r_n'x_n)$ in $M_1/\mathfrak{m}_K M_1$ fails to be in the $(R_1/\mathfrak{m}_K R_1)$-span of the images of $x_1, \ldots, x_n$. This yields a contradiction, so we must have $\lambda = 0$ and $r_i' = r_i \in R \cap \mathfrak{o}_F = R_1$ for $i = 1, \ldots, n$. Consequently, $x_1, \ldots, x_n$ form a basis of M with supremum norm $|\cdot|_M$, as desired. □

It is more difficult to deal with the case of $K[\![t]\!]_0$ with K not discretely valued. Although we will not use that case in what follows, for completeness we mention one result that applies to it.

Lemma 8.6.2. *Suppose that K is spherically complete with value group $\mathbb{R}$. Let M be a finite free module over $K[\![t/\beta]\!]_0$ for some $\beta > 0$. Let $|\cdot|_M$ be a supremum-equivalent norm on M compatible with $K[\![t/\beta]\!]_0$. For $j = 1, 2, \ldots$, let $|\cdot|_j$ be the quotient seminorm on $M/t^j M$ induced by $|\cdot|_M$. Suppose that for any $m \in M$,*

$$|m|_M = \lim_{j \to \infty} |m|_j. \tag{8.6.2.1}$$

Then $|\cdot|_M$ is the supremum norm for some basis of M.

Proof By rescaling, we reduce immediately to the case $\beta = 1$. Put $R = K[\![t]\!]_0$ and $R_1 = \mathfrak{o}_K[\![t]\!]$, so that R_1 is the set of $r \in R$ of norm at most 1. Let M_1 be the set of $x \in M$ with $|x|_M \leq 1$; then M_1 is a R_1-submodule of M. Since $|\cdot|_M$ is compatible with R, for any positive integer j, $M_1 \cap t^j M = t^j M_1$.

By hypothesis, we can find a basis $m'_1, \ldots, m'_n$ of M which determines a supremum norm equivalent to $|\cdot|_M$. That is, there exist $c_1, c_2 > 0$ such that for any $a_1, \ldots, a_n \in R$,

$$c_1 \max_i \{|a_i|_R\} \leq |a_1 m'_1 + \cdots + a_n m'_n|_M \leq c_2 \max_i \{|a_i|_R\}.$$

As in the proof of Lemma 1.3.5, it follows that for any positive integer j and any $a'_1, \ldots, a'_n \in R/t^j R$,

$$c_1 \max_i \{|a'_i|_{R/t^j R}\} \leq |a'_1 m'_1 + \cdots + a'_n m'_n|_j \leq c_2 \max_i \{|a'_i|_{R/t^j R}\}. \tag{8.6.2.2}$$

In particular, each $|\cdot|_j$ is a norm.

By Lemma 1.5.5, $|\cdot|_1$ is the supremum norm defined by some basis $m_{1,1}, \ldots, m_{n,1}$ of M/tM. By Lemma 1.5.4 (applied by viewing each $M/t^j M$ as a K-vector space), for $i = 1, \ldots, n$ and $j = 1, 2, \ldots$, we can construct $m_{i,j+1} \in M/t^{j+1} M$ lifting $m_{i,j} \in M/t^j M$ and satisfying $|m_{i,j+1}|_{j+1} = |m_{i,j}|_j$.

For each i, the inverse limit of the $m_{i,j}$ determines an element m_i of $M \otimes_R K[\![t]\!]$. However, by (8.6.2.2), if we write m_i as a $K[\![t]\!]$-linear combination of $m'_1, \ldots, m'_n$, then each coefficient is a power series bounded in norm by c_1^{-1}, so $m_i \in M$. By (8.6.2.1), $|m_i|_M = 1$.

For any $x \in M$ with $|x|_M \leq 1$, in each $M/t^j M$ we can write $x = a'_{1,j} m_1 + \cdots + a'_{n,j} m_n$ with $a'_{1,j}, \ldots, a'_{n,j} \in R/t^j R$. We prove by induction on j that $|a'_{1,j}|, \ldots, |a'_{n,j}| \leq 1$. This is true for $j = 1$ because the $m_1, \ldots, m_n$ were chosen so that the supremum norm they define matches $|\cdot|_1$. Given the claim for $j - 1$, lift each $a'_{i,j-1} \in R_1/t^{j-1} R_1$ to $b_{i,j} \in R_1/t^j R_1$ by making the last coefficient zero, so that the norm is preserved. Then $t^{1-j}(x - \sum_i b_{i,j} m_i)$ is an element of M/tM of norm at most 1, so it is an $\mathfrak{o}_K$-linear combination of $m_1, \ldots, m_n$. This completes the induction.

We thus obtain a representation $x = a_1 m_1 + \cdots + a_n m_n$ with $a_1, \ldots, a_n \in R_1$. Since any element of M can be written as an element of K times an element of M of norm at most 1, we deduce that $m_1, \ldots, m_n$ is a basis of M, and that the supremum norm defined by this basis coincides with $|\cdot|_M$. □

Question 8.6.3. *Is condition* (8.6.2.1) *necessary?*

Notes

The Hadamard three-circle theorem (Proposition 8.2.3(c)) is a special case of the fact that the *Shilov boundary* of the annulus $\alpha \le |t| \le \beta$ consists of the two circles $|t| = \alpha$ and $|t| = \beta$. For much amplification of this remark, including a full-blown theory of harmonic functions on Berkovich analytic curves, see [389]. For an alternate presentation, restricted to the Berkovich projective line but otherwise more detailed, see [31].

Proposition 8.3.2 can also be proved and generalized in the context of affinoid subspaces of the Berkovich projective line. See Proposition 26.2.1.

The gluing lemma (Lemma 8.3.6) is a special case of the gluing property of coherent sheaves on affinoid rigid analytic spaces, i.e., the theorems of Kiehl and Tate [69, Theorems 8.2.1/1 and 9.4.2/3]. The factorization argument in the proof, however, is older still; it is the nonarchimedean version of what is called a *Birkhoff factorization* over an archimedean field. Similarly, Definition 8.4.3 corresponds to the definition of a locally free coherent sheaf on the corresponding rigid or Berkovich analytic space. Such a sheaf is only guaranteed to be freely generated by global sections in case K is spherically complete [233, Theorem 3.14]; in fact, a previous result of Lazard [289] implies that this property, even when restricted to modules of rank 1, is in fact equivalent to the spherical completeness of K.

Lemma 8.5.8(a) is also a glueing statement for coherent sheaves, but this time on the adic space associated to the ring $K\langle \alpha/t, t/\delta \rrbracket_{\mathrm{an}}$; for the analogue of Kiehl's theorem for noetherian adic spaces, see [268, Theorem 2.3.3]. As far as we know, Lemma 8.5.8(b) does not admit a similar interpretation.

We again thank Liang Xiao for his help with the proof of Lemma 8.6.1.

Exercises

8.1 Prove Lemma 8.0.4. (Hint: for each point of I, find an open neighborhood covered by one or two elements of S. Then reduce to the usual compactness property of a bounded closed interval.)

8.2 Verify the assertions of Remark 8.2.4. (Hint: a typical example where the piecewise affinity fails is $\sum_{n=1}^{\infty} p^{1/n} t^n$.)

8.3 Prove Proposition 8.2.5. (Hint: it may be easiest to first construct the limit using a single $\rho \in [\alpha, \beta]$, then show that it must also work for the other ρ.)

8.4 Prove that if K is discretely valued, then for any $f \in K\langle \alpha/t, t/\beta \rrbracket_0$, there exists a polynomial $P \in K[t]$ and a unit $g \in K\langle \alpha/t, t/\beta \rrbracket_0^{\times}$ such that $f = Pg$. (Hint: the Newton polygon has finite width in this case, so one may argue as in Proposition 8.3.2.)

8.5 Prove that the ring $K\{t\}$ is not noetherian. (Hint: pick a sequence of points in the open unit disc converging to the boundary, and consider the ideal of functions vanishing on all but finitely many of these points.)

8.6 Prove that if K is not trivially or discretely valued, then $K[\![t]\!]_0$ is not noetherian. (Hint: as in the previous exercise, but choose the points so that the Newton polygon of a function vanishing on all of the points has finite height.)

8.7 Prove that if K is discretely valued, then $\mathfrak{o}_K \langle t \rangle = \mathfrak{o}_K [\![t]\!] \cap K\langle t \rangle$ is noetherian. It isn't otherwise, because then $\mathfrak{o}_K$ itself is not noetherian.

8.8 Prove that each maximal ideal of $\mathfrak{o}_K \langle t \rangle$ is generated by $\mathfrak{m}_K$ together with some $P \in \mathfrak{o}_K[t]$ whose reduction modulo $\mathfrak{m}_K$ is irreducible in $\kappa_K[t]$.

8.9 Adapt the proof of Lemma 8.3.4 to prove that for any invertible $n \times n$ matrix A over $K[\![\rho/t, t/\rho]\!]_{\mathrm{an}}$, there exists $U \in \mathrm{GL}_n(K(t))$ such that $|UA - I_n|_\rho < 1$.

8.10 Using the previous exercise, show that Lemma 8.5.8(b) remains true if we take $\gamma = \delta$.

8.11 State and prove an analogue of Lemma 8.3.6 for gluing together finite free modules over $K[\![1/t, t]\!]_{\mathrm{an}}$ and $K[\![t]\!]_0$, using an isomorphism over the completion of $K[\![t]\!]_0 \otimes_{K[t]} K(t)$ under $|\cdot|_1$, to obtain a module over $K[\![t]\!]_{\mathrm{an}}$.

8.12 Prove that if the additive value group of K is not equal to $\mathbb{R}$, then one can find two elements $f, g \in \mathfrak{o}_K[\![t]\!]$ such that the intersection of the ideals (f) and (g) is not finitely generated. (Hint: arrange the Newton polygons of f and g in such a way that the infimum of the valuations of the constant terms of elements of $(f) \cap (g)$ is not in the value group of K.) This argument is due to Anderson–Watkins [14, §3, Theorem]; it implies that the ring $o_K[\![t]\!]$ is not coherent (a *coherent* ring being one in which every finitely generated ideal is finitely presented). The presentation in [14] does not use the language of Newton polygons; for a similar argument that does use this language, see [257, Theorem 1.2].

9

Radius and generic radius of convergence

In this chapter, we begin to approach a fundamental question peculiar to the study of nonarchimedean differential modules. It was already pointed out in Chapter 0 that a differential module over a nonarchimedean disc can fail to have horizontal sections even in the absence of singularities. The radius of convergence of local horizontal sections is thus an important numerical invariant, the control of which is a key factor in the production of solutions of p-adic differential equations.

Unfortunately, the radius of convergence is often difficult to compute directly. One of Dwork's fundamental insights is that one can get much better control over the radius of convergence around a so-called *generic point*. The properties of the *generic radius of convergence* can then be used to infer information about the actual convergence of horizontal sections. For instance, Dwork's *transfer theorem* asserts that the radius of convergence of a differential module over a nonarchimedean disc is no less than the generic radius of convergence at the boundary of the disc.

However, both the radius and generic radius of convergence are rather coarse invariants. Just as the notion of spectral radius is refined by the notion of the full spectrum, we can introduce *subsidiary radii of convergence* and *subsidiary generic radii of convergence* which detect whether some local horizontal sections at a point converge further than others. We will devote much effort in the remainder of Part III to analyzing the behavior of these refined invariants.

Hypothesis 9.0.1. Throughout this chapter, we will view the rings $K\langle \alpha/t, t/\beta \rangle$, $K[\![\alpha/t, t/\beta]\!]_{\mathrm{an}}$, and so forth as differential rings with derivation $d = \frac{d}{dt}$, the formal differentiation in the variable t, unless otherwise specified (as in Section 9.6).

9.1 Differential modules have no torsion

We start with some simple but critical observations about the categories of differential modules over the rings of power series introduced in Chapter 8.

Lemma 9.1.1. *Let R be one of the following rings:* $K\langle t/\beta\rangle$, $K[\![t/\beta]\!]_{an}$, $K\langle\alpha/t,t/\beta\rangle$, $K\langle\alpha/t,t/\beta]\!]_{an}$, $K[\![\alpha/t,t/\beta]\!]_{an}$, *or (if K is discretely valued)* $K[\![t/\beta]\!]_0$ *or* $K\langle\alpha/t,t/\beta]\!]_0$. *Then R has no nonzero proper differential ideal.*

Proof If I is a nonzero differential ideal, then by Proposition 8.3.2 (or Proposition 8.5.2), I contains a nonzero element $P \in K[t]$. But then I also contains $d^{\deg(P)}(P)$, which is a nonzero element of K (because K is of characteristic 0), so I must be the unit ideal. □

Proposition 9.1.2. *For R as in Lemma 9.1.1, any finite differential module over R is free. Consequently, the finite free differential modules over R form an abelian category.*

Proof Let M be a finite differential module over R. If $m \in M$ is annihilated by the nonzero element $r \in R$, then $0 = D(rm) = rD(m) + d(r)m$, and so $D(m)$ is annihilated by r^2. Consequently, the torsion submodule T of M is also a differential module. Also, T is finite over R because R is a principal ideal domain (by either Proposition 8.3.2 or Proposition 8.5.2) and hence noetherian. Hence the annihilator I of T is a nonzero ideal of R. It is also a differential ideal: if $r \in I$, then for any $m \in T$, $0 = D(rm) = rD(m) + d(r)m = d(r)m$, so $d(r) \in I$. By Lemma 9.1.1, I must be the trivial ideal. Hence $T = 0$, so M is torsion-free; since R is a principal ideal domain, M must also be free. □

Corollary 9.1.3. *For R as in Lemma 9.1.1, let M be a finite differential module over R. Then any finite set of horizontal sections which is linearly independent over K forms part of a basis of M.*

Proof Let S be a finite set of horizontal sections which is linearly independent over K. By Lemma 5.1.5, S is also linearly independent over R. In this case, S determines an injective morphism from a trivial differential module to M. By Proposition 9.1.2, the image of this map must be a direct summand of M as an R-module; this yields the desired result. □

Corollary 9.1.4. *For R as in Lemma 9.1.1, let M be a finite differential module over R of rank n, admitting a set S of n horizontal sections linearly independent over K. Then M is trivial and $H^0(M)$ is the K-span of S.*

Since we also wish to deal with open discs and annuli, we must formally define differential modules on them.

Definition 9.1.5. Consider the region $|t| \in I$ where $I \subseteq [0, \infty)$ is an interval; this could be an open or closed disc, or a closed, open, or half-open annulus. By a *finite differential module* M on this region, we will mean a coherent locally free module in the sense of Definition 8.4.3, in which each of the modules M_i carries the structure of a differential module over $K\langle \alpha_i/t, t/\beta_i \rangle$, and each of the isomorphisms $M_{i+1} \otimes K\langle \alpha_i/t, t/\beta_i \rangle \cong M_i$ is horizontal. By Proposition 9.1.2, we see that these again form an abelian category.

9.2 Antidifferentiation

One initially surprising fact about p-adic analysis is that the relationship between differentiation and boundary convergence is reversed from the archimedean case: whereas d carries $K\langle \alpha/t, t/\beta \rangle$ into itself, antidifferentiation does not. Instead, one has the following.

Lemma 9.2.1. *For any* $x = \sum_i x_i t^i \in K\{\alpha/t, t/\beta\}$ *with* $x_{-1} = 0$*, there exists* $y \in K\{\alpha/t, t/\beta\}$ *for which* $d(y) = x$.

Proof Exercise. □

Corollary 9.2.2. *Let* M *be the trivial differential module over* $K\langle t/\beta \rangle$ *(resp. over* $K\langle \alpha/t, t/\beta \rangle$*). Then for any* δ *with* $0 \le \delta < \beta$ *(resp. any* γ, δ *with* $\alpha < \gamma \le \delta < \beta$*), the map* $H^1(M) \to H^1(M \otimes K\langle t/\delta \rangle)$ *(resp.* $H^1(M) \to H^1(M \otimes K\langle \gamma/t, t/\delta \rangle)$*) is the zero map.*

This gives us an explicit description of the unipotent differential modules. (Recall that these are the successive extensions of trivial modules.)

Lemma 9.2.3. *Let* M *be a finite unipotent differential module for the derivation* $t\frac{d}{dt}$ *over* $K\langle \alpha/t, t/\beta \rangle$ *with* $0 < \alpha < \beta$*. Then for any* γ, δ *with* $\alpha < \gamma \le \delta < \beta$*,* $M \otimes K\langle \gamma/t, t/\delta \rangle$ *admits a basis on which the matrix of action of* D *is nilpotent with entries in* K.

Proof Let $v_1, \ldots, v_n$ be a basis of M such that, letting M_i denote the span of $v_1, \ldots, v_i$, we have that M_i is stable under D and M_i/M_{i-1} is trivial for $i = 1, \ldots, n$. We proceed by induction on n; that being said, we may assume that the matrix of action of D on $v_1, \ldots, v_{n-1}$ is upper triangular nilpotent with entries in K.

We now write $D(v_n) = c_1 v_1 + \cdots + c_{n-1} v_{n-1}$. If $c_{n-1} \notin K$, we can choose α', β' with $\alpha < \alpha' < \gamma \le \delta < \beta' < \beta$, then (by Lemma 9.2.1) choose $x \in K\langle \alpha'/t, t/\beta' \rangle$ such that $c_{n-1} - t\frac{dx}{dt} \in K$. If we replace v_n by $v_n' = v_n - x v_{n-1}$, then $D(v_n') = c_1' v_1 + \cdots + c_{n-1}' v_{n-1}$ with $c_{n-1}' \in K$. Similarly, if $c_{n-1}, \ldots, c_{n-i+1} \in K$

but $c_{n-i} \notin K$, we can change basis to put c_{n-i} into K while possibly changing $c_1, \ldots, c_{n-i-1}$ but not $c_{n-i+1}, \ldots, c_{n-1}$. Repeating this process, we ultimately obtain a basis of the desired form. □

9.3 Radius of convergence on a disc

Definition 9.3.1. Let M be a finite differential module over $K\langle t/\beta \rangle$, $K[\![t/\beta]\!]_{\mathrm{an}}$, or $K[\![t/\beta]\!]_0$. Define the *radius of convergence* of M around 0, denoted $R(M)$, to be the supremum of the set of $\rho \in (0, \beta)$ such that $M \otimes K\langle t/\rho \rangle$ has a basis of horizontal elements; we refer to those elements as *local horizontal sections* of M. For M a finite differential module on the open disc of radius β around $t = 0$, define $R(M)$ as the supremum of $R(M \otimes K\langle t/\gamma \rangle)$ over all $\gamma < \beta$. For $\gamma \leq \beta$, note that

$$R(M \otimes K\langle t/\gamma \rangle) = \begin{cases} \gamma & \gamma \leq R(M) \\ R(M) & \gamma > R(M). \end{cases}$$

Example 9.3.2. In general, if $p > 0$, then it is possible to have $R(M) < \beta$; that is, there is no p-adic analogue of the fundamental theorem of ordinary differential equations (as was noted in Example 0.4.1). For instance, consider the module $M = K\langle t/\beta \rangle v$ with $D(v) = v$; for $\beta > p^{-1/(p-1)}$, $R(M) = p^{-1/(p-1)}$ because the latter is the radius of convergence of the exponential series. (This is essentially Example 0.4.1 again.)

That said, the local form of the fundamental theorem of ordinary differential equations has the following analogue.

Proposition 9.3.3 (p-adic Cauchy theorem). *Let M be a finite differential module over $K\langle t/\beta \rangle$, $K[\![t/\beta]\!]_{\mathrm{an}}$, or $K[\![t/\beta]\!]_0$. Then $R(M) > 0$.*

Proof By shrinking β, we reduce to the case over $K\langle t/\beta \rangle$. One can give a direct proof of this, but instead we will deduce this from Dwork's transfer theorem (Theorem 9.6.1). We will give a direct proof of a slightly stronger result later (Proposition 18.1.1); see also the notes. □

Here are some direct consequences of the definition of the radius of convergence; note the parallels with properties of the spectral radius (Lemma 6.2.8).

Lemma 9.3.4. *Let M, M_1, M_2 be finite differential modules over $K\langle t/\beta \rangle$, $K[\![t/\beta]\!]_{\mathrm{an}}$, or $K[\![t/\beta]\!]_0$.*

(a) If $0 \to M_1 \to M \to M_2 \to 0$ is exact, then

$$R(M) = \min\{R(M_1), R(M_2)\}.$$

(b) We have

$$R(M^\vee) = R(M).$$

(c) We have

$$R(M_1 \otimes M_2) \geq \min\{R(M_1), R(M_2)\},$$

with equality when $R(M_1) \neq R(M_2)$.

Proof For (a), it is clear that $R(M) \leq \min\{R(M_1), R(M_2)\}$; we must check that equality holds. Choose $\lambda < \min\{R(M_1), R(M_2)\}$, so that $M_1 \otimes K\langle t/\lambda\rangle$ and $M_2 \otimes K\langle t/\lambda\rangle$ are both trivial. For any $\lambda' < \lambda$, the map $H^1(M_2^\vee \otimes M_1) \to H^1((M_2^\vee \otimes M_1) \otimes K\langle t/\lambda'\rangle)$ is zero by Corollary 9.2.2, so $M \otimes K\langle t/\lambda'\rangle$ is trivial by Lemma 5.3.3. Since we can make λ and λ' as close to $\min\{R(M_1), R(M_2)\}$ as we like, we find that $R(M) \geq \min\{R(M_1), R(M_2)\}$.

For (b), we obtain $R(M^\vee) \geq R(M)$ from the fact that if $M \otimes K\langle t/\lambda\rangle$ is trivial, then so is its dual $M^\vee \otimes K\langle t/\lambda\rangle$. Since M and $M^\vee$ enter symmetrically, we get $R(M^\vee) = R(M)$.

For (c), the inequality is clear from the fact that the tensor product of two trivial modules over $K\langle t/\lambda\rangle$ is also trivial. The last assertion follows as in the proof of Lemma 6.2.8(c). □

Example 9.3.5. Assume that $p > 0$ and let M be the differential module of rank 1 over $K\langle t/\beta\rangle$ defined by $D(v) = \lambda v$ with $\lambda \in K$. Then it is an exercise to show that

$$R(M) = \min\{\beta, p^{-1/(p-1)}|\lambda|^{-1}\}$$

This includes an important example of Dwork; see Example 17.1.4.

9.4 Generic radius of convergence

In general, the radius of convergence of a differential module is difficult to compute. We are thus led to introduce a related but simpler invariant.

Definition 9.4.1. For $\rho > 0$, let F_ρ be the completion of $K(t)$ under the ρ-Gauss norm $|\cdot|_\rho$. Put $d = \frac{d}{dt}$ on $K(t)$; then d extends by continuity to F_ρ. For M a module over a ring R which maps to F_ρ in an apparent way, we will often write M_ρ as shorthand for the base extension $M \otimes_R F_\rho$. When substituting a specific value for ρ, say 1, for clarity we will write $M_{\rho=1}$ instead of M_1.

By embedding $K(t)$ into the field (by Corollary 8.5.3) $\bigcup_{\alpha \in (0,\rho)} K\langle \alpha/t, t/\rho\rangle_{\text{an}}$

and expressing elements of the latter as Laurent series, we compute that

$$\left|\frac{d^n}{n!}\right|_{F_\rho} = \rho^{-n}. \tag{9.4.1.1}$$

It follows that

$$|d|_{F_\rho} = \rho^{-1}, \qquad |d|_{\mathrm{sp},F_\rho} = \lim_{n\to\infty} |n!|^{1/n}\rho^{-1} = \begin{cases} \rho^{-1} & p = 0 \\ p^{-1/(p-1)}\rho^{-1} & p > 0. \end{cases}$$

(See Proposition 9.10.2 for a refinement of this assertion.) It is a common convention to define

$$\omega = \begin{cases} 1 & p = 0 \\ p^{-1/(p-1)} & p > 0, \end{cases}$$

so that we may write $|d|_{\mathrm{sp},F_\rho} = \omega\rho^{-1}$.

Remark 9.4.2. Note that F_ρ coincides with the ring $K[\![\rho/t, t/\rho]\!]_{\mathrm{an}}$ of analytic elements on the circle $|t| = \rho$ (see Definition 8.5.1). As a result, F_ρ is commonly known as the *field of analytic elements* of norm ρ.

We will also make a related construction in case $\rho = 1$. For a way to unify the two constructions, see Definition 9.10.1.

Definition 9.4.3. Put $\mathcal{E} = K\langle 1/t, t]\!]_0$, or in other words, let $\mathcal{E}$ be the completion of $\mathfrak{o}_K((t)) \otimes_{\mathfrak{o}_K} K$ for the 1-Gauss norm $|\cdot|_1$. If K is discretely valued, the supremum in the Gauss norm is achieved; consequently, $\mathcal{E}$ is a field and its residue field is equal to $\kappa_K((t))$. However, none of this applies if K is not discretely valued. (This is the same issue that arises in Remark 8.2.4.) In any case, $\mathcal{E}$ is complete under $|\cdot|_1$, there is an isometric map $F_1 \to \mathcal{E}$ carrying t to t, and the supremum is achieved for elements of $\mathcal{E}$ in the image of that map; this at least gives an embedding of $\kappa_K((t))$ into the quotient $\{x \in \mathcal{E}: |x|_1 \leq 1\}/\{x \in \mathcal{E}: |x|_1 < 1\}$.

Definition 9.4.4. Let (V, D) be a nonzero finite differential module over F_ρ or $\mathcal{E}$. We define the *generic radius of convergence* (or for short, the *generic radius*) of V to be

$$R(V) = \omega|D|_{\mathrm{sp},V}^{-1};$$

note that $R(V) > 0$. We will see later (Proposition 9.7.5) that this does indeed compute the radius of convergence of horizontal sections of V on a generic disc. (For V the zero module, set $R(V) = \rho$.)

Remark 9.4.5. Note that the map $F_1 \to \mathcal{E}$ is isometric, and $|d|_{\text{sp},\mathcal{E}} = \omega = |d|_{\text{sp},F_1}$. Consequently, for any finite differential module V over F_1, Corollary 6.2.7 implies that $|D|_{\text{sp},V \otimes \mathcal{E}} = |D|_{\text{sp},V}$.

We can translate some basic properties of the spectral radius (Lemma 6.2.8) into properties of generic radii of convergence, leading to the following analogue of Lemma 9.3.4. Alternatively, one can first check Proposition 9.7.5 and then simply invoke Lemma 9.3.4 itself around a generic point.

Lemma 9.4.6. *Let V, V_1, V_2 be nonzero finite differential modules over F_ρ.*

(a) For an exact sequence $0 \to V_1 \to V \to V_2 \to 0$,

$$R(V) = \min\{R(V_1), R(V_2)\}.$$

(b) We have

$$R(V^\vee) = R(V).$$

(c) We have

$$R(V_1 \otimes V_2) \geq \min\{R(V_1), R(V_2)\},$$

with equality when $R(V_1) \neq R(V_2)$.

Definition 9.4.7. In some situations, it is more natural to consider the *intrinsic generic radius of convergence*, or for short the *intrinsic radius*, defined as

$$IR(V) = \rho^{-1} R(V) = \frac{|d|_{\text{sp},F_\rho}}{|D|_{\text{sp},V}} \in (0,1].$$

(The upper bound of 1 comes from Lemma 6.2.4.) To emphasize the difference, we may refer to the unadorned generic radius of convergence defined earlier as the *extrinsic* generic radius of convergence. (See Proposition 9.7.6 and the chapter notes for some reasons why the intrinsic radius deserves such a name.)

Remark 9.4.8. For I an interval, M a differential module on the annulus $|t| \in I$, and $\rho \in I$, it is unambiguous to refer to the generic radius of convergence $R(M_\rho)$ of M at radius ρ. This is defined by first making a base change to $K\langle\alpha/t, t/\beta\rangle$ for some closed subinterval $[\alpha, \beta]$ of I containing ρ. Since the resulting module M_ρ does not depend on the choice of $[\alpha, \beta]$, neither does its generic radius.

Remark 9.4.9. Let F' be a nonarchimedean field containing F, and define F'_ρ by analogy with F_ρ. Let V be a nonzero differential module over F_ρ and set $V' = V \otimes_{F_\rho} F'_\rho$. One may check directly that $|D|_{\text{sp},V} = |D|_{\text{sp},V'}$, and therefore $R(V) = R(V')$. However, the same logic will not directly apply to subsidiary radii; see Remark 9.8.4.

9.5 Some examples in rank 1

Assume $p > 0$ for these examples. See also Example 9.9.3.

Example 9.5.1. In Example 9.3.5, $D^s(v) = \lambda^s v$ for all nonnegative integers s. By Lemma 6.2.5,

$$R(M_\beta) = \min\{\beta, \omega|\lambda|^{-1}\} = R(M).$$

Another important class of examples is given as follows.

Example 9.5.2. For $\lambda \in K$, let V_λ be the differential module of rank 1 over F_ρ defined by $D(v) = \lambda t^{-1} v$. It is an exercise to show that $IR(V_\lambda) = 1$ if and only if $\lambda \in \mathbb{Z}_p$.

We can further classify Example 9.5.2 as follows.

Proposition 9.5.3. *We have $V_\lambda \cong V_{\lambda'}$ if and only if $\lambda - \lambda' \in \mathbb{Z}$.*

Proof Note that $V_\lambda \cong V_{\lambda'}$ if and only if $V_{\lambda-\lambda'}$ is trivial, so we may reduce to the case $\lambda' = 0$. By Example 9.5.2, V_λ is nontrivial whenever $\lambda \notin \mathbb{Z}_p$; by direct inspection, V_λ is trivial whenever $\lambda \in \mathbb{Z}$.

It remains to deduce a contradiction assuming that V_λ is trivial, $\lambda \in \mathbb{Z}_p$, and $\lambda \notin \mathbb{Z}$. There is no harm in enlarging K now, so we may assume that K contains a scalar of norm ρ; by rescaling, we may reduce to the case $\rho = 1$. We now have $f \in F_1^\times$ such that $t\frac{df}{dt} = \lambda f$; by multiplying by an element of $K^\times$, we can force $|f|_1 = 1$.

Let λ_1 be an integer such that $\lambda \equiv \lambda_1 \pmod{p}$. Then

$$\left|\frac{d(ft^{-\lambda_1})}{dt}\right|_1 = |(\lambda - \lambda_1)ft^{-\lambda_1-1}|_1 \le p^{-1}.$$

Using the embedding $F_1 \hookrightarrow \mathcal{E}$, we may expand $f = \sum_{i\in\mathbb{Z}} f_i t^i$ with $\max_i\{|f_i|\} = 1$. The previous calculation then forces $|f_i| \le p^{-1}$ unless $i \equiv \lambda_1 \equiv \lambda \pmod{p}$.

By considering the reduction of f modulo p^n and arguing similarly, we find that $|f_i| \le p^{-1}$ unless $i \equiv \lambda \pmod{p^n}$ for all n. But since $\lambda \notin \mathbb{Z}$, this means that the image of f in $\kappa_K((t))$ cannot have any terms at all, contradiction. $\square$

9.6 Transfer theorems

A fundamental relationship between radius of convergence and generic radius of convergence is provided by the following result. In the language of Dwork, this is a *transfer theorem*, because it transfers convergence information from one disc to another. (Note that Proposition 9.3.3, which asserts that $R(M) > 0$, is an immediate corollary.)

Theorem 9.6.1 (Dwork). *For any nonzero finite differential module M over $K\langle t/\beta\rangle$ or $K[\![t/\beta]\!]_{\mathrm{an}}$, $R(M) \geq R(M_\beta)$. That is, the radius of convergence is at least the generic radius.*

Proof Suppose $\lambda < \beta$ and $\lambda < \omega |D|_{\mathrm{sp},M}^{-1}$. Fix a supremum norm $|\cdot|_M$ on M compatible with $|\cdot|_\lambda$. We claim that for any $x \in M$, the Taylor series

$$y = \sum_{i=0}^{\infty} \frac{(-t)^i}{i!} D^i(x) \tag{9.6.1.1}$$

converges under $|\cdot|_M$. To see this, pick $\epsilon > 0$ such that $\lambda\omega^{-1}(|D|_{\mathrm{sp},M} + \epsilon) < 1$. By Lemma 6.1.8, there exists $c > 0$ such that $|D^i(x)|_M \leq c(|D|_{\mathrm{sp},M} + \epsilon)^i |x|_M$ for all i. The i-th term of the sum defining y thus has norm at most $\lambda^i \omega^{-i} c(|D|_{\mathrm{sp},M} + \epsilon)^i |x|_M$, which tends to 0 as $i \to \infty$.

By differentiating the series expression as in Remark 5.8.4, we find that $D(y) = 0$, so y is a horizontal section of $M \otimes K\langle t/\lambda\rangle$. If we run this construction over a basis of M, we obtain horizontal sections of $M \otimes K\langle t/\lambda\rangle$ whose reductions modulo t also form a basis; these sections are thus K-linearly independent and so form a basis of $M \otimes K\langle t/\lambda\rangle$ by Proposition 9.1.2. This proves the claim. $\square$

Here is a simple example of how one may apply the transfer theorem, modulo a forward reference to Theorem 11.3.2(a).

Example 9.6.2. Recall the hypergeometric differential equation

$$y'' + \frac{(c - (a + b + 1)z)}{z(1-z)}y' - \frac{ab}{z(1-z)}y = 0$$

considered in Chapter 0. In general, one solution in the ring $K((z))$ is given by the hypergeometric series

$$F(a, b; c; z) = \sum_{i=0}^{\infty} \frac{a(a+1)\cdots(a+i)b(b+1)\cdots(b+i)}{c(c+1)\cdots(c+i)i!} z^i.$$

Let us now restrict to the case $a, b, c \in \mathbb{Z}_p \cap \mathbb{Q}$; assume in addition that c is not a positive integer. Let m be the denominator of c. In the ring $K((z^{1/m}))$, the general solution is

$$AF(a, b; c; z) + Bz^{1-c}F(a+1-c, b+1-c; 2-c; z) \qquad (A, B \in K), \tag{9.6.2.1}$$

and it converges for $|z| < 1$ (see the exercises for Chapter 0).

We now pass to the associated differential module M of rank 2, defined over the ring $K\langle\alpha/z, z\rangle_{\mathrm{an}}$ for any $\alpha > 0$. From (9.6.2.1), we see that for any $\beta \in (0, 1)$,

$M \otimes K\langle \alpha/z, z/\beta \rangle$ has a filtration in which one quotient is a trivial module and the other has the form $D(w) = \lambda z^{-1} w$ for some $\lambda \in \mathbb{Z}_p \cap \mathbb{Q}$. From Example 9.5.2 and Lemma 9.4.6, we deduce that $R(M_\rho) = 1$ for $\rho \in (0, 1)$. By Theorem 11.3.2(a), $R(M_\rho)$ is continuous at $\rho = 1$, so we also have $R(M_{\rho=1}) = 1$.

We now expand in power series around another value $z_0 \in \mathfrak{o}_K$ which is not congruent to 0 or 1 modulo $\mathfrak{m}_K$. We get another differential module N over $K[\![z - z_0]\!]_{\mathrm{an}}$ such that $M_{\rho=1} \cong N_{\rho=1}$, so in particular $R(N_{\rho=1}) = 1$. By Theorem 9.6.1, $R(N) = 1$; that is, the general solution of the hypergeometric differential equation at $z = z_0$ converges in the disc $|z - z_0| < 1$.

Remark 9.6.3. In Example 9.6.2, one cannot directly prove that $R(N_{\rho=1}) = 1$ using Theorem 6.5.3 because of the limitation on slopes therein. An alternate approach that does work is to construct a Frobenius structure on N; see Part V.

9.7 Geometric interpretation

As promised, here is a construction that explains the terminology "generic radius of convergence". We will expand upon this point in Chapter 25.

Definition 9.7.1. Let L be a complete extension of K. A *generic point* of L of norm ρ relative to K is an element $t_\rho \in L$ with $|t_\rho| = \rho$, such that there is no $z \in K^{\mathrm{alg}}$ with $|z - t_\rho| < \rho$. For instance, t itself is a generic point of F_ρ of norm ρ.

Remark 9.7.2. If $t_\rho \in L$ is a generic point, then evaluation at t_ρ gives an isometry $K[t] \to L$ for the ρ-Gauss norm on $K[t]$. To see this, it suffices to check after replacing K by a completed algebraic closure, then enlarging L to contain this enlarged K. Then any $P \in K(t)$ can be factored as $c \prod_i (t - z_i)$ for some $z_i \in K$, and for each i, $|z_i - t_\rho| \geq \rho$ because t_ρ is a generic point. Consequently,

$$|P(t_\rho)| = |c| \prod_i |t_\rho - z_i|$$
$$= |c| \prod_i \max\{\rho, |z_i|\}$$
$$= |c| \prod_i |t - z_i|_\rho$$
$$= |P|_\rho.$$

Definition 9.7.3. Let L be a complete extension of K. For any $t_\rho \in L$ for which $|t_\rho| = \rho$, the substitution $t \mapsto t_\rho + (t - t_\rho)$ induces an isometric map $K[t] \to L\langle(t - t_\rho)/\rho\rangle$. However, if (and only if) t_ρ is a generic point, then the composition of this map with the reduction modulo $t - t_\rho$ is again an isometry, by Remark 9.7.2. Hence in this case, the map $K[t] \to L\langle(t - t_\rho)/\rho\rangle$ extends to an isometry $F_\rho \to L[\![(t - t_\rho)/\rho]\!]_{\mathrm{an}}$.

Remark 9.7.4. In Berkovich's theory of nonarchimedean analytic geometry, the geometric interpretation of the above construction is that the analytic space corresponding to F_ρ is obtained from the closed disc of radius ρ by removing the open disc of radius ρ centered around each point of K^{alg}. As a result, it still contains any open disc of radius ρ that does not meet K^{alg}.

Proposition 9.7.5. *Let V be a finite differential module over F_ρ, and put $V' = V \otimes_{F_\rho} L[\![(t - t_\rho)/\rho]\!]_{\mathrm{an}}$. Then the generic radius of convergence of V is equal to the radius of convergence of V'.*

Proof Let G_λ be the completion of $L(t - t_\rho)$ for the λ-Gauss norm; then the map $F_\rho \to G_\lambda$ is an isometry for any $\lambda < \rho$. By Corollary 6.2.7,

$$|D|_{\mathrm{sp},V \otimes_{F_\rho} G_\lambda} = \max\{|d|_{\mathrm{sp},G_\lambda}, |D|_{\mathrm{sp},V}\} = \max\{\omega\lambda^{-1}, |D|_{\mathrm{sp},V}\}.$$

On one hand, we may deduce from this that $R(V) \leq R(V')$ by applying Theorem 9.6.1 to $V \otimes_{F_\rho} L\langle(t - t_\rho)/\lambda\rangle$ for a sequence of values of λ converging to ρ.

On the other hand, pick any $\lambda < R(V')$; then $V \otimes_{F_\rho} G_\lambda$ is a trivial differential module, so the spectral radius of D on it is $\omega\lambda^{-1}$. We thus have

$$|D|_{\mathrm{sp},V} \leq \omega\lambda^{-1},$$

so $R(V) \geq \lambda$. This yields $R(V) \geq R(V')$. $\qquad\square$

Here is an example illustrating both the use of the geometric interpretation and a good transformation property of the intrinsic normalization.

Proposition 9.7.6. *Let m be a nonzero integer not divisible by p, and let $f_m \colon F_\rho \to F_{\rho^{1/m}}$ be the substitution $t \mapsto t^m$. Then for any finite differential module V over F_ρ, $IR(V) = IR(f_m^*(V))$.*

Proof Let ζ_m be a primitive m-th root of unity in K^{alg}. The claim then follows from the geometric interpretation plus the fact that

$$|t - t_\rho \zeta_m^i| < c\rho \text{ for some } i \in \{0, \dots, m-1\} \Leftrightarrow |t^m - t_\rho^m| < c\rho^m \qquad (c \in (0,1)), \tag{9.7.6.1}$$

whose proof is left as an exercise. $\qquad\square$

Remark 9.7.7. Beware that in Proposition 9.7.6, we are not quite performing the standard base change. In explicit terms, if d_m denotes the derivation with respect to t on $F_{\rho^{1/m}}$, and d_1 denotes the derivation with respect to t on F_ρ extended via f_m, then

$$d_m = mt^{m-1}d_1.$$

We will encounter this issue again when we perform Frobenius pullback and pushforward in Chapter 10. One may view it as an argument in favor of a coordinate-free perspective (Proposition 6.3.1).

Remark 9.7.8. A similar construction can be made for $\mathcal{E}$. Let L be the completion of $\mathfrak{o}_K((t_1)) \otimes_{\mathfrak{o}_K} K$ for the 1-Gauss norm. Then the substitution $t \mapsto t_1 + (t - t_1)$ induces an isometry $\mathfrak{o}_K((t)) \to \mathfrak{o}_L[\![t - t_1]\!]$ for the 1-Gauss norm; this map extends to an isometric embedding of $\mathcal{E}$ into the completion of $\mathfrak{o}_L[\![t - t_1]\!] \otimes_{\mathfrak{o}_L} L$ for the 1-Gauss norm.

9.8 Subsidiary radii

It is sometimes important to consider not only the generic radius of convergence, but also some secondary invariants.

Definition 9.8.1. Let V be a finite differential module over F_ρ. Let $V_1, \ldots, V_m$ be the Jordan–Hölder constituents of V. We define the multiset of *subsidiary generic radii of convergence*, or for short the *subsidiary radii*, to consist of $R(V_i)$ with multiplicity $\dim V_i$ for $i = 1, \ldots, m$. We list these as $s_1, \ldots, s_n$ in increasing order, so that $R(V) = s_1 \leq \cdots \leq s_n$. We also have *intrinsic subsidiary generic radii of convergence*, obtained by multiplying the subsidiary radii by ρ^{-1}.

Remark 9.8.2. If we replace F_ρ by $\mathbb{C}((z))$ in the definition of intrinsic subsidiary generic radii of convergence, then the negative logarithm of the product of the radii equals the irregularity of V. Thus our analysis of the variation of subsidiary radii, in the remainder of this part, will also imply results about variation of irregularity. See [250] for an application of this.

Remark 9.8.3. It is not immediate from the definition how to interpret the subsidiary radii in terms of the convergence of solutions of differential equations. We will give such an interpretation in Theorem 11.9.2.

Remark 9.8.4. With notation as in Remark 9.4.9, it is not apparent that the subsidiary radii of V and V' coincide, because constituents of V may not remain irreducible upon base extension from F_ρ to F'_ρ. We will only be able to deduce this coincidence later; see Proposition 10.6.6.

9.9 Another example in rank 1

We introduce one more important example in rank 1.

Definition 9.9.1. For $p > 0$, the *Artin–Hasse exponential series* is the formal power series

$$E_p(t) = \exp\left(\sum_{i=0}^{\infty} \frac{t^{p^i}}{p^i}\right).$$

Proposition 9.9.2. We have $E_p(t) \in (\mathbb{Z}_p \cap \mathbb{Q})[\![t]\!]$.

Proof This follows from the formal identity

$$E_p(x) = \prod_{n \geq 1, p \nmid n} (1 - x^n)^{-\mu(n)/n} \tag{9.9.2.1}$$

in which $\mu(n)$ is the Möbius function: $\mu(n)$ equals $(-1)^e$ if n is the product of $e \geq 0$ distinct primes and 0 otherwise (exercise). □

Example 9.9.3 (Matsuda). Let h be a nonnegative integer, and suppose that K contains a primitive p^{h+1}-st root of unity ζ. Let M_h be the differential module of rank 1 on the whole t-line for which the action of D on a generator v is given by

$$D(v) = \sum_{i=0}^{h} (\zeta^{p^i} - 1)t^{p^i-1}v;$$

note that M_0 is isomorphic to the module of Example 9.3.5 for $\lambda = \zeta - 1$. On the open disc $|t| < 1$, M_h admits the horizontal section

$$\frac{E_p(t)}{E_p(\zeta t)}v = \exp\left(\sum_{i=0}^{h} \frac{1 - \zeta^{p^i}}{p^i}t^{p^i}\right)v \tag{9.9.3.1}$$

by Proposition 9.9.2, so $R(M_h) \geq 1$. Since this horizontal section is bounded, it also gives a horizontal section in the open unit disc around a generic point of norm 1, so $R(M_{h,\rho=1}) = 1$. (This can also be seen by arguing that M_h is trivial on any disc of the form $|t| < \rho$ for $\rho < 1$, so $R(M_{h,\rho}) = 1$, then using continuity of the generic radius of convergence as in Example 9.6.2.) By Proposition 9.7.6, the same remains true if we pull back along the map $t \mapsto ct^m$ for any positive integer m not divisible by p and any $c \in \mathfrak{o}_K^\times$. (This generalizes further; see the chapter notes.)

Remark 9.9.4. Note that

$$\left(\frac{E_p(t)}{E_p(\zeta t)}\right)^P = \exp((1-\zeta)pt)\frac{E_p(t^P)}{E_p(\zeta^P t^P)}.$$

Consequently, on some disc of radius greater than 1, $M_h^{\otimes p}$ is isomorphic to the pullback of M_{h-1} along $t \mapsto t^P$.

Remark 9.9.5. It is possible to prove directly that for $\rho \in [1, \infty)$ sufficiently close to 1,

$$IR(M_{h,\rho}) = \rho^{-p^h} \qquad (9.9.5.1)$$

(exercise). We can also do it using variational properties of the generic radius; see the exercises for Chapter 11.

9.10 Comparison with the coordinate-free definition

It is worth reconciling our definition of generic radius of convergence with the coordinate-free formula in Proposition 6.3.1. In the process, we can give a more unified treatment of F_ρ and $\mathcal{E}$.

Definition 9.10.1. Let F be a complete nonarchimedean field of characteristic 0 equipped with a bounded derivation d. For $u \in F$, we say that d is of *rational type* for the parameter u if the following conditions hold.

(a) We have $d(u) = 1$ and $|ud|_F = 1$.
(b) For each positive integer n, $|d^n/n!|_F \le |d|_F^n$.

For instance, this holds for $F = F_\rho$ with $d = \frac{d}{dt}$ and $u = t$ by (9.4.1.1), and similarly for $F = \mathcal{E}$ if K is discretely valued. Note also that condition (b) is vacuously true if κ_F is of characteristic 0.

Proposition 9.10.2. *Set notation as in Definition 9.10.1. Then for any nonnegative integer n and any $c_0, \ldots, c_n \in F$,*

$$\left|\sum_{i=0}^n c_i \frac{u^i}{i!} d^i\right|_F = \max_i\{|c_i|\}. \qquad (9.10.2.1)$$

Proof The left side of (9.10.2.1) is less than or equal to the right side because by hypothesis $|u^i d^i/i!|_F \le |u|^i|d|_F^i \le 1$ for all i. Conversely, let $j \in \{0, \ldots, n\}$ be the minimal index for which $|c_j| = \max_i\{|c_i|\}$. Since

$$c_i \frac{u^i}{i!} d^i(u^j) = \binom{j}{i} c_i u^j$$

and F is of characteristic 0,

$$\left| c_i \frac{u^i}{i!} d^i(u^j) \right| \begin{cases} = |c_j u^j| & (i = j) \\ < |c_j u^j| & (i < j) \\ = 0 & (i > j). \end{cases}$$

Hence $\left| \sum_{i=0}^n c_i u^i d^i(u^j)/i! \right| = |c_j u^j|$, so the left side of (9.10.2.1) is greater than or equal to the right side. □

Corollary 9.10.3. *Set notation as in Definition 9.10.1. Then for any nonzero finite differential module V over F,*

$$|D|_{\mathrm{sp},V} = |d|_{\mathrm{sp},F} \lim_{s \to \infty} |D_s|^{1/s}$$

where D_s is as in Proposition 6.3.1.

Proof This follows from Proposition 6.3.1, since (6.3.1.2) holds by virtue of Proposition 9.10.2. □

9.11 An explicit convergence estimate

The following estimate is due to Dwork and Robba [151, Theorem 3.6]; we follow the proof of [149, Theorem IV.2.1]. We will pick up this thread again in Chapter 18.

Lemma 9.11.1. *Choose $u_1, \ldots, u_n \in K\{t/\beta\}$ for which the Wronskian matrix $W(u_1, \ldots, u_n)$ (Definition 5.6.4) is nonsingular. For $i = 0, 1, \ldots$, define the vector G_i of length n by*

$$G_i = \frac{1}{i!} \Big(d^i(u_1) \quad \cdots \quad d^i(u_n) \Big) W(u_1, \ldots, u_n)^{-1}.$$

Then

$$\limsup_{\alpha \to \beta^-} |G_i|_\alpha \le \beta^{-i} \sup\{|\lambda_1 \cdots \lambda_{n-1}|^{-1} : 1 \le \lambda_1 < \cdots < \lambda_{n-1} \le i\}$$

Proof There is no harm in assuming that $\beta = 1$. We prove the same claim for $u_1, \ldots, u_n \in \operatorname{Frac} K\{t\}$ by induction on n. For the base case $n = 1$, the claim is that $\limsup_{\alpha \to 1^-} |G_i|_\alpha \le 1$ for all i; this holds if $u_1 \in K(t)$ because we may embed u_i into F_1 and invoke (9.4.1.1), and follows in the general case by continuity.

For the induction step, write u for the tuple $u_1, \ldots, u_n$ and let $G(u)$ be the infinite matrix with rows $G_0, G_1, \ldots$. Define the infinite matrices $U = U(u)$

and $P = P(v)$ as in Lemma 5.1.8. Write u' for the tuple $u'_1, \ldots, u'_{n-1}$ where $u'_i = u_{i+1}/u_1$. Let P_n be the $n \times n$ lower triangular matrix with

$$(P_n)_{ij} = \binom{i}{j} d^{i-j}(u_1);$$

then as in Lemma 5.1.8, we compute the block decomposition

$$W(u) = P_n \begin{pmatrix} 1 & u' \\ 0 & W(u') \end{pmatrix}.$$

Let Δ be the infinite diagonal matrix with entries $1, \frac{1}{2}, \frac{1}{3}, \ldots$. By the previous computations plus Lemma 5.1.8,

$$\begin{aligned} G(u) &= U(u)W(u)^{-1} \\ &= P(u_1)U(1, u')W(u)^{-1} \\ &= P(u_1) \begin{pmatrix} 1 & u' \\ 0 & \Delta U(u') \end{pmatrix} W(u)^{-1} \\ &= P(u_1) \begin{pmatrix} 1 & 0 \\ 0 & \Delta G(u') \end{pmatrix} P_n^{-1}. \end{aligned}$$

By (9.4.1.1), $\limsup_{\alpha \to 1^-} |P(u_1)|_\alpha, |P_n^{-1}|_\alpha \leq 1$; we may thus deduce the claim from the induction hypothesis. $\qquad \square$

Notes

As noted in [149, Appendix III] (which see for more information), the p-adic Cauchy theorem (Proposition 9.3.3) was originally proved by Lutz [298]. (Note that this result long predates Dwork's work.) See Proposition 18.1.1 for a related result.

The idea of restricting a p-adic differential module to a generic disc has its origin in the work of Dwork [141], although in retrospect, the base change involved is quite natural in Berkovich's framework of nonarchimedean analytic geometry. Our definition of the generic radius of convergence is taken from Christol and Dwork [96]. The interpretation in terms of Berkovich geometry (as adopted by Baldassarri and Di Vizio in [42] and by Baldassarri in [41]) will figure prominently in Part VII.

The intrinsic generic radius of convergence (original terminology) was introduced in [245], where it is called the "toric normalization" in light of Proposition 9.7.6.

The subsidiary radii (original terminology) have not been studied much in prior literature, outside of the work of Young [415]. We will give Young's interpretation of the subsidiary radii as radii of convergence of certain horizontal sections, in a refined form, as Theorem 11.9.2.

Our description of the Artin–Hasse exponential follows Robert [355, §7.2]. Matsuda's example, from [306], is an explicit instance of a general construction introduced by Robba [353]. In turn, Matsuda's example can be greatly generalized, as shown by Pulita [343] building on work of Chinellato; one obtains an analogous construction from any Lubin–Tate group over $\mathbb{Q}_p$, with Matsuda's example arising from the multiplicative group. (The introduction to [343] provides a much more detailed historical discussion.)

The notion of a derivation of rational type was introduced in [271, §1.4] as a way to isolate those features of the field F_ρ which are needed to carry out the spectral theory of differential modules. For instance, it is shown in [271] that rational type is preserved under unramified or tamely ramified field extensions. The coordinate-free interpretation of the generic radius of convergence (Corollary 9.10.3) is useful in the study of the irregularity of higher-dimensional flat meromorphic connections [250].

Exercises

9.1 Prove Lemma 9.2.1. Then exhibit an example showing that the cokernel of $\frac{d}{dt}$ on $K\langle \alpha/t, t/\beta \rangle$ is not spanned over K by t^{-1}. That is, antidifferentiation with respect to t is not well-defined on $K\langle \alpha/t, t/\beta \rangle$.

9.2 Prove Example 9.3.5. (Hint: use the exponential series to construct a horizontal section.)

9.3 Prove Example 9.5.2. (Hint: consider the cases $\lambda \in \mathbb{Z}_p$, $\lambda \in \mathfrak{o}_K \setminus \mathbb{Z}_p$, and $\lambda \notin \mathfrak{o}_K$ separately.)

9.4 In Example 9.5.2, give an explicit formula for $IR(V_\lambda)$, in terms of ρ and the minimum distance from λ to an integer.

9.5 Prove (9.7.6.1). (Hint: you may find it easier to start with the cases where $m > 0$ and where $m = -1$, then deduce the general result from these.)

9.6 The following considerations illustrate the pitfalls of using $t\frac{d}{dt}$ instead of $\frac{d}{dt}$ in the p-adic setting.

(a) Verify that $|t\frac{d}{dt}|_{\mathrm{sp},F_\rho} \neq |t|_\rho |\frac{d}{dt}|_{\mathrm{sp},F_\rho}$.

(b) Show that the inequality (6.3.1.1) of Proposition 6.3.1 can be strict for $F = F_\rho$ and $d = t\frac{d}{dt}$. (Hint: use Example 9.5.2.)

9.7   With notation as in Proposition 9.7.6, show that all of the intrinsic subsidiary radii of V match those of $f_m^*(V)$, not just the generic radius.

9.8   Here is an "off-center" analogue of Proposition 9.7.6 suggested by Liang Xiao (compare with Theorem 10.8.2). Let m be a nonzero integer not divisible by p. Given $\rho \in (0, 1]$, let $f_m \colon F_\rho \to F_\rho$ be the map $t \mapsto (t + 1)^m - 1$. Then for any finite differential module V over F_ρ, $R(V) = R(f_m^*(V))$. (As in the previous exercise, one also obtains equality for the other subsidiary radii.)

9.9   Prove (9.9.2.1). (Hint: take logarithms. See also [355, §VII.2.2].)

9.10   Prove (9.9.5.1) by analyzing an explicit horizontal section around a generic point. A similar argument is [85, Proposition 1.5.1]. (Hint: use the equality $|1 - \zeta| = p^{-p^{-h+1}/(p-1)}$ from Example 2.1.6.)

10

Frobenius pullback and pushforward

In this chapter, we introduce Dwork's technique of descent along Frobenius in order to analyze the generic radius of convergence and subsidiary radii of a differential module, primarily in the range where Newton polygons do not apply. In one direction, we introduce a somewhat refined form of the *Frobenius antecedents* introduced by Christol and Dwork. These fail to apply in an important boundary case; we remedy this by introducing the new notion of *Frobenius descendants*.

Using these results, we are able to improve a number of results from Chapter 6 in the special case of differential modules over F_ρ. For instance, we get a full decomposition by spectral radius, extending the visible decomposition theorem (Theorem 6.6.1) and the refined visible decomposition theorem (Theorem 6.8.2). We will use these results again to study variation of subsidiary radii, and decomposition by subsidiary radii, in the remainder of Part III.

Notation 10.0.1. Throughout this chapter, we retain Hypothesis 9.0.1. We also continue to use F_ρ to denote the completion of $K(t)$ for the ρ-Gauss norm, viewed as a differential field with respect to $d = \frac{d}{dt}$ unless otherwise specified.

Notation 10.0.2. Throughout this chapter, we also assume $p > 0$ unless otherwise specified. Let μ_p denote the group of p-th roots of unity in K^{alg}. Following Definition 9.4.1, we write

$$\omega = \begin{cases} 1 & p = 0 \\ p^{-1/(p-1)} & p > 0. \end{cases}$$

10.1 Why Frobenius?

Remark 10.1.1. It may be helpful to review the current state of affairs, to clarify why Frobenius descent is needed.

Let V be a finite differential module over F_ρ. Then the possible values of the spectral radius $|D|_{\mathrm{sp},V}$ are the real numbers greater than or equal to $|d|_{\mathrm{sp},F_\rho} = \omega\rho^{-1}$, corresponding to generic radii of convergence less than or equal to ρ. However, using the Newton polygon of a twisted polynomial, we cannot distinguish among values of the spectral radius less than or equal to the operator norm $|d|_{F_\rho} = \rho^{-1}$. In particular, we cannot use this technique to prove a decomposition theorem for differential modules that separates components of spectral radius between $\omega\rho^{-1}$ and ρ^{-1}.

One might try to overcome this difficulty by considering not d but a high power of d, particularly a p^n-th power. The trouble with this is that iterating a derivation does not give another derivation, but something much more complicated. We will instead differentiate with respect to t^{p^n} rather than t, in order to increase the spectral radius into the range where Newton polygons become useful.

10.2 p-th powers and roots

We first make some calculations in answer to the following question: if two p-adic numbers are close together, how close are their p-th powers or their p-th roots? (See also [149, §V.6] and [89, Proposition 4.6.4].)

Remark 10.2.1. We observed previously, in slightly different notation (see (9.7.6.1)), that when m is a positive integer coprime to p, for ζ_m a primitive m-th root of unity,

$$|t - \eta\zeta_m^i| < \lambda|\eta| \text{ for some } i \in \{0, \dots, m-1\} \Leftrightarrow |t^m - \eta^m| < \lambda|\eta|^m \quad (\lambda \in (0,1)).$$

This breaks down for $m = p$, because $|1 - \zeta_p| = p^{-1/(p-1)} < 1$ by Example 2.1.6. Instead, we have the following bounds.

Lemma 10.2.2. *Pick* $t, \eta \in K$.

(a) *For* $\lambda \in (0,1)$, *if* $|t - \eta| \leq \lambda|\eta|$, *then*

$$|t^p - \eta^p| \leq \max\{\lambda^p, p^{-1}\lambda\}|\eta^p| = \begin{cases} \lambda^p|\eta^p| & \lambda \geq \omega \\ p^{-1}\lambda|\eta^p| & \lambda \leq \omega. \end{cases}$$

(b) Suppose $\zeta_p \in K$. If $|t^p - \eta^p| \leq \lambda |\eta^p|$, then there exists $m \in \{0, \ldots, p-1\}$ such that

$$|t - \zeta_p^m \eta| \leq \min\{\lambda^{1/p}, p\lambda\}|\eta| = \begin{cases} \lambda^{1/p}|\eta| & \lambda \geq \omega^p \\ p\lambda|\eta| & \lambda \leq \omega^p. \end{cases}$$

Moreover, if $\lambda \geq \omega^p$, we may always take $m = 0$.

We will use repeatedly, and without comment, the fact that

$$\lambda \mapsto \max\{\lambda^p, p^{-1}\lambda\}, \qquad \lambda \mapsto \min\{\lambda^{1/p}, p\lambda\}$$

are strictly increasing functions from $[0, 1]$ to $[0, 1]$ that are inverse to each other.

Proof There is no harm in assuming $\zeta_p \in K$ for both parts. For (a), factor $t^p - \eta^p$ as $(t - \eta) \prod_{m=1}^{p-1}(t - \eta\zeta_p^m)$, and write

$$t - \eta\zeta_p^m = (t - \eta) + \eta(1 - \zeta_p^m).$$

If $|t - \eta| \geq p^{-1/(p-1)}|\eta|$, then $t - \eta$ is the dominant term, otherwise $\eta(1 - \zeta_p^m)$ dominates. This gives the claimed bounds.

For (b), consider the Newton polygon of

$$t^p - \eta^p - c = \sum_{i=0}^{p-1} \binom{p}{i} \eta^i (t - \eta)^{p-i} - c$$

viewed as a polynomial in $t - \eta$. Suppose $|c| = \lambda|\eta^p|$. If $\lambda \geq p^{-p/(p-1)}$, then the terms $(t - \eta)^p$ and c dominate, and all roots have norm $\lambda^{1/p}|\eta|$. Otherwise, the terms $(t - \eta)^p$, $p(t - \eta)\eta^{p-1}$, and c dominate, so one root has norm $p\lambda|\eta|$ and the others are larger. Repeating with η replaced by $\zeta_p^m \eta$ for $m = 0, \ldots, p-1$ gives p distinct roots, which accounts for all of the roots. □

Corollary 10.2.3. *Let $T : K[\![t^p - \eta^p]\!] \to K[\![t - \eta]\!]$ be the substitution*

$$t^p - \eta^p \mapsto \sum_{i=0}^{p-1} \binom{p}{i} \eta^i (t - \eta)^{p-i}.$$

(a) If $f \in K\langle (t^p - \eta^p)/(\lambda|\eta^p|) \rangle$ for some $\lambda \in (0, 1)$, then $T(f) \in K\langle (t - \eta)/(\lambda'|\eta|) \rangle$ for $\lambda' = \min\{\lambda^{1/p}, p\lambda\}$.

(b) If $T(f) \in K\langle (t - \eta)/(\lambda|\eta|) \rangle$ for some $\lambda \in (p^{-1/(p-1)}, 1)$, then $f \in K\langle (t^p - \eta^p)/(\lambda'|\eta^p|) \rangle$ for $\lambda' = \lambda^p$.

(c) Suppose K contains a primitive p-th root of unity ζ_p. For $m = 0, \ldots, p - 1$, let $T_m : K[\![t^p - \eta^p]\!] \to K[\![t - \zeta_p^m \eta]\!]$ be the substitution $t^p - \eta^p \mapsto \sum_{i=0}^{p-1} \binom{p}{i} \zeta_p^{im} \eta^i (t - \zeta_p^m \eta)^{p-i}$. If for some $\lambda \in (0, \omega]$ one has $T_m(f) \in$

$K\langle(t - \zeta_p^m \eta)/(\lambda|\eta|)\rangle$ *for* $m = 0, \ldots, p-1$, *then* $f \in K\langle(t^p - \eta^p)/(\lambda'|\eta^p|)\rangle$
for $\lambda' = p^{-1}\lambda$.

10.3 Moving along Frobenius

We now define Frobenius pullback and pushforward operations, and show how they affect generic radius of convergence.

Definition 10.3.1. Let F'_ρ be the completion of $K(t^p)$ for the ρ^p-Gauss norm, viewed as a subfield of F_ρ and equipped with the derivation $d' = \frac{d}{d(t^p)}$. We then have

$$d = \frac{d(t^p)}{dt}d' = pt^{p-1}d'.$$

Given a finite differential module (V', D') over F'_ρ, we may view $\varphi^*V' = V' \otimes_{F'_\rho} F_\rho$ as a differential module over F_ρ with

$$D(v \otimes f) = D'(v) \otimes pt^{p-1}f + v \otimes d(f).$$

Note that this is *not* the usual base-change operation, because the restriction of d to F'_ρ is not d'; this is the same situation as in Remark 9.7.7.

Lemma 10.3.2. *Let* (V', D') *be a finite differential module over* F'_ρ. *Then*

$$IR(\varphi^*V') \geq \min\{IR(V')^{1/p}, pIR(V')\}.$$

Proof For any $\lambda < IR(V')$, any complete extension L of K, and any generic point $t_\rho \in L$ relative to K of norm ρ, $V' \otimes_{F'_\rho} L\langle(t^p - t_\rho^p)/(\lambda\rho^p)\rangle$ admits a basis of horizontal sections. By Corollary 10.2.3(a), the same holds for $(\varphi^*V') \otimes_{F'_\rho} L\langle(t - t_\rho)/(\min\{\lambda^{1/p}, p\lambda\}\rho)\rangle$. □

Here is a important example for which Lemma 10.3.2 gives a strict inequality.

Definition 10.3.3. For $m = 0, \ldots, p-1$, let W_m be the differential module over F'_ρ with one generator v, such that

$$D(v) = \frac{m}{p}t^{-p}v.$$

(The generator v is meant to behave like $t^m = (t^p)^{m/p}$.) From the Newton polygon associated to v, we may read off $IR(W_m) = \omega^p$ for $m \neq 0$ since this is within the range of applicability of Theorem 6.5.3. Note that the inequality of Lemma 10.3.2 is strict in this case, since φ^*W_m is trivial and so has intrinsic radius 1.

Definition 10.3.4. For V a differential module over F_ρ, define the *Frobenius descendant* of V as the module $\varphi_* V$ obtained from V by restriction along $F'_\rho \to F_\rho$. We view $\varphi_* V$ as a differential module over F'_ρ with differential $D' = p^{-1} t^{1-p} D$. Note that this operation commutes with duals, but not with tensor products because of a rank mismatch: the rank of $\varphi_* V$ is p times that of V. However, see Lemma 10.3.6(f).

Remark 10.3.5. The definition of the Frobenius descendant extends to differential modules over $K\langle \alpha/t, t/\beta \rangle$ for $\alpha > 0$, but not for $\alpha = 0$, since we must divide by a power of t to express D' in terms of D. The underlying problem is that (in geometric terms) the map φ ramifies at the point $t = 0$. We will see one way to deal with this problem in the discussion of off-center Frobenius descendants (Section 10.8).

The operation φ_* enjoys a number of useful properties, whose verifications we leave to the reader.

Lemma 10.3.6. *Let V be a differential module over F_ρ.*

(a) There are canonical isomorphisms

$$\iota_m : (\varphi_* V) \otimes W_m \cong \varphi_* V \qquad (m = 0, \ldots, p - 1).$$

(b) A submodule U of $\varphi_* V$ is itself the Frobenius descendant of a submodule of V if and only if $\iota_m(U \otimes W_m) = U$ for $m = 0, \ldots, p - 1$.

(c) For V' a differential module over F'_ρ, there is a canonical isomorphism

$$\varphi_* \varphi^* V' \cong \bigoplus_{m=0}^{p-1} (V' \otimes W_m).$$

(d) Suppose that $\mu_p \subset K$. Define the pullback $\zeta^*(V, D)$ as the differential module $(\zeta^* V, D')$ with

$$\zeta^* V = V \otimes_{F_\rho, \zeta} F_\rho, \qquad D'(v \otimes f) = D(v) \otimes \zeta f + v \otimes d(f).$$

Then there is a canonical isomorphism

$$\varphi^* \varphi_* V \cong \bigoplus_{\zeta \in \mu_p} \zeta^* V.$$

More precisely, the map $\zeta^* V \to \varphi^* \varphi_* V$ takes $v \otimes 1$ to $\frac{1}{p} \sum_{i=0}^{p-1} (t/\zeta)^i v \otimes t^{-i}$.

(e) There are canonical bijections

$$H^i(V) \cong H^i(\varphi_* V) \qquad (i = 0, 1).$$

(f) Suppose that $\mu_p \subset K$. For differential modules V_1, V_2 over F_ρ, there is a canonical isomorphism

$$\varphi_* V_1 \otimes \varphi_* V_2 \cong \bigoplus_{\zeta \in \mu_p} \varphi_* (V_1 \otimes \zeta^* V_2).$$

Proof Exercise.   □

10.4 Frobenius antecedents

An important counterpart to the construction of Frobenius descendants is the construction of Frobenius antecedents; this operation inverts the pullback operation φ^* when the intrinsic radius is sufficiently large.

Definition 10.4.1. Let (V, D) be a finite differential module over F_ρ. A *Frobenius antecedent* of V is a differential module (V', D') over F'_ρ such that $IR(V') > p^{-p/(p-1)}$, together with an isomorphism $V \cong \varphi^* V'$. By Lemma 10.3.2, a necessary condition for existence of a Frobenius antecedent is that $IR(V) > \omega$; Theorem 10.4.2 will imply that this condition is also sufficient.

Theorem 10.4.2 (after Christol–Dwork). *Let (V, D) be a finite differential module over F_ρ such that $IR(V) > \omega$. Then there exists a unique Frobenius antecedent V' of V. Moreover, $IR(V') = IR(V)^p$.*

Proof We may assume $\zeta_p \in K$, as otherwise we may check everything by adjoining ζ_p and then performing a Galois descent at the end.

We first check existence. Since $|D|_{\mathrm{sp},V} < \rho^{-1}$, we may define an action of $\mathbb{Z}/p\mathbb{Z}$ on V using Taylor series:

$$\zeta_p^m(x) = \sum_{i=0}^{\infty} \frac{(\zeta_p^m t - t)^i}{i!} D^i(x) \qquad (x \in V, m \in \mathbb{Z}/p\mathbb{Z}).$$

Note that the maps $P_j : V \to V$ defined by

$$P_j(v) = \frac{1}{p} \sum_{i=0}^{p-1} \zeta_p^{-ij} \cdot \zeta_p^i(v) \qquad (j = 0, \ldots, p-1)$$

are F'_ρ-linear projectors onto the generalized eigenspaces for the characters of $\mathbb{Z}/p\mathbb{Z}$. Note also that these eigenspaces are permuted by multiplication by t, so they all have the same dimension. We may conclude that the fixed space V' is an F'_ρ-subspace of V and the natural map $\varphi^* V' \to V$ is an isomorphism. (This calculation amounts to a simple instance of Noether's extension of Hilbert's Theorem 90, or of Galois descent.)

By applying the $\mathbb{Z}/p\mathbb{Z}$-action to a basis of horizontal sections of V in the generic disc $|t - t_\rho| \leq \lambda\rho$ for some $\lambda \in (\omega, IR(V))$, then invoking Corollary 10.2.3(b), we may construct horizontal sections of V' in the generic disc $|t^p - t_\rho^p| \leq \lambda^p \rho^p$. Hence $IR(V') \geq IR(V)^p > \omega^p$.

To check uniqueness, suppose $V \cong \varphi^*V' \cong \varphi^*V''$ with $IR(V'), IR(V'') > \omega^p$. By Lemma 10.3.6(c),

$$\varphi_*V \cong \bigoplus_{m=0}^{p-1}(V' \otimes W_m) \cong \bigoplus_{m=0}^{p-1}(V'' \otimes W_m). \qquad (10.4.2.1)$$

For $m = 1, \ldots, p - 1$, $IR(W_m) = \omega^p$. For any Jordan–Hölder constituent X' of $V' \otimes W_m$, write $X' = X \otimes W_m$ where $X = X' \otimes W_{-m}$ is a constituent of V'. By Lemma 9.4.6(a), $IR(X) \geq IR(V') > \omega^p = IR(W_m)$; hence by Lemma 9.4.6(c), $IR(X') = \omega^p$. Returning to (10.4.2.1), we deduce that the factor $V'' \otimes W_0$ cannot share a constituent with $V' \otimes W_m$ for any $m \in \{1, \ldots, p - 1\}$, so we must have $V'' \otimes W_0 \subseteq V' \otimes W_0$ and vice versa. Hence $V' \cong V''$.

For the last assertion, note that the proof of existence gives $IR(V') \geq IR(V)^p$, whereas Lemma 10.3.2 gives the reverse inequality. □

Corollary 10.4.3. *Let V' be a differential module over F'_ρ such that $IR(V') > \omega^p$. Then V' is the Frobenius antecedent of φ^*V', so $IR(V') = IR(\varphi^*V')^p$.*

Proof By Lemma 10.3.2, $IR(\varphi^*V') \geq IR(V')^{1/p} > \omega$, so φ^*V' has a unique Frobenius antecedent by Theorem 10.4.2. Since $IR(V') > \omega^p$, V' is that antecedent. □

The construction of Frobenius antecedents carries over to discs and annuli as follows.

Theorem 10.4.4. *Let M be a finite differential module over $K\langle\alpha/t, t/\beta\rangle$ (we may allow $\alpha = 0$) such that $IR(M_\rho) > \omega$ for $\rho \in [\alpha, \beta]$. (It will follow from Theorem 11.3.2(e) that we only need to check this condition for $\rho = \alpha$ and $\rho = \beta$.) Then there exists a unique differential module M' over $K\langle\alpha^p/t^p, t^p/\beta^p\rangle$ such that $M = M' \otimes_{K\langle\alpha^p/t^p, t^p/\beta^p\rangle} K\langle\alpha/t, t/\beta\rangle$ and*

$$IR(M' \otimes_{K\langle\alpha^p/t^p, t^p/\beta^p\rangle} F'_\rho) > \omega^p \qquad (\rho \in [\alpha, \beta]);$$

this M' also satisfies

$$IR(M' \otimes_{K\langle\alpha^p/t^p, t^p/\beta^p\rangle} F'_\rho) = IR(M_\rho)^p \qquad (\rho \in [\alpha, \beta]).$$

(A similar statement holds with $K\langle\alpha/t, t/\beta\rangle$, $K\langle\alpha^p/t^p, t^p/\beta^p\rangle$ replaced with $K[\![\alpha/t, t/\beta]\!]_{an}$, $K[\![\alpha^p/t^p, t^p/\beta^p]\!]_{an}$, respectively, by a similar proof.)

Proof   Define the projectors P_j as in the proof of Theorem 10.4.2, and let M' be the image of P_0. By arguing as in Theorem 10.4.2, we may show that the map

$$M' \otimes_{K\langle \alpha^p/t^p, t^p/\beta^p \rangle} K\langle \alpha/t, t/\beta \rangle \to M \qquad (10.4.4.1)$$

becomes an isomorphism after inverting t. The kernel and cokernel of this map are differential modules, so they cannot have any nontrivial t-torsion by Proposition 9.1.2. Hence (10.4.4.1) is an isomorphism, and we may continue as in Theorem 10.4.2.   □

10.5 Frobenius descendants and subsidiary radii

We saw in Lemma 10.3.2 that we can only weakly control the behavior of generic radius of convergence under Frobenius pullback. Under Frobenius pushforward, we can do much better; we can control not only the generic radius of convergence, but also the subsidiary radii. This will lead to a refinement of Lemma 10.3.2; see Corollary 10.5.4. (See Example 11.7.2 for an explicit instance.)

Theorem 10.5.1. *Let V be a finite differential module over F_ρ with intrinsic subsidiary radii $s_1 \leq \cdots \leq s_n$. Then the intrinsic subsidiary radii of $\varphi_* V$ comprise the multiset*

$$\bigcup_{i=1}^{n} \begin{cases} \{s_i^p, \omega^p \ (p-1 \ times)\} & s_i > \omega \\ \{p^{-1} s_i \ (p \ times)\} & s_i \leq \omega. \end{cases}$$

(To clarify, in the case $s_i > \omega$ the element ω^p is to be included $p-1$ times for each value of i, while in the case $s_i \leq \omega$ the element $p^{-1} s_i$ is to be included p times for each value of i.) In particular,

$$IR(\varphi_* V) = \min\{p^{-1} IR(V), \omega^p\}.$$

To clarify the exposition, we first prove Theorem 10.5.1 under the additional hypothesis that $\mu_p \subset K$; we will relax this hypothesis in the next section. Until then, all results in this section must be read as including the hypothesis.

Proof of Theorem 10.5.1 when $\mu_p \subset K$ It suffices to consider V irreducible. First suppose $IR(V) > \omega$. Let V' be the Frobenius antecedent of V (as per Theorem 10.4.2); note that V' is also irreducible. By Lemma 10.3.6(c), $\varphi_* V \cong \bigoplus_{m=0}^{p-1} (V' \otimes W_m)$. Since each W_m has rank 1, $V' \otimes W_m$ is also irreducible. Since $IR(V') = IR(V)^p$ by Theorem 10.4.2, and $IR(V' \otimes W_m) = \omega^p$ for $m \neq 0$ by Lemma 9.4.6(c), we have the claim.

Next suppose $IR(V) \leq \omega$. We first show that

$$IR(\varphi_* V) \geq p^{-1} IR(V) = \max\{IR(V)^p, p^{-1} IR(V)\}.$$

For t_ρ a generic point of radius ρ and $\lambda \in (0, \omega)$, the module $\varphi_* V \otimes_{F'_\rho} L\langle(t^p - t_\rho^p)/(p^{-1}\lambda\rho^p)\rangle$ splits as the direct sum of $V \otimes_{F_\rho} L\langle(t - \zeta_p^m t_\rho)/(\lambda\rho)\rangle$ over $m = 0, \ldots, p - 1$. If $\lambda < IR(V)$, by applying Corollary 10.2.3(c), we obtain $IR(\varphi_* V) \geq p^{-1}\lambda$.

Next, let W' be any irreducible subquotient of $\varphi_* V$; then $IR(W') \geq IR(\varphi_* V)$, so on one hand Lemma 10.3.2 gives

$$IR(\varphi^* W') \geq \min\{IR(W')^{1/p}, pIR(W')\}$$

$$\geq \min\{IR(\varphi_* V)^{1/p}, pIR(\varphi_* V)\} \geq IR(V). \qquad (10.5.1.1)$$

On the other hand, $\varphi^* W'$ is a subquotient of $\varphi^* \varphi_* V$; since we are assuming that $\mu_p \subseteq K$, we may apply Lemma 10.3.6(d) to see that $\varphi^* \varphi_* V$ is isomorphic to $\bigoplus_{\zeta \in \mu_p} \zeta^* V$. By Corollary 6.2.7, $IR(\zeta^* V) = IR(V)$ for each $\zeta \in \mu_p$. Since each $\zeta^* V$ is irreducible, each Jordan–Hölder constituent of $\varphi^* W'$ must be isomorphic to $\zeta^* V$ for some $\zeta \in \mu_p$, yielding $IR(\varphi^* W') = IR(V)$. That forces each inequality in (10.5.1.1) to be an equality; in particular, $IR(W')$ and $IR(\varphi_* V)$ have the same image under the injective map $s \mapsto \min\{s^{1/p}, ps\}$. We conclude that $IR(W') = IR(\varphi_* V) = p^{-1} IR(V)$, proving the claim. $\square$

Remark 10.5.2. One might be tempted to think that the argument for the inequality $IR(\varphi_* V) \geq p^{-1} IR(V)$ in the proof of Theorem 10.5.1 should carry over to the case $IR(V) > \omega$, in which case it would lead to the false conclusion $IR(\varphi_* V) \geq IR(V)^p$. What breaks down in this case is that pushing forward a basis of local horizontal sections of V only gives you $(\dim V)$ local horizontal sections of $\varphi_* V$; what they span is precisely the Frobenius antecedent of V.

Corollary 10.5.3. *Let $s_1 \leq \cdots \leq s_n$ be the intrinsic subsidiary radii of V.*

(a) *For i such that $s_i \leq \omega$, the product of the pi smallest intrinsic subsidiary radii of $\varphi_* V$ is equal to $p^{-pi} s_1^p \cdots s_i^p$.*

(b) *For i such that either $i = n$ or $s_{i+1} \geq \omega$, the product of the $pi + (p-1)(n-i)$ smallest intrinsic subsidiary radii of $\varphi_* V$ is equal to $p^{-ni} s_1^p \cdots s_i^p$.*

In particular, the product of the intrinsic subsidiary radii of $\varphi_* V$ is $p^{-np} s_1^p \cdots s_n^p$.

Note that both conditions apply when $s_i = \omega$; this will be important later (Remark 11.6.2).

As promised, we now obtain a refined version of Lemma 10.3.2.

Corollary 10.5.4. *Let V' be a differential module over F'_ρ such that $IR(V') \neq$*

ω^p. Then $IR(\varphi^*V') = \min\{IR(V')^{1/p}, pIR(V')\}$. (*This fails when $IR(V') = \omega^p$, e.g., for $V' = W_m$.*)

Proof In case $IR(V') > \omega^p$, this holds by Corollary 10.4.3. Otherwise, by Lemma 10.3.6(c), $\varphi_*\varphi^*V' \cong \bigoplus_{m=0}^{p-1}(V' \otimes W_m)$ and $IR(V' \otimes W_m) = IR(V')$ since $IR(V') < IR(W_m)$. Hence by Theorem 10.5.1,

$$IR(V') = IR(\varphi_*\varphi^*V') = \min\{p^{-1}IR(\varphi^*V'), \omega^p\}.$$

We get a contradiction if the right side equals ω^p, so we must have $IR(V') = p^{-1}IR(\varphi^*V') \leq \omega^p$, proving the claim. $\qquad\square$

10.6 Decomposition by spectral radius

As our first application of descent along Frobenius, we extend the visible decomposition theorem, and its refined form, in the special case of differential module over F_ρ. We then deduce some consequences analogous to the consequences of the visible decomposition theorem. To do this, we must use Frobenius descendants to cross the bound of $|d|_{F_\rho}$ on the spectral radius. This cannot be done using Frobenius antecedents alone, as they give no information in the boundary case $IR(V) = \omega$.

In this section, we may suppress the hypothesis $p > 0$, since the case $p = 0$ is already covered by the original decomposition theorems. In case $p > 0$, however, we must temporarily add the hypothesis that $\mu_p \subset K$ because Theorem 10.5.1 was originally proved under this added hypothesis. After proving Theorem 10.6.2, we will establish Theorem 10.5.1 for general K, at which point all results of the previous section and this section will become valid without the added hypothesis.

Proposition 10.6.1. *Let V_1, V_2 be irreducible finite differential modules over F_ρ with $IR(V_1) \neq IR(V_2)$. Then $H^1(V_1 \otimes V_2) = 0$.*

Proof We may assume that $IR(V_2) > IR(V_1)$; note that $IR(V_1^\vee) = IR(V_1)$ by Lemma 9.4.6(b). If $p = 0$, or if $p > 0$ and $IR(V_1) < \omega$, then any short exact sequence $0 \to V_2 \to V \to V_1^\vee \to 0$ splits by the visible decomposition theorem (Theorem 6.6.1), yielding the desired vanishing by Lemma 5.3.3. We may thus assume hereafter that $p > 0$.

Suppose that $p > 0$ and $IR(V_1) = \omega$. Let V_2' be the Frobenius antecedent of V_2; it is also irreducible, and $IR(V_2') = IR(V_2)^p > \omega^p$ by Theorem 10.4.2. By Theorem 10.5.1, each irreducible subquotient W of φ_*V_1 satisfies $IR(W) = \omega^p$; hence $H^1(W \otimes V_2') = 0$ by the previous case, so $H^1(\varphi_*V_1 \otimes V_2') = 0$ by the snake lemma.

By Lemma 10.3.6(a), (c),

$$\varphi_*V_1 \otimes \varphi_*V_2 \cong \bigoplus_{m=0}^{p-1} (\varphi_*V_1 \otimes W_m \otimes V_2')$$

$$\cong (\varphi_*V_1 \otimes V_2')^{\oplus p}.$$

This yields $H^1(\varphi_*V_1 \otimes \varphi_*V_2) = 0$. By Lemma 10.3.6(f), $\varphi_*(V_1 \otimes V_2)$ is a direct summand of $\varphi_*V_1 \otimes \varphi_*V_2$, so $H^1(\varphi_*(V_1 \otimes V_2)) = 0$. By Lemma 10.3.6(e), $H^1(V_1 \otimes V_2) = H^1(\varphi_*(V_1 \otimes V_2)) = 0$.

In the general case, $1 \geq IR(V_2) > IR(V_1)$. If $IR(V_1) > \omega$, then Theorem 10.4.2 implies that V_1, V_2 have Frobenius antecedents V_1', V_2'. In addition, for any extension $0 \to V_1 \to V \to V_2^\vee \to 0$, the module V satisfies $IR(V) > \omega$ by Lemma 9.4.6, so Theorem 10.4.2 implies that the whole sequence is itself is the pullback of an extension $0 \to V_1' \to V' \to (V_2')^\vee \to 0$. To show that V always splits, it suffices to do so for V'; that is, we may reduce from V_1, V_2 to V_1', V_2'. By repeating this enough times, we get to a situation where $IR(V_1) \leq \omega$. We may then apply the previous cases. □

From here, the proof of the following theorem is purely formal.

Theorem 10.6.2 (Strong decomposition theorem). *Let V be a finite differential module over F_ρ. Then there exists a decomposition*

$$V = \bigoplus_{s \in (0,1]} V_s$$

where every subquotient W_s of V_s satisfies $IR(W_s) = s$.

Proof We induct on $\dim V$; we need only consider reducible V. Choose a short exact sequence $0 \to U_1 \to V \to U_2 \to 0$ with U_2 irreducible. Split $U_1 = \bigoplus_{s \in (0,1]} U_{1,s}$ where every subquotient W_s of $U_{1,s}$ satisfies $IR(W_s) = s$. For each $s \neq IR(U_2)$, $H^1(U_2^\vee \otimes U_{1,s}) = 0$ by repeated application of Proposition 10.6.1 plus the snake lemma. By Lemma 5.3.3,

$$V = V' \oplus \bigoplus_{s \neq IR(U_2)} U_{1,s},$$

where $0 \to U_{1,IR(U_2)} \to V' \to U_2 \to 0$ is exact. □

We finally complete the proof of Theorem 10.5.1 by eliminating the hypothesis that $\mu_p \subset K$. Since that hypothesis was used nowhere else in this section or the previous one, this will imply that all results in these sections remain valid without the extra hypothesis.

Proof of Theorem 10.5.1   Given that we already proved Theorem 10.5.1 under the hypothesis that $\mu_p \subset K$, it will now suffice to check that for general K, the subsidiary radii of V and $V \otimes_K K(\mu_p)$ coincide. This will follow from Proposition 10.6.6 below, but to avoid circularity we must give a direct argument.

To this end, we may again assume that V is irreducible. By Remark 9.4.9, V and $V \otimes_K K(\mu_p)$ have the same intrinsic radius, but as per Remark 9.8.4 we cannot directly conclude that they have the same subsidiary radii. However, it is valid to apply Theorem 10.6.2 to $V \otimes_K K(\mu_p)$ to obtain a strong decomposition, which must be $\mathrm{Gal}(K(\mu_p)/K)$-stable by Remark 9.4.9 again. Consequently, this decomposition must descend to V; since V is irreducible, we conclude that $V \otimes_K K(\mu_p)$ is pure. This yields the needed equality of subsidiary radii.   □

As with the visible decomposition theorem, we obtain the following corollaries of Theorem 10.6.2 (now available with no extra hypotheses on K).

Corollary 10.6.3. *Let V be a finite differential module over F_ρ whose intrinsic subsidiary radii are all less than 1. Then $H^0(V) = H^1(V) = 0$.*

Proof   It is clear that $H^0(V)$ vanishes, as otherwise V would have a submodule with intrinsic generic radius of convergence equal to 1. To see that $H^1(V) = 0$, it suffices by Lemma 5.3.3 to show that any short exact sequence $0 \to V \to W \to X \to 0$ with X trivial is split. This follows by applying Theorem 10.6.2 to W: in the resulting decomposition, V and X must project into distinct summands, so $W \cong V \oplus X$.   □

Corollary 10.6.4. *With $V = \bigoplus_{s \in (0,1]} V_s$ as in Theorem 10.6.2, $H^i(V) = H^i(V_1)$ for $i = 0, 1$.*

This suggests that the difficulties in computing H^0 and H^1 arise in the case of intrinsic generic radius 1. We will pursue a closer study of this case in Chapter 13.

Using the strong decomposition theorem, we obtain a refined version of Corollary 6.2.9, extending Corollary 6.6.3.

Corollary 10.6.5. *If V_1, V_2 are irreducible finite differential modules over F_ρ and $IR(V_1) < IR(V_2)$, then every irreducible subquotient W of $V_1 \otimes V_2$ satisfies $IR(W) = IR(V_1)$.*

Proof   Decompose $V_1 \otimes V_2 = \bigoplus_{s \in (0,1]} V_s$ according to Theorem 10.6.2; by Lemma 9.4.6(c), $V_s = 0$ whenever $s < IR(V_1)$. If some V_s with $s > IR(V_1)$ were nonzero, then $V_1 \otimes V_2$ would have an irreducible submodule of intrinsic radius greater than $IR(V_1)$, in violation of Corollary 6.2.9.   □

We should also mention the following related result, extending Corollary 6.2.7 and Corollary 6.5.5.

Proposition 10.6.6. *Let $F_\rho \to E$ be an isometric embedding of complete nonarchimedean differential fields such that $|d|_{F_\rho} = |d|_E$ and $|d|_{\mathrm{sp},F_\rho} = |d|_{\mathrm{sp},E}$. Then for any finite differential module V over F_ρ, the full spectra of V and $V \otimes_{F_\rho} E$ are equal.*

Proof We may assume that V is irreducible with spectral radius s. We first check the case where $E = F'_\rho$ where F'/F is a finite Galois extension. Let W be any submodule of $V \otimes_{F_\rho} E$; then $V \otimes_{F_\rho} E$ is a quotient of the direct sum of the Galois conjugates of W, all of which have the same spectral radius. Hence W and $V \otimes_{F_\rho} E$ have the same spectral radius, which by Remark 9.4.9 is s. Moreover, every constituent of $V \otimes_{F_\rho} E$ is a conjugate of W and so has spectral radius s. This proves the claim.

In the general case, we may use the previous paragraph to add a primitive p-th root of unity ζ_p to F_ρ in case $p > 0$. If $p = 0$ or $s > \rho$, then Theorem 6.5.3 implies the desired result. Otherwise, we may apply Theorem 10.5.1, replacing F_ρ by F'_ρ and E by the fixed field under the action of $\mathbb{Z}/p\mathbb{Z}$ (given by Taylor series). Repeating this finitely many times, we get back to the case where Theorem 6.5.3 becomes applicable. □

We next extend the refined visible decomposition theorem (Theorem 6.8.2) to the full spectrum.

Theorem 10.6.7 (Refined strong decomposition theorem). *Let V be a finite differential module over F_ρ such that no subquotient of V has intrinsic radius equal to 1. Then for some finite tamely ramified extension E of F_ρ, $V \otimes_{F_\rho} E$ admits a (unique) direct sum decomposition*

$$V \otimes_{F_\rho} E = \bigoplus_i V_i$$

with each V_i refined, such that for $i \neq j$, $V_i \not\sim V_j$ in the sense of Definition 6.2.12.

Proof We may assume from the outset that $\mu_p \subset E$. By Theorem 10.6.2, we may reduce immediately to the case where V is pure and $IR(V) < 1$. If $p = 0$, or if $p > 0$ and $IR(V) < \omega$, then Theorem 6.8.2 gives the claim.

We next consider the case $IR(V) = \omega$; by Theorem 10.5.1, $\varphi_* V$ is again pure with $IR(\varphi_* V) = \omega^p$. By the previous paragraph, for some finite tamely extension E' of F'_ρ, we obtain a decomposition of $(\varphi_* V) \otimes_{F'_\rho} E'$ into nonequivalent refined submodules. For V_1, V_2 appearing in this decomposition, we declare V_1 and V_2 to be *weakly equivalent* if V_1 is equivalent to $V_2 \otimes_{E'} (W_m \otimes_{F'_\rho} E')$ for some m. This is again an equivalence relation; by Lemma 10.3.6(b), if we group the summands of $(\varphi_* V) \otimes_{F'_\rho} E'$ into weak equivalence classes, the resulting decomposition descends to $V \otimes_{F_\rho} E$ for $E = F_\rho \otimes_{F'_\rho} E'$.

Let X be a summand of $V \otimes_{F_\rho} E$ in this decomposition; it remains to check that X is refined. By Lemma 10.3.6(f),

$$(\varphi_* X^\vee) \otimes (\varphi_* X) \cong \bigoplus_{\zeta \in \mu_p} \varphi_* (X^\vee \otimes \zeta^* X).$$

Hence by the previous construction, $\varphi_* (X^\vee \otimes X)$ decomposes as a direct sum in which each summand can be twisted by a suitable $W_m \otimes_{F'_\rho} E'$ to raise its intrinsic radius above ω^p. However, by Lemma 10.3.6(a), $\varphi_* (X^\vee \otimes X)$ is isomorphic to its own twist by any $W_m \otimes_{F'_\rho} E'$. Hence its intrinsic subsidiary radii must comprise a multiset of $p \operatorname{rank}(X)^2$ elements in which exactly $\operatorname{rank}(X)^2$ of the elements are greater than ω^p. On the other hand, since $IR(X^\vee \otimes X) \geq IR(X) = \omega$ by Lemma 9.4.6(c), Theorem 10.5.1 implies that the intrinsic subsidiary radii of $\varphi_* (X^\vee \otimes X)$ include at most $\operatorname{rank}(X)^2$ elements greater than ω. In fact, equality occurs if and only if *all* of the intrinsic subsidiary radii of $X^\vee \otimes X$ are greater than ω. Hence X is refined.

Finally, we handle the case $\omega < IR(V) < 1$ by induction on the smallest integer h such that $IR(V) \leq \omega^{p^{-h}}$. The case $h = 0$ is handled by the arguments above. If $h > 0$, then by Theorem 10.4.2, V has a Frobenius antecedent W, for which $IR(W) = IR(V)^{1/p} \leq \omega^{p^{-h+1}}$. By the induction hypothesis, W admits a decomposition into nonequivalent refined submodules; pulling back by Frobenius then gives the desired decomposition of V by Corollary 10.5.4. □

10.7 Integrality of the generic radius

The relationship between generic radius of convergence and Newton polygons in the visible range (Theorem 6.5.3) suggests that the generic radius of convergence should satisfy some sort of integrality property. On one hand, we can infer such a property using Frobenius antecedents; on the other hand, a certain price must be paid.

Theorem 10.7.1. *Let V be a finite differential module over F_ρ with intrinsic subsidiary radii $s_1 \leq \cdots \leq s_n$. Let m be the largest integer such that $s_m = IR(V)$. Then for any nonnegative integer h,*

$$s_1 < \omega^{p^{-h}} \quad \Longrightarrow \quad s_1^m \in |K^\times|^{p^{-h}} \rho^{\mathbb{Z}}.$$

Proof For $h = 0$, we can read this off from a Newton polygon by invoking Theorem 6.5.3. To reduce from h to $h - 1$, if $IR(V) > \omega$, we replace V by its

Frobenius antecedent (Theorem 10.4.2); if $IR(V) = \omega$, we apply φ_* and invoke Corollary 10.5.3. $\qquad\qquad\qquad\square$

Here is an example to show that the exponent p^{-h} in the conclusion $s_1^m \in |K^\times|^{p^{-h}} \rho^{\mathbb{Z}}$ of Theorem 10.7.1 is not spurious.

Example 10.7.2. Suppose that $\pi \in K$ satisfies $|\pi| = \omega$. Pick $\lambda \in K^\times$ and $0 < \alpha \le \beta$ such that for $\rho \in [\alpha, \beta]$,

$$p^{1/(p-1)} < |\lambda| \rho^{-p} < p^{p/(p-1)}.$$

Let M be the differential module over $K\langle \alpha/t, t/\beta \rangle$ generated by a single element v satisfying $D(v) = -p\pi\lambda t^{-p-1} v$. Then $M \cong \varphi^* M'$, where M' is the differential module over $K\langle \alpha^p/t^p, t^p/\beta^p \rangle$ with generator w and $D'(w) = -\pi\lambda (t^p)^{-2} w$. We obtain

$$|D'|_{M' \otimes F'_\rho} = \omega |\lambda| \rho^{-2p} > \rho^{-p}.$$

Hence

$$IR(M' \otimes_{K\langle \alpha^p/t^p, t^p/\beta^p \rangle} F'_\rho) = |\lambda|^{-1} \rho^p$$
$$IR(M \otimes_{K\langle \alpha/t, t/\beta \rangle} F_\rho) = |\lambda|^{-1/p} \rho,$$

where the first equality follows by Theorem 6.5.3 and the second follows from the first by Corollary 10.4.3.

Remark 10.7.3. One can understand Example 10.7.2 in another way using the Dwork exponential series; see Definition 17.1.3.

Question 10.7.4. *What is the correct extension of Theorem 10.7.1 for the remaining subsidiary radii?*

10.8 Off-center Frobenius antecedents and descendants

Since pushing forward along Frobenius does not work well on a disc (Remark 10.3.5), we must also consider "off-center" Frobenius antecedents and descendants. Although this can be done rather more generally, we will stick to one case that will suffice for our purposes.

Definition 10.8.1. For $\rho \in (\omega, 1]$, let F''_ρ be the completion of $K((t+1)^p - 1)$ under the ρ^p-Gauss norm. Note that this coincides with the restriction of the ρ-Gauss norm on $K(t)$, because $|((t+1)^p - 1) - t^p|_\rho < |t^p|_\rho$. (One could allow $K((t+\mu)^p - \mu^p)$ for any $\mu \in K$ of norm 1, but there is no significant

loss of generality in rescaling t to reduce to the case $\mu = 1$.) For brevity, write
$u = (t + 1)^p - 1$. Equip F''_ρ with the derivation

$$d'' = \frac{d}{du} = \frac{1}{du/dt} d.$$

Given a differential module V'' over F''_ρ, we may view $\psi^* V'' = V'' \otimes_{F''_\rho} F_\rho$ as
a differential module over F_ρ. Given a differential module V over F_ρ, we may
view the restriction $\psi_* V$ of V along $F''_\rho \to F_\rho$ as a differential module over F''_ρ.

The main point here is that $du/dt \in K\langle t\rangle \cap K[\![t]\!]^\times_{an}$. Consequently, we can
extend both ψ^* and ψ_* not just to annuli but also to discs. This is needed to
establish the monotonicity property for subsidiary radii (Theorem 11.3.2(d)).

We may apply Lemma 10.2.2 with η replaced by $\eta + 1$, keeping in mind that
$|\eta + 1| = 1$ for $|\eta| < 1$. This has the net effect that everything that holds for φ
also holds for ψ, except that the intrinsic generic radius of convergence must be
replaced by the extrinsic one. Rather than rederive everything, we simply state
the analogues of Theorem 10.4.2 and Theorem 10.5.1 and leave the proofs as
exercises.

Theorem 10.8.2. *Let (V, D) be a finite differential module over F_ρ such that
$R(V) > \omega$. Then there exists a unique differential module (V'', D'') over F''_ρ
equipped with an isomorphism $V \cong \psi^* V''$ such that $R(V'') > \omega^p$. For this V'',
we have $R(V'') = R(V)^p$.*

Theorem 10.8.3. *Let V be a finite differential module over F_ρ with extrinsic
subsidiary radii $s_1, \ldots, s_n$. Then the extrinsic subsidiary radii of $\psi_* V$ comprise
the multiset*

$$\bigcup_{i=1}^n \begin{cases} \{s_i^p, \omega^p \ (p - 1 \ times)\} & s_i > \omega \\ \{p^{-1} s_i \ (p \ times)\} & s_i \leq \omega. \end{cases}$$

(The notation is to be interpreted as in Theorem 10.5.1.)

Remark 10.8.4. Note that one cannot expect Theorem 10.8.3 to hold for $\rho < \omega$,
as in that case ω^p is too large to appear as a subsidiary radius of $\psi_* V$.

Notes

Lemma 10.2.2 is taken from [233, §5.3] with some typos corrected.

The Frobenius antecedent theorem of Christol and Dwork [96, Théorème 5.4]
is slightly weaker than the one given here: it only applies for $IR(V) > p^{-1/p}$

rather than for $IR(V) > p^{-1/(p-1)}$. The discrepancy is created by the introduction of cyclic vectors, which create some regular singularities which can only eliminated under the stronger hypothesis. Much closer to the statement of Theorem 10.4.2 is [233, Theorem 6.13]; however, uniqueness is only asserted there when $IR(V') \geq IR(V)^p$.

The concept of the Frobenius descendant, and the results deduced using it, are original. This includes Theorem 10.5.1 and its off-center analogue (Theorem 10.8.3), the strong decomposition theorem (Theorem 10.6.2), and the refined strong decomposition theorem (Theorem 10.6.7). The latter was suggested by Liang Xiao; see the notes for Chapter 6 for motivation.

The notions of Frobenius antecedents and descendants extend to derivations of rational type, with one caveat: the assertion of Theorem 10.7.1 does not carry over. See [271, Theorem 1.4.21] for the correct statement.

Exercises

10.1 Prove Lemma 10.3.6.
10.2 Prove that for any finite differential module V' over F'_ρ with $IR(V') > \omega^p$, $H^0(V') = H^0(\varphi^*V')$. (The example $V' = W_m$ shows that the bound on $IR(V')$ cannot be relaxed.)
10.3 Derive Theorem 10.8.2 by imitating the proof of Theorem 10.4.2.
10.4 Derive Theorem 10.8.3 by imitating the proof of Theorem 10.5.1.

11

Variation of generic and subsidiary radii

In this chapter, we apply the tools developed in the preceding chapters to study the variation of the generic radius of convergence and the subsidiary radii associated to a differential module on a disc or annulus. We have already seen some instances where this study is needed to deduce consequences about convergence of solutions of p-adic differential equations (Example 9.6.2 and Example 9.9.3).

The statements we formulate are modeled on statements governing the variation of the Newton polygon of a polynomial over a ring of power series, as we vary the choice of a Gauss norm on the power series ring. The guiding principle is that in the visible spectrum, one should be able to relate variation of subsidiary radii to variation of Newton polygons via matrices of action of the derivation on suitable bases. This includes the relationship between subsidiary radii and Newton polygons for cyclic vectors (Theorem 6.5.3), but trying to use that approach directly creates no end of difficulties because cyclic vectors only exist in general for differential modules over fields. We implement the guiding principle in a somewhat more robust manner, using the work of Chapter 6 based on matrix inequalities.

To this principle, we must also add the techniques of descent along Frobenius introduced in Chapter 10, including the off-center variant. This allows us to overcome the limitation to the visible spectrum.

As corollaries of this analysis, we deduce some facts about the true radius of convergence of a differential module on a disc. We also establish a geometric interpretation of subsidiary radii in terms of convergence of local horizontal sections around a generic point, extending a result of Young. We will build further on this work when we discuss decomposition theorems in Chapter 12.

Throughout this chapter, we retain Notation 10.0.1, but we do not assume

$p > 0$. We will continue to use the convention from Definition 9.4.1:

$$\omega = \begin{cases} 1 & p = 0 \\ p^{-1/(p-1)} & p > 0. \end{cases}$$

11.1 Harmonicity of the valuation function

For $f \in K\langle\alpha/t, t/\beta\rangle$ and $r \in [-\log\beta, -\log\alpha]$, the function $r \mapsto v_r(f)$ is continuous, piecewise affine, and (by Proposition 8.2.3(b)) concave in r. However, one can make an even more precise statement; for simplicity, we only write this out explicitly for $r = 0$.

Definition 11.1.1. For $\overline{\mu}$ in some extension of k, let μ be a lift of $\overline{\mu}$ in some complete extension L of K. For $\alpha \leq 1 \leq \beta$, define the substitution

$$T_\mu : K\langle\alpha/t, t/\beta\rangle \to L[\![t]\!]_{\mathrm{an}}, \qquad t \mapsto t + \mu.$$

(This map extends to $K[\![\alpha/t, t/\beta]\!]_0$ if $\alpha < 1 < \beta$.) The function $r \mapsto v_r(T_\mu(f))$ on $[0, \infty)$ is continuous and piecewise affine; moreover, its right slope at $r = 0$ does not depend on the choice of the field L or the lift μ of $\overline{\mu}$. We call this slope $s_{\overline{\mu}}(f)$. For $1 < \beta$ (resp. $\alpha < 1$), define $s_\infty(f)$ (resp. $s_0(f)$) to be the left (resp. right) slope of the function $r \mapsto v_r(f)$ at $r = 0$.

We then have the following harmonicity property.

Proposition 11.1.2. *For $0 \leq \alpha < 1 < \beta$ and $f \in K\langle\alpha/t, t/\beta\rangle$ nonzero,*

$$s_\infty(f) = \sum_{\overline{\mu} \in \kappa_K^{\mathrm{alg}}} s_{\overline{\mu}}(f).$$

Proof Without loss of generality, we may assume that $|f|_1 = 1$. The quotient of $\mathfrak{o}_{F_1} \cap K\langle\alpha/t, t/\beta\rangle$ by the ideal generated by $\mathfrak{m}_K$ is isomorphic to $\kappa_K[t, t^{-1}]$; let $\overline{f}$ be the image of f in this quotient. Then $s_{\overline{\mu}}$ is the order of vanishing of $\overline{f}$ at $\overline{\mu}$, whereas s_∞ is the pole order of $\overline{f}$ at ∞. The desired equality then follows from the fact that a rational function has as many zeroes as poles (counted with multiplicity). $\qquad\square$

Remark 11.1.3. Note that $s_{\overline{\mu}}(f) \geq 0$ for $\overline{\mu} \neq 0$; thus Proposition 11.1.2 does indeed recover the concavity inequality $s_\infty \geq s_0$. Also, $s_{\overline{\mu}}(f) = 0$ if $\overline{\mu} \notin \kappa_K^{\mathrm{alg}}$, because the zeroes and poles of a rational function with coefficients in κ_K must be algebraic over κ_K.

Remark 11.1.4. For $f \in K\langle \alpha/t, t \rangle$ with $\alpha < 1$, we do not have a direct analogue of Proposition 11.1.2 because we cannot define $s_\infty(f)$. However, it is still true that $s_{\overline{\mu}}(f) \geq 0$ for $\overline{\mu} \in \kappa_K^{\mathrm{alg}}$, with equality for all but finitely many values. This remains true for $\alpha = 1$ if we skip over $\overline{\mu} = 0$.

11.2 Variation of Newton polygons

Before proceeding to differential modules, we study the variation of the Newton polygon of a polynomial over $K\langle \alpha/t, t/\beta \rangle$ or $K[\![\alpha/t, t/\beta]\!]_{\mathrm{an}}$ when measured with respect to different Gauss valuations. We begin with this both because it motivates the statements of the results for differential modules and because it will be used heavily in the proofs of those statements.

Theorem 11.2.1. *Let $P \in K[\![\alpha/t, t/\beta]\!]_{\mathrm{an}}[T]$ be a polynomial of degree n. For $r \in [-\log \beta, -\log \alpha]$, put $v_r(\cdot) = -\log | \cdot |_{e^{-r}}$. Let $\mathrm{NP}_r(P)$ be the Newton polygon of P under v_r. Let $f_1(P, r), \ldots, f_n(P, r)$ be the slopes of $\mathrm{NP}_r(P)$ (listed with multiplicity) in increasing order. For $i = 1, \ldots, n$, put $F_i(P, r) = f_1(P, r) + \cdots + f_i(P, r)$.*

(a) *(Linearity) For $i = 1, \ldots, n$, the functions $f_i(P, r)$ and $F_i(P, r)$ are continuous and piecewise affine in r. Moreover, even if $\alpha = 0$ there are only finitely many different slopes.*

(b) *(Integrality) If $i = n$ or $f_i(r_0) < f_{i+1}(r_0)$, then the slopes of $F_i(P, r)$ in some neighborhood of $r = r_0$ belong to $\mathbb{Z}$. Consequently, the slopes of each $f_i(P, r)$ and $F_i(P, r)$ belong to $\frac{1}{1}\mathbb{Z} \cup \cdots \cup \frac{1}{n}\mathbb{Z}$.*

(c) *(Superharmonicity) Suppose that $\alpha < 1 < \beta$. For $i = 1, \ldots, n$, let $s_{\infty,i}(P)$ and $s_{0,i}(P)$ be the left and right slopes of $F_i(P, r)$ at $r = 0$. For $\overline{\mu} \in (\kappa_K^{\mathrm{alg}})^\times$, let $s_{\overline{\mu},i}(P)$ be the right slope of $F_i(T_\mu(P), r)$ at $r = 0$. Then*

$$s_{\infty,i}(P) \geq \sum_{\overline{\mu} \in \kappa_K^{\mathrm{alg}}} s_{\overline{\mu},i}(P),$$

with equality if $i = n$ or $f_i(P, 0) < f_{i+1}(P, 0)$.

(d) *(Monotonicity) Suppose that P is monic and $\alpha = 0$. For $i = 1, \ldots, n$, the slopes of $F_i(P, r)$ are nonnegative.*

(e) *(Concavity) Suppose that P is monic. For $i = 1, \ldots, n$, the function $F_i(P, r)$ is concave.*

Proof Write $P = \sum_{i=0}^{n} P_i T^i$ with $P_i \in K[\![\alpha/t, t/\beta]\!]_{\mathrm{an}}$. By Proposition 8.5.2, the function $v_r(P_i)$ is continuous and concave in r and piecewise affine with slopes in $\mathbb{Z}$. Moreover, even if $\alpha = 0$ there are only finitely many different slopes.

For $s \in \mathbb{R}$ and $r \in [-\log \beta, -\log \alpha]$, put

$$v_{s,r}(P) = \min_i \{v_r(P_i) + is\};$$

that is, $v_{s,r}(P)$ is the y-intercept of the supporting line of $\mathrm{NP}_r(P)$ of slope s. Since $v_{s,r}(P)$ is the minimum of finitely many functions of the pair (r, s), each of which is continuous, piecewise affine with only finitely many different slopes, and concave, it also enjoys these properties. (Compare Remark 2.1.7.)

Note that $F_i(P, r)$ is the difference between the y-coordinates of the points of $\mathrm{NP}_r(P)$ of x-coordinates $i - n$ and $-n$. That is,

$$F_i(P, r) = \sup_s \{v_{s,r}(P) - (n - i)s\} - v_r(P_n). \tag{11.2.1.1}$$

The supremum in (11.2.1.1) is achieved by some s whose denominator is bounded by n. For any given r_0, for r in some neighborhood of r_0, there can only be finitely many values of s with denominator bounded by n achieving the supremum in (11.2.1.1) for at least one value of r in the neighborhood. Consequently, $F_i(P, r)$ is continuous and piecewise affine with only finitely many different slopes, proving (a).

If $i = n$ or $f_i(P, r_0) < f_{i+1}(P, r_0)$, then the point of $\mathrm{NP}_{r_0}(P)$ of x-coordinate $i - n$ is a vertex, and likewise for r in some neighborhood of r_0. In that case, for r near r_0,

$$F_i(P, r) = v_r(P_{n-i}) - v_r(P_n), \tag{11.2.1.2}$$

proving (b).

Assume that $\alpha < 1 < \beta$. Then Proposition 11.1.2 implies that

$$s_\infty(P_i) = \sum_{\bar{\mu} \in \kappa_K^{\mathrm{alg}}} s_{\bar{\mu}}(P_i) \qquad (i = 0, \ldots, n).$$

If $i = n$ or $f_i(P, 0) < f_{i+1}(P, 0)$, then this equation plus (11.2.1.2) yield that the desired inequality is an equality. Otherwise, let j, k be the least and greatest indices for which $f_j(P, 0) = f_i(P, 0) = f_k(P, 0)$; then $j \le i < k$, and the convexity of the Newton polygon implies

$$F_i(P, r) \le \frac{k - i}{k - j + 1} F_{j-1}(P, r) + \frac{i - j + 1}{k - j + 1} F_k(P, r), \tag{11.2.1.3}$$

with equality for $r = 0$. From this plus piecewise affinity, we deduce that

$$s_{\infty,i}(P) \geq \frac{k-i}{k-j+1} s_\infty(P_{n-j+1}) + \frac{i-j+1}{k-j+1} s_\infty(P_{n-k}),$$

$$s_{\overline{\mu},i}(P) \leq \frac{k-i}{k-j+1} s_{\overline{\mu}}(P_{n-j+1}) + \frac{i-j+1}{k-j+1} s_{\overline{\mu}}(P_{n-k}) \qquad (\overline{\mu} \in \kappa_K^{\mathrm{alg}}),$$

yielding (c).

Assume that $\alpha = 0$ and that P is monic. Then each $v_r(P_i)$ is a nondecreasing function of r, as then is each $v_{s,r}(P)$. Since $v_r(P_n) = 0$, $F_i(P,r)$ is nondecreasing by (11.2.1.1), proving (d).

To prove (e), one can reduce to working locally around $r = 0$ and then deduce the claim from (c) and (d) (because the latter implies that $s_{\overline{\mu},i}(P) \geq 0$ for $\overline{\mu} \neq 0$). However, one can also prove (e) directly as follows. Assume that P is monic, so that $P_n = 1$ and (11.2.1.1) reduces to

$$F_i(P,r) = \sup_s \{v_{s,r}(P) - (n-i)s\}.$$

It is not immediately clear from this that $F_i(P,r)$ is concave, since we are taking the supremum rather than the infimum of a collection of concave functions. To get around this, pick $r_1, r_2 \in [-\log\beta, -\log\alpha]$ and put $r_3 = ur_1 + (1-u)r_2$ for some $u \in [0,1]$. For $j \in \{1,2\}$, choose a value s_j of s which achieves the supremum in (11.2.1.1) for $r = r_j$. Put $s_3 = us_1 + (1-u)s_2$; using the concavity of $v_{s,r}(P)$ in both s and r, we have

$$\begin{aligned} F_i(P,r_3) &\geq v_{s_3,r_3}(P) - (n-i)s_3 \\ &\geq u(v_{s_1,r_1}(P) - (n-i)s_1) + (1-u)(v_{s_2,r_2}(P) - (n-i)s_2) \\ &= uF_i(P,r_1) + (1-u)F_i(P,r_2). \end{aligned}$$

This yields concavity for $F_i(P,r)$, proving (e). $\qquad\qquad\square$

Remark 11.2.2. A more geometric interpretation of the previous proof can be given by writing each P_i as $\sum_j P_{i,j} t^j$ and considering the lower convex hull of the set of points $\{(-i, -j, v(P_{i,j}))\}$ in $\mathbb{R}^3$. We leave elaboration of this point to the reader.

Remark 11.2.3. If $i = n$ or $f_i(P,r_0) < f_{i+1}(P,r_0)$, then (11.2.1.2) implies

$$f_1(P,r_0) + \cdots + f_i(P,r_0) \in v(K^\times) + \mathbb{Z}r_0.$$

This fact does not analogize to subsidiary radii, because one has to replace $v(K^\times)$ by its p-divisible closure. See Theorem 10.7.1 and Example 10.7.2.

Remark 11.2.4. The conclusions of Theorem 11.2.1 carry over if we replace

$f_i(P, r)$ by $\min\{f_i(P, r), ar + b\}$ for any fixed $a, b \in \mathbb{R}$ (except for (b), for which we need $a, b \in \mathbb{Z}$). This holds because

$$\sum_{j=1}^{i} \min\{f_i(P, r), ar + b\} = \sup_{s \leq ar+b} \{v_{s,r}(P) - (n - i)s\} - v_r(P_n);$$

in other words, the height of the relevant point is determined by the supporting lines of slopes less than or equal to $ar + b$, rather than all slopes. Note that in the notation of the proof of Theorem 11.2.1(e), the inequality $s_j \leq ar_j + b$ for $j = 1, 2$ implies the same for $j = 3$, so the proof of concavity goes through.

For bounded elements, we obtain a similar but slightly weaker conclusion.

Theorem 11.2.5. *Let $P \in K[\![\alpha/t, t/\beta]\!]_0[T]$ be a polynomial of degree n. Then the conclusions of Theorem 11.2.1 continue to hold except that if K is not discrete, in (a) the functions $f_i(P, r)$ and $F_i(P, r)$ are only piecewise affine on the interior of $[-\log\beta, -\log\alpha]$ (with possibly infinitely many slopes).*

Proof The revised statement of (a) holds by Remark 8.2.4. For the other parts, one may simply apply Theorem 11.2.1 over the ring $K\langle\gamma/t, t/\delta\rangle[T]$ for all γ, δ with $\alpha < \gamma \leq \delta < \beta$ (in case $\alpha \neq 0$) or $0 = \alpha = \gamma \leq \delta < \beta$ (in case $\alpha = 0$). $\square$

Remark 11.2.6. Picking up on Remark 11.1.4, for $f \in K\langle 1/t, t\rangle$, we have a weaker version of Theorem 11.2.1(c) asserting that $s_{\overline{\mu}, i}(P) = 0$ for all but finitely many $\overline{\mu} \in (\kappa_K^{\text{alg}})^\times$.

11.3 Variation of subsidiary radii: statements

In order to state the analogue of Theorem 11.2.1 for subsidiary radii of a differential module on a disc or annulus, we must set some corresponding notation.

Notation 11.3.1. Let M be a finite free differential module of rank n over $K\langle\alpha/t, t/\beta\rangle$, $K[\![\alpha/t, t/\beta]\!]_{\text{an}}$, or $K[\![\alpha/t, t/\beta]\!]_{\text{an}}$. For $\rho \in [\alpha, \beta]$, let

$$R_1(M, \rho), \ldots, R_n(M, \rho)$$

be the extrinsic subsidiary radii of $M_\rho = M \otimes F_\rho$ listed in increasing order, so that $R_1(M, \rho) = R(M_\rho)$ is the generic radius of convergence of M_ρ. For $r \in [-\log\beta, -\log\alpha]$, define

$$f_i(M, r) = -\log R_i(M, e^{-r}),$$

so that $f_i(M, r) \geq r$ for all r. Put $F_i(M, r) = f_1(M, r) + \cdots + f_i(M, r)$.

We now have the following results, whose proofs are distributed across the remainder of this chapter (Lemmas 11.5.1, 11.6.1, 11.6.3, and 11.7.1). Note that there is an overall sign discrepancy with Theorem 11.2.1, so that concavity becomes convexity and so forth. There are also some exceptions made in cases where $f_i(M, r) = r$.

Theorem 11.3.2. *Let M be a finite free differential module of rank n over $K\langle\alpha/t, t/\beta\rangle$, $K\langle\alpha/t, t/\beta\rangle_{\mathrm{an}}$, or $K[\![\alpha/t, t/\beta]\!]_{\mathrm{an}}$.*

(a) *(Linearity) For $i = 1, \dots, n$, the functions $f_i(M, r)$ and $F_i(M, r)$ are continuous and piecewise affine. Moreover, even if $\alpha = 0$ there are only finitely many different slopes.*

(b) *(Integrality) If $i = n$ or $f_i(M, r_0) > f_{i+1}(M, r_0)$, then the slopes of $F_i(M, r)$ in some neighborhood of r_0 belong to $\mathbb{Z}$. Consequently, the slopes of each $f_i(M, r)$ and $F_i(M, r)$ belong to $\frac{1}{1}\mathbb{Z} \cup \cdots \cup \frac{1}{n}\mathbb{Z}$.*

(c) *(Subharmonicity) Suppose that $\alpha < 1 < \beta$ and that $f_i(M, 0) > 0$. For $i = 1, \dots, n$, let $s_{\infty,i}(M)$ and $s_{0,i}(M)$ be the left and right slopes of $F_i(M, r)$ at $r = 0$. For $\overline{\mu} \in (\kappa_K^{\mathrm{alg}})^\times$, let $s_{\overline{\mu},i}(M)$ be the right slope of $F_i(T_\mu^*(M), r)$ at $r = 0$. Then*

$$s_{\infty,i}(M) \le \sum_{\overline{\mu} \in \kappa_K^{\mathrm{alg}}} s_{\overline{\mu},i}(M),$$

with equality if either $i = n$ and $f_n(M, 0) > 0$, or $i < n$ and $f_i(M, 0) > f_{i+1}(M, 0)$.

(d) *(Monotonicity) Suppose that $\alpha = 0$. For $i = 1, \dots, n$, for any point r_0 where $f_i(M, r_0) > r_0$, the slopes of $F_i(M, r)$ are nonpositive in some neighborhood of r_0. (Remember that if $\alpha = 0$, then $f_i(M, r) = r$ for r sufficiently large by Proposition 9.3.3; see also Proposition 11.8.1.)*

(e) *(Convexity) For $i = 1, \dots, n$, the function $F_i(M, r)$ is convex.*

Remark 11.3.3. Note that $f_i(M, r)$ and $F_i(M, r)$ are defined using the extrinsic normalization. However, if we switch to the intrinsic normalization, then everything in Theorem 11.3.2 stays the same except for (d), in which the upper bound on the slopes in a neighborhood of r_0 changes from 0 to -1.

Remark 11.3.4. Suppose instead that M is a finite free differential module of rank n over $K\langle\alpha/t, t/\beta\rangle_0$ (resp. over $K[\![\alpha/t, t/\beta]\!]_0$). We may then apply Theorem 11.3.2 to $M \otimes K\langle\gamma/t, t/\delta\rangle$ for all γ, δ with $\alpha < \gamma \le \delta < \beta$ (in case $\alpha \ne 0$ and we are working over $K[\![\alpha/t, t/\beta]\!]_0$) or $\alpha = \gamma \le \delta < \beta$ (otherwise). This implies that all of the conclusions of Theorem 11.3.2 continue to hold for M itself, except that in (a), the functions $f_i(M, r)$ and $F_i(M, r)$ are only defined on $(-\log\beta, -\log\alpha]$ (resp. on $(-\log\beta, -\log\alpha))$. One can also show that they extend continuously, and continue to be piecewise affine (with finitely many slopes) in case K is discrete, on the entirety of $[-\log\beta, -\log\alpha]$; see Remark 11.6.5.

Remark 11.3.5. Picking up on Remarks 11.1.4 and 11.2.6, for M defined over $K\langle 1/t, t\rangle$, we have a weaker version of Theorem 11.3.2(c) asserting that $s_{\overline{\mu},i}(M) = 0$ for all but finitely many $\overline{\mu} \in (\kappa_K^{\mathrm{alg}})^{\times}$.

11.4 Convexity for the generic radius

As a prelude to tackling Theorem 11.3.2, we give a quick proof of subharmonicity, monotonicity, and convexity (parts (c)–(e) of Theorem 11.3.2) for the function f_1 corresponding to the generic radius of convergence. This argument applies to both discs and annuli, and can be used in place of the full strength of Theorem 11.3.2 for many purposes; indeed, this is true for numerous results which predate Theorem 11.3.2. See the chapter notes for further details.

Proof of Theorem 11.3.2(c), (d), (e) for $i = 1$ Choose a basis of M and let D_s be the matrix via which D^s acts on this basis. Then recall from Lemma 6.2.5 that

$$R_1(M, \rho) = \min\{\rho, \omega \liminf_{s\to\infty} |D_s|_\rho^{-1/s}\}.$$

For each s, the function $r \mapsto -\log|D_s|_{e^{-r}}^{-1/s}$ is convex in r by Proposition 8.2.3(b). This implies the convexity of

$$f_1(M, r) = \max\{r, -\log\omega + \limsup_{s\to\infty}(-\log|D_s|_{e^{-r}}^{-1/s})\}.$$

Similarly, we deduce (c) by applying Proposition 11.1.2 to each D_s. If $\alpha = 0$, then the function $r \mapsto -\log|D_s|_{e^{-r}}^{-1/s}$ is nonincreasing, yielding (d). □

Remark 11.4.1. To improve upon this result, we will try to read off the generic radius of convergence, as well as the other subsidiary radii, from the Newton polygon of a cyclic vector. In so doing, we will encounter two obstructions.

(a) In general, one can only construct cyclic vectors for differential modules over differential fields, not over differential rings. (While Theorem 5.7.3 produces cyclic vectors over certain rings, it only does so locally for the Zariski topology, which appears to be insufficient for this purpose.)
(b) If $p > 0$, then some of the subsidiary radii may be greater than $\omega\rho$, in which case Newton polygons will not detect them.

The first problem will be addressed by using a cyclic vector over a fraction field to establish linearity, integrality and subharmonicity, then using a carefully chosen lattice to deduce monotonicity and convexity. The second problem will be addressed using Frobenius descendants.

11.5 Measuring small radii

In this section, we address concern (a) from Remark 11.4.1 using both cyclic vectors and matrix inequalities.

Lemma 11.5.1. *For any $i \in \{1, \ldots, n\}$ and any r_0 such that $f_i(M, r_0) > r_0 - \log \omega$, Theorem 11.3.2 holds in a neighborhood of r_0.*

Proof If M is defined over R, put $F = \text{Frac } R$. Choose a cyclic vector for $M \otimes_R F$ to obtain an isomorphism $M \otimes_R F \cong F\{T\}/F\{T\}P$ for some monic twisted polynomial P over F. We may then apply Corollary 6.5.4 and Theorem 11.2.1 to deduce (a), (b), (c) of Theorem 11.3.2.

To deduce (d), we may work in a right neighborhood of a single value r_0 of r. There is no harm in enlarging K (by Proposition 10.6.6), so we may assume $v(K^\times) = \mathbb{R}$. Then we may reduce to the case $r_0 = 0$ by replacing t by λt for some $\lambda \in K^\times$.

Since $v(K^\times) = \mathbb{R}$, we may pick $c_1, \ldots, c_n \in K$ such that

$$-\log |c_j| = \min\{-\log \omega - f_j(M, 0), 0\} \qquad (j = 1, \ldots, n).$$

Let $S \in F[U]$ be the untwisted polynomial with the same coefficients as P, and let $\mu_1, \ldots, \mu_n$ be the roots of S. By Corollary 6.5.4,

$$|c_i| = \max\{|\mu_i|, 1\} = \max\{\omega \exp(f_j(M, 0)), 1\} \qquad (i = 1, \ldots, n).$$

We now construct a good basis of $M \otimes_R F$ as in Theorem 6.5.3. Let B_0 be the basis of $M \otimes_R F$ given by

$$c_{n-1}^{-1} \cdots c_{n-j}^{-1} T^j \qquad (j = 0, \ldots, n-1).$$

Let N_0 be the matrix of action of D on B_0; it is a conjugated companion matrix of the form appearing in Proposition 4.3.10, corresponding to S. In particular, the singular values of N_0 are $|c_1|, \ldots, |c_{n-1}|, |\mu_1 \cdots \mu_n/(c_1 \cdots c_{n-1})|$. The latter equals $|c_n|$ if $|c_n| > 1$ and otherwise is less than or equal to 1.

By Lemma 8.6.1, the supremum norm defined by B_0 is also defined by some basis B_1 of $M \otimes_R K[[t]]_{\text{an}}$ (or for that matter $M \otimes_R K\langle t \rangle$ if the latter contains R). Let N_1 be the matrix of action of D on B_1. Theorem 6.7.4 implies that for r close to 0, the visible spectrum of $M \otimes_R F_{e^{-r}}$ is the multiset of those norms of eigenvalues of the characteristic polynomial of N_1 which exceed e^{-r}. We may then deduce (d) from Theorem 11.2.1(d). (Alternatively, one may replace K by a spherical completion, tensor M with $K[\![t]\!]_0$, and use Lemma 8.6.2.)

We may deduce (e) from (c) and (d) as noted in the proof of Theorem 11.2.1(e); however, it may also be proved directly as follows. We may again assume $r_0 = 0$. We may also assume $\alpha < 1 < \beta$, as otherwise there is nothing to check at r_0. Define B_0 as above. This time, apply Lemma 8.6.1 to construct a basis B_1' of $M \otimes_R K\langle 1/t, t \rangle$ defining the same supremum norm as B_0. We may approximate B_1' with a basis B_1 of $M \otimes_R K\langle \gamma/t, t/\delta \rangle$ for some $\alpha \leq \gamma < 1 < \delta \leq \beta$ defining the same supremum norm with respect to $| \cdot |_1$. (Here we use the facts that $K\langle \alpha/t, t/\beta \rangle$ is dense in $K\langle 1/t, t \rangle$ and that any element of $K\langle \alpha/t, t/\beta \rangle$ which becomes a unit in $K\langle 1/t, t \rangle$ is already a unit in $K\langle \gamma/t, t/\delta \rangle$ for some $\alpha \leq \gamma < 1 < \delta \leq \beta$. The latter holds because if the Newton polygon of an element of $K\langle \gamma/t, t/\delta \rangle$ has no slope equal to 0, then it also has no slopes in some neighborhood of 0.) Let N_1 be the matrix of action of D on B_1. Applying Theorem 6.7.4 to N_1, we may deduce (e) from Theorem 11.2.1(e). $\qquad\square$

11.6 Larger radii

We next address concern (b) from Remark 11.4.1, considering the cases $f_i(M, r_0) > r_0$ and $f_i(M, r_0) = r_0$ separately. We temporarily omit monotonicity, as it requires a slightly different argument. (We must also say a bit more about Theorem 11.3.2(a); see Remark 11.6.4.)

Lemma 11.6.1. *For any $i \in \{1, \ldots, n\}$ and any r_0 such that $f_i(M, r_0) > r_0$, the assertions (a), (b), (c), and (e) of Theorem 11.3.2 hold in a neighborhood of r_0.*

Proof This holds by Lemma 11.5.1 in case $p = 0$, so we may assume $p > 0$ throughout the proof. For each nonnegative integer j, we prove the claim for r_0 such that $f_i(M, r_0) > r_0 - p^{-j} \log \omega$, by induction on j; the base case $j = 0$ is precisely Lemma 11.5.1, so we may assume $j > 0$ hereafter. As in the proof of Lemma 11.5.1, we may reduce to the case $r_0 = 0$.

Let $R_1'(\rho^p), \ldots, R_{pn}'(\rho^p)$ be the subsidiary radii of $\varphi_* M \otimes F_\rho'$ in increasing order. (The normalization is chosen this way because the series variable in F_ρ' is t^p, which has norm ρ^p.) Put $g_i(pr) = -\log R_i'(e^{-pr})$. By Theorem 10.5.1, the list $g_1(pr), \ldots, g_{pn}(pr)$ consists of

$$\bigcup_{i=1}^{n} \begin{cases} \{p f_i(M, r), \ pr + \frac{p}{p-1} \log p \ (p-1 \text{ times})\} & f_i(M, r) \leq r - \log \omega \\ \{\log p + (p-1)r + f_i(M, r) \ (p \text{ times})\} & f_i(M, r) \geq r - \log \omega. \end{cases}$$

Thus we may deduce (a) from the induction hypothesis.

To check (b), (c), and (e), it suffices to handle cases where $i = n$ or $f_i(M, 0) > f_{i+1}(M, 0)$. (As in the proof of Theorem 11.2.1(c), we may linearly interpolate to establish convexity and subharmonicity in the other cases.) In these cases, as in Corollary 10.5.3, we have at least one of $f_i(M, 0) > -\log \omega$, in which case in some neighborhood of $r = 0$,

$$g_1(pr) + \cdots + g_{pi}(pr) = pF_i(M, r) + pi \log p + (p - 1)ipr; \qquad (11.6.1.1)$$

or $f_{i+1}(M, 0) < -\log \omega$ or $i = n$, in which case in some neighborhood of $r = 0$,

$$g_1(pr) + \cdots + g_{pi+(p-1)(n-i)}(pr) = pF_i(M, r) + pn \log p + (p - 1)npr. \qquad (11.6.1.2)$$

Moreover, $f_i(M, 0) > -p^{-j} \log \omega$ if and only if $g_{pi}(0) > -p^{-j+1} \log \omega$.

If $f_i(M, 0) > -\log \omega$, apply (11.6.1.1) and the induction hypothesis to write piecewise

$$\begin{aligned} F_i(M, r) &= p^{-1}(g_1(pr) + \cdots + g_{pi}(pr) - pi \log p - (p - 1)ipr) \\ &= p^{-1}(m(pr) + *) \\ &= mr + p^{-1}* \end{aligned}$$

for some $m \in \mathbb{Z}$. (Note that $*$ is not guaranteed to be in $pv(K^\times)$; this explains Example 10.7.2.) Otherwise, we may apply (11.6.1.2) to write piecewise

$$\begin{aligned} F_i(M, r) &= p^{-1}(g_1(pr) + \cdots + g_{pi+(p-1)(n-i)}(pr) - pn \log p - (p - 1)npr) \\ &= p^{-1}(m(pr) + *) \\ &= mr + p^{-1}* \end{aligned}$$

for some $m \in \mathbb{Z}$.   □

Remark 11.6.2. In the proof of Lemma 11.6.1, note the importance of the fact that the domains of applicability of (11.6.1.1) and (11.6.1.2) overlap: if $f_i(M, 0) = -\log \omega$, then (11.6.1.1) is valid for $r = 0$ but possibly not for nearby r values.

The case $f_i(M, r_0) = r_0$ remains inaccessible even using descent along Frobenius, so we make an *ad hoc* argument.

Lemma 11.6.3. *For any $i \in \{1, \ldots, n\}$ and any r_0 such that $f_i(M, r_0) = r_0$, Theorem 11.3.2 holds in a neighborhood of r_0.*

Proof As in the proof of Lemma 11.5.1, it suffices to consider the case $r_0 = 0$. We first check continuity. For this, note that the proofs of Lemma 11.5.1

and 11.6.1 show that for any $c > 0$, the function $\max\{f_i(M, r), r + c\}$ is continuous at $r = 0$. Consequently, for any $\epsilon > 0$, we can find $0 < \delta < \epsilon/2$ such that

$$|\max\{f_i(M, r), r + \epsilon/4\}| < \epsilon/2 \qquad (|r| < \delta).$$

For such r, $-\epsilon < -\delta < r \leq f_i(M, r) < \epsilon$; this yields continuity.

We next check convexity. Using Remark 11.2.4, the proofs of Lemma 11.5.1 and 11.6.1 show that for any $c > 0$, the function $\sum_{j=1}^{i} \max\{f_j(M, r), r + c\}$ is convex. (The key point is that the domain over which this holds does not depend on c.) Since this function tends to $F_i(M, r)$ as c tends to 0, we may deduce (e).

We next check piecewise affinity by induction on i. Given that the functions $f_1(M, r), \dots, f_{i-1}(M, r)$ are affine in a one-sided neighborhood of $r = 0$, say $[-\delta, 0]$, and given $f_i(M, 0) = 0$, it suffices to check linearity of $f_i(M, r) - r$ in some $[-\delta', 0]$. By (e), the set of $r \in [-\delta, 0]$ for which $f_i(M, r) - r \leq 0$ is connected. Since $f_i(M, r) - r \geq 0$ always, it follows that if $f_i(M, r_0) = 0$ for a single $r_0 \in [-\delta, 0]$, then $f_i(M, r) = 0$ for $r \in [-r_0, 0]$, so in particular $f_i(M, r) - r$ is linear in a one-sided neighborhood of 0. Otherwise, the slopes of $f_i(M, r) - r$ in $[-\delta, 0)$ form a sequence of discrete values which are negative and nondecreasing (by (e)). This sequence must then stabilize, so $f_i(M, r) - r$ is again linear in a one-sided neighborhood of 0. This proves (a).

To prove (b), note that when $f_i(M, 0) = 0$, the input hypothesis can only hold if $i = n$. Suppose we wish to check integrality of the right slope of $F_n(M, r)$ (the argument for the left slope is analogous). If $f_1(M, r) - r, \dots, f_n(M, r) - r$ are identically zero in a right neighborhood of 0, then we have nothing to check. Otherwise, let j be the greatest integer such that $f_j(M, r) - r$ is not identically zero in a right neighborhood of 0; we then deduce (b) by applying Lemma 11.6.1 with i replaced by j.

Since (c) and (d) make no assertion at $r = 0$ in case $f_i(0) = 0$, we are done. $\qquad \square$

Remark 11.6.4. Lemmas 11.6.1 and 11.6.3 fail to establish the last assertion of Theorem 11.3.2(a): if $\alpha = 0$, then each $f_i(M, r)$ has only finitely many different slopes. However, this follows from parts (b) and (e) of the theorem. Namely, each $F_i(M, r)$ has slopes which are discrete and nondecreasing, but also bounded above by i because $f_i(M, r) = r$ for r large (by Proposition 9.3.3). Hence each $F_i(M, r)$ has only finitely many different slopes, as does each $f_i(M, r)$.

Remark 11.6.5. Suppose that M is a finite free differential module of rank n over $K\langle \alpha/t, t/\beta \rrbracket_0$ (resp. over $K[\![\alpha/t, t/\beta]\!]_0$). As noted in Remark 11.6.5, Theorem 11.3.2 implies that the functions $f_i(M, r)$ and $F_i(M, r)$ are continuous and piecewise affine on $(-\log \beta, -\log \alpha]$ (resp. on $(-\log \beta, -\log \alpha)$). However, by imitating the proofs of Lemma 11.5.1, Lemma 11.6.1, and Lemma 11.6.3 using Theorem 11.2.5 in place of Theorem 11.2.1, we see that $f_i(M, r)$ and $F_i(M, r)$ extend continuously to $[-\log \beta, -\log \alpha]$. Moreover, if K is discretely valued and $\beta = 1$, then the limits of the quantities $\exp(-f_i(M, r))$ as $r \to 0^+$ are the subsidiary radii of $M \otimes \mathcal{E}$.

11.7 Monotonicity

To complete the proof of Theorem 11.3.2, we must prove (d) for $p > 0$ without the restriction $f_i(M, r_0) > r_0 - \log \omega$. The reason why we do not have (d) as part of Lemma 11.6.1 is that passing from M to $\varphi_* M$ introduces a singularity at $t = 0$ (Remark 10.3.5), so we cannot hope to infer monotonicity on $\varphi_* M$. To fix this, we must use off-center Frobenius descendants.

Lemma 11.7.1. *If $\alpha = 0$ and $f_i(M, r_0) > r_0$, then the slope of $f_i(M, r)$ in a right neighborhood of r_0 is nonpositive.*

Proof We may assume $p > 0$, as otherwise Lemma 11.5.1 implies this. We proceed as in the proof of Lemma 11.6.1, but using the off-center Frobenius ψ instead of φ. Again, we may assume $r_0 = 0$, and that $i = n$ or $f_i(M, 0) > f_{i+1}(M, 0)$ (reducing to the latter case by linear interpolation).

Let $R_1''(\rho^p), \ldots, R_n''(\rho^p)$ be the subsidiary radii of $\psi_* M \otimes F_n''$ in increasing order. Put $g_i(pr) = -\log R_i''(e^{-pr})$. By Theorem 10.8.3, if $f_i(M, 0) > -\log \omega$, then

$$g_1(pr) + \cdots + g_{pi}(pr) = pF_i(M, r) + pi \log p,$$

whereas if $f_{i+1}(M, 0) < -\log \omega$ or $i = n$, then

$$g_1(pr) + \cdots + g_{pi+(p-1)(n-i)}(pr) = pF_i(M, r) + pn \log p.$$

Moreover, $f_i(M, 0) > -p^{-j} \log \omega$ if and only if $g_{pi}(0) > -p^{-j+1} \log \omega$. We may thus proceed as in Lemma 11.6.1 to conclude. $\qquad \square$

Example 11.7.2. To see in action the discrepancy between the behavior of the centered and off-center Frobenius descendants, we consider an example suggested by Liang Xiao. (All verifications are left as an exercise.) Take $\beta > 1$, and let M be the differential module over $K\langle t/\beta \rangle$ with a single generator v satisfying $D(v) = t^{p-1} v$. Pick any $\alpha \in (0, 1)$, so that we may form $\varphi_* M$ on $K\langle \alpha/t^p, t^p/\beta \rangle$. Then $\varphi_* M$ splits as $\bigoplus_{m=0}^{p-1}(M' \otimes W_m)$, where M' has a single generator v' satisfying $D'(v') = p^{-1} v$, and W_m is defined as in Definition 10.3.3.

One then computes that for $m \neq 0$ and $\overline{\mu} \in \kappa_K^{\mathrm{alg}}$,

$$s_{\infty,1}(M') = 0$$
$$s_{\overline{\mu},1}(M') = 0$$
$$s_{\infty,1}(M' \otimes W_m) = 0$$
$$s_{0,1}(M' \otimes W_m) = 1$$
$$s_{-m,1}(M' \otimes W_m) = -1$$
$$s_{\overline{\mu},1}(M' \otimes W_m) = 0 \qquad (\overline{\mu} \neq 0, -m).$$

This yields

$$s_{\infty,p}(\varphi_* M) = 0$$
$$s_{0,p}(\varphi_* M) = p - 1$$
$$s_{\overline{\mu},p}(\varphi_* M) = -1 \qquad (\overline{\mu} \in \mathbb{F}_p^{\times})$$
$$s_{\overline{\mu},p}(\varphi_* M) = 0 \qquad (\overline{\mu} \notin \mathbb{F}_p)$$

and in turn

$$s_{\infty,1}(M) = -p + 1$$
$$s_{0,1}(M) = 0$$
$$s_{\overline{\mu},1}(M) = -1 \qquad (\overline{\mu} \in \mathbb{F}_p^{\times})$$
$$s_{\overline{\mu},1}(M) = 0 \qquad (\overline{\mu} \notin \mathbb{F}_p).$$

11.8 Radius versus generic radius

As promised, we can recover some information about the radius of convergence from the properties of the generic radius of convergence.

Proposition 11.8.1. *Let M be a finite differential module over $K\langle t/\beta \rangle$ or $K[\![t/\beta]\!]_{\mathrm{an}}$ for some $\beta > 0$. Then the radius of convergence of M equals e^{-r} for r the smallest value in $[-\log \beta, \infty)$ such that $f_1(r) = r$. Consequently, $f_1(r') = r'$ for all $r' \geq r$.*

Proof On one hand, by Theorem 9.6.1 the radius of convergence of M is at least the generic radius of convergence of $M_{\rho=e^{-r}}$, which by hypothesis equals e^{-r}. On the other hand, if $\lambda > e^{-r}$, then by hypothesis $f_1(-\log \lambda) > -\log \lambda$, or in other words $R(M_\lambda) < \lambda$. This means that $M \otimes K\langle t/\lambda \rangle$ cannot be trivial, so the radius of convergence cannot exceed λ. This proves the desired result. □

Corollary 11.8.2. *Let M be a finite differential module over $K\langle t/\beta \rangle$ or $K[\![t/\beta]\!]_{\mathrm{an}}$ for some $\beta > 0$. Then the radius of convergence of M belongs to the divisible closure of the multiplicative value group of K.*

Proof By Theorem 11.3.2(a), (b) and Theorem 10.7.1, the function $f_1(r)$ is piecewise of the form $ar+b$ with $a \in \mathbb{Q}$ and $b \in \mathbb{Q}v(K^\times)$. By Proposition 11.8.1, the radius of convergence of M equals e^{-r} for r the smallest value such that $f_1(r) = r$. To the left of this r, f_1 must be piecewise affine with slope $\neq 1$; by comparing the left and right limits at r, we deduce that $r = ar + b$ for some $a \neq 1$ rational and some $b \in \mathbb{Q}v(K^\times)$. Since this gives $r = b/(a-1)$, we deduce the claim. $\square$

It is not clear to what extent the implicit denominator in Corollary 11.8.2 can be controlled. Any answer would have to take the following example into account.

Example 11.8.3. Assume $p = 0$. Let M be the differential module over $K\langle t/\beta \rangle$ generated by a single element v satisfying $D(v) = \lambda t^n v$. Then

$$R(M) = \max\{\beta, |\lambda|^{-1/(n+1)}\},$$

which need not belong to $|K^\times|$.

We also have a criterion to detect when the radius of convergence equals the generic radius.

Corollary 11.8.4. *Let M be a finite differential module over $K\langle t/\beta \rangle$ or $K[\![t/\beta]\!]_{\mathrm{an}}$ for some $\beta > 0$, such that for some $\alpha \in (0, \beta)$, $R(M_\rho)$ is constant for $\rho \in [\alpha, \beta]$. Then $R(M) = R(M_\rho)$.*

Proof The hypothesis implies that $f_1(M, r)$ is constant in a right neighborhood of $r = -\log\beta$. By Theorem 11.3.2(d), $f_1(M, r)$ must remain constant until it becomes equal to r. By Proposition 11.8.1, we deduce the claim. $\square$

11.9 Subsidiary radii as radii of optimal convergence

The subsidiary generic radius of convergence can be interpreted as the radii of convergence of a well-chosen basis of local horizontal sections at a generic point. The argument is a variation on Corollary 11.8.4.

Definition 11.9.1. Let M be a differential module of rank n over $K\langle t/\beta \rangle$ or $K[\![t/\beta]\!]_{\mathrm{an}}$, or on the open disc of radius β. For $i = 1, \ldots, n$, the i-th *radius of optimal convergence* of M at 0 is the supremum of those $\lambda \in [0, \beta)$ for which there exist $n - i$ linearly independent horizontal sections of $M \otimes K\langle t/\lambda \rangle$. (Remember that by Corollary 9.1.3, it is equivalent to require linear independence of the horizontal sections over K or over $K\langle t/\lambda \rangle$.)

Note that there exists a basis of local horizontal sections $s_1, \ldots, s_n$ of M such that s_i has radius of convergence equal to the i-th radius of optimal radius of convergence of M at 0: once $s_{i+1}, \ldots, s_n$ have been chosen, there must be at least a one-dimensional space of choices left for s_i. Such a basis is sometimes called an *optimal basis* of local horizontal sections. (It might be more consistent with our earlier terminology to refer to the radii of optimal convergence as the *subsidiary radii of convergence*, but we have refrained from doing to avoid confusion with the subsidiary generic radii of convergence, which we commonly abbreviate to *subsidiary radii*.)

The following generalizes Proposition 9.7.5.

Theorem 11.9.2 (after Young). *Let (V, D) be a differential module over F_ρ of dimension n with subsidiary radii $r_1 \leq \cdots \leq r_n$. Let L be a complete extension of K, let t_ρ be a generic point of L relative to K of norm ρ, and put $V' = V \otimes_{F_\rho} L[\![(t - t_\rho)/\rho]\!]_{\mathrm{an}}$. Then the radii of optimal convergence of V' are also $r_1 \leq \cdots \leq r_n$.*

Proof We first produce a basis $s_1, \ldots, s_n$ for which s_i converges in the open disc of radius r_i around t_ρ for $i = 1, \ldots, n$. For this, we may apply Theorem 10.6.2 to decompose V into components each with a single subsidiary radius, and thus reduce to the case $r_1 = \cdots = r_n = r$. By the geometric interpretation of the generic radius (Proposition 9.7.5), each Jordan–Hölder constituent of V admits a basis of local horizontal sections on a generic disc of radius r. By Lemma 6.2.8(a), the same is true for V itself.

It remains to check that there cannot exist $n - i$ linearly independent local horizontal sections converging on a disc of radius strictly greater than r_i. We prove this by induction on n. Let m be the largest integer such that $r_1 = r_m$. Let V_1 be the component of V of subsidiary radius r_1, so that $\dim V_1 = m$. We will check that no local horizontal section of V_1 at t_ρ can have radius of convergence strictly greater than r_1.

Put

$$f_i(r) = f_i(V_1 \otimes_{F_\rho} L[\![(t - t_\rho)/\rho]\!]_{\mathrm{an}}, r) \qquad (i = 1, \ldots, m; \quad r \in [-\log \rho, \infty)).$$

By Theorem 11.3.2(c), the $f_i(r)$ are constant in a neighborhood of $r = -\log \rho$. By Theorem 11.3.2(c), (e) and induction on i,

$$f_i(r) = \begin{cases} -\log r_i & 0 < r \leq -\log r_i \\ r & r \geq -\log r_i. \end{cases}$$

However, if there were a local horizontal section of V_1 at t_ρ which converged on a closed disc of radius λ for some $\lambda \in (r_1, \rho)$, then $V_1 \otimes_{F_\rho} L\langle(t - t_\rho)/\lambda\rangle$ would have a trivial submodule, and so it would have λ as one of its subsidiary radii. This would force $f_n(r) = r$ for $r = -\log \lambda < -\log r_i$, contradiction.

We conclude that any local horizontal section of V that projects nontrivially onto V_1 has radius of convergence at most r_1. If $i \leq m$, we are done; otherwise, any linearly independent set of $n - i$ local horizontal sections converging in an open disc of radius greater than r_i must project to zero in V_1. We may thus reduce to applying the induction hypothesis to the complementary component.   □

Notes

The harmonicity property of functions on annuli (Proposition 11.1.2) may be best viewed inside a theory of subharmonic functions on one-dimensional Berkovich analytic spaces. See the discussion in Part VII.

For the function $f_1(M, r) = F_1(M, r)$ representing the generic radius of convergence, Christol and Dwork established convexity [96, Proposition 2.4] (using essentially the same short proof given here) and continuity at endpoints [96, Théorème 2.5] (see also [149, Appendix I]) in the case of a module over a full annulus. In the case of analytic elements, the continuity and piecewise affinity were conjectured by Dwork and proved by Pons [339, Théorème 2.2]. The analogous results for the higher $F_i(M, r)$ are original.

When restricted to intrinsic subsidiary radii less than ω, Theorem 11.9.2 is a result of Young [415, Theorem 3.1]. Young's proof is an explicit calculation using twisted polynomials and cyclic vectors.

As suggested earlier (see the notes for Chapter 9), Young's definition of radii of optimal convergence suggests an analogue in the framework of Baldassarri and Di Vizio. For instance, given a differential module on a closed disc of radius β, the radii of optimal convergence at a generic point of radius $\rho \in (0, \beta)$ should be defined in terms of an optimal basis of local horizontal sections, which are allowed to extend all the way across the disc of radius β rather than just the disc of radius ρ. Again, see Part VII for further development of this point of view.

Exercises

11.1   Give an example to show that in Theorem 11.2.1, f_2 need not be concave (even though f_1 and $f_1 + f_2$ are concave).

11.2   Verify Example 11.7.2.

11.3 Give another proof of (9.9.5.1) as follows. First find one value ρ_0 for which Theorem 6.5.3 implies that $IR(M_{\rho_0}) = \rho_0^{-b}$. Then use Theorem 11.3.2 to show that (9.9.5.1) holds for $\rho \in [1, \rho_0]$.

12

Decomposition by subsidiary radii

In Chapter 11, we established a number of important variational properties of the subsidiary radii of a differential module over a disc or annulus. In this chapter, we continue the analysis by showing that under suitable conditions, one can separate a differential module into components of different subsidiary radii. That is, we can globalize the decompositions by spectral radius provided by the strong decomposition theorem, in case a certain numerical criterion is satisfied.

As in Chapter 11, our discussion begins with some observations about power series, in this case identifying criteria for invertibility. We use these in order to set up a Hensel lifting argument to give the desired decompositions; again we must start with the visible case and then extend using Frobenius descendants. We end up with a number of distinct statements, covering open and closed discs and annuli, as well as analytic elements.

As a corollary of these results, we recover an important theorem of Christol and Mebkhout. That result gives a decomposition by subsidiary radii on an annulus in a neighborhood of a boundary radius at which the module is *solvable*, that is, all of the intrinsic subsidiary radii tend to 1. (It is not necessary to assume that the annulus is closed at this boundary.) One may view our results as a collection of quantitative refinements of the Christol–Mebkhout theorem.

Note that nothing is this chapter says anything useful in a case where the intrinsic subsidiary radii are *everywhere* equal to 1. We tackle this case in Chapter 13.

Throughout this chapter, besides Notation 10.0.1, we also retain Notation 11.3.1.

12.1 Metrical detection of units

One can identify the units in rings such as $K\langle\alpha/t, t/\beta\rangle$ in terms of power series coefficients (Lemma 8.2.6). However, for the present application, we need an alternate characterization based on metric data, i.e., Gauss norms.

Definition 12.1.1. For $f \in K\langle\alpha/t, t/\beta\rangle$ with $\alpha \leq 1 \leq \beta$, define the *discrepancy* of f at $r = 0$ as the sum

$$\mathrm{disc}(f, 0) = \sum_{\overline{\mu} \in (\kappa_K^{\mathrm{alg}})^{\times}} s_{\overline{\mu}}(f);$$

note that $\mathrm{disc}(f, 0) \geq 0$ because it is a sum of nonnegative terms. We define $\mathrm{disc}(f, r)$ for general $r \in [-\log\beta, -\log\alpha]$ by rescaling: assume without loss of generality that K contains a scalar c of norm e^{-r}, let $T_c : K\langle\alpha/t, t/\beta\rangle \to K\langle(\alpha e^r)/t, t/(\beta e^r)\rangle$ be the substitution $t \mapsto ct$, then put

$$\mathrm{disc}(f, r) = \mathrm{disc}(T_c(f), 0).$$

Lemma 12.1.2. *For $x \in K\langle t/\beta\rangle$ nonzero and $c \in K$ of norm β, x is a unit if and only if $s_0(T_c(x)) = \mathrm{disc}(x, -\log\beta) = 0$.*

Proof We may reduce to the case $\beta = 1$ and $|x|_1 = 1$. In this case, by Lemma 8.2.6, x is a unit if and only if its image modulo $\mathfrak{m}_K$ in $\kappa_K[t]$ is a unit. As noted in Proposition 11.1.2, the order of vanishing of this image at $\overline{\mu} \in \kappa_K^{\mathrm{alg}}$ is precisely $s_{\overline{\mu}}(x)$; this proves the claim. $\qquad\square$

There is also a simpler variant for bounded series and analytic elements.

Lemma 12.1.3. *For $x \in K[\![t/\beta]\!]_0$ or $x \in K[\![t/\beta]\!]_{\mathrm{an}}$ nonzero, x is a unit if and only if $s_0(T_c(x)) = 0$.*

Proof We may reduce again to the case $\beta = 1$ and $|x|_1 = 1$. Since $|x|_\rho$ is nonincreasing, $s_0(T_c(x)) = 0$ if and only if $|x|_\rho = 1$ for all $\rho \in (0, 1)$. This happens if and only if the constant term of x has norm 1, which happens if and only if x is a unit in $\mathfrak{o}_K[\![t]\!]$ or $\mathfrak{o}_K[\![t]\!] \cap F_1$ (since these are both local rings). $\quad\square$

For annuli, it is more convenient to prove a weak criterion first.

Lemma 12.1.4. *For $x \in \bigcup_{\alpha \in (0,\beta)} K\langle\alpha/t, t/\beta\rangle$ nonzero, x is a unit if and only if $\mathrm{disc}(x, -\log\beta) = 0$.*

Proof We may reduce again to the case $\beta = 1$ and $|x|_1 = 1$. In this case, by Lemma 8.2.6, x is a unit if and only if its image modulo $\mathfrak{m}_K$ in $\kappa_K[t, t^{-1}]$ is a unit. We then argue as in Lemma 12.1.2. $\qquad\square$

One may then deduce the following.

Lemma 12.1.5. *For $\alpha > 0$ and $x \in K\langle\alpha/t, t/\beta\rangle$ nonzero, x is a unit if and only if the function $r \mapsto v_r(x)$ is affine on $[-\log\beta, -\log\alpha]$, and $\mathrm{disc}(x, -\log\alpha) = \mathrm{disc}(x, -\log\beta) = 0$.*

Proof Note that by Proposition 11.1.2, $r \mapsto v_r(x)$ is affine on $[-\log\beta, -\log\alpha]$ if and only if $\mathrm{disc}(x, r) = 0$ for $r \in (-\log\beta, -\log\alpha)$. We may thus reformulate the desired result as follows: x is a unit if and only if $\mathrm{disc}(x, r) = 0$ for all $r \in [-\log\beta, -\log\alpha]$.

If x is a unit, then $\mathrm{disc}(x, r) = 0$ for all $r \in [-\log\beta, -\log\alpha]$ by Lemma 12.1.4. Conversely, given the latter condition, to check that x is a unit, it suffices (by Remark 8.1.3) to check that x is a unit in $K\langle\alpha_i/t, t/\beta_i\rangle$ for a finite collection of closed intervals $[\alpha_i, \beta_i]$ with union $[\alpha, \beta]$. However, Lemma 12.1.4 implies that one can cover a one-sided neighborhood of any element of $[\alpha, \beta]$ with such an interval; compactness of $[\alpha, \beta]$ (Lemma 8.0.4) then yields the claim. □

Remark 12.1.6. Another statement in this vein is the fact that the ring

$$\bigcup_{\alpha\in(0,\beta)} K\langle\alpha/t, t/\beta\rrbracket_{\mathrm{an}}$$

is a field (Corollary 8.5.3).

12.2 Decomposition over a closed disc

We consider decomposition by subsidiary radii first in the case of a closed disc. The numerical criterion in this case involves an analogue of the discrepancy function from the previous section.

Unfortunately, while the case of a closed disc is the most natural place to start from the point of view of the exposition, for technical reasons we will not complete its proof until after we treat some other cases. See Section 12.4 for the resolution of this issue.

Definition 12.2.1. Let M be a finite differential module over $K\langle\alpha/t, t/\beta\rangle$ with $\alpha \leq 1 \leq \beta$. Define the *i-th discrepancy of M at $r = 0$* as

$$\mathrm{disc}_i(M, 0) = -\sum_{\overline{\mu}\in(\kappa_K^{\mathrm{alg}})^\times} s_{\overline{\mu},i}(M);$$

it is always nonnegative by Theorem 11.3.2(d). Extend the definition to define $\mathrm{disc}_i(M, r)$ for general $r \in [-\log\beta, -\log\alpha]$ as in Definition 12.1.1.

Theorem 12.2.2. *Let M be a finite differential module over $K\langle t/\beta\rangle$ of rank n. Suppose that the following conditions hold for some $i \in \{1, \ldots, n-1\}$.*

(a) We have $f_i(M, -\log\beta) > f_{i+1}(M, -\log\beta)$.
(b) The function $F_i(M, r)$ is constant for r in a neighborhood of $-\log\beta$.
(c) We have $\mathrm{disc}_i(M, -\log\beta) = 0$.

Then there is a direct sum decomposition of M inducing, for each $\rho \in (0, \beta]$, the decomposition of M_ρ separating the first i subsidiary radii from the others.

Before proving Theorem 12.2.2, we record some observations which will simplify the proof.

Remark 12.2.3. To prove Theorem 12.2.2, it suffices to lift to M the decomposition of M_β separating the first i subsidiary radii from the rest. Namely, suppose $M_1 \oplus M_2$ is this decomposition and that $M_{1,\beta}$ accounts for the first i subsidiary radii of M_β. By Theorem 11.3.2(a), for r in a neighborhood of $-\log\beta$, $M_{1,e^{-r}}$ accounts for the first i subsidiary radii of $M_{e^{-r}}$. Consequently, for r in a neighborhood of $-\log\beta$, $f_j(M, r) = f_j(M_1, r)$ for $j = 1, \ldots, i$ and $f_j(M, r) = f_{j-i}(M_2, r)$ for $j = i+1, \ldots, n$.

By Theorem 11.3.2(d),(e), the function $F_i(M_1, r)$ is convex and nonincreasing as long as $f_i(M_1, r) > r$. For r in a neighborhood of $-\log\beta$, $F_i(M_1, r) = F_i(M, r)$ by the previous paragraph, and $F_i(M, r)$ is constant by hypothesis. Hence $F_i(M_1, r)$ must remain constant until the first value of r for which $f_i(M_1, r) = r$; denote this value by r_1. For $-\log\beta \le r \le r_1$, $f_i(M_1, r) = F_i(M_1, r) - F_{i-1}(M_1, r)$ is nondecreasing by Theorem 11.3.2(d), whereas for $-\log\beta \le r$, $f_1(M_2, r)$ is nonincreasing until it becomes equal to r and then stays equal to r thereafter. Hence $f_i(M_1, r) > f_1(M_2, r)$ for $-\log\beta \le r < r_1$ and $r_1 = f_i(M_1, r_1) \ge f_1(M_2, r_1) \ge r_1$; thus also $f_i(M_1, r) \ge r = f_1(M_2, r)$ for $r \ge r_1$. Consequently, $M_{1,e^{-r}}$ accounts for the first i subsidiary radii of $M_{e^{-r}}$ for all r.

We next make a simple but quite useful observation.

Lemma 12.2.4. *Let R, S, T be subrings of a common ring U with $S \cap T = R$. Let M be a finite free R-module. Then the intersection $(M \otimes_R S) \cap (M \otimes_R T)$ inside $M \otimes_R U$ is equal to M itself.*

This also holds when M is only locally free; see the exercises.

Remark 12.2.5. The immediate application of Lemma 12.2.4 is to replace K by a complete extension L in Theorem 12.2.2; inside the completion of $L(t)$ for the β-Gauss norm,

$$F_\beta \cap L\langle t/\beta \rangle = K\langle t/\beta \rangle$$

(exercise). Thus obtaining matching decompositions of M_β and $M \otimes_{K\langle t/\beta \rangle} L\langle t/\beta \rangle$ gives a corresponding decomposition of M itself.

For annuli, we will use the related fact that for any $\rho \in [\alpha, \beta]$, inside the completion of $L(t)$ for the ρ-Gauss norm,

$$F_\rho \cap L\langle \alpha/t, t/\beta \rangle = K\langle \alpha/t, t/\beta \rangle$$

(exercise).

We also need a lemma about polynomials over $K\langle t \rangle$.

Lemma 12.2.6. *Let $P = \sum_i P_i T^i$ and $Q = \sum_i Q_i T^i$ be polynomials over $K\langle t \rangle$ satisfying the following conditions.*

(a) We have $|P - 1|_1 < 1$.
(b) For $m = \deg(Q)$, Q_m is a unit and $|Q|_1 = |Q_m|_1$.

Then P and Q generate the unit ideal in $K\langle t \rangle[T]$.

Proof We may assume without loss of generality that $Q_m = 1$. The hypothesis that $|Q|_1 = |Q_m|_1 = 1$ implies that the Newton polygon of Q under $|\cdot|_1$ has all slopes nonnegative. Hence by Lemma 2.3.1, if $R \in K\langle t \rangle[T]$ and S is the remainder upon dividing R by Q, then $|S|_1 \leq |R|_1$.

Let S_i denote the remainder upon dividing $(1 - P)^i$ by Q. By the previous paragraph, the series $\sum_{i=0}^{\infty} S_i$ converges and its limit S satisfies $PS \equiv 1$ (mod Q). This proves the claim.   □

We are now ready to make the key step of establishing Theorem 12.2.2 in the visible range (Definition 6.5.1).

Lemma 12.2.7. *Theorem 12.2.2 holds if $f_i(M, -\log\beta) > -\log\omega - \log\beta$.*

Proof By invoking Remark 12.2.5 to justify enlarging K, then rescaling, we may reduce to the case $\beta = 1$. We may also assume that K has value group $\mathbb{R}$.

Set notation as in the section of the proof of Lemma 11.5.1 that addresses Theorem 11.3.2(d). Let $Q(T)$ be the characteristic polynomial of N_1, so that the Newton polygon of Q computes $f_1(M, r), \ldots, f_i(M, r)$ in a neighborhood of $r = 0$ (from the same section of the proof of Lemma 11.5.1). Condition (a) of Theorem 12.2.2 implies that these Newton polygons all have a vertex with x-coordinate $n - i$, whose position is determined by the coefficient of T^{n-i} in Q. Conditions (b) and (c) then imply that this coefficient satisfies the hypothesis of Lemma 12.1.2, and so is a unit in $K\langle t \rangle$. We can thus apply Theorem 2.2.1 to factor $Q = Q_2 Q_1$ in such a way that the roots of Q_1 are the i largest roots of Q under $|\cdot|_1$.

Use the basis B_1 to identify M with $K\langle t \rangle^n$, so that we may view N_1 as a $K\langle t \rangle$-linear endomorphism of M. By Lemma 12.2.6 applied after rescaling, Q_1 and Q_2 generate the unit ideal in $K\langle t \rangle[T]$. Hence M splits as a direct sum

$M_1' \oplus M_2'$ of modules (but not differential modules) with $M_i' = \ker(Q_i(N_1))$. We now invoke some of the approximation lemmas, as follows. Equip M with the supremum norm compatible with $|\cdot|_1$ defined by B_1, then equip M_1', M_2' with the induced quotient norms; these are both supremum-equivalent by Lemma 1.3.5. By Lemma 1.3.7, for any $c > 1$, we can approximate these norms to within a factor of at most c by supremum norms defined by bases of $M_1' \otimes_{K\langle t \rangle} F_1, M_2' \otimes_{K\langle t \rangle} F_1$. By Lemma 8.6.1, the latter norms are also defined by bases $B_{1,1}, B_{1,2}$ of M_1', M_2'.

Let U be the change-of-basis matrix from B_1 to $B_{1,1} \cup B_{1,2}$, so that $U^{-1}N_1U$ is a block diagonal matrix and $|U|_1, |U^{-1}|_1 \leq c$. Then the matrix of action of D on $B_{1,1} \cup B_{1,2}$ is $U^{-1}N_1U + U^{-1}d(U)$. If we write this in block form as

$$\begin{pmatrix} A & B \\ C & D \end{pmatrix},$$

then by taking c sufficiently close to 1, we may force the following conditions hold.

(a) The matrix A is invertible and $|A^{-1}|_1 \max\{|d|_1, |B|_1, |C|_1, |D|_1\} < 1$.
(b) The Newton slopes of A under $|\cdot|_1$ account for the first i subsidiary radii of $M_{\rho=1}$.

As in the proof of Lemma 6.7.3, we may now use Lemma 6.7.1 to produce a submodule of M accounting for the last $n - i$ subsidiary radii of $M_{\rho=1}$. By repeating this argument for $M^\vee$, we obtain a submodule of M accounting for the first i subsidiary radii of $M_{\rho=1}$. By Remark 12.2.3, this suffices to prove the desired result. $\qquad \square$

Remark 12.2.8. As in Lemma 11.5.1, one can prove Lemma 12.2.7 using Lemma 8.6.2 in place of Lemma 8.6.1, at the expense of some extra complications. First, one must replace K by a spherical completion with value group $\mathbb{R}$ and algebraically closed residue field (by invoking Theorem 1.5.3). One then only obtains the desired decomposition over $K[\![t]\!]_0$. By gluing with the original decomposition (defined over F_1), we recover a decomposition over $K[\![t]\!]_{\mathrm{an}}$. To remove poles in the disc $|t - \mu| < 1$, we apply the same argument with M replaced by $T_\mu^*(M)$ (in the sense of Definition 11.1.1).

Remark 12.2.9. At this point, we would ideally like to conclude the proof of Theorem 12.2.2 by using Frobenius descendants to extend the range of applicability of Lemma 12.2.7. Unfortunately, we cannot execute this strategy because both the centered and off-center Frobenius pushforwards introduce poles in the closed unit disc. Instead, we carry on to some other cases where this strategy does work, and then patch these results together to deal with closed discs in Section 12.4.

12.3 Decomposition over a closed annulus

Over a closed annulus, one has a decomposition theorem of a somewhat different shape than over a closed disc. Fortunately, we may follow a similar strategy as for Theorem 12.2.2 through to completion; we do not encounter the issue raised in Remark 12.2.9.

Theorem 12.3.1. *Let M be a finite differential module over $K\langle \alpha/t, t/\beta \rangle$ of rank n, for some $0 < \alpha \leq \beta$. Suppose that the following conditions hold for some $i \in \{1, \ldots, n - 1\}$.*

(a) We have $f_i(M, r) > f_{i+1}(M, r)$ for $-\log \beta \leq r \leq -\log \alpha$.
(b) The function $F_i(M, r)$ is affine for $-\log \beta \leq r \leq -\log \alpha$.
(c) We have $\mathrm{disc}_i(M, -\log \beta) = \mathrm{disc}_i(M, -\log \alpha) = 0$.

Then there is a direct sum decomposition of M inducing, for each $\rho \in \lfloor \alpha, \beta \rfloor$, the decomposition of M_ρ separating the first i subsidiary radii from the rest.

We first prove a lemma which looks somewhat more like Theorem 12.2.2.

Lemma 12.3.2. *Let M be a finite differential module over $K\langle \alpha/t, t/\beta \rangle$ of rank n. Suppose that the following conditions hold for some $i \in \{1, \ldots, n - 1\}$.*

(a) We have $f_i(M, -\log \beta) > f_{i+1}(M, -\log \beta)$.
(b) We have $\mathrm{disc}_i(M, -\log \beta) = 0$.

Then for some $\gamma \in [\alpha, \beta)$, there is a direct sum decomposition of $M \otimes_{K\langle \alpha/t, t/\beta \rangle} K\langle \gamma/t, t/\beta \rangle$ inducing, for each $\rho \in [\gamma, \beta]$, the decomposition of M_ρ separating the first i subsidiary radii from the rest.

Proof Using Remark 12.2.5 again, we may enlarge K and then reduce to the case $\beta = 1$. We first prove the lemma assuming that $f_i(M, 0) > -\log \omega$. Set notation as in the section of the proof of Lemma 11.5.1 that addresses Theorem 11.3.2(d), but take $\beta = 1$. As in the section of the proof of Lemma 11.5.1 that addresses Theorem 11.3.2(e), we can find a basis B_1 of $M \otimes_{K\langle \alpha/t, t \rangle} K\langle \gamma/t, t \rangle$ for some $\gamma \in [\alpha, 1)$ defining the same supremum norm as B_0. Let N_1 be the matrix of action of D on B_1. By conditions (a) and (b) of the lemma plus Lemma 12.1.4, the coefficient of T^{n-i} in the characteristic polynomial of N_1 is a unit in $K\langle \gamma/t, t \rangle$ for some $\gamma \in [\alpha, 1)$. We may thus continue as in the proof of Lemma 12.2.7.

To prove the lemma in general, choose $j > 0$ such that $f_i(M, 0) > -p^{-j} \log \omega$. Let $M_1' \oplus M_2'$ be the decomposition of $\varphi_* M$ separating the subsidiary radii less than or equal to $\exp(-p f_i(M, 0))$ into M_1' (which exists by the induction hypothesis). This might not be induced by a decomposition of M_1, because some factors of subsidiary radius ω^p that are needed in M_2' may instead be grouped into M_1'. To fix this, consider instead the decomposition

$$(\iota_0(M_1') \cap \cdots \cap \iota_{p-1}(M_1')) \oplus (\iota_0(M_2') + \cdots + \iota_{p-1}(M_2')),$$

where ι_m is defined as in Lemma 10.3.6(a). By Lemma 10.3.6(b), this decomposition is induced by a decomposition of M; by Theorem 10.5.1, it has the desired effect. □

To prove Theorem 12.3.1 from Lemma 12.3.2, we proceed as in the proof of Lemma 12.1.5. However, we must first give an alternate formulation of the hypotheses of the theorem.

Remark 12.3.3. In the statement of Theorem 12.3.1, given condition (a), we may reformulate conditions (b) and (c) together as the following condition.

(b′) We have $\mathrm{disc}_i(M, r) = 0$ for $-\log \beta \leq r \leq -\log \alpha$.

To see this, note that if $\alpha < 1 < \beta$, then condition (a) implies that equality holds in Theorem 11.3.2(c), so $s_{0,i}(M) - s_{\infty,i}(M) = \mathrm{disc}_i(M, 0)$. Consequently, $F_i(M, r)$ is affine in a neighborhood of 0 if and only if $\mathrm{disc}_i(M, 0) = 0$. By rescaling, we obtain the desired equivalence. (Compare the proof of Lemma 12.1.5.)

Proof of Theorem 12.3.1 We first treat the case $\alpha = \beta$. In this case, we may argue as in Lemma 12.2.7 if $f_i(M, -\log \beta) > -\log \omega - \log \beta$, then argue as in Lemma 12.3.2 to deduce the general case.

Assume hereafter that $\alpha < \beta$. By Remark 12.3.3, if M satisfies the given hypothesis, then so does $M \otimes_{K \langle \alpha/t, t/\beta \rangle} K \langle \gamma/t, t/\delta \rangle$ for each closed subinterval $[\gamma, \delta] \subseteq [\alpha, \beta]$. For each $\rho \in (\alpha, \beta]$, Lemma 12.3.2 implies that for some $\gamma \in [\alpha, \rho)$, $M \otimes_{K \langle \alpha/t, t/\beta \rangle} K \langle \gamma/t, t/\rho \rangle$ admits a decomposition with the desired property. Similarly, for each $\rho \in [\alpha, \beta)$, for some $\gamma \in (\rho, \beta]$, $M \otimes_{K \langle \alpha/t, t/\beta \rangle} K \langle \rho/t, t/\gamma \rangle$ admits a decomposition with the desired property.

By the compactness of $[\alpha, \beta]$ (Lemma 8.0.4), we can cover $[\alpha, \beta]$ with finitely many intervals $[\gamma_i, \delta_i]$ for which $M \otimes_{K \langle \alpha/t, t/\beta \rangle} K \langle \gamma_i/t, t/\delta_i \rangle$ admits a decomposition with the desired property. Since the decomposition of $M \otimes_{K \langle \alpha/t, t/\beta \rangle} K \langle \gamma_i/t, t/\delta_i \rangle$ is uniquely determined by the induced decomposition over F_ρ for any single $\rho \in [\gamma_i, \delta_i]$, these decompositions agree on overlaps of the covering intervals. By the gluing lemma (Lemma 8.3.6), we obtain a decomposition of M itself. □

Remark 12.3.4. As in Remark 12.2.8, to prove Lemma 12.3.2 one may use Lemma 8.6.2 in place of Lemma 8.6.1. Again, one enlarges K to be spherically complete with value group $\mathbb{R}$ and algebraically closed residue field. One then notes that $F = \bigcup_{\gamma<1} K\langle \gamma/t, t \rrbracket_{\text{an}}$ is a field (Corollary 8.5.3), so one can approximate the basis B_0 of $M_{\rho=1}$ with a basis B_1 of $M \otimes_{K\langle \alpha/t, t \rangle} F$ defining the same supremum norm. One then obtains a decomposition of $M \otimes_{K\langle \alpha/t, t \rangle} K\langle \gamma/t, t \rrbracket_{\text{an}}$ for some γ, and one can remove the unwanted poles as in Remark 12.2.8.

12.4 Partial decomposition over a closed disc or annulus

We next state decomposition theorems which apply to a closed disc or annulus without requiring a discrepancy condition. The price one must pay is that one must work with analytic elements. However, in this context we can salvage the proof for a disc; this will in turn enable us to treat the case of a closed disc (Theorem 12.2.2).

Theorem 12.4.1. *Let M be a finite differential module of rank n over $K\llbracket t/\beta \rrbracket_{\text{an}}$. Suppose that the following conditions hold for some $i \in \{1, \ldots, n-1\}$.*

(a) *The function $F_i(M, r)$ is constant in a neighborhood of $r = -\log\beta$.*
(b) *We have $f_i(M, -\log\beta) > f_{i+1}(M, -\log\beta)$.*

Then M admits a direct sum decomposition separating the first i subsidiary radii of M_ρ for $\rho \in (0, \beta]$.

Proof We first establish an analogue of Lemma 12.2.7 in this situation by replacing Lemma 12.1.3 with Lemma 12.1.2. We then complete the proof by arguing as in Lemma 12.3.2, but replacing the centered Frobenius pushforward $\varphi_* M$ with the off-center Frobenius pushforward $\psi_* M$ (using Theorem 10.8.3 in place of Theorem 10.5.1). □

Proof of Theorem 12.2.2 As in Remark 12.2.3, from the hypothesis of Theorem 10.6.2 we deduce that for $\alpha \in (0, \beta)$ sufficiently close to β, $M \otimes_{K\langle t/\beta \rangle} K\langle \alpha/t, t/\beta \rangle$ satisfies the hypothesis of Theorem 12.3.1. We may now deduce Theorem 12.2.2 by applying Theorem 12.3.1 to $M \otimes_{K\langle t/\beta \rangle} K\langle \alpha/t, t/\beta \rangle$ and Theorem 12.4.1 to $M \otimes_{K\langle t/\beta \rangle} K\llbracket t/\beta \rrbracket_{\text{an}}$. By Lemma 8.5.8(b), these decompositions glue together to give the desired result. □

For annuli, we get two different statements depending on whether we lift the discrepancy condition on one or both endpoints.

Theorem 12.4.2. *Let M be a finite differential module of rank n over the ring*

$K\langle\alpha/t, t/\beta\rangle_{an}$ for some $\alpha < \beta$. Suppose that the following conditions hold for some $i \in \{1, \ldots, n-1\}$.

(a) The function $F_i(M, r)$ is affine for $-\log\beta \leq r \leq -\log\alpha$.
(b) We have $f_i(M, r) > f_{i+1}(M, r)$ for $-\log\beta \leq r \leq -\log\alpha$.
(c) We have $\mathrm{disc}_i(M, -\log\alpha) = 0$.

Then M admits a direct sum decomposition separating the first i subsidiary radii of M_ρ for $\rho \in [\alpha, \beta]$.

Proof To obtain a decomposition of $M \otimes_{K\langle\alpha/t, t/\beta\rangle_{an}} K\langle\gamma/t, t/\beta\rangle_{an}$ for some $\gamma \in (\alpha, \beta)$, we may proceed as in Theorem 12.3.1 except with Lemma 12.1.4 replaced by Remark 12.1.6. We may then do likewise with α and β interchanged, fill in the middle using Theorem 12.3.1, and using Lemma 8.5.8(a) to conclude. $\quad\square$

Theorem 12.4.3. *Let M be a finite differential module of rank n over the ring $K[\![\alpha/t, t/\beta]\!]_{an}$ for some $\alpha < \beta$. Suppose that the following conditions hold for some $i \in \{1, \ldots, n-1\}$.*

(a) *The function $F_i(M, r)$ is affine for $-\log\beta \leq r \leq -\log\alpha$.*
(b) *We have $f_i(M, r) > f_{i+1}(M, r)$ for $-\log\beta \leq r \leq -\log\alpha$.*

Then M admits a direct sum decomposition separating the first i subsidiary radii of M_ρ for $\rho \in [\alpha, \beta]$.

Proof As in Theorem 12.4.2. $\quad\square$

Remark 12.4.4. Readers familiar with affinoid algebras should be able to extend the results of this section to cases where M is defined over the ring of analytic elements for a disc contained in a one-dimensional affinoid space. (We leave even the definition of this ring as an unstated exercise.)

If K is discrete, one can also extend to modules defined over $K[\![t/\beta]\!]_0$ or $K\langle\alpha/t, t/\beta\rangle_0$. However, if K is not discretely valued, one runs into various difficulties associated with the fact that $\mathcal{E}$ is no longer a field; compare Remark 8.2.4.

Remark 12.4.5. If $p = 0$, it should be possible to get a decomposition theorem over $K[\![\alpha/t, t/\beta]\!]_0$ even without assuming that $f_i(M, r) > f_{i+1}(M, r)$ for $r = -\log\beta, -\log\alpha$; a statement along these lines (for K discretely valued) appears in [250]. However, this fails completely if $p > 0$; one can generate numerous counterexamples using the theory of isocrystals (Appendix B).

12.5 Decomposition over an open disc or annulus

As for rings of analytic elements, when working over open discs we obtain a decomposition theorem without any discrepancy condition.

Theorem 12.5.1. *Let M be a finite differential module of rank n over the open disc of radius β. Suppose that the following conditions hold for some $i \in \{1, \ldots, n-1\}$ and some $\gamma \in (0, \beta)$.*

(a) The function $F_i(M, r)$ is constant for $-\log \beta < r \le -\log \gamma$.
(b) We have $f_i(M, r) > f_{i+1}(M, r)$ for $-\log \beta < r \le -\log \gamma$.

Then M admits a unique decomposition separating the first i subsidiary radii of M_ρ for $\rho \in [\gamma, \beta)$.

Proof As in Remark 12.3.3, note that combining (a) with subharmonicity (Theorem 11.3.2(c)) yields that $\mathrm{disc}_i(M, \delta) = 0$ for $\delta \in (\gamma, \beta)$. Thus for any such δ, we may apply Theorem 12.2.2 to $M \otimes K\langle t/\delta \rangle$; doing so for all such δ (or a sequence increasing to β) yields the desired result. □

Similarly, for open annuli, we obtain a decomposition theorem without a discrepancy condition at endpoints.

Theorem 12.5.2. *Let M be a finite differential module of rank n over the open annulus of inner radius α and outer radius β. Suppose that the following conditions hold for some $i \in \{1, \ldots, n-1\}$.*

(a) The function $F_i(M, r)$ is affine for $-\log \beta < r < -\log \alpha$.
(b) We have $f_i(M, r) > f_{i+1}(M, r)$ for $-\log \beta < r < -\log \alpha$.

Then M admits a unique decomposition separating the first i subsidiary radii of M_ρ for any $\rho \in (\alpha, \beta)$.

Proof By Remark 12.3.3, the conditions of Theorem 12.3.1 are satisfied by $M \otimes K\langle \gamma/t, t/\delta \rangle$ whenever $\alpha < \gamma \le \delta < \beta$. Gluing together the resulting decompositions yields the desired result. □

Remark 12.5.3. One can also obtain a decomposition theorem for a half-open annulus by covering the half-open annulus with an open annulus and a closed annulus, and then gluing together the decompositions given by Theorem 12.3.1 and Theorem 12.3.1. Similarly, one can obtain decomposition theorems on more exotic subspaces of the affine line by gluing; the reader knowledgeable enough to be interested in such statements should at this point have no trouble formulating and deriving them. See also Remark 12.4.4.

12.6 Modules solvable at a boundary

One of the most important special cases of our decomposition theorems occurs in the following setting, which occurs frequently in applications.

Definition 12.6.1. Let M be a finite differential module on the half-open annulus with closed inner radius α and open outer radius β. We say M is *solvable at* β if $R(M_\rho) \to \beta$ as $\rho \to \beta^-$, or equivalently if $IR(M_\rho) \to 1$ as $\rho \to \beta^-$. (One can also make a similar definition with the roles of the inner and outer radius reversed; we will not refer to that definition here, but see Remark 27.2.3.)

Lemma 12.6.2. *Let M be a finite differential module on the half-open annulus with closed inner radius α and open outer radius β, which is solvable at β. There exist $b_1 \geq \cdots \geq b_n \in [0, \infty)$ such that for $\rho \in [\alpha, \beta)$ sufficiently close to β, the intrinsic subsidiary radii of M_ρ are $(\rho/\beta)^{b_1}, \ldots, (\rho/\beta)^{b_n}$. Moreover, if $i = n$ or $b_i > b_{i+1}$, then $b_1 + \cdots + b_i \in \mathbb{Z}$.*

Proof For $r \to (-\log \beta)^+$, $F_i(M, r) - ir$ is a convex function (by Theorem 11.3.2(e)) with slopes in a discrete subset of $\mathbb{R}$ (by Theorem 11.3.2(a), (b)). Moreover, it is nonnegative and its limit is 0; this implies that the slopes are all nonnegative. Hence these slopes must eventually stabilize; that is, each $f_i(M, r)$ is linear in a neighborhood of $-\log \beta$. This provides the existence of $b_1, \ldots, b_n$; by Theorem 11.3.2(b), if $i = n$ or $b_i > b_{i+1}$, then $b_1 + \cdots + b_i \in \mathbb{Z}$. □

Definition 12.6.3. Let M be a finite differential module on the half-open annulus with closed inner radius α and open outer radius β, which is solvable at β. The quantities $b_1, \ldots, b_n$ defined by Lemma 12.6.2 will be called the *differential slopes* of M at β. (They are also called *ramification numbers*; the reason for this will become clear when we consider quasiconstant differential modules in Chapter 19. See specifically Theorem 19.4.1.)

We now recover a decomposition theorem of Christol and Mebkhout; see the notes for further discussion. We will see several applications of this result later in the book.

Theorem 12.6.4 (Christol–Mebkhout). *Let M be a finite differential module on the half-open annulus with closed inner radius α and open outer radius β. Suppose that M is solvable at β. Then for any sufficiently large $\gamma \in [\alpha, \beta)$, the restriction of M to the open annulus with inner radius γ and outer radius β splits uniquely as a direct sum $\bigoplus_{b \in [0, \infty)} M_b$ such that for each $b \in [0, \infty)$, for all $\rho \in [\gamma, \beta)$, the intrinsic subsidiary radii of $M_{b,\rho}$ are all equal to $(\rho/\beta)^b$.*

Proof By Lemma 12.6.2, we are in a case where Theorem 12.5.2 may be applied. □

Remark 12.6.5. For some differential modules for which one has fairly explicit series expansions for local horizontal sections, one may be able to establish solvability at a boundary by explicit estimates. (A related strategy appears in Example 9.6.2.) However, it is more common for solvability to be established by proving the existence of a Frobenius structure; this notion will be introduced in Chapter 17.

Remark 12.6.6. Beware that in Theorem 12.6.4, if M is the restriction of a differential module over the closed annulus with inner radius α and outer radius β, or over the ring of analytic elements on the open annulus, the decomposition of M given by the theorem need not descend back to this original structure. Compare Remark 12.4.5.

12.7 Solvable modules of rank 1

We now give a partial classification of modules of rank 1 on an open annulus which are solvable at a boundary. Throughout this section, assume $p > 0$.

Definition 12.7.1. Fix a coherent system of p-power roots of 1 in K^{alg}; that is, for each h, the chosen p^{h+1}-st root of unity should be a p-th root of the chosen p^h-th root of unity. For $c \in \mathfrak{o}_K^\times$ and n a positive integer, write $n = mp^h$ with m coprime to p, and let $M_{n,c}$ be the pullback of the module M_h defined in Example 9.9.3 along the map $t \mapsto ct^{-m}$ (using the chosen p^{h+1}-st root of unity); this module is solvable at 1.

Theorem 12.7.2. *Assume that K contains the p^h-th roots of unity for all $h \leq \log_p b$. Let M denote a finite differential module of rank 1 on a half-open annulus with open outer radius 1, which is solvable at 1 with differential slope b. Then there exist $c_1, \ldots, c_b \in \{0\} \cup \mathfrak{o}_K^\times$ and nonnegative integers $j_1, \ldots, j_b$ such that*

$$M \otimes (\varphi^{j_1})^*(M_{1,c_1}) \otimes \cdots \otimes (\varphi^{j_b})^*(M_{b,c_b})$$

has differential slope 0.

We will refine this statement a bit later by eliminating the Frobenius pullbacks (Theorem 17.1.6).

Proof By Theorem 11.3.2(b), $b \in \mathbb{Z}$. It thus makes sense to proceed by induction on b; we may assume $b > 0$. Pick $0 < \alpha < \beta < 1$ such that for some nonnegative integer j,

$$p^{-p^{-j+1}/(p-1)} < IR(M_\alpha) = \alpha^b < IR(M_\beta) = \beta^b < p^{-p^j/(p-1)}.$$

By Theorem 10.4.4, M admits a j-fold Frobenius antecedent N over the ring $K\langle \alpha^{p^j}/t^{p^j}, t^{p^j}/\beta^{p^j} \rangle$. Pick a generator v of N and put $D(v) = nv$. Then $|n|_{\rho^{p^j}} = p^{-1/(p-1)} \rho^{-p^j(b+1)}$ for $\rho \in [\alpha, \beta]$, so in this range n is dominated by a term of the form $n_{-b-1}(t^{p^j})^{-b-1}$ with $|n_{b-1}| = p^{-1/(p-1)}$. Let ζ be the chosen p^{h+1}-st root of unity, write $b = p^h m$ with m coprime to p, and take

$$c_b = \frac{n_{-b-1}}{m(\zeta - 1)};$$

then

$$IR((N \otimes M_{b,c_b})_{\rho^{p^j}}) > \rho^{p^j b} \qquad (\rho \in [\alpha, \beta]).$$

We may use the log-concavity of the intrinsic radius (Theorem 11.3.2(e)) to deduce that the differential slope of $N \otimes M_{b,c_b}$ is strictly less than b. Thus the induction hypothesis gives the desired result. $\qquad\square$

Corollary 12.7.3. *Let M denote a finite differential module of rank 1 on a half-open annulus with open outer radius 1, which is solvable at 1 with differential slope $b > 0$. If $M^{\otimes p}$ has differential slope 0, then b is not divisible by p.*

Proof There is no harm in enlarging K, so that the hypotheses of Theorem 12.7.2 are satisfied; we may thus assume $M \cong (\varphi^{j_1})^*(M_{1,c_1}) \otimes \cdots \otimes (\varphi^{j_b})^*(M_{b,c_b})$. By Remark 9.9.4 and Theorem 10.4.2, for $c_b \in \mathfrak{o}_K^\times$:

- if b is coprime to p, then $M_{b,c_b}^{\otimes p}$ has differential slope 0;
- if b is divisible by p, then $M_{b,c_b}^{\otimes p}$ has differential slope b/p.

Moreover, these differential slopes are preserved under Frobenius pullback by Theorem 10.4.2. This implies the desired result as follows. If b were divisible by p, then $(\varphi^{j_b})^*(M_{b,c_b})^{\otimes p}$ would have differential slope b/p, whereas $(\varphi^{j_1})^*(M_{1,c_1})^{\otimes p} \otimes \cdots \otimes (\varphi^{j_{b-1}})^*(M_{b-1,c_{b-1}})^{\otimes p}$ would have differential slope strictly less than b/p. By Lemma 9.4.6(c), $M^{\otimes p}$ would have differential slope $b/p > 0$, contradiction. $\qquad\square$

12.8 Clean modules

It is reasonable to ask whether the refined strong decomposition theorem (Theorem 10.6.7) admits an analogue over a disc or annulus. The following discussion provides one possible answer to this question, though not a definitive one.

Definition 12.8.1. Let M be a finite differential module of rank n over the ring $K[\![\alpha/t, t/\beta]\!]_{an}$. By a *spectral decomposition* (resp. a *refined decomposition*) of M, we will mean a direct sum decomposition $\bigoplus_i M_i$ of M such that for each $\rho \in [\alpha, \beta]$, the decomposition $M_\rho \cong \bigoplus_i M_{i,\rho}$ is the minimal decomposition given by Theorem 10.6.2 (resp. Theorem 10.6.7).

Lemma 12.8.2. *Let M be a finite differential module of rank n over $K[\![\alpha/t, t/\beta]\!]_{an}$. Suppose that $F_n(M, r)$ is affine on $[-\log\beta, -\log\alpha]$. Let $j \in \{0, \ldots, n\}$ be the smallest integer such that $f_i(M, r)$ is affine on $[-\log\beta, -\log\alpha]$ for all $i > j$. If $0 < j < n$, then $f_j(M, r) > f_{j+1}(M, r)$ for all $r \in (-\log\beta, -\log\alpha)$; if $j = n$, then $f_j(M, r) > r$ for all $r \in (-\log\beta, -\log\alpha)$.*

Proof There is nothing to check if $j = 0$, and $j = 1$ is impossible, so we may assume $j > 1$. Note that $F_j(M, r) = F_n(M, r) - f_{j+1}(M, r) - \cdots - f_n(M, r)$ is affine by hypothesis, and $F_{j-1}(M, r)$ is convex by Theorem 11.3.2(e), so $f_j(M, r) = F_j(M, r) - F_{j-1}(M, r)$ is concave. Moreover, $f_j(M, r)$ is bounded below by the affine function $f_{j+1}(M, r)$ if $j < n$, or by the affine function r if $j = n$. Hence if this inequality becomes an equality for any one interior point of $[-\log\beta, -\log\alpha]$, then it must hold identically, contrary to the choice of j. This proves the claim. □

Definition 12.8.3. Let M be a finite differential module of rank n over the ring $K[\![\alpha/t, t/\beta]\!]_{an}$. We say that M is *clean* if $F_n(M, r)$ and $F_{n^2}(M^\vee \otimes M, r)$ are both affine on $[-\log\beta, -\log\alpha]$.

Theorem 12.8.4. *Let M be a finite clean differential module of rank n over $K[\![\alpha/t, t/\beta]\!]_{an}$.*

(a) *For $i \in \{1, \ldots, n\}$, $F_i(M, r)$ is affine (and so $f_i(M, r)$ is affine).*
(b) *For $i \in \{1, \ldots, n-1\}$, we either have $f_i(M, r) = f_{i+1}(M, r)$ for all $r \in (-\log\beta, -\log\alpha)$ or $f_i(M, r) > f_{i+1}(M, r)$ for all $r \in (-\log\beta, -\log\alpha)$. In addition, either $f_n(M, r) = r$ for all $r \in (-\log\beta, -\log\alpha)$ or $f_i(M, r) > r$ for all $r \in (-\log\beta, -\log\alpha)$.*
(c) *There exists a spectral decomposition of M.*

Proof To check (a), we may replace K by a field with algebraically closed residue field and value group $\mathbb{R}$. Take j as in Lemma 12.8.2 and suppose by way of contradiction that $j > 0$ (which in turn forces $j > 1$). Pick a point $r_0 \in (-\log\beta, -\log\alpha)$ at which $f_j(M, r)$ fails to be affine, or equivalently where $F_{j-1}(M, r)$ fails to be affine; we may rescale to reduce to the case $r_0 = 0$. Then by Theorem 11.3.2(c), we must then have $s_{\overline{\mu}, j-1}(M) < 0$ for some $\overline{\mu} \in \kappa_K^\times$.

On the other hand, $f_j(M, 0) > 0$ if $j > 0$ (by Lemma 12.8.2) and hence

$s_{\overline{\mu},j}(M) = 0$ (by Theorem 11.3.2(c) again). Similarly, if $h \in \{0, \ldots, n^2\}$ denotes the smallest index for which $f_i(M^\vee \otimes M, r) = r$ identically for $i > h$, then $f_h(M^\vee \otimes M, 0) > 0$ if $h > 0$ and $s_{\overline{\mu},h}(M^\vee \otimes M) = 0$.

Choose a lift μ of $\overline{\mu}$ in K, and put $N = T_\mu^*(M)$ as a differential module over $K[\![t]\!]_{\text{an}}$. Note that $F_h(N^\vee \otimes N, r)$ is constant for r in a right neighborhood of 0, $f_h(N^\vee \otimes N, 0) > 0$ if $h > 0$, and $f_i(N^\vee \otimes N, r) = r$ for $i > h$. By Theorem 12.4.1, there exists a direct sum decomposition $P_0 \oplus P_1$ of $N^\vee \otimes N$ such that for each $\rho \in (0, 1)$, $P_{0,\rho}$ accounts for the first h subsidiary radii of $(N^\vee \otimes N)_\rho$. By Theorem 9.6.1, P_1 restricts to a trivial differential module over the open unit disc.

For any $\rho \in (0, 1)$, any direct sum decomposition of N_ρ is defined by projectors which are horizontal elements of $(N^\vee \otimes N)_\rho$. For ρ sufficiently close to 1, the subsidiary radii of $P_{0,\rho}$ are all strictly less than ρ (by Lemma 12.8.2 again), so the projectors must belong to $P_{1,\rho}$. Since P_1 is trivial on the open unit disc, the projectors must extend to horizontal elements of $N^\vee \otimes N$ over the open unit disc. That is, they define a direct sum decomposition of N over the open unit disc.

It follows that over the open unit disc, N admits a direct sum decomposition $\bigoplus_i N_i$ in which for each i and each $\rho \in (0, 1)$ sufficiently close to 1, $N_{i,\rho}$ has only a single subsidiary radius. Namely, given any decomposition not satisfying this condition, we can apply Theorem 10.6.2 and then the previous paragraph to obtain a finer decomposition. (This can only be repeated as many times as the rank of N.)

Let S be the set of indices i for which $\lim_{r \to 0^+} f_1(N_i, r) \geq f_j(M, 0)$. Since $f_j(M, 0) > 0$, in a neighborhood of $r = 0$, $f_1(N_i, r)$ is affine and nonincreasing for each $i \in S$ by Theorem 11.3.2(a),(d), and $F_j(N, r)$ is a positive linear combination of these functions. However, the right slope of $F_j(N, r)$ at $r = 0$ is $s_{\overline{\mu},j}(M) = 0$, so $f_1(N_i, r)$ must be constant in a neighborhood of $r = 0$ for each $i \in S$. In a neighborhood of $r = 0$, $F_{j-1}(N, r)$ is a nonnegative linear combination of the $f_1(N_i, r)$ for $i \in S$, so it also has right slope 0 at $r = 0$. But this contradicts the fact that $s_{\overline{\mu},j-1}(M) < 0$.

This contradiction leads to the conclusion that $j = 0$, which implies (a). Given (a), we deduce (b) from the fact that $f_i(M, r)$ and $f_{i+1}(M, r)$ are affine functions on $[-\log\beta, -\log\alpha]$ satisfying $f_i(M, r) \geq f_{i+1}(M, r)$ (or in case $i = n$, the same argument with $f_{i+1}(M, r)$ replaced by r). Given (a) and (b), the hypothesis of (c) implies that $f_i(M, r) > f_{i+1}(M, r)$ for all $r \in [-\log\beta, -\log\alpha]$, so the claim follows from Theorem 12.4.3. □

For refined decompositions, we obtain a slightly weaker result where we lose control of the endpoints of the interval.

Lemma 12.8.5. *Let M be a finite clean differential module of rank n over $K[\![\alpha/t, t/\beta]\!]_{\mathrm{an}}$. Suppose that for some $\rho \in (\alpha, \beta)$, M_ρ admits a refined decomposition $\bigoplus_i V_i$ as per Theorem 10.6.7. Then for each i for which $IR(V_i) < 1$, there exist a finite extension K' of K and a differential module N_i over $K'[\![\alpha/t, t/\beta]\!]_{\mathrm{an}}$ of rank 1 such that*

$$IR(V_i^\vee \otimes N_{i,\rho}) > IR(V_i).$$

Moreover, if κ_K is perfect, we can take $K' = K$.

Proof As in the proof of Theorem 10.6.7, we may use Frobenius antecedents (Theorem 10.4.4) and Frobenius descendants (Theorem 10.5.1) to reduce to the case where $IR(V) < \omega\rho$. Apply Theorem 5.4.2 to choose a cyclic vector of $M \otimes_{K[\![\alpha/t, t/\beta]\!]_{\mathrm{an}}} \mathrm{Frac}(K[\![\alpha/t, t/\beta]\!]_{\mathrm{an}})$, then set notation as in the proof of Theorem 6.8.2 with $F = F' = F_\rho$. By comparing the information from Corollary 6.5.4 with the assumption that M is clean, we see that for each root λ of Q_i, the function $r \mapsto -\log|\lambda|_{e^{-r}}$ is affine in a neighborhood of $-\log\rho$. By considering Newton polygons, we deduce that there exist $\mu \in K'$ and $j \in \mathbb{Z}$ such that $|\lambda - \mu t^{j-1}|_\rho < |\lambda|_\rho$; moreover, $j \neq 0$ because $IR(V_i) < 1$. Let N_i be the differential module over $K[\![\alpha/t, t/\beta]\!]_{\mathrm{an}}$ on a single generator v satisfying $D(v) = j^{-1}t^j v$; then N has the desired property. $\qquad\square$

Theorem 12.8.6. *Let M be a finite clean differential module of rank n over $K[\![\alpha/t, t/\beta]\!]_{\mathrm{an}}$. Then for any α', β' with $\alpha < \alpha' \leq \beta' < \beta$, there exist a finite tamely extension K' of K and a positive integer m not divisible by p such that for the map $K[\![\alpha/t, t/\beta]\!]_{\mathrm{an}} \to K'[\![(\alpha')^{1/m}/t, t/(\beta')^{1/m}]\!]_{\mathrm{an}}$ given by $t \mapsto t^m$, $M \otimes_{K[\![\alpha/t, t/\beta]\!]_{\mathrm{an}}} K'[\![(\alpha')^{1/m}/t, t/(\beta')^{1/m}]\!]_{\mathrm{an}}$ admits a refined decomposition.*

Proof It suffices to prove the claim locally around a single value of $\rho \in (\alpha, \beta)$. Put $r_0 = -\log\rho$. By Theorem 10.6.7, there exists a finite tamely ramified extension E of F_ρ such that $M \otimes_{K[\![\alpha/t, t/\beta]\!]_{\mathrm{an}}} E$ admits a refined strong decomposition. Since the same remains true after enlarging E, by Proposition 3.3.6, we can choose K' and m so that $E = F_\rho \otimes_{K[t]} K'[t^{1/m}]$ works. To simplify notation hereafter, we may assume that $K' = K$ and $m = 1$ (as enlarging K and replacing t as the series parameter by $t^{1/m}$ do not disturb the cleanness hypothesis); we will then prove the claim for these values of K' and m. By Theorem 12.8.4, we may also assume that M admits a one-term spectral decomposition.

Let $\bigoplus_i V_i$ be the refined strong decomposition of M_ρ and construct differential modules N_i of rank 1 corresponding to V_i as in Lemma 12.8.5. For some neighborhood $[\alpha', \beta']$ of ρ, $(N_i^\vee \otimes M) \otimes_{K[\![\alpha/t, t/\beta]\!]_{\mathrm{an}}} K[\![\alpha'/t, t/\beta']\!]_{\mathrm{an}}$ admits a spectral decomposition by Theorem 12.4.2; contracting this decomposition with N_i exposes a summand M_i of $M \otimes_{K[\![\alpha/t, t/\beta]\!]_{\mathrm{an}}} K'[\![\alpha'/t, t/\beta']\!]_{\mathrm{an}}$ whose base extension to $F_\rho \otimes_K K'$ is equivalent to $V_i \otimes_K K'$.

This gives the desired result except that we have replaced K with a finite extension K' which is not necessarily tame; indeed, it is obtained by adjoining p-power roots of some elements of K so as to achieve some purely inseparable extension of κ_K. But this means that the resulting refined decomposition admits Galois descent for the Galois closure of K' over K, so we get the decomposition of the originally stated form. $\qquad\square$

Notes

Our results on modules solvable at a boundary are originally due to Christol and Mebkhout [99, 100]. In particular, Lemma 12.6.2 for the generic radius is [99, Théorème 4.2.1], and the decomposition theorem (which implies Lemma 12.6.2 in general) is [100, Corollaire 2.4–1]. However, the proof technique of Christol and Mebkhout is significantly different from ours: they construct the desired decomposition by exhibiting convergent sequences for a certain topology on the ring of differential operators. This does not appear to give quantitative results; that is, one does not control the range over which the decomposition occurs, although we are not sure whether this is an intrinsic limitation of the method. (Keep in mind that the approach here crucially uses our method of Frobenius descendants, which was not available when [99, 100] were written.)

Note also that Christol and Mebkhout work directly with a differential module on an open annulus as a ring-theoretic object; this requires a freeness result of the following form. If K is spherically complete, any coherent locally free module on the half-open annulus with closed inner radius α and open outer radius β is induced by a finite free module over the ring $\bigcap_{\rho\in[\alpha,\beta)} K\langle\alpha/t, t/\rho\rangle$. (That is, any coherent locally free sheaf on this annulus is freely generated by global sections.) See the notes for Chapter 8 for further discussion.

A partial extension of the work of Christol and Mebkhout can be found in the paper of Pons [339]. However, in the absence of a theory of Frobenius descendants, Pons was thus forced to impose somewhat awkward hypotheses in order to avoid the exceptional value $\omega\rho$ for the generic radius of convergence.

For the attribution of Theorem 12.7.2, see the notes for Chapter 17.

The notion of a clean differential module is original, motivated partly by discussions with Liang Xiao. It is a first attempt to model in p-adic differential theory a higher-rank analogue of Kato's notion of cleanness for a rank-1 étale sheaf [222]. Similar considerations in residual characteristic 0 appear in [250]. Lemma 12.8.5 is loosely based on [413, §5].

Similar results can be found in an unpublished manuscript of Christol [93].

Exercises

12.1   Prove the analogue of Lemma 12.2.4 in which M is only required to be locally free (in the sense of Remark 5.3.4).

12.2   Let L be a complete extension of K. Prove that for any $\beta > 0$, within the completion of $L(t)$ for the β-Gauss norm,

$$F_\beta \cap L\langle t/\beta \rangle = K\langle t/\beta \rangle, \qquad F_\beta \cap L\langle \beta/t, t/\beta \rangle = K\langle \beta/t, t/\beta \rangle.$$

(Hint: use Remark 8.5.5.)

12.3   Use the previous exercise to deduce that for any $\rho \in [\alpha, \beta]$, within the completion of $L(t)$ for the ρ-Gauss norm,

$$F_\rho \cap L\langle \alpha/t, t/\beta \rangle = K\langle \alpha/t, t/\beta \rangle.$$

12.4   (a)   Prove that for $x \in K[\![t]\!]_0$ nonzero, the following conditions are equivalent.

   (i) The element x is a unit.

   (ii) The function $v_r(x)$ is constant in a neighborhood of $r = 0$.

   (iii) The function $v_r(x)$ is constant for all $r \in [0, \infty)$.

(b)   Prove that for $x \in \bigcup_{\alpha \in (0,1)} K\langle \alpha/t, t]\!]_0$ nonzero, x is a unit if and only if the function $r \mapsto v_r(x)$ is affine in some neighborhood of $r = 0$. In fact, this always happens if K is discretely valued, since the Newton polygon of any $x \in K\langle \alpha/t, t]\!]_0$ has finite width in this case (see the exercises for Chapter 8). Hence in this case, the ring $\bigcup_{\alpha \in (0,1)} K\langle \alpha/t, t]\!]_0$ is a field. We will encounter this field again under the name of the *bounded Robba ring*; see Definition 15.1.2.

13

p-adic exponents

In this chapter, we study *p*-adic differential modules in a situation left untreated by our preceding analysis, namely when the intrinsic generic radius of convergence is equal to 1 everywhere. (This condition is commonly called the *Robba condition*.) This setting is loosely analogous to the study of regular singularities of formal meromorphic differential modules considered in Chapter 7; in particular, there is a meaningful theory of *p-adic exponents* in this setting.

However, some basic considerations indicate that *p*-adic exponents must necessarily be more complicated than the exponents considered in Chapter 7. For instance, the *p*-adic analogue of the Fuchs theorem (Theorem 7.3.8) can fail unless we impose a further condition: the difference between exponents must not be *p*-adic Liouville numbers.

With this in mind, we may proceed to construct *p*-adic exponents for differential modules satisfying the Robba condition. Such modules carry an action of the group of *p*-power roots of unity via Taylor series; under favorable circumstances, the module splits into isotypical components for the characters of this group. We may identify these characters with elements of $\mathbb{Z}_p$, and these give the exponents.

Throughout this chapter, retain Notation 10.0.1 and assume $p > 0$.

13.1 *p*-adic Liouville numbers

Definition 13.1.1. For $\lambda \in K$, the *type* of λ, denoted type(λ), is the radius of convergence of the power series

$$\sum_{m=0, m \neq \lambda}^{\infty} \frac{t^m}{\lambda - m}.$$

This cannot exceed 1, as there are infinitely many m for which $|\lambda - m| = \max\{|\lambda|, 1\}$. (For instance, we may take $m \in p\mathbb{Z}_p$ if $|\lambda| \geq 1$ and $m \in 1 + p\mathbb{Z}_p$ if $|\lambda| < 1$.) Moreover, if $\lambda \notin \mathbb{Z}_p$, then $|\lambda - m|$ is bounded below, so $\mathrm{type}(\lambda) = 1$. We will thus focus on the attention on the case $\lambda \in \mathbb{Z}_p$.

Definition 13.1.2. We say that $\lambda \in K$ is a *p-adic Liouville number* if either λ or $-\lambda$ has type less than 1, and a *p-adic non-Liouville number* otherwise. Note that it is necessary to consider both λ and $-\lambda$ because they may have different types (exercise).

The following alternate characterization of the type may be helpful.

Definition 13.1.3. For $\lambda \in \mathbb{Z}_p$, let $\lambda^{(m)}$ be the unique integer in $\{0, \ldots, p^m - 1\}$ congruent to λ modulo p^m.

Proposition 13.1.4. *For $\lambda \in \mathbb{Z}_p$ not a nonnegative integer,*

$$-\frac{1}{\log_p \mathrm{type}(\lambda)} = \liminf_{m \to \infty} \frac{\lambda^{(m)}}{m}. \tag{13.1.4.1}$$

In particular, λ has type 1 if and only if $\lambda^{(m)}/m \to \infty$ as $m \to \infty$.

Proof It suffices to check that for $0 < \eta < 1$,

$$\limsup_{m \to \infty}(m + \lambda^{(m)} \log_p \eta) = -\infty \tag{13.1.4.2}$$

when $\eta < \mathrm{type}(\lambda)$ and

$$\limsup_{m \to \infty}(m + \lambda^{(m)} \log_p \eta) = \infty \tag{13.1.4.3}$$

when $\eta > \mathrm{type}(\lambda)$. Namely, (13.1.4.2) implies $m + \lambda^{(m)} \log_p \eta \leq 0$ for all large m, so $\liminf_{m \to \infty} \frac{\lambda^{(m)}}{m} \geq -1/(\log_p \eta)$, whereas (13.1.4.3) implies $m + \lambda^{(m)} \log_p \eta \geq 0$ for infinitely many m, so $\liminf_{m \to \infty} \frac{\lambda^{(m)}}{m} \leq -1/(\log_p \eta)$.

Suppose first that $\mathrm{type}(\lambda) > \eta > 0$; then as $s \to \infty$, $\eta^s/|\lambda - s| \to 0$ or equivalently $v_p(\lambda - s) + s \log_p \eta \to -\infty$. (Here v_p denotes the renormalized valuation with $v_p(p) = 1$.) Since λ is not a nonnegative integer, $\lambda^{(m)} \to \infty$ as $m \to \infty$, so

$$v_p(\lambda - \lambda^{(m)}) + \lambda^{(m)} \log_p \eta \to -\infty.$$

The left side does not increase if we replace $v_p(\lambda - \lambda^{(m)})$ by m, so we may deduce (13.1.4.2).

Suppose next that $\mathrm{type}(\lambda) < \eta < 1$; then we may choose a sequence s_j such that as $j \to \infty$, $v_p(\lambda - s_j) + s_j \log_p \eta \to \infty$. Put $m_j = v_p(\lambda - s_j)$, so that $s_j \geq \lambda^{(m_j)}$. Then

$$m_j + \lambda^{(m_j)} \log_p \eta \to \infty,$$

yielding (13.1.4.3). □

The alternate characterization is convenient for such verifications as the fact that rational numbers are non-Liouville (exercise), or this stronger result [149, Proposition VI.1.1], whose proof we omit.

Proposition 13.1.5. *Any element of $\mathbb{Z}_p$ algebraic over $\mathbb{Q}$ is non-Liouville.*

We will later encounter the p-adic Liouville property in yet another apparently different form. (See the exercises for an alternate proof of this lemma.)

Lemma 13.1.6. *For λ not a nonnegative integer, in $K[\![x]\!]$ we have*

$$\sum_{m=0}^{\infty} \frac{x^m}{\lambda(1-\lambda)(2-\lambda)\cdots(m-\lambda)} = e^x \sum_{m=0}^{\infty} \frac{(-x)^m}{m!} \frac{1}{\lambda - m}.$$

Proof The coefficient of x^m on the right side is a sum of the form $\sum_{i=0}^{m} c_i/(i-\lambda)$ for some $c_i \in \mathbb{Q}$. It is thus a rational function of λ of the form $P(\lambda)/(\lambda(1-\lambda)\cdots(m-\lambda))$, where P has coefficients in $\mathbb{Q}$ and degree at most m. To check that $P(\lambda) = 1$ identically, we need only check this for $\lambda = 0, \ldots, m$.

In other words, to check the original identity, it suffices to check after multiplying both sides by $\lambda - i$ and evaluating at $\lambda = i$, for each nonnegative integer i. On the left side, we obtain

$$\sum_{m=i}^{\infty} \frac{-x^m}{(-1)^{i-1} i! (m-i)!}.$$

On the right side, we obtain

$$e^x \frac{(-x)^i}{i!},$$

which is the same thing. □

Corollary 13.1.7. *If $\lambda \in K$ is not a nonnegative integer, and* $\text{type}(\lambda) = 1$, *then the series*

$$\sum_{m=0}^{\infty} \frac{x^m}{\lambda(1-\lambda)(2-\lambda)\cdots(m-\lambda)}$$

has radius of convergence at least $p^{-1/(p-1)}$.

13.2 *p*-adic regular singularities

We now consider a *p*-adic analogue of Theorem 7.3.8. Recall that in this theorem, the eigenvalues of the constant matrix N_0 are assumed to be *weakly prepared*, meaning that no two of them differ by a nonzero integer. Here, we will need to impose a more restrictive condition.

Definition 13.2.1. We say that a finite set is *p-adic non-Liouville* if its elements are *p*-adic non-Liouville numbers. We say the set has *p-adic non-Liouville differences* if the difference between any two elements of the set is a *p*-adic non-Liouville number.

Theorem 13.2.2 (*p*-adic Fuchs theorem for discs). *Let $N = \sum_{i=0}^{\infty} N_i t^i$ be an $n \times n$ matrix over $K\langle t/\beta \rangle$ for some $\beta > 0$. Assume that N_0 has weakly prepared eigenvalues with p-adic non-Liouville differences. Then there exists $\gamma > 0$ such that the fundamental solution matrix for N, which exists and is unique by Proposition 7.3.6, has entries in $K\langle t/\gamma \rangle$. (The same conclusion then holds for the inverse of the fundamental solution matrix.)*

Proof Recall that the fundamental solution matrix U is computed by the recursion (7.3.6.1):

$$N_0 U_i - U_i N_0 + i U_i = - \sum_{j=1}^{i} N_j U_{i-j} \qquad (i > 0).$$

There is no harm in enlarging K to include the eigenvalues $\lambda_1, \ldots, \lambda_n$ of N_0. By Lemma 7.3.5, the map $X \mapsto i + N_0 X - X N_0$ has eigenvalues $\lambda_g - \lambda_h + i$ for $g, h \in \{1, \ldots, n\}$. View the operator $X \mapsto N_0 X - X N_0$ as a linear transformation T on the space V of $n \times n$ matrices over K. The matrix of action of $(i + T)^{-1}$ on some basis of V can be written as a matrix of cofactors of $i + T$ divided by the determinant of $i + T$. If we fix the basis of V, then the entries of $i + T$ are bounded independently of i, as are the cofactors; we thus obtain a bound of the form

$$|U_i|\beta^i \leq c|N|_\beta \max_{j<i}\{|U_j|\beta^j\} \prod_{\substack{g,h=1}}^{n} |\lambda_g - \lambda_h + i|^{-1}.$$

for some $c > 0$ not depending on i. (For a somewhat more careful argument in this vein, see Proposition 18.1.1.)

Thus to conclude the theorem, it suffices to verify that for each $g, h \in \{1, \ldots, n\}$, the quantity $\lambda = \lambda_g - \lambda_h$ has the property that

$$\prod_{i=1}^{m} \max\{1, |\lambda - i|^{-1}\}$$

grows at worst exponentially. If $\lambda \notin \mathbb{Z}_p$, then $|\lambda - i|^{-1}$ is bounded above and the claim is verified. Otherwise, Corollary 13.1.7 and the hypothesis that λ is a p-adic non-Liouville number give the desired estimate. $\qquad\square$

By a slight modification of the argument (which we omit), one may obtain the following result of Clark [102, Theorem 3], which may be viewed as a p-adic analogue of Corollary 7.3.11.

Theorem 13.2.3 (Clark). *Let M be a finite differential module over $K\langle t/\beta\rangle$ for the derivation $t\frac{d}{dt}$, with a regular singularity at 0 whose exponents have p-adic non-Liouville differences. Then for any $x \in M$ and $y \in M \otimes_{K\langle t/\beta\rangle} K[\![t]\!]$ such that $D(y) = x$, $y \in M \otimes_{K\langle t/\beta\rangle} K\langle t/\rho\rangle$ for some $\rho > 0$.*

Remark 13.2.4. The conclusion of Theorem 13.2.2 remains true, with the same proof, if it is assumed only that the pairwise differences between eigenvalues of N_0 all have type greater than 0. However, it is possible for the conclusion to fail otherwise. To construct a counterexample, put $a = b = 1$, choose $c \in \mathbb{Z}_p$ with type$(-c) = 0$, and consider the differential module associated to the hypergeometric differential equation (0.3.2.2). Then the eigenvalues of N_0 are 0 and c, whose difference has type 0; correspondingly, the hypergeometric series (0.3.2.1), i.e.,

$$F(a, b; c; z) = \sum_{i=0}^{\infty} \frac{a(a+1)\cdots(a+i-1)b(b+1)\cdots(b+i-1)}{c(c+1)\cdots(c+i-1)i!}z^i,$$

gives rise to a formal horizontal section with radius of convergence 0.

13.3 The Robba condition

Given a finite differential module on an annulus for the derivation $t\frac{d}{dt}$, we would like to be able to tell whether it extends over a disc with a regular singularity at $t = 0$. It turns out that when the exponents of that singularity are constrained to lie in $\mathbb{Z}_p$ (e.g., if they are rational numbers with denominators prime to p), one gets a strong necessary condition from the generic radius of convergence.

Definition 13.3.1. Let M be a finite differential module on the disc/annulus $|t| \in I$, for I an interval. We say that M satisfies the *Robba condition* if $IR(M_\rho) = 1$ for all nonzero $\rho \in I$.

Proposition 13.3.2. *Let M be a finite differential module on the open disc of radius β for the derivation $t\frac{d}{dt}$. Suppose that M satisfies the Robba condition in some annulus. Then the eigenvalues of the action of D on M/tM belong to $\mathbb{Z}_p$.*

Proof Let $N = \sum_{i=0}^{\infty} N_i t^i$ be the matrix of action of D on some basis of M. Suppose N_0 has an eigenvalue $\lambda \notin \mathbb{Z}_p$; there is no harm in enlarging K to force $\lambda \in K$. Choose $v \in M$ such that the image of v in M/tM is a nonzero eigenvector of N_0 of eigenvalue λ. Let D' be the derivation corresponding to $d' = \frac{d}{dt}$ instead of $t\frac{d}{dt}$. Then with notation as in Example 9.5.2, for any $\rho < \beta$,

$$\max\left\{ \limsup_{s \to \infty} |(D')^s v|_\rho^{1/s}, |d'|_{\mathrm{sp},F_\rho} \right\} \geq |D'|_{\mathrm{sp},V_\lambda,\rho} > p^{-1/(p-1)}\rho,$$

so $IR(M_\rho) < 1$ by Lemma 6.2.5. $\qquad\square$

We will establish a partial converse to Proposition 13.3.2 later (Theorem 13.7.1). In the interim, we mention the following result.

Proposition 13.3.3. *Let M be a finite differential module on the open disc of radius β for the derivation $t\frac{d}{dt}$. Suppose that the matrix of action N_0 of D on some basis of M has entries in K. Then M satisfies the Robba condition if and only if N_0 has eigenvalues in $\mathbb{Z}_p$.*

Proof Exercise, or see [149, Corollary IV.7.6]. $\qquad\square$

13.4 Abstract *p*-adic exponents

In the previous section, we considered a finite differential module on an annulus for the derivation $t\frac{d}{dt}$, and saw that the Robba condition was necessary for extending the module over a disc with a regular singularity at $t = 0$. Moreover, the exponents of that regular singularity must belong to $\mathbb{Z}_p$. We may then ask whether it is possible to identify these exponents by looking only at the original annulus.

The answer to this question is complicated by the fact that the exponents are only well-defined as elements of the quotient $\mathbb{Z}_p/\mathbb{Z}$. This means we cannot hope to identify them using purely *p*-adic considerations; in fact, we must use *archimedean* considerations to identify them. Here are those considerations.

Definition 13.4.1. We will say that $A, B \in \mathbb{Z}_p^n$ are *equivalent* (or sometimes *strongly equivalent*) if there exists a permutation σ of $\{1, \ldots, n\}$ such that $A_i - B_{\sigma(i)} \in \mathbb{Z}$ for $i = 1, \ldots, n$. This is evidently an equivalence relation.

Definition 13.4.2. We say that $A, B \in \mathbb{Z}_p^n$ are *weakly equivalent* if there exists a constant $c > 0$, a sequence $\sigma_1, \sigma_2, \ldots$ of permutations of $\{1, \ldots, n\}$, and signs $\epsilon_{i,m} \in \{\pm 1\}$ such that

$$(\epsilon_{i,m}(A_i - B_{\sigma_m(i)}))^{(m)} \leq cm \qquad (i = 1, \ldots, n; m = 1, 2, \ldots).$$

In other words, $A_i - B_{\sigma_m(i)}$ has a representative modulo p^m of size at most cm. Again, this is clearly an equivalence relation, and equivalence implies weak equivalence.

Lemma 13.4.3. *If $A, B \in \mathbb{Z}_p$ (regarded as 1-tuples) are weakly equivalent, then they are equivalent.*

Proof For some $c > 0$,

$$|\epsilon_{1,m+1}(\epsilon_{1,m+1}(A - B))^{(m+1)} - \epsilon_{1,m}(\epsilon_{1,m}(A - B))^{(m)}| \leq 2cm + c,$$

and the left side is an integer divisible by p^m. For m large enough, $p^m > 2cm + c$ and so

$$\epsilon_{1,m+1}(\epsilon_{1,m+1}(A - B))^{(m+1)} = \epsilon_{1,m}(\epsilon_{1,m}(A - B))^{(m)}.$$

Hence for m large enough, $\epsilon_{1,m}$ is constant and $\epsilon_{1,m}(A - B)$ is a constant nonnegative integer. □

Corollary 13.4.4. *Suppose $A \in \mathbb{Z}_p^n$ is weakly equivalent to hA for some positive integer h. Then $A \in (\mathbb{Z}_p \cap \mathbb{Q})^n$.*

Proof We are given that for some $c > 0$, some permutations σ_m, and some signs $\epsilon_{i,m}$,

$$(\epsilon_{i,m}(A_i - hA_{\sigma_m(i)}))^{(m)} \leq cm.$$

The order of σ_m divides $n!$, so

$$(\pm(A_i - h^{n!}A_i))^{(m)} \leq n!cm$$

for some choice of sign (depending on i, m). That is, for each i, the 1-tuple consisting of $(h^{n!} - 1)A_i$ is weakly equivalent to zero. By Lemma 13.4.3, $(h^{n!} - 1)A_i \in \mathbb{Z}$, so $A_i \in \mathbb{Z}_p \cap \mathbb{Q}$. □

Proposition 13.4.5. *Suppose that $A, B \in \mathbb{Z}_p^n$ are weakly equivalent and that B has p-adic non-Liouville differences. Then A and B are equivalent.*

Proof There is no harm in replacing B by an equivalent tuple in which $B_i - B_j \in \mathbb{Z}$ if and only if $B_i = B_j$. For any $c > 0$, by Proposition 13.1.4 there exists m_0 such that for $1 \leq j, k \leq n$ and $m \geq m_0$,

$$B_j \neq B_k \Rightarrow (B_j - B_k)^{(m)} > 3cm.$$

However, for some c and σ_m, for all m,

$$(\pm(A_i - B_{\sigma_m(i)}))^{(m)} \leq cm$$
$$(\pm(A_i - B_{\sigma_{m+1}(i)}))^{(m+1)} \leq c(m + 1)$$

and so

$$(\pm(B_{\sigma_m(i)} - B_{\sigma_{m+1}(i)}))^{(m)} \leq 2cm + c.$$

For m sufficiently large, taking $j = \sigma_m(i), k = \sigma_{m+1}(i)$ now yields a contradiction unless $B_{\sigma_m(i)} = B_{\sigma_{m+1}(i)}$. That is, for m large we may take $\sigma_m = \sigma$ for some fixed σ, so

$$(\pm(A_i - B_{\sigma(i)}))^{(m)} \leq cm \qquad (m = 1, 2, \dots).$$

By Lemma 13.4.3, $A_i - B_{\sigma(i)} \in \mathbb{Z}$, so A and B are equivalent. $\qquad\square$

Example 13.4.6. Here is an example of Dwork (from [295, Exemple 2.3.2]) of sets which are weakly equivalent but not equivalent. Let γ be an increasing function on the nonnegative integers, such for some $\epsilon > 0$, $\gamma(i+1)p^{-\gamma(i)} \geq \epsilon$ for all $i \geq 0$. Put

$$\alpha = \sum_{i=0}^{\infty} p^{\gamma(2i)}, \qquad \beta = \sum_{i=0}^{\infty} p^{\gamma(2i+1)}.$$

One verifies (exercise) that the pairs $(\alpha, -\beta)$ and $(\alpha-\beta, 0)$ are weakly equivalent but not equivalent. In this case, $\alpha, \beta, \alpha+\beta, \alpha-\beta$ are all p-adic Liouville numbers.

13.5 Exponents for annuli

We now give the abstract definition of the p-adic exponents of a differential module on an annulus satisfying the Robba condition, after a motivating remark.

Remark 13.5.1. Let N be a differential module over $K\langle t/\beta \rangle$ for the derivation $t\frac{d}{dt}$, such that the matrix of action N_0 of D on N has entries in K and eigenvalues in $\mathbb{Z}_p$. By Proposition 13.3.3, N satisfies the Robba condition. Our best hope at this point would be to prove that any differential module M over $K\langle \alpha/t, t/\beta \rangle$ satisfying the Robba condition is isomorphic to such an N. This turns out to be too much to ask for, on account of difficulties with Liouville numbers (not to mention the closed boundary), but for the moment let us postulate the existence of an isomorphism $N \otimes_{K\langle t/\beta \rangle} K\langle \alpha/t, t/\beta \rangle \cong M$ and see what it tell us.

Assume (for clarity) that K contains the group μ_{p^∞} of all p-power roots of unity in K^{alg}. Since M and N satisfy the Robba condition, the Taylor series construction gives an action of μ_{p^∞}; recall that we used the p-power action already in the construction of Frobenius antecedents (Theorem 10.4.2). The K-valued characters of μ_{p^∞} may be naturally identified with $\mathbb{Z}_p$ by identifying $a \in \mathbb{Z}_p$ with the map carrying $\zeta \in \mu_{p^\infty}$ to ζ^a; we can use the structure of N to decompose M into character spaces for this action. Namely, perform a shearing transformation if needed (Proposition 7.3.10) to ensure that N_0 has prepared eigenvalues. Then choose the basis of N so that the matrix N_0 splits into blocks corresponding to individual eigenspaces V_j for the action of D (so in particular, the K-span of the chosen basis of N equals $\bigoplus_j V_j$). Let $\lambda_j \in \mathbb{Z}_p$ be the eigenvalue corresponding to V_j. We may then identify M with $(\bigoplus_j V_j) \otimes_K K\langle \alpha/t, t/\beta \rangle$; under this identification, $t^i V_j$ is an eigenspace with character $\lambda_j + i$.

The strategy for constructing p-adic exponents is to turn this argument on its head, by first constructing the eigenspaces for the actions of the finite subgroups of μ_{p^∞}, then extracting elements of these which stabilize at a basis of M. One might expect an obstruction to arise because the full action of $1 + \mathfrak{m}_K$ by Taylor series need not be semisimple, but this does not occur; see the exercises for a statement that may help to explain this.

Following the strategy discussed in the previous remark, we now introduce the definition of exponents.

Definition 13.5.2. Let M be a finite differential module of rank n over $K\langle \alpha/t, t/\beta \rangle$ satisfying the Robba condition. For $m > 0$, let μ_{p^m} denote the group of p^m-th power roots of unity in K^{alg}, and put $\mu_{p^\infty} = \bigcup_{m>0} \mu_{p^m}$. For $\zeta \in \mu_{p^\infty}$, define the action of ζ on $M \otimes_{K\langle \alpha/t,t/\beta \rangle} K(\zeta)\langle \alpha/t, t/\beta \rangle$ via the Taylor series

$$\zeta(x) = \sum_{i=0}^{\infty} (\zeta - 1)^i \binom{D}{i}(x);$$

this series converges because of the Robba condition. Fix a basis $e_1, \ldots, e_n$ of M. An *exponent* for M is an element $A \in \mathbb{Z}_p^n$ for which there exist a sequence $\{S_{m,A}\}_{m=1}^{\infty}$ of $n \times n$ matrices over $K\langle \alpha/t, t/\beta \rangle$ satisfying the following conditions.

(a) For $j = 1, \ldots, n$ and $m = 1, 2, \ldots$, put $v_{m,A,j} = \sum_i (S_{m,A})_{ij} e_i$. Then for all $\zeta \in \mu_{p^m}$, $\zeta(v_{m,A,j}) = \zeta^{A_j} v_{m,A,j}$.
(b) For some $k > 0$, $|S_{m,A}|_\rho \le p^{km}$ for all m and all $\rho \in [\alpha, \beta]$. (This condition will be used to obtain some sort of convergence for the $S_{m,A}$.)
(c) We have $\left| \det(S_{m,A}) \right|_\rho \ge 1$ for all m and all $\rho \in [\alpha, \beta]$. (This condition prevents the $S_{m,A}$ from converging to zero.)

We make the following observations.

(i) The property of being an exponent does not depend on the choice of the basis, although the matrices S_m do depend on this choice.
(ii) If A is an exponent for M, then so is any $B \in \mathbb{Z}_p^n$ equivalent to A.

(iii) To obtain (b), it is enough to have $|S_{m,A}|_\rho \leq cp^{km}$ for some $c, k > 0$, as we can enlarge k to absorb c. It is even enough to check this just for $\rho = \alpha, \beta$, by the Hadamard three-circle theorem (Proposition 8.2.3(c)).

(iv) Choose any α', β' with $\alpha < \alpha' < \beta' < \beta$. The following argument shows that for m sufficiently large (depending on α' and β'), $\det(S_{m,A})$ is a unit in $K\langle \alpha'/t, t/\beta' \rangle$ and so $S_{m,A}$ is invertible over this ring. (Compare [148, Lemma 1.2] and the use of "leverage" in the proof of Theorem 13.5.6.)

Put $\gamma = \sqrt{\alpha\beta}$ and $\eta = \sqrt{\beta/\alpha} = \gamma/\alpha = \beta/\gamma$. Write $\det(S_{m,A}) = \sum_i x_i t^i$, let j be the largest index for which $|x_j|\gamma^j = |\det(S_{m,A})|_\gamma$, and let $[\alpha'', \beta'']$ be the largest subinterval of $[\alpha, \beta]$ containing γ for which $|x_j|\rho^j = |\det(S_{m,A})|_\rho$ for all $\rho \in [\alpha'', \beta'']$. Since the set of zeroes of $\det(S_{m,A})$ is stable under multiplication by μ_{p^m}, by Proposition 2.1.5 the width of each segment of the Newton polygon of $\det(S_{m,A})$ is a multiple of p^m. Combining this observation with (c) and Proposition 8.2.3(c) yields

$$|\det(S_{m,A})|_\alpha \eta^j (\alpha''/\alpha)^{-p^m} \geq |\det(S_{m,A})|_\gamma \geq 1$$
$$|\det(S_{m,A})|_\beta \eta^{-j} (\beta/\beta'')^{-p^m} \geq |\det(S_{m,A})|_\gamma \geq 1.$$

Adding (b), we see that

$$p^{2knm} \geq |\det(S_{m,A})|_\alpha |\det(S_{m,A})|_\beta \geq (\alpha''/\alpha)^{p^m} (\beta/\beta'')^{p^m}$$

or equivalently

$$2kn \log p \frac{m}{p^m} \geq \log(\alpha''/\alpha) + \log(\beta/\beta'').$$

Since each quantity on the right-hand side is positive, each tends to 0 as $m \to \infty$. Consequently, for m sufficiently large we have $\alpha'' < \alpha' < \beta' < \beta''$, proving the claim.

Before constructing exponents in general, we note the following extension of Remark 13.5.1.

Proposition 13.5.3. *Let M be a differential module of rank n over $K\langle t/\beta \rangle$ for the derivation $t\frac{d}{dt}$, such that the eigenvalues of D on M/tM are in $\mathbb{Z}_p$. Then for any $\alpha \in (0, \beta)$, these eigenvalues form an exponent for $M \otimes_{K\langle t/\beta \rangle} K\langle \alpha/t, t/\beta \rangle$ (which satisfies the Robba condition by Proposition 13.3.3).*

Proof By applying shearing transformations (Proposition 7.3.10) as needed, we may reduce to the case where the eigenvalues of D on M/tM are prepared. Let $e_1, \ldots, e_n$ be a basis of M reducing modulo t to a basis of M/tM consisting of bases for the generalized eigenspaces of D. For $j = 1, \ldots, n$ and $m =$

$1, 2, \ldots$, let A_j be the eigenvalue corresponding to e_j, and put

$$v_{m,A,j} = p^{-m} \sum_{\zeta \in \mu_{p^m}} \zeta^{-A_j} \cdot \zeta(e_j).$$

Then the resulting matrices $S_{m,A}$ have the desired property (exercise). □

We now give the general construction of exponents, using the discrete Fourier transform for the group $\mathbb{Z}/p^m\mathbb{Z}$. (Compare [148, Lemma 3.1, Corollary 3.3] or [295, Proposition 3.1.1].)

Lemma 13.5.4. *Let V be a finite differential module of rank n over F_ρ for the derivation $\frac{d}{dt}$, such that $IR(V) = 1$. Choose a basis of V, and for $i = 1, 2, \ldots$, let N_i be the matrix of action of $D^i/i!$ on this basis. Then for some $c > 0$,*

$$|N_i|_\rho \rho^i \leq c p^{(n-1)\lfloor \log_p i \rfloor} \qquad (i = 1, 2, \ldots).$$

Proof With notation as in Theorem 11.9.2, the differential module V' over $L[\![(t - t_\rho)/\rho]\!]_{an}$ is trivial. We may thus apply Lemma 9.11.1 to a basis of horizontal sections to deduce the claim. (We will give an alternate proof in Chapter 18; see Corollary 18.2.5.) □

Theorem 13.5.5. *Let M be a finite differential module of rank n over $K\langle \alpha/t, t/\beta \rangle$ satisfying the Robba condition. Then there exists an exponent for M.*

Proof Fix a basis $e_1, \ldots, e_n$ of M. For any $A \in \mathbb{Z}_p^n$ and any positive integer m, we wish to define the matrix $S_{m,A}$ so that $v_{m,A,j}$ corresponds to the projection of e_j into the eigenspace of μ_{p^m} for the character A_j. To achieve this, we let $E(\zeta)$ be the matrix of action of $\zeta \in \mu_{p^\infty}$ and put

$$S_{m,A} = p^{-m} \sum_{\zeta \in \mu_{p^m}} E(\zeta) \operatorname{Diag}(\zeta^{-A_1}, \ldots, \zeta^{-A_n}).$$

This matrix is invariant under $\operatorname{Gal}(K(\mu_{p^\infty})/K)$, and so has entries in $K\langle \alpha/t, t/\beta \rangle$. By the vector interpretation, it satisfies condition (a) of Definition 13.5.2. (Another way to see this is to check the identity

$$E(\zeta)\zeta(S_{m,A}) = S_{m,A} \operatorname{Diag}(\zeta^{A_1}, \ldots, \zeta^{A_n}) \qquad (\zeta \in \mu_{p^m}),$$

and use the formula for change of basis in a difference module. See Remark 14.1.3.)

We check using Lemma 13.5.4 that $S_{m,A}$ satisfies condition (b) of Definition 13.5.2. For each $\zeta \in \mu_{p^m}$, $|\zeta - 1| \leq p^{-p^{-m+1}/(p-1)}$ by Example 2.1.6. If we write

$$E(\zeta) = \sum_{i=0}^{\infty} (\zeta - 1)^i t^i N_i,$$

then under $|\cdot|_\rho$, the i-th summand is bounded under $|\cdot|_\rho$ by $\max\{1, |N_1|_\rho^{n-1}\} p^{c(m,i)}$ for

$$c(m,i) = (n-1)\log_p i - \frac{ip^{-m+1}}{p-1} = (n-1)m + (n-1)\log_p(ip^{-m}) - \frac{ip^{-m}}{p(p-1)}.$$

Here $c(m,i) - (n-1)m$ is a function of ip^{-m} which is continuous on $(0,\infty)$ and tends to $-\infty$ as ip^{-m} tends to either 0 or ∞, so it is bounded above independent of i and m. Hence $|E(\zeta)|_\rho \le p^{km}$ for some $k > 0$, implying a similar bound for $S_{m,A}$ (with a larger k).

We next choose A to satisfy condition (c) of Definition 13.5.2. Note that

$$v_{m,A,1} \wedge \cdots \wedge v_{m,A,n} = \det(S_{m,A}) e_1 \wedge \cdots \wedge e_n$$

and that for $b \in \{0, \ldots, p-1\}^n$,

$$v_{m,A,j} = \sum_{b=0}^{p-1} v_{m+1, A+p^m b, j} \qquad (j = 1, \ldots, n). \tag{13.5.5.1}$$

(In words, (13.5.5.1) states that each eigenspace for the action of μ_{p^m} is the direct sum of p eigenspaces for the action of $\mu_{p^{m+1}}$.) Hence for any $A \in \mathbb{Z}_p^n$ and any $m \ge 0$,

$$\det(S_{m,A}) = \sum_{b \in \{0,\ldots,p-1\}^n} \det(S_{m+1, A+p^m b}). \tag{13.5.5.2}$$

Write $\det(S_{m,A}) = \sum_{i \in \mathbb{Z}} s_{m,A,i} t^i$. By (13.5.5.2), we can choose b so that $|s_{m+1, A+p^m b, 0}| \ge |s_{m,A,0}|$. Since $S_{0,A}$ is the identity matrix, this allows us to make a choice of A for which the matrices $S_{m,A}$ satisfy $|s_{m,A,0}| \ge |s_{0,A,0}| = 1$ for all m. Since $\left|\det(S_{m,A})\right|_\rho \ge |s_{m,A,0}|$ for all $\rho \in [\alpha, \beta]$, this yields condition (c) of Definition 13.5.2. $\qquad\square$

We also have the following limited uniqueness result for the exponents; here we see the first appearance of a non-Liouville condition. We also must begin to assume that the annulus has positive width. (Compare [148, Theorem 4.4].)

Theorem 13.5.6. *Assume $\alpha < \beta$. Let M be a finite differential module of rank n over $K\langle\alpha/t, t/\beta\rangle$ satisfying the Robba condition. Then any two exponents for M are weakly equivalent. In particular, if M admits an exponent A with non-Liouville differences, then (by Lemma 13.4.3) any other exponent for M is equivalent to A.*

Proof Fix a basis for M, let A, B be two exponents for M, and let $S_{m,A}$, $S_{m,B}$ be the corresponding sequences of matrices (for the same constant $k > 0$). By shrinking the interval $[\alpha, \beta]$ slightly, we may ensure that the matrices $S_{m,A}$ are invertible (see Definition 13.5.2(iv)). For $j = 1, \ldots, n$, put $v_{m,j,A} =$

$\sum_i (S_{m,A})_i e_i$ and $v_{m,j,B} = \sum_i (S_{m,B})_i e_i$. Let $T_m = S_{m,A}^{-1} S_{m,B}$ be the change-of-basis matrix between the bases $v_{m,j,A}$ and $v_{m,j,B}$ of M. Since $|S_{m,A}^{-1}|_\rho \le |S_{m,A}|_\rho^{n-1} |\det(S_{m,A})|_\rho^{-1}$ (from the description of the inverse matrix using cofactors),

$$|T_m|_\rho \le |S_{m,B}|_\rho |S_{m,A}|_\rho^{n-1} |\det(S_{m,A})|_\rho^{-1} \le p^{nkm} \qquad (\rho \in [\alpha, \beta]).$$
(13.5.6.1)

We will make progress by leveraging (in a strong metaphorical sense) the upper bound (13.5.6.1) against various lower bounds involving T_m, using the log-convexity of the Gauss norm as a function of the radius, i.e., the Hadamard three-circle theorem (Proposition 8.2.3(c)).

To begin with,

$$|\det(T_m)|_\rho \ge |\det(S_{m,A})|_\rho^{-1} \ge |S_{m,A}|_\rho^{-n} \ge p^{-nkm} \qquad (\rho \in [\alpha, \beta]).$$

Put $\gamma = \sqrt{\alpha\beta}$. From the additive formula for the determinant of T_m, there must be a permutation σ_m of $\{1, \dots, n\}$ such that

$$\prod_{i=1}^n |(T_m)_{i, \sigma_m(i)}|_\gamma \ge p^{-nkm}.$$

We now apply "leverage" by combining with (13.5.6.1) to isolate a single factor, yielding

$$|(T_m)_{i, \sigma_m(i)}|_\gamma \ge p^{-nkm} \prod_{j \ne i} |(T_m)_{j, \sigma_m(j)}|_\gamma^{-1} \ge p^{-n^2 km} \qquad (i = 1, \dots, n).$$

(13.5.6.2)

Write $(T_m)_{i, \sigma_m(i)} = \sum_{j \in \mathbb{Z}} T_{i,j} t^j$ with $T_{i,j} \in K$; we can then choose $j = j(i, m)$ so that $|T_{i,j} t^j|_\gamma = |(T_m)_{i, \sigma_m(i)}|_\gamma$. We now apply "leverage" again to limit the size of j. Put $\eta = \sqrt{\beta/\alpha} > 1$. In case $j \ge 0$, combine (13.5.6.2) with the case $\rho = \beta$ of (13.5.6.1) to get

$$\eta^j = (\beta/\gamma)^j = \frac{|T_{i,j} t^j|_\beta}{|T_{i,j} t^j|_\gamma} \le p^{(n^2+n)km};$$

in case $j \le 0$, combine (13.5.6.2) with the case $\rho = \alpha$ of (13.5.6.1) to get

$$\eta^{-j} = (\gamma/\alpha)^{-j} = \frac{|T_{i,j} t^j|_\alpha}{|T_{i,j} t^j|_\gamma} \le p^{(n^2+n)km}.$$

In either case, we deduce that

$$|j| \log_p \eta \le (n^2 + n)km.$$

Finally, note that

$$\mathrm{Diag}(t^{B_1^{(m)}}, \dots, t^{B_n^{(m)}}) T_m \, \mathrm{Diag}(t^{-A_1^{(m)}}, \dots, t^{-A_n^{(m)}})$$

must have entries in $K\langle \alpha^{p^m}/t^{p^m}, t^{p^m}/\beta^{p^m}\rangle$ by condition (a) of Definition 13.5.2. Thus the integer $j = j(i, m)$ is a representative of $B_i - A_{\sigma_m(i)}$ modulo p^m which is bounded in size by a constant times m. This implies that A and B are weakly equivalent, as desired.   □

Corollary 13.5.7. *Assume $\alpha < \beta$. Let M be a finite differential module of rank n over $K\langle \alpha/t, t/\beta\rangle$ satisfying the Robba condition. Suppose that for some α, β with $\alpha \le \alpha' < \beta' \le \beta$, $M \otimes_{K\langle \alpha/t, t/\beta\rangle} K\langle \alpha'/t, t/\beta'\rangle$ admits an exponent A with p-adic non-Liouville differences. Then A is also an exponent for M.*

Proof By Theorem 13.5.5, there exists an exponent B for M. By Theorem 13.5.6, A and B are weakly equivalent; since A has p-adic non-Liouville differences, we may conclude from Lemma 13.4.3 that A and B are equivalent. Hence A is also an exponent for M.   □

In general, it is quite difficult to compute the p-adic exponents of a differential module. However, one can at least check the following compatibility, which will lead to an important instance in which the exponent of a differential module can be controlled. See Corollary 13.6.2.

Lemma 13.5.8. *Let M be a finite differential module of rank n over $K\langle \alpha/t, t/\beta\rangle$ satisfying the Robba condition, and let $\varphi : K\langle \alpha/t, t/\beta\rangle \rightarrow K\langle \alpha^{1/q}/t, t/\beta^{1/q}\rangle$ be the substitution $t \mapsto t^q$. If A is an exponent of M, then qA is an exponent of $\varphi^* M$.*

13.6 The p-adic Fuchs theorem for annuli

We now come to the question of whether one can really invert the construction of Remark 13.5.1, i.e., whether any differential module satisfying the Robba condition is isomorphic to a differential module over a disc. We can only hope to treat this question in case the module has an exponent with p-adic non-Liouville differences: in this case the exponent is unique up to equivalence by Theorem 13.5.6, so there is no ambiguity about how to fill in the hole in the annulus. The fact that no other conditions are necessary is the content of the following remarkable theorem of Christol and Mekhout.

Theorem 13.6.1 (Christol–Mekhout). *Let M be a finite differential module on an open annulus for the derivation $t\frac{d}{dt}$ satisfying the Robba condition, admitting an exponent on some closed subannulus of positive width with p-adic non-Liouville differences. (By Corollary 13.5.7, the same exponent then serves for every such subannulus.) Then M admits a basis on which the matrix of action* *of D has entries in K and eigenvalues representing the exponent of M (and hence belonging to $\mathbb{Z}_p$). Consequently, M admits a canonical decomposition*

$$M = \bigoplus_{\lambda \in \mathbb{Z}_p/\mathbb{Z}} M_\lambda$$

in which each M_λ has exponent identically equal to λ.

The proof is loosely analogous to that of Theorem 13.5.6, with a twist: instead of comparing two different sequences of matrices, we compare one sequence with itself.

Proof Let α, β be the inner and outer radii of the original annulus. It suffices to construct a basis of the desired form over the closed annulus $\alpha' \leq |t| \leq \beta'$ for every pair α', β' with $\alpha < \alpha' < \beta' < \beta$. Choose α'', β'' with $\alpha < \alpha'' < \alpha' < \beta' < \beta'' < \beta$ and $\alpha'/\alpha'' = \beta''/\beta' = \eta > 1$, so that M is represented by a finite free module over $K\langle \alpha''/t, t/\beta'' \rangle$.

By hypothesis, M admits an exponent A on some closed subannulus of positive width, having p-adic non-Liouville differences; by replacing A with an equivalent exponent, we can also force it to have no nonzero integer differences. By Corollary 13.5.7, A is also an exponent for M on any closed subannulus of positive width, including $\alpha'' \leq |t| \leq \beta''$. We may thus fix a basis of $M \otimes K\langle \alpha''/t, t/\beta'' \rangle$ and then define a sequence of invertible matrices $S_{m,A}$ as in Definition 13.5.2. As in the proof of Theorem 13.5.6, for all m,

$$|S_{m,A}|_\rho \leq p^{km}, \quad |S_{m,A}^{-1}|_\rho \leq p^{(n-1)km} \qquad (\rho \in [\alpha'', \beta'']).$$

For $m' \geq m$, set $T_{m',m} = S_{m',A} S_{m,A}^{-1}$. As in (13.5.6.1),

$$|T_{m',m}|_\rho, |T_{m',m}^{-1}|_\rho \leq p^{nkm'} \qquad (\rho \in [\alpha'', \beta'']).$$

Choose $\lambda \in (0, 1)$ and $c > 0$ so that $p^{8nk} \eta^{-c} \leq \lambda$. Since A has p-adic non-Liouville differences, there exists $m_0 > 0$ such that for $m \geq m_0$, the congruence $h \equiv A_i - A_j \pmod{p^m}$ forces either $h = A_i - A_j = 0$ or $|h| \geq cm$. By enlarging m_0 if needed, we may also ensure that $A_i \equiv A_j \pmod{p^{m_0}}$ if and only if $A_i = A_j$.

The strategy is to renormalize the matrices $S_{m,A}$ to obtain a convergent sequence. To this end, we will construct invertible matrices R_m over K for $m \geq m_0$ such that $R_{m_0} = I_n$, $(R_m)_{ij} = 0$ whenever $A_i \neq A_j$ (or equivalently $A_i \not\equiv A_j \pmod{p^m}$), and

$$|I_n - R_m S_{m,A}^{-1} S_{m+1,A} R_{m+1}^{-1}|_\rho \leq \lambda^m \qquad (\rho \in [\alpha', \beta'], m \geq m_0).$$

This will imply that for $m \geq m_0$ and $\rho \in [\alpha', \beta']$, $|I_n - S_{m_0,A}^{-1} S_{m,A} R_m^{-1}|_\rho < 1$ and $|S_{m_0,A}^{-1} S_{m,A} R_m^{-1} - S_{m_0,A}^{-1} S_{m+1,A} R_m^{-1}|_\rho \leq \lambda^m$. Consequently, the sequence

$S_{m_0,A}^{-1} S_{m,A} R_m^{-1}$ for $m = m_0, m_0 + 1, \dots$ will converge to an invertible matrix U over $K\langle \alpha'/t, t/\beta' \rangle$ which will have the desired effect (more on this below).

The construction of the R_m proceeds recursively as follows. Given the matrices $R_{m_0}, \dots, R_m$, we first verify that

$$|R_m|, |R_m^{-1}| \le p^{nkm}.$$

This is clear for $m = m_0$, so we may assume $m > m_0$. Choose any $\rho \in [\alpha', \beta']$. As noted above, $|I_n - S_{m_0,A}^{-1} S_{m,A} R_m^{-1}|_\rho < 1$, so $|S_{m_0,A}^{-1} S_{m,A} R_m^{-1}|_\rho = |R_m S_{m,A}^{-1} S_{m_0,A}|_\rho = 1$. We then deduce the claim using the bounds on the quantities $|S_{m,A}|_\rho, |S_{m,A}^{-1}|_\rho$ from above.

Next, put $T_m = R_m S_{m,A}^{-1} S_{m+1,A}$; we then have $|T_m|_\rho, |T_m^{-1}|_\rho \le p^{2nk(m+1)}$ for all $\rho \in [\alpha'', \beta'']$. If we write $T_m = \sum_{h \in \mathbb{Z}} T_{m,h} t^h$, then $(T_{m,h})_{ij}$ can only be nonzero if $h \equiv A_i - A_j \pmod{p^m}$, which forces either $h = 0$ or $|h| \ge cm$. If $h > 0$, then

$$|(T_{m,h})_{ij} t^h|_{\alpha'} \le |(T_{m,h})_{ij} t^h|_{\beta'} \le |(T_{m,h})_{ij} t^h|_{\beta''} \eta^{-cm} \le p^{2nk(m+1)} \eta^{-cm},$$

while if $h < 0$, then

$$|(T_{m,h})_{ij} t^h|_{\beta'} \le |(T_{m,h})_{ij} t^h|_{\alpha'} \le |(T_{m,h})_{ij} t^h|_{\alpha''} \eta^{-cm} \le p^{2nk(m+1)} \eta^{-cm}.$$

We may now take $R_{m+1} = T_{m,0}$, because for $\rho \in [\alpha', \beta']$,

$$|I_n - R_{m+1} T_m^{-1}|_\rho \le |T_m^{-1}|_\rho |T_m - T_{m,0}|_\rho \le p^{2nk(m+1)} p^{2nk(m+1)} \eta^{-cm} \le \lambda^m < 1$$

and so $|I_n - T_m R_{m+1}^{-1}|_\rho \le \lambda^m$. (Note that indeed $(R_m)_{ij} = 0$ whenever $A_i \ne A_j \pmod{p^m}$, and that the latter condition is equivalent to $A_i \ne A_j$ by our choice of m_0.) This completes the construction of the R_m.

At this point, we have exhibited a invertible matrix U over $K\langle \alpha'/t, t/\beta' \rangle$ such that $S_{m_0,A} U$ is the change-of-basis matrix to a basis $e_1, \dots, e_n$ of $M \otimes K\langle \alpha'/t, t/\beta' \rangle$ with the property that

$$\zeta(e_i) = \zeta^{A_i} e_i \qquad (\zeta \in \mu_{p^\infty}; i = 1, \dots, n).$$

We next check that the matrix of action N of D on $e_1, \dots, e_n$ has entries in K. To this end, note that the actions of ζ and D commute; writing $N_{ij} = \sum_{n \in \mathbb{Z}} N_{ijn} t^n$, we have

$$\sum_i \left(\zeta^{A_j} \sum_{n \in \mathbb{Z}} N_{ijn} t^n \right) e_i = D(\zeta^{A_j} e_j)$$

$$= (D \circ \zeta)(e_j) = (\zeta \circ D)(e_j)$$

$$= \sum_i \left(\sum_{n \in \mathbb{Z}} \zeta^{n+A_i} N_{ijn} t^n \right) e_i.$$

Since we previously enforced the assumption that A has no nonzero integer differences, it follows that $N_{ijn} = 0$ whenever $n \neq 0$ or $A_i \neq A_j$.

From the previous paragraph, we see that M decomposes into summands, each corresponding to a single value in A. It remains to check that the multiset of eigenvalues of N equals A; for this, we first reduce to the case where A consists of a single element, and then to the case where that single element equals 0. In this case, if b occurs as an eigenvalue of N, then by Example 9.5.2 we must have $b \in \mathbb{Z}_p$, and there must exist a nonzero eigenvector v of N with eigenvalue b of the form $c_1 e_1 + \cdots + c_n e_n$ for some $c_1, \ldots, c_n \in K$. However, we then have

$$\zeta^b v = \zeta(v) = \zeta(c_1 e_1 + \cdots + c_n e_n) = c_1 e_1 + \cdots + c_n e_n = v,$$

a contradiction if $b \neq 0$. □

The hypothesis of Theorem 13.6.1 is difficult to verify in general, because of the indirect nature of the definition of exponents. However, we do have the following important case where the condition can be verified.

Corollary 13.6.2. *Let M be a finite differential module on the open annulus with inner radius α and outer radius β, satisfying the Robba condition. Let q be a power of p, and suppose that the intervals $(\alpha^{1/q}, \beta^{1/q})$ and (α, β) have nonempty intersection. Suppose moreover that on some annulus there exists an isomorphism $\varphi_K^* \varphi^* M \cong M$, where $\varphi_K : K \to K$ is an isometry and φ is the substitution $t \mapsto t^q$. Then any exponent for M consists of rational numbers; consequently, the conclusion of Theorem 13.6.1 applies.*

Proof Choose α', β' with $\alpha \leq \alpha' < \beta' \leq \beta$ in such a way that the intervals $((\alpha')^{1/q}, (\beta')^{1/q})$ and (α', β') have nonempty intersection. On the intersection, by Theorem 13.5.5 M admits some exponent A; by Corollary 13.5.7, A is also an exponent of both M and $\varphi_K^* \varphi^* M$ on their entire domains of definition. The rationality of A holds by Lemma 13.5.8 and Corollary 13.4.4. By Theorem 13.5.6 and Lemma 13.4.3, any exponent for M is equivalent to A. We may thus apply Theorem 13.6.1. □

Remark 13.6.3. Corollary 13.6.2 is critical in what follows; it gives rise to a quasiunipotence result (Lemma 21.2.1) which can be used to establish the p-adic local monodromy theorem (Theorem 20.1.4).

One other consequence of Theorem 13.6.1 that can be stated without reference to exponents is the following. (In fact, it can also be proved without exponents; see Remark 13.7.3.)

Corollary 13.6.4. *Let M be a finite differential module on an open annulus for* *the derivation $t\frac{d}{dt}$ satisfying the Robba condition. Suppose that the restriction M' of M to some smaller open annulus is either trivial or unipotent. Then the same is true for M.*

Proof If M' is unipotent, then it admits the exponent 0 by Remark 13.5.1. However, any exponent A of M restricts to an exponent of M', so is weakly equivalent to 0 by Proposition 13.5.6. By Lemma 13.4.3, A is equivalent to 0, so Theorem 13.6.1 implies that M is unipotent. If now M' is trivial, then M is forced to be trivial also. $\square$

Remark 13.6.5. Let M be a finite differential module of rank 1 on an open annulus for the derivation $t\frac{d}{dt}$ satisfying the Robba condition. By Theorem 13.5.5, M admits an exponent $A \in \mathbb{Z}_p$ which automatically has p-adic non-Liouville differences. Consequently, we may apply Theorem 13.6.1 to deduce that M is free on a single generator v satisfying $D(v) = Av$ (as in Example 9.5.2). However, it is also possible to give a direct proof of this; see the exercises.

13.7 Transfer to a regular singularity

As an application of Theorem 13.6.1, we obtain a transfer theorem in the presence of a regular singularity, in the spirit of Theorem 9.6.1 but with a somewhat weaker estimate. For the necessity of weakening the estimate, see Example 13.7.2.

Theorem 13.7.1. *Let M be a finite differential module of rank n on the open unit disc for the derivation $t\frac{d}{dt}$, with a regular singularity at $t = 0$ whose exponents are in $\mathbb{Z}_p$ and have p-adic non-Liouville differences. Then on the open disc of radius $\lim_{\rho \to 1^-} R(M_\rho)^n$, M admits a basis on which the matrix of action of D has entries in $\mathbb{Z}_p$. In particular, if M is solvable at 1, then this basis is defined over the whole open unit disc.*

Proof Using shearing transformations (as in Proposition 7.3.10), we may reduce to the case where the exponents of M at the regular singularity $t = 0$ are prepared. Note that for any $\rho \in (0, 1)$, $M \otimes K\langle t/\rho \rangle$ admits a basis. By Theorem 13.2.2, the fundamental solution matrix in terms of this basis has entries defined over some disc of positive radius. From this and Proposition 13.3.3, it follows that $R(M_\rho) = \rho$ for $\rho \in (0, 1)$ sufficiently small.

Let λ be the supremum of $\rho \in (0, 1)$ for which $R(M_\rho) = \rho$. Note that the function $f_1(r) = -\log R(M_{e^{-r}})$ is convex by Theorem 11.3.2(e), is equal to r for r sufficiently large by the previous paragraph, and is also equal to r for

$r = -\log \lambda$ by continuity (Theorem 11.3.2(a)). Consequently, $f_1(r) = r$ for all $r \geq -\log \lambda$.

Choose $\alpha, \beta \in (0, \lambda)$ with $\alpha < \beta$, such that the fundamental solution matrix of M (with respect to some basis on the closed disc of radius β) converges in the open disc of radius β. Let N_0 be the matrix of action of D on the basis B_0 given by the columns of the fundamental solution matrix; by construction, N_0 has entries in K whose eigenvalues are prepared and form a tuple $A \in \mathbb{Z}_p^n$ with p-adic non-Liouville differences. On the open annulus of inner radius α and outer radius β, M admits the exponent A by Remark 13.5.1. By Corollary 13.5.7, A is also an exponent for M on the open annulus of inner radius α and outer radius λ. By Theorem 13.6.1, on that annulus we obtain another basis B_1 of M on which D acts via a matrix N_1 with entries in K and eigenvalues equal to A.

Let U be the change-of-basis matrix from B_0 to B_1, which is an invertible $n \times n$ matrix over $K\{\alpha/t, t/\beta\}$; it satisfies the equation

$$U^{-1} N_0 U + U^{-1} t \frac{d}{dt}(U) = N_1,$$

or $N_0 U - U N_1 + t \frac{d}{dt}(U) = 0$. Since N_0, N_1 have entries in K, if we write $U = \sum_i U_i t^i$, then $N_0 U_i - U_i N_1 + i U_i = 0$ for each $i \in \mathbb{Z}$. By Lemma 7.3.5, the map $U_i \mapsto N_0 U_i - U_i N_1 + i U_i$ on $n \times n$ matrices over K has eigenvalues each of which is i plus a difference between two elements of A. In particular, since A is prepared, these eigenvalues can never vanish for $i \neq 0$, so $U = U_0$ has entries in K. (Compare the uniqueness argument in Proposition 7.3.6.) It follows that B_0 is a basis of M on the open disc of radius λ, on which D acts via a matrix over K. Since that matrix has eigenvalues in $\mathbb{Z}_p$, we can conjugate over K to put the matrix in Jordan normal form, in which case the entries will belong to $\mathbb{Z}_p$.

It remains to give a lower bound for λ. By Theorem 11.3.2, for $r \in [0, -\log \lambda]$, the function f_1 is continuous, convex, and piecewise affine, with slopes belonging to $\frac{1}{1}\mathbb{Z} \cup \cdots \cup \frac{1}{n}\mathbb{Z}$. Since the slope for $r > -\log \lambda$ is equal to 1, the slopes for $r \leq -\log \lambda$ cannot exceed 1. Moreover, there cannot be a slope equal to 1 in this range, as otherwise it would occur as the left slope at $r = -\log \lambda$, so there would exist $\rho > \lambda$ for which $R(M_\rho) = \rho$, contrary to the definition of λ. Consequently, f_1 has all slopes less than or equal to $(n-1)/n$ for $r \in [0, -\log \lambda]$, yielding

$$-\log \lambda = f_1(-\log \lambda) \leq f_1(0) + \frac{n-1}{n}(-\log \lambda).$$

From this we deduce $\lambda \geq \lim_{\rho \to 1^-} R(M_\rho)^n$, as desired. $\qquad \square$

Example 13.7.2. The following example shows that the exponent n in Theorem 13.7.1 cannot be replaced by 1. Pick $\lambda \in K^\times$ with $|\lambda| > 1$. Let M be the differential module of rank 2 over the open unit disc for the operator $t\frac{d}{dt}$, whose action of D is given on a basis by

$$\begin{pmatrix} 0 & \lambda \\ t & 0 \end{pmatrix}.$$

Since the reduction of this matrix modulo t is nilpotent, the exponents of M are both zero; in particular, they are in $\mathbb{Z}_p$ with p-adic non-Liouville differences.

For $\rho \in (|\lambda|^{-1}, 1)$, we may apply Theorem 6.5.3 to compute $|t^{-1}D|_{sp,M_\rho} = |\lambda|^{1/2}\rho^{-1/2}$, so $R(M_\rho) = p^{-1/(p-1)}|\lambda|^{-1/2}\rho^{1/2}$. To compute $R(M_\rho)$ for $\rho \le |\lambda|^{-1}$, we note that the function $f_1(M, r)$ is affine of slope $1/2$ for $0 \le r \le \log|\lambda|$ and of slope 1 for r large. By Theorem 11.3.2(a),(b),(e), f_1 is piecewise affine, and its slopes on $[0, \infty)$ must lie in $[1/2, 1] \cap (\mathbb{Z} \cup \frac{1}{2}\mathbb{Z})$. Consequently, $f_1(M, r)$ must have slope $1/2$ until it reaches a point at which $f_1(M, r) = r$, and slope 1 thereafter. In other words,

$$R(M_\rho) = \begin{cases} p^{-1/(p-1)}|\lambda|^{-1/2}\rho^{1/2} & \rho > p^{-2/(p-1)}|\lambda|^{-1} \\ \rho & \rho \le p^{-2/(p-1)}|\lambda|^{-1}. \end{cases}$$

In particular, the fundamental solution matrix of M in the given basis converges on the open disc of radius $p^{-2/(p-1)}|\lambda|^{-1} = \lim_{\rho\to 1^-} R(M_\rho)^2$ but not on any larger disc.

Remark 13.7.3. The final assertion of Theorem 13.7.1 (if M is solvable at 1, then the fundamental solution matrix converges in the open unit disc) can also be proved without relying on the p-adic Fuchs theorem for annuli (Theorem 13.6.1); see the chapter notes. This in turn can be used to give a proof of Corollary 13.6.4 without relying on Theorem 13.6.1, as follows.

Let M be a finite differential module on the open annulus $\alpha < |t| < \beta$ for the derivation $t\frac{d}{dt}$ satisfying the Robba condition. Suppose that the restriction M' of M to some smaller open annulus $\gamma < |t| < \delta$ is unipotent. Then on this smaller annulus, by Lemma 9.2.3, we obtain a basis of M on which $t\frac{d}{dt}$ acts via a nilpotent matrix over K. This defines a differential module on the disc $|t| < \delta$ with a nilpotent regular singularity at $t = 0$; by gluing with M over the annulus $\gamma < |t| < \delta$, we obtain a differential module on the disc $|t| < \beta$ with a nilpotent regular singularity at $t = 0$, which is solvable at the boundary of the disc. By the final assertion of Theorem 13.7.1, this module admits a basis on which $t\frac{d}{dt}$ acts via a nilpotent matrix over K. This gives us such a basis of M over the annulus $\gamma < |t| < \beta$; by a similar argument, we get a similar basis over the annulus $\alpha < |t| < \delta$. These bases can only differ by a K-linear transformation (as in the proof of Theorem 13.7.1), so each one gives a basis of M itself on which $t\frac{d}{dt}$ acts via a nilpotent matrix over K. Hence M is unipotent.

13.8 Liouville partitions

For completeness, we mention a refinement of the theorem of Christol and Mebkhout, based on the following definition from [255, Definition 3.4.4].

Definition 13.8.1. Let $n_1, \ldots, n_k$ be positive integers and let $A_i \in \mathbb{Z}_p^{n_i}$ be a tuple for $i = 1, \ldots, k$. Let $A \in \mathbb{Z}_p^{n_1 + \cdots + n_k}$ be a permutation of the concatenation of $A_1, \ldots, A_k$. We say that $A_1, \ldots, A_k$ form an *integer partition* (resp. a *Liouville partition*) of A if there do not exist distinct values $g, h \in \{1, \ldots, k\}$ and elements $a_g \in A_g, a_h \in A_h$ such that $a_g - a_h$ is an integer (resp. an integer or a p-adic Liouville number). This implies in particular that A_g and A_h are disjoint, justifying the use of the term "partition".

Note that A always admits a maximal integer partition, namely the partition into $\mathbb{Z}$-cosets. This partition is a Liouville partition if and only if A has p-adic non-Liouville differences.

In this language, we may upgrade Proposition 13.4.5 following [255, Proposition 3.4.5].

Proposition 13.8.2. *Let $A_1, \ldots, A_k$ be a Liouville partition of $A \in \mathbb{Z}_p^n$.*

(a) *For $g = 1, \ldots, k$, let B_g be a tuple weakly equivalent to A_g. Then $B_1, \ldots, B_k$ form a Liouville partition of their concatenation; in particular, $B_1, \ldots, B_k$ are pairwise disjoint.*

(b) *Suppose that $B \in \mathbb{Z}_p^n$ is weakly equivalent to A. Then B admits a Liouville partition $B_1, \ldots, B_k$ such that B_g is weakly equivalent to A_g for $g = 1, \ldots, k$.*

Proof For $g = 1, \ldots, k$, let n_g denote the length of the tuples A_g and B_g. For any $c > 0$, by Proposition 13.1.4 there exists m_0 such that for $g, h \in \{1, \ldots, k\}$ distinct, $i_g \in \{1, \ldots, n_g\}$, $i_h \in \{1, \ldots, n_g\}$, and $m \geq m_0$,

$$((A_g)_{i_g} - (A_h)_{i_h})^{(m)} > (3c + 1)m. \tag{13.8.2.1}$$

To prove (a), let $g, h \in \{1, \ldots, k\}$ be distinct and choose $j_g \in \{1, \ldots, n_g\}, j_h \in \{1, \ldots, n_h\}$; we must show that $(B_g)_{j_g} - (B_h)_{j_h}$ is not a integer or a p-adic Liouville number. By hypothesis, for some $c > 0$, for each m we can find $i_g \in \{1, \ldots, n_g\}, i_h \in \{1, \ldots, n_g\}$ (depending on m) such that for suitable signs (also depending on m),

$$\pm((A_g)_{i_g} - (B_g)_{j_g})^{(m)} \leq cm, \qquad \pm((A_h)_{i_h} - (B_h)_{j_h})^{(m)} \leq cm.$$

Combining with (13.8.2.1) yields

$$((B_g)_{j_g} - (B_h)_{h_g})^{(m)} > (c + 1)m,$$

which implies the desired conclusion.

To prove (b), we first make a notational shift: we write the elements of A as $a_1, \ldots, a_n$ and the elements of B as $b_1, \ldots, b_n$ to avoid confusion with the partitions. By hypothesis, for each m, there exist a permutation σ_m of $\{1, \ldots, n\}$ and some signs $\epsilon_{i,m}$ such that

$$(\epsilon_{i,m}(a_{\sigma_m(i)} - b_i))^{(m)} \le cm \qquad (i = 1, \ldots, n).$$

In particular,

$$(\epsilon_{i,m}(a_{\sigma_m(i)} - a_{\sigma_{m+1}(i)}))^{(m)} \le (2c+1)m \qquad (i = 1, \ldots, n).$$

Combining this with (13.8.2.1), we see that $\sigma_m^{-1} \circ \sigma_{m+1}$ must respect the partition of A into $A_1, \ldots, A_k$; we thus obtain a partition $B_1, \ldots, B_k$ of B for which A_g and B_g are weakly equivalent for $g = 1, \ldots, k$. By (a), $B_1 \ldots, B_k$ form a Liouville partition of B. □

The following corollary is [255, Corollary 3.4.6].

Corollary 13.8.3. *Suppose that $A, B \in \mathbb{Z}_p^n$ are weakly equivalent. Then A contains an integer or a p-adic Liouville number if and only if B does.*

Proof Note that A contains an integer or a p-adic Liouville number if and only if $\{0\}$ and A fail to form a Liouville partition of their union. The claim thus follows by applying Proposition 13.8.2(a) to $\{0\}, A$ and $\{0\}, B$. □

The following corollary is [255, Corollary 3.4.9].

Corollary 13.8.4. *Suppose that $A, B \in \mathbb{Z}_p^n$ are weakly equivalent and A contains no p-adic Liouville numbers. Then there exist Liouville partitions A_1, A_2 of A and B_1, B_2 of B satisfying the following conditions.*

(a) *The multisets A_1, B_1 consist entirely of integers.*
(b) *The multisets A_2, B_2 are weakly equivalent and contain no integers or p-adic Liouville numbers.*

In particular, B also contains no p-adic Liouville numbers.

Proof Let A_1, A_2 be a partition of A for which A_1 contains exactly the integers in A; by hypothesis, this is a Liouville partition of A. By Proposition 13.8.2(b), B admits a Liouville partition B_1, B_2 in which B_i is weakly equivalent to A_i for $i = 1, 2$. Since A_1 consists only of integers, by Lemma 13.4.3 the same holds for B_1. Since A_2 contains no integers or p-adic Liouville numbers, by Corollary 13.8.3 the same holds for B_2. This proves the claim. □

The following is [255, Corollary 3.4.10].

Corollary 13.8.5. *Choose $A \in \mathbb{Z}_p^n$ and set*

$$A - A = (A_i - A_j : i, j = 1, \ldots, n) \in \mathbb{Z}_p^{n^2}$$

(using any bijection of $\{1, \ldots, n\} \times \{1, \ldots, n\}$ with $\{1, \ldots, n^2\}$). If $A - A$ is weakly equivalent to a tuple containing no p-adic Liouville numbers, then A has p-adic non-Liouville differences.

Proof By Corollary 13.8.4, $A - A$ is itself p-adic non-Liouville, as desired. □

We now state without proof a theorem giving a partial form of the Christol–Mebkhout decomposition under a weaker condition on the exponent.

Theorem 13.8.6. *Let M be a finite differential module on an open annulus for the derivation $t\frac{d}{dt}$ satisfying the Robba condition. Suppose that on some closed subannulus of positive width, M admits an exponent A admitting the Liouville partition $A_1, \ldots, A_k$. Then M admits a unique direct sum decomposition $M = M_1 \oplus \cdots \oplus M_k$ such that for $g = 1, \ldots, k$, on any closed subannulus of positive width, M_g admits an exponent weakly equivalent to A_g.*

Proof See the corrected proof of [255, Theorem 3.4.20] given in [269]. □

This in turn implies a refined form of Corollary 13.6.4.

Corollary 13.8.7. *Let M be a finite differential module on an open annulus I for the derivation $t\frac{d}{dt}$ satisfying the Robba condition. Suppose that on some closed subannulus of positive width, M admits an exponent A containing no p-adic Liouville numbers. Then for any open subannulus J of I, the restriction M' of M to J has the property that the restriction map $H^0(M) \to H^0(M')$ is an isomorphism.*

Proof The hypothesis on A implies that the partition of A given by $A \cap \mathbb{Z}, A \setminus \mathbb{Z}$ is a Liouville partition. By Theorem 13.8.6, we may thus reduce to the case where A consists entirely of integers. We may thus apply Theorem 13.6.1 to show that M is unipotent, from which the claim follows at once. □

Remark 13.8.8. The example of Remark 13.2.4 can be modified (by varying c) to give cases where the radius of convergence of the hypergeometric series takes any specified value in $(0, 1)$. Such cases show that the hypothesis on the exponent in Corollary 13.8.7 cannot be relaxed.

Notes

The definition of a *p*-adic Liouville number was introduced by Clark [102]; our presentation follows [149, §VI.1].

The cited theorem of Clark [102, Theorem 3] is actually somewhat stronger than Theorem 13.2.3, as it allows differential operators of possibly infinite order.

The example in Remark 13.2.4 is loosely inspired by an example of Monsky (a counterexample against a slightly different assertion); see [150, §7] or [149, §IV.8] for discussion.

Proposition 13.3.2 is originally due to Christol; compare [149, Proposition IV.7.7].

The theory of exponents for differential modules on a *p*-adic annulus satisfying the Robba condition was originally developed by Christol and Mebkhout [98, §4–5]; in particular, Theorem 13.6.1 appears therein as [98, Théorème 6.2–4]. A somewhat more streamlined development was later given by Dwork [148], in which Theorem 13.6.1 appears as [148, Theorem 7.1]. (Dwork asserts that he did not verify the equivalence between the two constructions; we do not recommend losing any sleep over this.) Our treatment is essentially that of Dwork with a few technical simplifications. Another treatment appears in the manuscript of Christol [93]. Besides all of these, there is also an expository article by [295] that summarizes the proofs (using Dwork's approach) and provides some further context, including the formulation of the *p-adic index theorem*. (The treatment in [255], including the new concept of Liouville partitions, was closely modeled on the treatment given here.)

It is claimed in [295, §3.2] that the converse of Theorem 13.5.6 holds, i.e., if A is an exponent of a differential module M and B is weakly equivalent to A, then B is also an exponent of M. We do not know of a proof of this; [148, Remark 4.5] suggests that this may not be entirely trivial.

A somewhat more elementary treatment of Theorem 13.7.1 than the one given here is given in [149, §6]; it does not rely on the *p*-adic Fuchs theorem for annuli (Theorem 13.6.1). However, it gives a weaker result: it only establishes convergence of the fundamental solution matrix in the open disc of radius $\lim_{\rho \to 1^-} R(M_\rho)^{n^2}$. That weaker result is due to Christol, who presented it himself in [89, Théorème 6.4.7] and [90].

The weaker result implies that if $\lim_{\rho \to 1^-} R(M_\rho) = 1$, then the fundamental solution matrix converges in the open unit disc. In case of a nilpotent singularity, one can show this even more directly; see [239, Lemma 3.6.2]. As noted in Remark 13.7.3, this can be used to derive Corollary 13.6.4 without using Theorem 13.6.1.

An intriguing archimedean analogue to the theory of p-adic exponents appears in the local analytic theory of q-difference equations in the case $|q| = 1$. This seems to extend an analogy between q-difference equations with $|q| \neq 1$ and p-adic differential modules with intrinsic subsidiary radii strictly less than 1. See [133] for part of this story.

Exercises

13.1 Prove that rational numbers are p-adic non-Liouville numbers.

13.2 Define the sequence of integers $c_0, c_1, \ldots$ by $c_0 = 0$ and $c_{n+1} = p^{c_n}$ for $n \geq 0$. Show that $\sum_{n=0}^{\infty} c_n$ is a p-adic Liouville number. (Hint: apply Proposition 13.1.4 with $m = c_{n+1} - 1$.)

13.3 Give another proof of Lemma 13.1.6 (as in [149, Lemma VI.1.2]) by first verifying that both sides of the desired equation have the same coefficients of x^0 and x^1, and are killed by the second-order differential operator $\frac{d}{dx}(\frac{d}{dx} - \lambda - x)$.

13.4 Show that Theorem 13.2.2 can be deduced from Theorem 13.2.3. (Hint: show that if $H^0(M) \neq 0$, then 0 must occur as an eigenvalue of N_0.)

13.5 Prove that there exists $a \in \mathbb{Z}_p$ satisfying

$$\text{type}(a) = 1, \qquad \text{type}(-a) < 1.$$

13.6 Prove Proposition 13.3.3.

13.7 Verify Example 13.4.6.

13.8 (a) Prove that the set of $A \in \mathbb{Z}_n^p$ which are weakly equivalent to 0 is a subgroup of $\mathbb{Z}_n^p$.

(b) Give a counterexample against the claim that if A_i is weakly equivalent to B_i for $i = 1, 2$, then $A_1 + A_2$ is weakly equivalent to $B_1 + B_2$. (Hint: give an example using rational numbers, in which case weak equivalence implies equivalence by an earlier exercise.)

13.9 Complete the proof of Proposition 13.5.3.

13.10 Prove that if a finite differential module M over $K\langle \alpha/t, t/\beta \rangle$ admits a basis on which $t\frac{d}{dt}$ acts via a nilpotent matrix over K, then the K-span of the basis is fixed by the Taylor series action of μ_{p^∞}, but not necessarily by the full action of $1 + \mathfrak{m}_K$. (Hint: reduce to the fact that μ_{p^∞} is in the kernel of the logarithm map defined using the power series $\log(1 - z)$, which converges for $|z| < 1$.)

13.11 Let M be a finite differential module on an open annulus satisfying the Robba condition and admitting an exponent with p-adic non-Liouville differences. Use Theorem 13.6.1 to prove that $H^0(M)$ and $H^1(M)$ are both finite-dimensional of the same dimension. In other words, the index of M exists and equals 0.

13.12 Show that Corollary 13.6.2 can already be deduced from the special case of Theorem 13.6.1 in which the exponent A is identically zero. (Hint: pull back along a map of the form $t \mapsto t^m$.)

13.13 Give a direct proof (not using any results from this chapter) of the assertion of Remark 13.6.5.

Part IV

Difference Algebra and Frobenius Modules

14

Formalism of difference algebra

We now step away from differential modules for a little while, to study the related subject of *difference algebra*. This is the theory of algebraic structures enriched not with a derivation but with an endomorphism of rings. Our treatment of difference algebra will run largely in parallel with what we did for differential algebra, but in a somewhat abridged fashion; our goal is to say just enough to be able to use difference algebra to say nontrivial things about p-adic differential equations. We will begin to do that in Part V.

In this chapter, we introduce the formalism of difference rings, fields, and modules, and the associated notion of twisted polynomials. We then study briefly the analogue of algebraic closure for a difference field. Finally, we make a detailed study of difference modules over a complete nonarchimedean field, culminating with a classification of difference modules for the Frobenius automorphism of a complete unramified p-adic field with algebraically closed residue field (the Dieudonné–Manin classification).

Hypothesis 14.0.1. Throughout Parts IV and V, we retain Notation 8.0.1 with $p > 0$ (so that K is a complete nonarchimedean field of characteristic 0 and residual characteristic $p > 0$), but unless otherwise specified we will require K to be *discretely valued*. This is necessary to avoid a number of technical complications, some of which we will point out as we go along. (We will make almost no reference to K in this particular chapter; we only include this hypothesis here in order to place it at the beginning of Part IV.)

14.1 Difference algebra

We start with the central definition. See the notes for some explanation of the term "difference" in this context.

Definition 14.1.1. A *difference ring* or *difference field* is a ring or field R equipped with an endomorphism φ. A *difference module* over R is an R-module M equipped with a map $\Phi\colon M \to M$ which is additive and φ-semilinear; the latter means that

$$\Phi(rm) = \varphi(r)\Phi(m) \qquad (r \in R, m \in M).$$

This data determines an R-linear map from $\varphi^* M = M \otimes_{R,\varphi} R$ to M sending $m \otimes 1$ to $\Phi(m)$; conversely, any R-linear map $\varphi^* M \to M$ induces the structure of a difference module on M. A free difference module is *trivial* if it admits a Φ-invariant basis; when we refer to "the trivial difference module", we mean the module R itself with $\Phi = \varphi$. A difference submodule of R is also called a *difference ideal*; if I is a difference ideal of R, then the quotient ring R/I is again a difference ring for the induced action of φ.

Morphisms of difference rings or modules are required to be φ-equivariant. Tensor products behave as expected: if M, N are difference modules over the same difference ring R, then $M \otimes_R N$ acquires the structure of a difference module with

$$\Phi(m \otimes n) = \Phi(m) \otimes \Phi(n) \qquad (m \in M, n \in N).$$

In particular, if $R \to S$ is a morphism of difference rings, we can perform base change from difference modules over R to difference modules over S.

As in differential algebra, difference modules can often be described in terms of matrices of action.

Definition 14.1.2. If M is a finite difference module over R freely generated by $e_1, \ldots, e_n$, then we can recover the action of Φ from the $n \times n$ matrix A defined by

$$\Phi(e_j) = \sum_i A_{ij} e_i.$$

Namely, if we use the basis to identify M with the space of column vectors of length n over R, then

$$\Phi(v) = A\varphi(v).$$

In parallel with the differential case, we call A the *matrix of action* of Φ on the basis $e_1, \ldots, e_n$.

We record the effect of a change of basis on a matrix of action.

Remark 14.1.3. Let M be a finite free difference module over R and let $e_1, \ldots, e_n$ and $e'_1, \ldots, e'_n$ be two bases of M. Recall that the change-of-basis matrix U from the first basis to the second is the $n \times n$ matrix U satisfying

$e'_j = \sum_i U_{ij} e_i$. If A is the matrix of action of Φ on the e_i, then the matrix of action of Φ on the e'_i is

$$A' = U^{-1} A \varphi(U).$$

It is important to consider dual modules, but the description in difference algebra is a bit more complicated than in differential algebra. We only introduce it in the free case. (One can extend the definition to locally free modules; compare Remark 5.3.4.)

Definition 14.1.4. Let M be a finite free difference module over R. We say M is *dualizable* if the map $\varphi^*(M) \to M$ induced by Φ is an isomorphism. Given a basis of M, it is equivalent to check that the matrix of action A of Φ on that basis is invertible (the same then holds for any other basis). However, even in this case, the action of Φ on M itself is not invertible unless the action of φ on R is (i.e., unless R is *inversive* in the sense of Definition 14.1.5).

If M is dualizable, then there is a unique way to equip the module-theoretic dual $M^\vee = \mathrm{Hom}_R(M, R)$ with a Φ-action so that

$$\Phi(f)(\Phi(m)) = f(m) \qquad (f \in M^\vee, m \in M).$$

In terms of a basis, the matrix of action on the dual basis is given by the inverse transpose A^{-T}. We call the resulting difference module the *dual* of M.

There is also a notion of an opposite difference ring, but only under an extra hypothesis.

Definition 14.1.5. We say that the difference ring R is *inversive* if φ is an automorphism. In this case, we can define the *opposite difference ring* R^{opp} to be R again, but now equipped with the endomorphism φ^{-1}. If R is inversive and M is a finite free difference module over R, we define the *opposite module* M^{opp} of M to be M equipped with the inverse of Φ; that is, if A is the matrix of action of Φ on some basis, then the matrix of action of Φ^{-1} is given by $\varphi^{-1}(A^{-1})$.

Definition 14.1.6. For M a difference module, write

$$H^0(M) = \ker(\mathrm{id} - \Phi), \qquad H^1(M) = \mathrm{coker}(\mathrm{id} - \Phi).$$

If M_1, M_2 are difference modules with M_1 dualizable, then $H^0(M_1^\vee \otimes M_2)$ computes φ-equivariant morphisms from M_1 to M_2, and $H^1(M_1^\vee \otimes M_2)$ computes extensions $0 \to M_2 \to M \to M_1 \to 0$ of difference modules (exercise). That is, we have natural identifications

$$H^0(M_1^\vee \otimes M_2) = \mathrm{Hom}(M_1, M_2), \qquad H^1(M_1^\vee \otimes M_2) = \mathrm{Ext}(M_1, M_2)$$

of groups (see the proof of Lemma 5.3.3 for the group structure on extensions).

14.2 Twisted polynomials

As in differential algebra, there is a relevant notion of twisted polynomials.

Definition 14.2.1. For R a difference ring, we define the *twisted polynomial ring* $R\{T\}$ as the set of finite formal sums $\sum_{i=0}^{\infty} r_i T^i$. As for the twisted polynomial ring over a difference ring (Definition 5.5.1) the set $R\{T\}$ admits a not necessarily commutative ring structure with addition given by formal addition of formal sums and multiplication characterized by the identity $Tr = \varphi(r)T$. For any $P \in R\{T\}$, the quotient $R\{T\}/R\{T\}P$ is a difference module; if M is a difference module, we say $m \in M$ is a *cyclic vector* if there is an isomorphism $M \cong R\{T\}/R\{T\}P$ carrying m to 1.

Definition 14.2.2. If R is inversive, we again have a formal adjoint construction: given $P \in R\{T\}$, its formal adjoint is obtained by pushing the coefficients to the right side of T. This may then be viewed as an element of the opposite ring of $R\{T\}$, which we may identify with $R^{\mathrm{opp}}\{T\}$.

It is not completely straightforward to analogize the cyclic vector theorem to difference modules; see the exercises for one attempt to do so. Instead, we will use only the following simple observation.

Lemma 14.2.3. *Any irreducible finite difference module over a difference field contains a cyclic vector.*

Proof If F is a difference field, M is a finite difference module over F, and $m \in M$ is nonzero, then $m, \Phi(m), \ldots$ generate a nonzero difference submodule of M. If M is irreducible, this submodule must be the entirety of M. $\square$

Definition 14.2.4. If φ is isometric for a norm $|\cdot|$ on F, then we have the usual definition of Newton polygons and slopes for twisted polynomials, with the natural analogue of additivity of slopes (i.e., Proposition 2.1.2). If R is inversive, then a twisted polynomial and its adjoint have the same Newton polygon.

As in the differential case, we may apply the master factorization theorem (Theorem 2.2.2), as in the proof of Theorem 2.2.1, to get a factorization result. However, the result is inherently asymmetric; see Remark 14.2.6.

Theorem 14.2.5. *Let F be a difference field complete for a norm* $| \cdot |$ *under which* φ *is isometric. Then any monic twisted polynomial* $P \in F\{T\}$ *admits a unique factorization*

$$P = P_{r_1} \cdots P_{r_m}$$

for some $r_1 < \cdots < r_m$, *where each* P_{r_i} *is monic with all slopes equal to* r_i. *(The same holds with the factors in the opposite order if F is inversive, but not always otherwise; see Remark 14.2.6 and the exercises.)*

Remark 14.2.6. It is worth clarifying why the conditions of Theorem 2.2.2 can be satisfied by $F\{T\}$ but not necessarily by its opposite ring. The key condition is (b), which states that with

$$U = \{P \in F[T] : \deg(P) \leq \deg(S) - m\}$$
$$V = \{P \in F[T] : \deg(P) \leq m - 1\}$$
$$W = \{P \in F[T] : \deg(P) \leq \deg(S)\}$$
$$a = 1$$
$$b = T^m,$$

the map $(u, v) \mapsto av + ub$ surjects $U \times V$ onto W. This holds because any element of $F\{T\}$ whose coefficients of $1, T, \ldots, T^{m-1}$ vanish is divisible by T^m on the right. On the other hand, if $r \notin \varphi(F)$, then rT^m is not divisible on the left by T, let alone by T^m.

14.3 Difference-closed fields

We briefly study the difference-theoretic analogue of the algebraic closure property of ordinary fields.

Definition 14.3.1. We will say that a difference field F is *weakly difference-closed* if every dualizable finite difference module over F is trivial. We say F is *strongly difference-closed* if F is inversive and weakly difference-closed.

Remark 14.3.2. Note that the property that F is weakly difference-closed includes the fact that short exact sequences of dualizable finite difference modules over F always split. By contrast, if for instance φ is the identity map, then this is never true even if F is algebraically closed, because linear transformations need not be semisimple.

Here is a useful criterion for checking for difference closure.

Lemma 14.3.3. *The difference field F is weakly difference-closed if and only if the following conditions all hold.*

(a) *Every nonconstant monic twisted polynomial $P \in F\{T\}$ factors as a product of linear factors.*

(b) *For every $c \in F^\times$, there exists $x \in F^\times$ with $\varphi(x) = cx$.*

(c) *For every $c \in F^\times$, there exists $x \in F^\times$ with $\varphi(x) - x = c$.*

Proof We first suppose that F is weakly difference-closed. To prove (a), it suffices to check that if $P \in F\{T\}$ is nonconstant monic with nonzero constant term, then P factors as $P_1 P_2$ with P_2 linear. The nonzero constant term implies that $M = F\{T\}/F\{T\}P$ is a dualizable finite difference module over F, so must be trivial by the hypothesis that F must be weakly difference-closed. In particular, there exists a short exact sequence $0 \to M_1 \to M \to M_2 \to 0$ with M_2 trivial of dimension 1; this corresponds to a factorization $P = P_1 P_2$ with P_2 linear.

To prove (b), note that $F\{T\}/F\{T\}(T - c^{-1})$ must be trivial. Since each element of this difference module is represented by some $x \in F$, we must have $x \in F^\times$ such that $Tx - x$ is divisible on the right by $T - c^{-1}$. By comparing degrees, we see that $Tx - x = y(T - c^{-1})$ for some $y \in F$. Then $y = \varphi(x)$ and $yc^{-1} = x$, proving the claim.

To prove (c), form the φ-module V corresponding to the matrix $\begin{pmatrix} 1 & c \\ 0 & 1 \end{pmatrix}$. By construction, we have a short exact sequence $0 \to V_1 \to V \to V_2 \to 0$ with V_1, V_2 trivial; since V must also be trivial, this extension must split. In other words, we can find $x \in F$ such that

$$\begin{pmatrix} 1 & x \\ 0 & 1 \end{pmatrix}\begin{pmatrix} 1 & c \\ 0 & 1 \end{pmatrix}\begin{pmatrix} 1 & -\varphi(x) \\ 0 & 1 \end{pmatrix} = \begin{pmatrix} 1 & 0 \\ 0 & 1 \end{pmatrix},$$

that is, $\varphi(x) - x = c$.

Conversely, suppose that (a)–(c) hold. Every nonzero dualizable finite difference module over F admits an irreducible quotient. This quotient admits a cyclic vector by Lemma 14.2.3, and so admits a quotient of dimension 1 by (a). That quotient in turn is trivial by (b). By induction, we deduce that every dualizable finite difference module over F admits a filtration whose successive quotients are trivial of dimension 1. This filtration splits by (c). □

The notion of difference closure is particularly simple for the absolute Frobenius endomorphism (the p-power map) on a field of characteristic $p > 0$.

Proposition 14.3.4. *Let F be a separably (resp. algebraically) closed field of* *characteristic $p > 0$ equipped with a power of the absolute Frobenius. Then F is weakly (resp. strongly) difference-closed.*

Proof For $P = \sum_{i=0}^{m} P_i T^m \in F\{T\}$ with $m > 0$, $P_m = 1$, and $P_0 \neq 0$, the polynomial $Q(x) = \sum_{i=0}^{m} P_i x^{q^i}$ has degree $q^m \geq 2$, and $x = 0$ occurs as a root only with multiplicity 1. Moreover, the formal derivative of P is a constant polynomial, so has no common roots with P; hence P is a separable polynomial. Since F is separably closed, there must exist a nonzero root x of Q; this implies criteria (a) and (b) of Lemma 14.3.3. To deduce (c), note that for $c \in F^\times$, the polynomial $x^q - x - c$ is again separable, and so has a root in F. $\quad\square$

Remark 14.3.5. At this point, it is natural to ask whether there is something analogous to an "algebraic closure" of a difference field F. If F is of characteristic p and φ acts as a power of the absolute Frobenius, then Proposition 14.3.4 implies that the usual algebraic closure of F is strongly difference-closed. In general, one can use a formal argument to show that F can be enlarged to a strongly difference-closed difference field (see exercises). One can then ask whether a minimal such extension is unique up to (noncanonical) isomorphism; this leads to the theory of *Picard–Vessiot extensions* [374, Chapter 1].

14.4 Difference algebra over a complete field

Hypothesis 14.4.1. For the rest of this chapter, let F be a difference field complete for a nonarchimedean norm $|\cdot|$ with respect to which φ is isometric. (For short, we will say that F is an *isometric* complete nonarchimedean difference field.) We do not assume that F is inversive; if not, then we can embed F into an inversive difference field by forming the completion F' of the direct limit of the system

$$ F \xrightarrow{\varphi} F \xrightarrow{\varphi} \cdots . $$

We sometimes call F' the *perfection*, or more precisely the *φ-perfection*, of F. For V a difference module over F, write F' as shorthand for $V \otimes_F F'$.

As in the differential case, we would like to classify finite difference modules over F by the spectral radius of Φ; this turns out to be somewhat easier in the difference case because there is no analogue of the distinction between the visible and full spectra. A related fact is that the use of matrix inequalities here is much closer to what we saw in Chapter 4 than in Chapter 6; the main difference is that we do not begin with a meaningful theory of eigenvalues or eigenvectors.

We first note the following analogue of Lemma 6.2.5.

Lemma 14.4.2. *Let V be a nonzero finite difference module over F. Let $e_1, \ldots, e_n$ be a basis of V. For each nonnegative integer s, let A_s be the matrix of action of Φ^s on $e_1, \ldots, e_n$. Then*

$$|\Phi|_{\mathrm{sp},V} = \lim_{s \to \infty} |A_s|^{1/s}.$$

Proof It suffices to observe that using the supremum norm defined by the basis $e_1, \ldots, e_n$, the operator norm of Φ^s is precisely $|A_s|$. Thus the limit on the right side exists and matches the definition of the left side. □

The following basic properties will help with our classification, as long as we are mindful of the discrepancies between the differential and difference cases. For instance, the difference case has a simpler rule for tensor products, but a less simple rule for duals. (It may be helpful to keep in mind the case where φ is the identity map, and relate the following observations to eigenvalues of linear transformations.)

Lemma 14.4.3. *Let V, V_1, V_2 be nonzero finite difference modules over F.*

(a) For a short exact sequence $0 \to V_1 \to V \to V_2 \to 0$,

$$|\Phi|_{\mathrm{sp},V} = \max\{|\Phi|_{\mathrm{sp},V_1}, |\Phi|_{\mathrm{sp},V_2}\}.$$

(b) We have

$$|\Phi|_{\mathrm{sp},V_1 \otimes V_2} = |\Phi|_{\mathrm{sp},V_1} |\Phi|_{\mathrm{sp},V_2}.$$

(c) For any complete extension E of F to which φ extends isometrically,

$$|\Phi|_{\mathrm{sp},V} = |\Phi|_{\mathrm{sp},V \otimes_F E}.$$

(d) For V dualizable,

$$|\Phi|_{\mathrm{sp},V} |\Phi|_{\mathrm{sp},V^\vee} \geq 1.$$

Proof Exercise. □

Lemma 14.4.4. *If $V \cong F\{T\}/F\{T\}P$ and P has only one slope $r < \infty$ in its Newton polygon, then V admits a basis on which*

$$|\Phi|_V = \mathrm{e}^{-r}, \qquad |\Phi^{-1}|_{V'} = \mathrm{e}^r.$$

Consequently,

$$|\Phi|_{\mathrm{sp},V} = \mathrm{e}^{-r}, \qquad |\Phi^\vee|_{\mathrm{sp},V} = \mathrm{e}^r,$$

and if F is inversive, then also

$$|\Phi^{-1}|_{\mathrm{sp},V} = \mathrm{e}^r.$$

Proof Put $n = \deg(P)$, and define a norm on V by

$$|a_0 + \cdots + a_{n-1}T^{n-1}| = \max_i\{|a_i|e^{-ri}\};$$

then evidently

$$|\Phi|_V = e^{-r}, \qquad |\Phi^{-1}|_{V'} = |\Phi|_{V^\vee} = e^r.$$

We deduce that

$$|\Phi|_{\mathrm{sp},V} \le e^{-r}, \qquad |\Phi^{-1}|_{\mathrm{sp},V'}, |\Phi|_{V^\vee} \le e^r;$$

since

$$1 \le |\Phi|_{\mathrm{sp},V}|\Phi^{-1}|_{\mathrm{sp},V'} \le e^{-r}e^r, \qquad 1 \le |\Phi|_{\mathrm{sp},V}|\Phi|_{\mathrm{sp},V^\vee} \le e^{-r}e^r,$$

we obtain the desired equalities. □

Corollary 14.4.5. *For any nonzero finite difference module V over F, either $|\Phi|_{\mathrm{sp},V} = 0$ or there exists an integer $m \in \{1, \ldots, \dim_F V\}$ such that $|\Phi|_{\mathrm{sp},V}^m \in |F^\times|$.*

Definition 14.4.6. Let V be a nonzero finite difference module over F. We say that V is *pure of norm s* if all of the Jordan–Hölder constituents of V have spectral radius s. Note that V is pure of norm 0 if and only if $\Phi^{\dim_F V} = 0$. If V is pure of norm 1, we also say that V is *étale* or *unit-root*.

Remark 14.4.7. It is more common to classify pure modules using additive notation, i.e., in terms of the *slope* $(-\log s)$ instead of the norm s. (Correspondingly, pure modules are also called *isoclinic* modules.) For better or worse, we have decided here to keep the notation consistent with the multiplicative terminology used in the differential case. We will switch to the additive language only in the next section, in order to talk about Hodge and Newton polygons.

Proposition 14.4.8. *Let V be a nonzero finite difference module over F. Then V is pure of norm $s > 0$ if and only if*

$$|\Phi|_{\mathrm{sp},V'} = s, \qquad |\Phi^{-1}|_{\mathrm{sp},V'} = s^{-1}; \qquad (14.4.8.1)$$

in this case, $V^\vee$ is pure of norm s^{-1}.

Proof If V is pure of norm s, then (14.4.8.1) holds by Lemma 14.4.3 (parts (a) and (c)) and Lemma 14.4.4. Conversely, if (14.4.8.1) holds and W is an irreducible subquotient of V, then by Lemma 14.4.3(a),

$$|\Phi|_{\mathrm{sp},W'} \le |\Phi|_{\mathrm{sp},V'}, \qquad |\Phi^{-1}|_{\mathrm{sp},W'} \le |\Phi^{-1}|_{\mathrm{sp},V'}.$$

We thus have

$$1 \le |\Phi|_{\mathrm{sp},W'}|\Phi^{-1}|_{\mathrm{sp},W'} \le ss^{-1} = 1,$$

which forces $|\Phi|_{\mathrm{sp},W} = |\Phi|_{\mathrm{sp},W'} = s$ and $|\Phi^{-1}|_{\mathrm{sp},W'} = s^{-1}$. In particular, W is pure of norm s. Moreover, by Lemma 14.2.3, there is an isomorphism $W \cong F\{T\}/F\{T\}P$ for some twisted polynomial P; by Theorem 14.2.5, P has only one slope in its Newton polygon. By Lemma 14.4.4, that slope must equal $-\log s$, and so $W^\vee$ must be pure of norm s^{-1}, as then is $V^\vee$.   □

Corollary 14.4.9. *Let V_1, V_2 be nonzero finite difference modules over F which are pure of respective norms s_1, s_2. Then $V_1 \otimes V_2$ is pure of norm $s_1 s_2$.*

Note that this does not follow immediately from Lemma 14.4.3(b), because the tensor product of two irreducibles need not be irreducible.

Proof   If $s_1 s_2 = 0$, then $V_1 \otimes V_2$ is pure of norm 0. Otherwise, Lemma 14.4.3(b) plus one direction of Proposition 14.4.8 yields

$$|\Phi|_{\mathrm{sp},(V_1 \otimes V_2)'} = |\Phi|_{\mathrm{sp},V_1'}|\Phi|_{\mathrm{sp},V_2'} = s_1 s_2,$$

$$|\Phi^{-1}|_{\mathrm{sp},(V_1 \otimes V_2)'} = |\Phi^{-1}|_{\mathrm{sp},V_1'}|\Phi^{-1}|_{\mathrm{sp},V_2'} = s_1^{-1} s_2^{-1},$$

so the other direction of Proposition 14.4.8 implies that $V_1 \otimes V_2$ is pure of norm $s_1 s_2$.   □

Corollary 14.4.10. *Let V be a nonzero finite difference module over F. Then for any positive integer d, V is pure of norm s if and only if V becomes pure of norm s^d when viewed as a difference module over (F, φ^d).*

Proposition 14.4.11. *Let V be a nonzero finite difference module over F. Suppose that either*

(a) $|\Phi|_{\mathrm{sp},V} < 1$, or
(b) F is inversive and $|\Phi^{-1}|_{\mathrm{sp},V} < 1$.

Then $H^1(V) = 0$.

Proof   In case (a), given $v \in V$, the series

$$w = \sum_{i=0}^{\infty} \Phi^i(v)$$

converges to a solution of $w - \Phi(w) = v$. In case (b), the series

$$w = -\sum_{i=0}^{\infty} \Phi^{-i-1}(v)$$

does likewise.   □

Corollary 14.4.12. *Let V_1, V_2 are nonzero finite differential modules over F which are pure of respective norms s_1, s_2. If either*

(a) $s_1 < s_2$; or
(b) F is inversive and $s_1 > s_2$;

then any exact sequence $0 \to V_1 \to V \to V_2 \to 0$ splits.

Proof If $s_2 > 0$, then by Proposition 14.4.8 and Corollary 14.4.9, $V_2^\vee \otimes V_1$ is pure of norm s_1/s_2, so Proposition 14.4.11 gives the desired splitting. Otherwise, we must be in case (b), so we can pass to the opposite ring to make the same conclusion. □

If F is inversive, we again get a decomposition theorem.

Theorem 14.4.13. *Suppose that F is inversive. Let V be a finite difference module over F. Then there exists a unique direct sum decomposition*

$$V = \bigoplus_{s \geq 0} V_s$$

of difference modules, in which each nonzero V_s is pure of norm s. (Note that V is dualizable if and only if $V_0 = 0$.)

Proof This follows at once from Corollary 14.4.12. □

Remark 14.4.14. Note that if φ is the identity map on F, Theorem 14.4.13 simply reproduces the decomposition of V in which V_s consists of the generalized eigenspaces for all eigenvalues of norm s.

If F is not inversive, we only get a filtration instead of a decomposition.

Theorem 14.4.15. *Let V be a finite difference module over F. Then there exists a unique filtration*

$$0 = V_0 \subset V_1 \subset \cdots \subset V_l = V$$

of difference modules, such that each successive quotient V_i/V_{i-1} is pure of some norm s_i, and $s_1 > \cdots > s_l$. (Note that V is dualizable if and only if $V = 0$ or $s_l > 0$.)

Proof Start with any filtration $0 = V_0 \subset V_1 \subset \cdots \subset V_l = V$ with irreducible successive quotients. Let s_i be the radius of V_i/V_{i-1}. In case $s_i < s_{i+1}$, we may apply Corollary 14.4.12 to choose another difference module V_i' containing V_{i-1} and contained in V_{i+1}, such that $V_i'/V_{i-1}, V_{i+1}/V_i'$ are pure of slopes s_{i+1}, s_i, respectively.

By repeating this process, we eventually reach the case where $s_1 \geq \cdots \geq s_l$.

We obtain existence of a good filtration from the given filtration by simply omitting V_i whenever $s_i = s_{i-1}$. Uniqueness follows by tensoring with F' and invoking the uniqueness in Theorem 14.4.13. □

The following alternate characterization of pureness may be useful in some situations.

Proposition 14.4.16. *Let V be a finite difference module over F, and choose $\lambda \in F^\times$. Then V is pure of norm $|\lambda|$ if and only if there exists a basis of V on which Φ acts via λ times an element of $\mathrm{GL}_n(\mathfrak{o}_F)$.*

Proof If such a basis exists, then Proposition 14.4.8 implies that V is pure of norm $|\lambda|$. Conversely, if V is irreducible of spectral radius $|\lambda|$, then Lemma 14.4.4 provides a basis of the desired form. Otherwise, we proceed by induction on $\dim_F V$. Suppose we are given a short exact sequence $0 \to V_1 \to V \to V_2 \to 0$ in which V_1, V_2 admit bases of the desired form. Let $e_1, \ldots, e_m \in V$ form such a basis for V_1, and let $e_{m+1}, \ldots, e_n \in V$ lift such a basis for V_2. Then for $\mu \in F$ of sufficiently small norm,

$$e_1, \ldots, e_m, \mu e_{m+1}, \ldots, \mu e_n$$

will form a basis of V of the desired form. □

Remark 14.4.17. Note that whenever V is pure of positive norm, we can apply Proposition 14.4.16 after replacing Φ by some power of itself, thanks to Corollary 14.4.5.

14.5 Hodge and Newton polygons

The theory of Hodge and Newton polygons, which we introduced when studying nonarchimedean matrix inequalities in Chapter 4, admits a close analogue when considering difference algebra over a complete nonarchimedean field. Throughout this section, we continue to retain Hypothesis 14.4.1.

Definition 14.5.1. Let V be a finite difference module over F equipped with the supremum norm with respect to some basis. Let A be the matrix of action of Φ on this basis; define the *Hodge polygon* of V as the Hodge polygon of the matrix A (see Definition 4.3.3). Given the choice of the norm on V, this definition is independent of the choice of the basis: we can only change basis by a matrix $U \in \mathrm{GL}_n(\mathfrak{o}_F)$ to replace A by $U^{-1}A\varphi(U)$, and the fact that φ is an isometry ensures that $\varphi(U) \in \mathrm{GL}_n(\mathfrak{o}_F)$ also. As in the linear case, we list the Hodge slopes $s_{H,1}, \ldots, s_{H,n}$ in increasing order.

Definition 14.5.2. Let V be a finite difference module over F. Define the *Newton polygon* of V to have slopes $s_{N,1}, \ldots, s_{N,n}$ such that r appears with multiplicity equal to the dimension of the quotient of norm e^{-r} in Theorem 14.4.15.

Lemma 14.5.3. *Let V be a finite difference module over F. Then*

$$s_{H,1} + \cdots + s_{H,i} = -\log |\Phi|_{\wedge^i V} \quad (i = 1, \ldots, n)$$
$$s_{N,1} + \cdots + s_{N,i} = -\log |\Phi|_{\mathrm{sp}, \wedge^i V} \quad (i = 1, \ldots, n).$$

Proof The first assertion follows from the corresponding fact in the linear case (analogous to Lemma 4.1.9). The second assertion reduces to the fact that if V is irreducible of dimension n and spectral radius s, then $\wedge^i V$ has spectral radius s^i for $i = 1, \ldots, n$; this follows from the fact that in the basis given by Lemma 14.4.4, $\left|\wedge^i \Phi\right|_{\wedge^i V} = s^i$. $\qquad\square$

Corollary 14.5.4 (Newton above Hodge). *We have*

$$s_{N,1} + \cdots + s_{N,i} \geq s_{H,1} + \cdots + s_{H,i} \quad (i = 1, \ldots, n)$$

with equality for $i = n$.

As in the linear case (Theorem 4.3.11), we have a Hodge–Newton decomposition theorem.

Theorem 14.5.5. *Let V be a finite difference module over F equipped with a basis. If for some $i \in \{1, \ldots, n-1\}$ we have*

$$s_{N,i} < s_{N,i+1}, \qquad s_{N,1} + \cdots + s_{N,i} = s_{H,1} + \cdots + s_{H,i},$$

then we can change basis by a matrix in $\mathrm{GL}_n(\mathfrak{o}_F)$ so that the matrix of action of Φ becomes block upper triangular, with the top left block accounting for the first i Hodge and Newton slopes of V. Moreover, if F is inversive and $s_{H,i} > s_{H,i+1}$, we can ensure that the matrix of action of Φ becomes block diagonal.

Proof We first change basis by a matrix in $\mathrm{GL}_n(\mathfrak{o}_F)$ to ensure that the F-span of the first i basis vectors equals the step of the filtration of Theorem 14.4.15 consisting of norms greater than or equal to $\exp(-s_{N,i})$. We may then proceed as in Theorem 4.3.11. $\qquad\square$

Remark 14.5.6. Beware that the Newton polygon, unlike the Hodge polygon, cannot be directly read off from a matrix of action of Φ; see the exercises for a counterexample. However, this recipe does give the correct answer if the matrix of action of Φ is a companion matrix; this is a restatement of the following fact.

Proposition 14.5.7. *If $V \cong F\{T\}/F\{T\}P$, then the Newton polygon of V coincides with that of P.*

Proof This reduces to Lemma 14.4.4 using Theorem 14.2.5. □

We also obtain the following analogue of Proposition 4.4.11.

Proposition 14.5.8. *Let V be a finite difference module over F, equipped with the supremum norm for some basis. For $i = 1, 2, \ldots,$ let $s_{H,1,i}, \ldots, s_{H,n,i}$ be the Hodge slopes of V as a difference module over (F, φ^i). Then as $i \to \infty$, the quantities $i^{-1}s_{H,1,i}, \ldots, i^{-1}s_{H,n,i}$ converge to the Newton slopes of V.*

Proof The claim is independent of the choice of basis, so by Theorem 14.4.15 we may reduce to considering a pure module. In that case, Lemma 14.4.2 implies that $i^{-1}s_{H,1,i} \to s_{N,1}$ as $i \to \infty$.

Note that $s_{N,1} = \cdots = s_{N,n}$ by purity and that $i^{-1}s_{H,1,i} \geq i^{-1}s_{H,j,i}$ for $j \geq 1$. Also, the Newton slopes of V as a difference module over (F, φ^i) are $is_{N,1}, \ldots, is_{N,n}$; by the equality case of Corollary 14.5.4, we obtain $i^{-1}s_{H,1,i} + \cdots + i^{-1}s_{H,n,i} = s_{N,1} + \cdots + s_{N,n}$. All of this plus the fact that $i^{-1}s_{H,1,i} \to s_{N,1}$ as $i \to \infty$ yield the desired convergence. □

Proposition 14.5.9. *Suppose that $F = K$ (i.e., F is discretely valued) and $R = F[\![t]\!]_0$ carries the structure of an isometric complete nonarchimedean difference ring for the norm $|\cdot|_1$. Let M be a finite free difference module over R with least Newton slope c. Then there exists a basis of M with respect to which for each positive integer m, the least Hodge slope of Φ^m is the least element of $v(F^\times)$ greater than or equal to cm.*

Proof Recall that if K is discrete, then $\mathcal{E}$ is the completed fraction field of R (see Definition 9.4.3). Construct a suitable basis of $M \otimes_R \mathcal{E}$ by imitating the proof of Lemma 4.3.13, then apply Lemma 8.6.1. □

14.6 The Dieudonné–Manin classification theorem

We next prove a difference-algebraic analogue of the fact that linear transformations over an algebraically closed field can be put into Jordan normal form. The analogous statement turns out to be even better: in this setting, all objects are semisimple. We continue to retain Hypothesis 14.4.1.

We first define some standard difference modules of a particularly simple form, to be used as building blocks.

Definition 14.6.1. For $\lambda \in F$ and d a positive integer, let $V_{\lambda,d}$ be the difference module over F with basis $e_1, \ldots, e_d$ such that

$$\Phi(e_1) = e_2, \quad \ldots, \quad \Phi(e_{d-1}) = e_d, \quad \Phi(e_d) = \lambda e_1.$$

Lemma 14.6.2. *Suppose $\lambda \in F^\times$ and the positive integer d are such that there is no $i \in \{1, \ldots, d-1\}$ such that $|\lambda|^{i/d} \in |F^\times|$. Then $V_{\lambda,d}$ is irreducible.*

Proof Note that

$$\Phi^d(e_i) = \varphi^{i-1}(\lambda)e_i \qquad (i = 1, \ldots, d).$$

Hence by Proposition 14.4.16, $V_{\lambda,d}$ is pure of norm $\lambda^{1/d}$, as then is any submodule. But if the submodule were proper and nonzero, we would have a violation of Corollary 14.4.5. $\qquad\square$

We show next that if F has sufficiently large residue field, one can classify all dualizable finite difference modules in terms of the standard modules $V_{\lambda,d}$. One can always enlarge F to reach this case; see the exercises.

Theorem 14.6.3. *Let F be a complete discretely valued field equipped with an isometric endomorphism φ, such that κ_F is strongly difference-closed. Then every dualizable finite difference module over F can be split (non-uniquely) as a direct sum of submodules, each of the form $V_{\lambda,d}$ for some λ, d. Moreover, given any generator π of $\mathfrak{m}_F$, we can force each λ to be a power of π.*

Proof We first check that if V is pure of norm 1, then V is trivial; for this step, we only need κ_F to be weakly difference-closed. We must show that for any $A \in \mathrm{GL}_n(\mathfrak{o}_F)$, there exists a convergent sequence $U_1, U_2, \ldots \in \mathrm{GL}_n(\mathfrak{o}_F)$ such that

$$U_m^{-1} A \varphi(U_m) \equiv I_n \pmod{\pi^m}.$$

Specifically, we will insist that $U_{m+1} \equiv U_m \pmod{\pi^m}$. Finding U_1 amounts to trivializing a dualizable difference module of dimension n over κ_F, which is possible because κ_F is assumed to be weakly difference-closed. For $m > 1$, given U_m, we wish to set $U_{m+1} = U_m(I_n + \pi^m X_m)$ for some X_m, in such a way that

$$(I_n + \pi^m X_m)^{-1}(U_m^{-1} A \varphi(U_m))(I_n + \varphi(\pi)^m \varphi(X_m)) \equiv I_n \pmod{\pi^{m+1}}.$$

Since already $U_m^{-1} A \varphi(U_m) \equiv I_n \pmod{\pi^m}$, this amounts to solving

$$-X_m + \pi^{-m}(U_m^{-1} A \varphi(U_m) - I_n) + \varphi(\pi)^m \pi^{-m} \varphi(X_m) \equiv 0 \pmod{\pi},$$

which is guaranteed to be possibly thanks to criterion (c) from Lemma 14.3.3.

We next check that φ is surjective on $\mathfrak{o}_F$, which implies that F is inversive. Given $x \in \mathfrak{o}_F$, it suffices to exhibit a sequence $y_0, y_1, y_2, \ldots$ with $y_m \equiv y_{m+1} \pmod{\pi^m}$, such that $\varphi(y_m) \equiv x \pmod{\pi^m}$. We start with $y_0 = 0$; given y_m, solving the equation

$$\varphi(\pi^{-m}(y_{m+1} - y_m)) \equiv \varphi(\pi)^{-m}(x - \varphi(y_m)) \pmod{\pi}$$

is possible because κ_F is inversive.

We next check that if V is trivial, then $H^1(V) = 0$. Given $x \in \mathfrak{o}_F$, it suffices to exhibit a sequence $y_0, y_1, y_2, \ldots$ with $y_m \equiv y_{m+1} \pmod{\pi^m}$, such that $\varphi(y_m) - y_m \equiv x \pmod{\pi^m}$. Again, we start with $y_0 = 0$; given y_m, solving the equation

$$\pi^{-m}\varphi(\pi^m)\varphi(\pi^{-m}(y_{m+1}-y_m)) - \pi^{-m}(y_{m+1}-y_m) \equiv \pi^{-m}(x-\varphi(y_m)+y_m) \pmod{\pi}$$

is possible by criteria (b) and (c) from Lemma 14.3.3. (We must use (b) to remove the leading coefficient on the first term before applying (c).)

At this point, we may apply Theorem 14.4.13 to reduce the desired result to the case where V is pure of norm $s > 0$. Let d be the smallest positive integer such that $s^d = |\pi^m|$ for some integer m. Then the first paragraph implies that $\pi^{-m}\Phi^d$ fixes some nonzero element of V; this gives us a nonzero map from $V_{\pi^m, d}$ to V. By Lemma 14.6.2, this map must be injective. Repeating this argument, we write V as a successive extension of copies of $V_{\pi^m, d}$. However, $V_{\pi^m, d}^\vee \otimes V_{\pi^m, d}$ is pure of norm 1, so has trivial H^1 as above. Thus V splits as a direct sum of copies of $V_{\pi^m, d}$, as desired. □

By Proposition 14.3.4, Theorem 14.6.3 has the following immediate corollary.

Corollary 14.6.4. *Let F be a complete discretely valued field such that κ_F is algebraically closed of characteristic $p > 0$. Let $\varphi: F \to F$ be an isometric automorphism lifting a power of the absolute Frobenius on κ_F. Then every dualizable finite difference module over F can be split (non-uniquely) as a direct sum of difference submodules, each of the form $V_{\lambda, d}$ for some $\lambda \in F^\times$ and some positive integer d coprime to the valuation of λ. Moreover, given any generator π of $\mathfrak{m}_F$, we can force each λ to be a power of π.*

Remark 14.6.5. Let k be an algebraically closed field of characteristic $p > 0$, let $W(k)$ be the ring of p-typical Witt vectors (i.e., the unique complete discrete valuation ring with residue field k and maximal ideal generated by p), and put $F = \mathrm{Frac}(W(k))$. Then for each power q of p, there is a unique Frobenius lift φ on $W(k)$, namely the Witt vector Frobenius. For this data, Corollary 14.6.4 with $\pi = p$ is precisely the *Dieudonné–Manin theorem*, i.e., the classification theorem of rational Dieudonné modules over k. (For more on Witt vectors, see the chapter notes and the exercises.)

Corollary 14.6.6. *Let F be a complete discretely valued field equipped with an isometric endomorphism φ, such that κ_F is strongly difference-closed. Then for any finite difference module V over F, $H^1(V) = 0$.*

Proof By Theorem 14.4.13, this reduces to the case where V is pure of some norm $s \geq 0$. If $s > 0$, then V is dualizable and Theorem 14.6.3 implies the claim. If $s = 0$, then the action of Φ on V is nilpotent, so we may check the claim for V by checking the claim for the kernel and image of V. We may thus reduce to the case where the action of Φ on V is zero, in which case this reduces to the fact that F is inversive (because φ is isometric). $\qquad\square$

Notes

The parallels between difference and differential algebra are quite close, enough so that a survey of references for difference algebra strongly resembles its differential counterpart. An established but rather dry reference is [105]; a somewhat more lively modern reference, which develops difference Galois theory under somewhat restrictive conditions, is [374]. We again mention [15] as a useful unifying framework for difference and differential algebra.

The choice of the modifier "difference" in the phrase "difference algebra" is motivated by the following basic example. Consider the automorphism on the ring of entire functions $f: \mathbb{C} \to \mathbb{C}$ given by $\varphi(f)(z) = f(z + q)$ for some q. The difference operator

$$\Delta(f) = \frac{\varphi(f) - f}{q}$$

obeys the twisted Leibniz rule

$$\Delta(fg) = \varphi(f)\Delta(g) + \Delta(f)(g),$$

and its limit as $q \to 1$ is the usual derivation with respect to z. We note in passing that forming a similar operator when φ is a Frobenius lift (and $q = p$) leads to the *p-derivations* of Joyal [218, 219] and Buium [73, 74], which have recently resurfaced in the Bhatt–Scholze construction of *prismatic cohomology* [64].

Proposition 14.3.4 appears in SGA7 [129, Exposé XXII, Corollaire 1.1.10], wherein Katz attributes it to Lang. Indeed, it is a special case of the nonabelian Artin–Schreier theory associated to an algebraic group over a field of positive characteristic (in our case GL_n), via the *Lang torsor*; see [285]. It is also the basis for the theory of (φ, Γ)-modules, which we will mention briefly in Appendix C.

Suppose that k is a perfect field of characteristic $p > 0$, and that C_k is a complete discrete valuation ring with residue field k and maximal ideal generated by p (e.g., $k = \mathbb{F}_p$ and $C_k = \mathbb{Z}_p$). It was originally noticed by Teichmüller that each element $x \in k$ has a unique lift $[x] \in C_k$ admitting p^n-th roots for all n; these are also the unique system of lifts forming a multiplicative subset. One can use these *multiplicative lifts*[1] as digits to form a canonical base p expansion of any element of C_k; the arithmetic operations can be expressed using certain polynomials in the p-power roots of these digits. These polynomials were used by Witt to prove that C_k exists, is unique, and is functorial in k. More generally, Witt gave a functor for each p defined on arbitrary rings (the functor of *p-typical Witt vectors*), and associating k as above to C_k; he also gave a natural functor through which all of the functors for different choices of p factor (the functor of *big Witt vectors*). These have applications far beyond their origins, in fields as diverse as arithmetic geometry, algebraic topology, and combinatorics. See [200] for a comprehensive summary.

In the special case of the difference field $\mathrm{Frac}(W(k))$, with k perfect of characteristic $p > 0$ and φ equal to the functorial lift of the absolute p-power Frobenius, a number of results in this chapter appear (in marginally less generality) in [224], such as the following.

- Corollary 14.5.4 reproduces Mazur's [224, Theorem 1.4.1].
- Theorem 14.5.5 is [224, Theorem 1.6.1].
- Proposition 14.5.8 reproduces [224, Corollary 1.4.4] without requiring non-negative Hodge slopes (as Katz does in his "basic slope estimate" [224, 1.4.3]).
- Proposition 14.5.9 reproduces (and slightly generalizes) one case of [224, Theorem 2.6.1].

For the original classification of rational Dieudonné modules over an algebraically closed field, see Manin's original paper [302] or the book of Demazure [131]. We do not have a prior reference for Theorem 14.6.3, but nonetheless we do not consider it to be original.

An equal-characteristic analogue of Remark 14.6.5 is to take $F = k((z))$ for k an algebraically closed field of characteristic $p > 0$, with φ acting as Frobenius on k and trivially on z. This special case of Corollary 14.6.4 is due to Laumon [288, Theorem 2.4.5].

As in Chapter 4, one can interpret what we have done here as the special case for GL_n of a construction for any reductive algebraic group (compare the discussion of Proposition 14.3.4 earlier in these notes). This point of view was originally introduced by Kottwitz [277, 278]; a full development of the analogy was subsequently pursued by Kottwitz [279] and Csima [120].

---

[1] In the existing literature, multiplicative lifts are more commonly called *Teichmüller lifts*. We recommend changing this terminology on account of Teichmüller's well-documented complicity with the Nazi regime.

Exercises

14.1 For M_1, M_2 difference modules over a difference ring R with M_1 dualizable, give a canonical identification of $H^1(M_1^\vee \otimes M_2)$ with the Yoneda extension group $\mathrm{Ext}(M_1, M_2)$. (Hint: as in Lemma 5.3.3, you may wish to first reduce to the case $M_1 = R$.)

14.2 Prove Lemma 14.4.3. (Hint: Lemma 14.4.2 may be helpful, particularly for (c) and (d).)

14.3 Let F be a difference field (of arbitrary characteristic) containing an element x such that $\varphi(x) = \lambda x$ for some λ fixed by φ which is not a root of unity. Prove that every finite difference module for M admits a cyclic vector. (Hint: imitate the proof of Theorem 5.4.2. At the key step, instead of getting a polynomial in s, you should get a polynomial in λ^s; the root of unity condition forces λ^s to take infinitely many values.)

14.4 Let F be the completion of $\mathbb{Q}_p(t)$ for the 1-Gauss norm, viewed as a difference field for φ equal to the substitution $t \mapsto t^p$. Let V be the difference module corresponding to the matrix

$$A = \begin{pmatrix} 1 & t \\ 0 & p \end{pmatrix}.$$

Prove that there is a nonsplit short exact sequence $0 \to V_1 \to V \to V_2 \to 0$ with V_1, V_2 pure of norms s_1, s_2 with $s_1 < s_2$.

14.5 Here is a beautiful example from [224, §1.3] (attributed to B. Gross). Let p be a prime congruent to 3 modulo 4, put $F = \mathbb{Q}_p(i)$ with $i^2 = -1$, and let φ be the automorphism $i \mapsto -i$ of F over $\mathbb{Q}_p$. Define a difference module V of rank 2 over F using the matrix

$$A = \begin{pmatrix} 1-p & (p+1)i \\ (p+1)i & p-1 \end{pmatrix}.$$

Compute the Newton polygons of A and V and verify that they do not coincide. (Hint: find another basis of V on which Φ acts diagonally.)

14.6 (a) Prove that every difference field can be embedded into a strongly difference-closed field. (Hint: use your favorite equivalent of the axiom of choice, e.g., Zorn's lemma or transfinite induction.)

(b) Prove that every complete isometric discretely valued difference field can be embedded into a complete isometric difference field with the same value group having strongly difference-closed residue field.

14.7 Prove that Theorem 14.6.3 continues to hold if the hypothesis that F is discretely valued is replaced by one of the following hypotheses.

Formalism of difference algebra

    (a) There exists a constant $\epsilon > 0$ with the following property: for $c \in \{0, 1\}$ and $x \in \mathfrak{o}_F$, there exists $y \in \mathfrak{o}_F$ with $|\varphi(y) - cy - x| < \epsilon |x|$. (Hint: use ϵ to replace the π-adically convergent sequences in the proof of Theorem 14.6.3.)

    (b) The field F is spherically complete. (Hint: attempt to form convergent sequences as in (a), then invoke spherical completeness if the sequences fail to converge.)

14.8   This exercise is related to the construction of Witt vectors. Let k, ℓ be perfect fields of characteristic $p > 0$. Suppose C_k, C_ℓ are complete discrete valuation rings with residue fields identified with k, ℓ, respectively, whose maximal ideals are generated by p. Prove that any homomorphism $k \to \ell$ lifts uniquely to a homomorphism $C_k \to C_\ell$. (Hint: read the description of Witt vectors in the notes.)

14.9   Let F be a difference field on which the action of φ is via the identity map. Prove that for any strongly difference-closed field F' containing F, the action of φ on F' is *not* via the identity map.

15

Frobenius modules

Having introduced the general formalism of difference algebra, and made a more careful study over a complete nonarchimedean field, we specialize to the sort of power series rings over which we studied differential algebra. Most of the rings are ones we have seen before, but we encounter a couple of new variations, notably the *Robba ring*. This ring consists of power series convergent on some annulus of outer radius 1, but with unspecified inner radius which may vary with the choice of the series. This may seem to be a strange construction at first, but it is rather natural from the point of view of difference algebra: the endomorphisms we will consider (Frobenius lifts) do not preserve the region of convergence of an individual series, but do act on the Robba ring as a whole.

This chapter serves mostly to set definitions and notation for what follows. That said, one nontrivial result here is the behavior of the Newton polygon under specialization. Remember that Hypothesis 14.0.1 is in force, so the field K will always be discretely valued.

15.1 A multitude of rings

One can talk about Frobenius structures on a variety of rings; for convenience, we review the definitions of some of the rings introduced in Chapter 8, then add some special notations that will be useful later.

Remark 15.1.1. The following rings were defined in Chapter 8:

$$K\langle\alpha/t, t/\beta\rangle = \left\{\sum_{i\in\mathbb{Z}} c_i t^i : c_i \in K, \lim_{i\to-\infty} |c_i||\alpha^i = 0, \lim_{i\to\infty} |c_i||\beta^i = 0\right\}$$

$$K[\![t]\!]_0 = \left\{\sum_{i=0}^{\infty} c_i t^i : c_i \in K, \sup_i\{|c_i|\} < \infty\right\}$$

$$K\{t\} = \left\{\sum_{i=0}^{\infty} c_i t^i : c_i \in K, \lim_{i\to\infty} |c_i||\rho^i = 0 \quad (\rho \in (0,1))\right\}$$

$$K\langle\alpha/t, t]\!]_0 = \left\{\sum_{i\in\mathbb{Z}} c_i t^i : c_i \in K, \lim_{i\to-\infty} |c_i||\alpha^i = 0, \sup_i\{|c_i|\} < \infty\right\}$$

$$K\langle\alpha/t, t\} = \left\{\sum_{i\in\mathbb{Z}} c_i t^i : c_i \in K, \lim_{i\to-\infty} |c_i||\alpha^i = 0, \lim_{i\to\infty} |c_i||\rho^i = 0 \quad (\rho \in (0,1))\right\}.$$

Definition 15.1.2. For later use, we give special notations to certain rings appearing in this framework. We already have defined $\mathcal{E}$ to be the completion of $\mathfrak{o}_K((t)) \otimes_{\mathfrak{o}_K} K = K[\![t]\!]_0[t^{-1}]$ for the 1-Gauss norm; that is, $\mathcal{E}$ consists of formal sums $\sum_i c_i t^i$ which have bounded coefficients and satisfy $|c_i| \to 0$ as $i \to -\infty$. Since we are assuming that K is discretely valued, this is a complete nonarchimedean field with residue field $\kappa_K((t))$. We next set

$$\mathcal{E}^\dagger = \bigcup_{\alpha\in(0,1)} K\langle\alpha/t, t]\!]_0;$$

that is, $\mathcal{E}^\dagger$ consists of formal sums $\sum c_i t^i$ which have bounded coefficients and converge in some range $\alpha \le |t| < 1$. This ring is sometimes called the *bounded Robba ring*, since it consists of the bounded elements of the Robba ring (see Definition 15.1.4). Note that it can also be written as

$$\mathcal{E}^\dagger = \bigcup_{\alpha\in(0,1)} K[\![\alpha/t, t]\!]_0.$$

Lemma 15.1.3. *The ring $\mathcal{E}^\dagger$ has the following properties.*

(a) *The ring $\mathcal{E}^\dagger$ is a field.*
(b) *Under the 1-Gauss norm $|\cdot|_1$, the valuation ring $\mathfrak{o}_{\mathcal{E}^\dagger}$ is a local ring with maximal ideal $\mathfrak{m}_K \mathfrak{o}_{\mathcal{E}^\dagger}$.*
(c) *The field $\mathcal{E}^\dagger$, equipped with $|\cdot|_1$, is henselian (see Remark 3.0.2).*

This last property implies that finite separable extensions of $\kappa_{\mathcal{E}^\dagger} = \kappa_K((t))$ lift functorially to finite étale extensions of $\mathfrak{o}_{\mathcal{E}^\dagger}$ (see Definition 19.4.6) and thus give rise to unramified extensions of $\mathcal{E}^\dagger$. In particular, the maximal unramified extension $\mathcal{E}^{\dagger,\mathrm{unr}}$ carries an action of the absolute Galois group of $\kappa_K((t))$.

Proof To check (a), note that any nonzero $x \in \mathcal{E}^\dagger$ has Newton polygon of finite width (since K is discretely valued). We can thus choose α so that $x \in K\langle\alpha/t, t]\!]_0$ has no slopes between 0 and $-\log\alpha$. In this case, x is a unit in $K\langle\alpha/t, t]\!]_0$ by Lemma 8.2.6(c), yielding (a). We deduce (b) as an immediate corollary of (a).

To prove (c), it suffices to check the criterion of Remark 3.0.2: for any monic polynomial $P(x) \in \mathfrak{o}_{\mathcal{E}^\dagger}[x]$ and any simple root $\bar{r} \in \kappa_{\mathcal{E}^\dagger}$ of $\overline{P} \in \kappa_{\mathcal{E}^\dagger}[x]$, there exists a unique root $r \in \mathfrak{o}_{\mathcal{E}^\dagger}$ of P lifting $\bar{r}$. In fact, there exists a unique such root $r \in \mathfrak{o}_{\mathcal{E}}$ by Hensel's lemma (Remark 2.2.3), so it suffices to produce such a root in $\mathfrak{o}_{\mathcal{E}^\dagger}$.

We may rescale P by a unit in $\mathfrak{o}_{\mathcal{E}^\dagger}$ to force $\overline{P}'(\bar{r}) = 1$. Choose $r_0 \in \mathfrak{o}_{\mathcal{E}^\dagger}$ lifting $\bar{r}$. We can then choose $\alpha \in (0, 1)$ so that $K\langle\alpha/t, t]\!]_0$ contains both r_0 and the coefficients of P, and in addition $|P(r_0)|_\alpha, |1 - P'(r_0)|_\alpha < 1$. Since $|P(r_0)|_1, |1-P'(r_0)|_1 < 1$, this implies $|P(r_0)|_\rho, |1-P'(r_0)|_\rho < 1$ for $\rho \in [\alpha, 1)$ by Proposition 8.2.3(c).

We finish with a standard application of Newton's method. Given r_0 as defined above, for $i \geq 0$ set $r_{i+1} = r_i - P(r_i)/P'(r_i)$. We find by induction on i (exercise) that $|P(r_{i+1})|_\rho \leq |P(r_i)|_\rho^2$ and $|1 - P'(r_i)|_\rho < 1$ for $\rho \in [\alpha, 1)$. Consequently, the r_i form a Cauchy sequence under $|\cdot|_\rho$ for each $\rho \in [\alpha, 1)$; since $K\langle\alpha/t, t]\!]_0$ is Fréchet complete for these norms (Proposition 8.2.5), the sequence has a limit r of the desired form. This yields (c). □

We next confect another novel but important ring.

Definition 15.1.4. Put

$$\mathcal{R} = \bigcup_{\alpha \in (0,1)} K\langle\alpha/t, t\};$$

that is, $\mathcal{R}$ consists of formal sums $\sum_i c_i t^i$ which converge in some range $\alpha \leq |t| < 1$, but need not be bounded. The ring $\mathcal{R}$ is commonly known as the *Robba ring* with coefficients in K.

Remark 15.1.5. Beware that since $\mathcal{R}$ consists of series with unbounded coefficients, the 1-Gauss norm $|\cdot|_1$ is not defined on all of $\mathcal{R}$. We will conventionally write $|x|_1 = \infty$ if $x \in \mathcal{R}$ has unbounded coefficients. A related issue is that $\mathcal{R}$ is not a field; indeed, it is not even noetherian, by a similar argument as for $K\{t\}$ (see the exercises for Chapter 8). We will return to this and related ring-theoretic issues concerning $\mathcal{R}$ later (see Section 16.2).

15.2 Substitutions and Frobenius lifts

We next define a class of endomorphisms on these rings.

Lemma 15.2.1. *Let $u \in \mathfrak{o}_{\mathcal{E}^\dagger}$ be an element whose image in the residue field of $\kappa_{\mathcal{E}} = \kappa_K((t))$ has t-adic valuation equal to some positive integer q.*

(a) There exist $\epsilon \in (0,1)$, $\eta \in (0,1)$, and $c \in \mathfrak{o}_K^\times$ such that $u \in K\langle\epsilon/t, t\rrbracket_0$ and

$$|ut^{-q} - c|_{\rho^{1/q}} \leq \eta \qquad (\rho \in [\epsilon, 1]).$$

(b) For $\beta, \gamma \in [\epsilon, 1)$ with $\beta \leq \gamma$,

$$\sum_i f_i t^i \in K\langle\beta/t, t/\gamma\rangle \implies \sum_i f_i u^i \in K\langle\beta^{1/q}/t, t/\gamma^{1/q}\rangle$$

and

$$\left|\sum_i f_i t^i\right|_\beta = \left|\sum_i f_i u^i\right|_{\beta^{1/q}}.$$

Proof   Write $u = \sum_i u_i t^i$ with $u_i \in \mathfrak{o}_K$ and take $c = u_i$. Since $|ut^{-q} - c|_1 < 1$, by continuity we can choose $\epsilon \in (0,1)$ and $\eta \in (0,1)$ satisfying (a). This inequality implies $|u^i t^{-qi} - c^i|_{\rho^{1/q}} \leq \eta$ for all $i \in \mathbb{Z}$. To verify (b), note that for $\beta, \gamma \in [\epsilon, 1)$ with $\beta \leq \gamma$ and $f = \sum_i f_i t^i \in K\langle\beta/t, t/\gamma\rangle$,

$$|f|_\beta = \left|\sum_i f_i c^i t^{qi}\right|_{\beta^{1/q}} \geq \eta \left|\sum_i f_i (c^i t^{qi} - u^i)\right|_{\beta^{1/q}}. \qquad \square$$

Corollary 15.2.2. *Let $u \in \mathfrak{o}_{\mathcal{E}}$ be an element whose image in the residue field of $\kappa_{\mathcal{E}} = \kappa_K((t))$ is a nonzero element of $t\kappa_K[\![t]\!]$.*

(a) There exists a unique K-linear endomorphism $\varphi_u \colon \mathfrak{o}_{\mathcal{E}} \to \mathfrak{o}_{\mathcal{E}}$ given by

$$\sum_i c_i t^i \mapsto \sum_i c_i u^i,$$

where the sum on the right converges for the t-adic topology modulo each power of p. (That is, it converges for the "weak topology"; see Definition C.1.5.) This map extends uniquely to an endomorphism of $\mathcal{E}$.

(b) If $u \in \mathcal{E}^\dagger$, then for each $\alpha \in (0,1)$, there exists $\alpha' \in (0,1)$ such that φ_u carries $K\langle\alpha/t, t\rrbracket_0$ into $K\langle\alpha'/t, t\rrbracket_0$. In particular, φ_u carries $\mathcal{E}^\dagger$ into itself (and likewise for $\mathfrak{o}_{\mathcal{E}^\dagger}$).

(c) If $u \in \mathcal{E}^\dagger$, then for each $\alpha \in (0,1)$, there exists $\alpha' \in (0,1)$ such that φ_u extends continuously (for the Fréchet topology) to an endomorphism from $K\langle\alpha/t, t\}_0$ into $K\langle\alpha'/t, t\}_0$. In particular, φ_u carries $\mathcal{R}$ into itself.

(d) If $u \in \bigcup_\alpha K\langle\alpha/t, t\rangle$, then for each $\alpha \in (0, 1)$, there exists $\alpha' \in (0, 1)$ such that φ_u extends continuously (for the Fréchet topology) to an endomorphism from $K\langle\alpha/t, t\rangle$ into $K\langle\alpha'/t, t\rangle$. In particular, φ_u carries $\mathcal{R}$ into itself.

(e) If $u \in K[\![t]\!]_0$, then φ_u carries $K[\![t]\!]_0$ into itself. Moreover, the extension of φ_u to $\mathcal{R}$ from (c) carries $K\{t\}$ into itself.

(f) If $u \in F_1$, then φ_u carries F_1 into itself.

Proof We may deduce (b), (c), (d) directly from Lemma 15.2.1. To deduce (a), note that to check convergence for the p-adic topology modulo p^n for some fixed n, we may replace u with an element of $\mathfrak{o}_{\mathcal{E}^\dagger}$ congruent to u modulo p^n. From (a), both (e) and (f) are immediate. $\qquad\square$

Definition 15.2.3. Let R be one of the following rings:

- $K\langle t\rangle$, $K[\![t]\!]_0$, or $K\{t\}$;
- the union of $K\langle\alpha/t, t\rangle$, $K\langle\alpha/t, t\rangle_0$, or $K\langle\alpha/t, t\}$ over all $\alpha \in (0, 1)$;
- F_1, the completion of $K(t)$ for the 1-Gauss norm;
- $\mathcal{E}$, the completion of $K[\![t]\!]_0[t^{-1}]$ for the 1-Gauss norm;
- $\mathcal{R}$, the Robba ring (see Definition 15.1.4).

By a *K-linear substitution* on R, we will mean an endomorphism φ_u of the form given by Corollary 15.2.2. Note that the element u can be recovered as $\varphi_u(t)$. More generally, for $\varphi_K : K \to K$ an isometry, by a *φ_K-semilinear substitution* on R, we will mean an endomorphism which is the composition of a K-linear substitution with the map $\sum_i c_i t^i \mapsto \sum_i \varphi_K(c_i) t^i$.

Let q be a power of p. By a *q-power Frobenius lift* on R, we will mean a φ_K-semilinear substitution φ on R for some isometry φ_K with the property that $|\varphi(t) - t^q|_1 < 1$. The most important case is when φ is *absolute*, i.e., when φ_K lifts the q-power Frobenius on κ_K. (These are not the most permissive conditions possible; see the chapter notes.)

Remark 15.2.4. Note that unless φ is absolute, the property that it is a Frobenius lift depends on the implicit choice of the series parameter t; that is, it is not invariant under isometric automorphisms of R.

Remark 15.2.5. Note that in Definition 15.2.3, one cannot define a Frobenius lift on an individual ring like $K\langle\alpha/t, t\rangle_0$; for instance, the simple substitution $t \to t^q$ carries $K\langle\alpha/t, t\rangle_0$ to $K\langle\alpha^{1/q}/t, t\rangle_0$. One can make this remark more quantitative, as in the following lemma.

Definition 15.2.6. Let φ be a Frobenius lift on $K[\![t]\!]_0$. We say that φ is *centered* if there exists $\lambda \in \mathfrak{m}_K$ such that

$$\varphi(t - \lambda) \equiv 0 \pmod{t - \lambda}.$$

We call such λ a *center* of φ; it will follow from Lemma 15.2.7 that λ is unique if it exists, and it always exists if φ_K is inversive (but not always otherwise; see the exercises). We say that φ is *zero-centered* if its center is equal to 0, i.e., if $\varphi(t) \equiv 0 \pmod{t}$.

Lemma 15.2.7. *Suppose that φ is a Frobenius lift on $K[\![t]\!]_0$ and that $\varphi_K : K \to K$ is inversive. Then φ is centered.*

Proof Exercise. $\square$

15.3 Generic versus special Frobenius

For difference modules over $K[\![t]\!]_0$, there are two natural Newton polygons and an important relationship between them.

Definition 15.3.1. Let M be a finite free difference module over $K[\![t]\!]_0$, for φ a centered Frobenius lift. Define the *generic Newton polygon* of M to be the Newton polygon of $M \otimes_{K[\![t]\!]_0} \mathcal{E}$. Define the *special Newton polygon* of M to be the Newton polygon of $M/(t - \lambda)M$, for λ the center of φ.

The following result is sometimes called the *semicontinuity theorem* for Newton polygons.

Theorem 15.3.2 (Grothendieck, Katz). *Let M be a finite free difference module over $K[\![t]\!]_0$, for φ a centered Frobenius lift. Then the special Newton polygon lies on or above the generic Newton polygon, with the same endpoints.*

Proof Choose a basis of M, and use it to define supremum norms on $M \otimes_{K[\![t]\!]_0} \mathcal{E}$ and $M/(t - \lambda)M$. Then it is evident that for any positive integer n, the Hodge polygon of Φ^n acting on $M/(t - \lambda)M$ lies on or above the Hodge polygon of Φ^n acting on $M \otimes_{K[\![t]\!]_0} \mathcal{E}$, with the same endpoints. If we divide all slopes by n and take limits as $n \to \infty$, then Proposition 14.5.8 implies that the generic (resp. special) Hodge slopes converge to the generic (resp. special) Newton slopes. $\square$

As in the comparison of Hodge and Newton polygons, one gets a decomposition result in case the special and generic Newton polygons touch somewhere.

Theorem 15.3.3. *Let M be a finite free difference module of rank n over $K[\![t]\!]_0$, for φ a centered Frobenius lift. Let $s_{g,1} \le \cdots \le s_{g,n}$ and $s_{s,1} \le \cdots \le s_{s,n}$ be the generic and special Newton slopes, respectively. Suppose that for some $i \in \{1, \ldots, n-1\}$,*

$$s_{s,i} < s_{s,i+1}, \qquad s_{g,1} + \cdots + s_{g,i} = s_{s,1} + \cdots + s_{s,i}.$$

Then there exists a difference submodule N of M with M/N free whose generic and special Newton slopes are $s_{g,1}, \ldots, s_{g,i}$ and $s_{s,1}, \ldots, s_{s,i}$, respectively. Moreover, if $s_{g,i} < s_{g,i+1}$, then N is unique.

We will relax the hypothesis for uniqueness later (Corollary 16.4.7).

Proof We may assume φ is zero-centered. Uniqueness in case $s_{g,i} < s_{g,i+1}$ follows from the uniqueness in $M \otimes_{K[\![t]\!]_0} \mathcal{E}$, as in Theorem 14.4.15. For existence, we first replace φ by a suitable power to ensure that all of the slopes are in the additive value group of K; we then apply Proposition 14.4.16 followed by Lemma 8.6.1 to change basis in M, so as to ensure that the generic Hodge slopes of M are also equal to $s_{g,1}, \ldots, s_{g,n}$.

If $s_{s,H,1}, \ldots, s_{s,H,n}$ denote the special Hodge slopes in this basis, then

$$s_{s,1} + \cdots + s_{s,i} \geq s_{s,H,1} + \cdots + s_{s,H,i}$$

by Corollary 14.5.4, but also

$$s_{s,H,1} + \cdots + s_{s,H,i} \geq s_{g,1} + \cdots + s_{g,i}$$

as in the proof of Theorem 15.3.2 (since $s_{g,1}, \ldots, s_{g,i}$ match the generic Hodge slopes of M). Consequently, $s_{s,1} + \cdots + s_{s,i} = s_{s,H,1} + \cdots + s_{s,H,i}$; that is, for this basis, the condition of Theorem 14.5.5 is also satisfied by M/tM.

We can thus change basis over $\mathfrak{o}_K[\![t]\!]$ to obtain a new basis of M on which the action of Φ is via a block matrix

$$A_0 = \begin{pmatrix} B_0 & C_0 \\ D_0 & E_0 \end{pmatrix}$$

in which modulo t, B_0 accounts for the first i Hodge and Newton slopes of M/tM, E_0 accounts for the remaining Hodge and Newton slopes of M/tM, and D_0 vanishes. In particular, $\det(B_0 \pmod{t})$ has valuation $s_{s,1} + \cdots + s_{s,i}$. The valuation of $\det(B_0)$ itself cannot be any greater, but it must be at least the sum of the first i generic Hodge slopes of M, which we also know to be $s_{s,1} + \cdots + s_{s,i}$. Consequently, $\det(B_0)$ is a unit in $K[\![t]\!]_0$, and it has minimal valuation among all $i \times i$ minors of A_0. Then by Cramer's rule (as in the proof of Theorem 4.3.11), $D_0 B_0^{-1}$ and $B_0^{-1} C_0$ have entries in $\mathfrak{o}_K[\![t]\!]$. Note also that the least Hodge slope of B_0^{-1} is $-s_{g,i}$, since $\det(B_0)$ is a unit and the entries of $\det(B_0)B_0^{-1}$ are the cofactors of B_0.

Conjugating by the block lower triangular unipotent matrix U_0 with off-diagonal block $D_0 B_0^{-1}$, we obtain

$$A_1 = U_0^{-1} A_0 \varphi(U_0) = \begin{pmatrix} B_0 + C_0 \varphi(D_0 B_0^{-1}) & C_0 \\ E_0 \varphi(D_0 B_0^{-1}) - D_0 B_0^{-1} C_0 \varphi(D_0 B_0^{-1}) & E_0 - D_0 B_0^{-1} C_0 \end{pmatrix}.$$

Since $D_0 B_0^{-1}$ has entries in $t \mathfrak{o}_K [\![t]\!]$, $A_1 \equiv A_0 \pmod{t}$. We can thus iterate the construction to obtain a sequence of block matrices

$$A_l = \begin{pmatrix} B_l & C_l \\ D_l & E_l \end{pmatrix} \qquad (l = 0, 1, \ldots)$$

for which $A_{l+1} = U_l^{-1} A_l \varphi(U_l)$ with $U_l \in \mathrm{GL}_n(\mathfrak{o}_K [\![t]\!])$ equal to the block lower triangular unipotent matrix with off-diagonal block $D_l B_l^{-1}$. Note that $B_{l+1} = B_l(I + B_l^{-1} C_l \varphi(D_l B_l^{-1}))$ in which $B_l^{-1} C_l \varphi(D_l B_l^{-1})$ has entries in the ideal $(t, \mathfrak{m}_K)$, so the Hodge slopes of B_l are preserved. In particular, the least Hodge slope of B_l^{-1} equals $s_{g,i}$.

To complete the proof, we must show that the U_l converge to the identity for the $(t, \mathfrak{m}_K)$-adic topology. Since the least Hodge slope of B_l^{-1} equals $s_{g,i}$, it is enough to check that the D_l converge to zero for the same topology. Moreover, it suffices to check that for each positive integer m, $D_l \pmod{t^m}$ converges to zero for the $\mathfrak{m}_K$-adic topology.

We do this by induction on m, the case $m = 1$ being clear because $D_l \equiv 0 \pmod{t}$. Suppose that the convergence is known modulo t^m. Write

$$D_{l+1} = E_l \varphi(D_l) \varphi(B_l^{-1}) - D_l(B_l^{-1} C_l) \varphi(D_l B_l^{-1}).$$

Setting $D_l = \sum_h D_{l,h} t^h$, we then have

$$D_{l+1,m} t^m \equiv E_l \varphi(D_{l,m} t^m) \varphi(B_l^{-1}) \; D_{l,m} t^m (B_l^{-1} C_l) \varphi(D_l B_l^{-1}) + \cdots \pmod{t^{m+1}},$$
$$(15.3.3.1)$$

where the ellipsis represents terms depending on $D_{l,0}, \ldots, D_{l,m-1}$ which are already known to converge to 0 as $l \to \infty$.

On one hand, the sum of the least Hodge slopes of the reductions of E_l and $\varphi(B_l^{-1})$ modulo t equals $s_{g,i+1} - s_{g,i} \geq 0$. Thus the reduction of $E_l \varphi(D_{l,m} t^m) \varphi(B_l^{-1})$ modulo t^{m+1} has p-adic valuation no less than that of the reduction of $\varphi(D_{l,m} t^m)$, which in turn has p-adic valuation strictly greater than that of the reduction of $D_{l,m} t^m$. On the other hand, $B_l^{-1} C_l$ has entries in $\mathfrak{o}_K [\![t]\!]$, and $\varphi(D_l B_l^{-1})$ has entries in $(t, \mathfrak{m}_K)$ since D_l is divisible by t. Thus the reduction of $D_{l,m} t^m (B_l^{-1} C_l) \varphi(D_l B_l^{-1})$ modulo t^{m+1} has valuation strictly greater than that of the reduction of $D_{l,m} t^m$. We may conclude that for any $c > 0$, for l sufficiently large (so that the terms represented by the ellipsis in (15.3.3.1) all have norm less than c), $|D_{l+1,m}| < \max\{|D_{l,m}|, c\}$. This yields the desired convergence. $\qquad \square$

Theorem 15.3.4. *Let M be a finite free difference module of rank n over $K[\![t]\!]_0$, for φ a zero-centered Frobenius lift. Suppose that the generic and special Frobenius slopes of M are all equal to a single value r. Then there is a canonical isomorphism $M \cong (M/tM) \otimes_K K[\![t]\!]_0$ of difference modules.*

Proof First suppose that $r = 0$. By Lemma 8.6.1, we can choose a basis for which the generic Hodge slopes are all equal to 0. Let A be the matrix of action of Φ on this basis. We wish to construct an $n \times n$ matrix $U = \sum_{i=0}^{\infty} U_i t^i$ over $\mathfrak{o}_K [\![t]\!]$ with $U_0 = I_n$ such that $U^{-1} A \varphi(U) = A_0$, or equivalently $U = A \varphi(U) A_0^{-1}$. Since the map $U \mapsto A \varphi(U) A_0^{-1}$ is contractive for the $(t, \mathfrak{m}_K)$-adic topology on $I_n + t M_{n \times n}(\mathfrak{o}_K [\![t]\!])$, it has a unique fixed point by the contraction mapping theorem, which gives the desired isomorphism.

If $r \in v(K^\times)$, we may apply the above argument after twisting by a scalar. Otherwise, we may replace φ by a power, then twist and apply the above argument. $\qquad \square$

15.4 A reverse filtration

Since $\mathcal{E}^\dagger$ is not complete, we cannot apply Theorem 14.4.15 to filter a finite difference module over $\mathcal{E}^\dagger$ with pure quotients of decreasing norms. It was originally observed by de Jong that one can get a filtration with pure quotients of *increasing* norms, at the expense of replacing $\mathcal{E}^\dagger$ with its φ-perfection.

Definition 15.4.1. Let φ be a Frobenius lift on $\mathcal{E}^\dagger$. Let $\mathcal{E}_\varphi$ denote the φ-perfection of $\mathcal{E}$, that is, the completion of the direct limit $R_0 \to R_1 \to \cdots$ with $R_i = \mathcal{E}$ and the transition map $R_i \to R_{i+1}$ being φ. Choose ϵ as in Lemma 15.2.1. For $\alpha \in [\epsilon, 1)$, we may define $| \cdot |_\alpha$ on $\mathcal{E}_\varphi$ as the function $| \cdot |_{\alpha^{1/q^i}}$ on R_i; this is consistent because Lemma 15.2.1 guarantees that the transition maps are isometries. Let $\mathcal{E}_\varphi^\dagger$ be the subring of $x \in \mathcal{E}_\varphi$ such that for some $\alpha \in [\epsilon, 1)$ depending on x, $|x|_\beta < \infty$ for $\beta \in [\alpha, 1]$. Note that within $\mathcal{E}_\varphi$,

$$\mathcal{E} \cap \mathcal{E}_\varphi^\dagger = \mathcal{E}^\dagger.$$

We obtain the following analogue of Lemma 15.1.3.

Lemma 15.4.2. *The ring $\mathcal{E}_\varphi^\dagger$ has the following properties.*

(a) The ring $\mathcal{E}_\varphi^\dagger$ is a field.
(b) Under the 1-Gauss norm $| \cdot |_1$, the valuation ring $\mathfrak{o}_{\mathcal{E}_\varphi^\dagger}$ is a local ring with maximal ideal $\mathfrak{m}_K \mathfrak{o}_{\mathcal{E}_\varphi^\dagger}$.
(c) The field $\mathcal{E}_\varphi^\dagger$, equipped with $| \cdot |_1$, is henselian.

Proof Exercise. $\qquad \square$

Example 15.4.3. Suppose φ_K is the identity and $\varphi(t) = t^q$. Then $\mathcal{E}_\varphi$ may be identified with the set of formal sums $\sum_{i \in \mathbb{Z}[1/p]} c_i t^i$ with the property that for each $\epsilon > 0$, the set of $i \in \mathbb{Z}[1/p]$ for which $|c_i| \geq \epsilon$ is both bounded below and has bounded denominators. In this interpretation, $\mathcal{E}_\varphi^\dagger$ consists of those sums $\sum_i c_i t^i$ for which there exists $\alpha \in (0, 1)$ for which $|c_i|\alpha^i \to 0$ as $i \to -\infty$. (Equivalently, there exists $\alpha \in (0, 1)$ for which $|c_i|\alpha^i$ remains bounded as $i \to -\infty$.)

Theorem 15.4.4. *Let V be a finite free dualizable difference module over $\mathcal{E}_\varphi^\dagger$. Then there exists a unique filtration*

$$0 = V_0 \subset V_1 \subset \cdots \subset V_l = V$$

of difference modules, such that each successive quotient V_i/V_{i-1} is pure of some norm s_i, and $s_1 < \cdots < s_l$.

Proof By Theorem 14.4.15, we obtain a decomposition of $V' = V \otimes_{\mathcal{E}_\varphi^\dagger} \mathcal{E}_\varphi$ as a direct sum $\bigoplus_s V_s'$ of difference modules, in which each nonzero V_s' is pure of norm s. By replacing φ with a power, we can ensure that each s for which $V_s' \neq 0$ appears in $|F^\times|$. It suffices to check that the summand of V' of least norm descends to a submodule of V, as we may then repeat the argument after quotienting by this submodule.

By twisting and then approximating a suitable basis of V', we obtain a basis of V on which the matrix of action of Φ^{-1} is via a block matrix

$$\begin{pmatrix} A & B \\ C & D \end{pmatrix}$$

in which A is invertible over $\mathfrak{o}_{\mathcal{E}_\varphi^\dagger}$, and B, C, D have entries in $\mathfrak{m}_K \mathfrak{o}_{\mathcal{E}_\varphi^\dagger}$. Put $X = CA^{-1}$, then change basis by the block lower triangular unipotent matrix with lower left block X; we obtain the new matrix of action

$$\begin{pmatrix} A' & B' \\ C' & D' \end{pmatrix} = \begin{pmatrix} A + B\varphi^{-1}(X) & B \\ (D - XB)\varphi^{-1}(X) & D - XB \end{pmatrix}.$$

Repeating this operation yields a p-adically convergent sequence of changes of basis; in the limit, we get a block upper triangular matrix in which the first block corresponds to the summand of V' of least norm. To prove that this summand descends to a submodule of V, it suffices to check that the change-of-basis matrices are bounded in norm by 1 under $|\cdot|_\alpha$ for some $\alpha \in (0, 1)$.

At the first stage, for α sufficiently close to 1,

$$|A^{-1}|_\alpha \max\{|B|_\alpha, |C|_\alpha, |D|_\alpha\} < 1,$$

which implies that $|X|_\alpha < 1$. Since also $|X|_1 < 1$, by Proposition 8.2.3(c),

$|X|_{\alpha^{1/q}} < 1$; by Lemma 15.2.1, $|\varphi^{-1}(X)|_\alpha = |X|_{\alpha^{1/q}} < 1$. Hence

$$|A^{-1}A' - I_n|_\alpha < 1$$
$$\max\{|B'|_\alpha, |C'|_\alpha, |D|_\alpha\} \leq \max\{|B|_\alpha, |C|_\alpha, |D|_\alpha\}.$$

We may thus use the same α for the next stage, so $|X|_\alpha < 1$ at all stages. □

In order to use the reverse filtration later, we will need the following projection construction.

Lemma 15.4.5. *There exists a $\mathcal{E}$-linear map $\lambda \colon \mathcal{E}_\varphi \to \mathcal{E}$ sending $\mathcal{E}_\varphi^\dagger$ to $\mathcal{E}^\dagger$, such that $\lambda(x) = x$ for all $x \in \mathcal{E}$.*

Proof Suppose first that κ_K is inversive. In this case, the residue field of $\mathcal{E}_\varphi$ can be written as $\bigcup_{i=0}^\infty \kappa_K((t^{1/q^i}))$. Each element of this union can uniquely be written as a finite sum $\sum_{i=0}^\infty \varphi^{-1}(x_i)$ with $x_i \in \kappa_K((t))$, such that the coefficient of t^j in x_i vanishes whenever $i > 0$ and j is divisible by q.

This implies by induction that each $x \in \mathcal{E}_\varphi$ can be written uniquely as a convergent (for the $\mathfrak{m}_K$-adic topology) sum $\sum_{i=0}^\infty \varphi^{-i}(x_i)$ with $x_i \in \mathcal{E}$, such that the coefficient of t^j in x_i vanishes whenever $i > 0$ and j is divisible by q. In this presentation, for ϵ as in Lemma 15.2.1, $|\varphi^{-i}(x_i)|_\alpha \leq |x|_\alpha$ for $\alpha \in [\epsilon, 1)$ (exercise). We may thus take $\lambda(x) = x_0$.

In the general case, let K' be the φ-perfection of K, and put $\tilde{\mathcal{E}} = K'\langle 1/t, t]]_0$. (That is, $\tilde{\mathcal{E}}$ is the analogue of $\mathcal{E}$ with base field K' instead of K.) Argue as above to construct a map $\mathcal{E}_\varphi \to \tilde{\mathcal{E}}$. Then choose any continuous linear map $K' \to K$ whose composition with the inclusion $K \to K'$ is the identity, and use it to define a projection $\tilde{\mathcal{E}} \to \mathcal{E}$. The composition $\mathcal{E}_\varphi \to \tilde{\mathcal{E}} \to \mathcal{E}$ has the desired effect. □

Corollary 15.4.6. *The multiplication map $\mu \colon \mathcal{E} \otimes_{\mathcal{E}^\dagger} \mathcal{E}_\varphi^\dagger \to \mathcal{E}_\varphi$ is injective.*

Proof Exercise. □

Remark 15.4.7. The characterization of Theorem 15.4.4 as a "reverse filtration" arises from the following consideration. For V a finite free dualizable difference module over $\mathcal{E}_\varphi^\dagger$, we may apply Theorem 14.4.13 to decompose $V \otimes_{\mathcal{E}_\varphi^\dagger} \mathcal{E}_\varphi$ into pure submodules, and then assemble these into a filtration as per Theorem 14.4.15, in which the norms occur in descending order. On the other hand, we can also assemble the terms into a filtration in which the norms occur in ascending order; this gives the filtration from Theorem 15.4.4 tensored over $\mathcal{E}_\varphi^\dagger$ with $\mathcal{E}_\varphi$.

15.5 Substitution maps in the Robba ring

We summarize some theorems and conjectures of Berger [50].

Theorem 15.5.1. *Let φ be a K-linear substitution on $\mathcal{R}$ with $\varphi(t) \in K[\![t]\!]_0$. For any $\mu \in K$, we have $(\operatorname{Frac}\mathcal{R})^{\varphi=\mu} \subset K\{t\}^{\times}$.*

Proof In the terminology of [50], the restriction on φ is that it is *of finite height*. Consequently, the claim is a reformulation of [50, Theorem A].   □

Theorem 15.5.2. *Let φ be a K-linear substitution on $\mathcal{R}$ with $\varphi(t) \in K[\![t]\!]_0$.*

(a) *We have $K\{t\}^{\varphi=1} = K$.*
(b) *Suppose that $h \in K\{t\}^{\varphi=\mu}$ for some $\mu \in K \setminus \{1\}$. Then $c = \left.\frac{d\varphi(t)}{dt}\right|_{t=0}$ is a nonzero element of K and there exists a positive integer k such that $\mu = c^k$.*
(c) *With notation as in (b), there exists a power series $\log_{\varphi} \in tK\{t\}$ such that $\varphi(\log_{\varphi}) = c\log_{\varphi}$, and for this series we have $h \in K\log_{\varphi}^k$.*

Proof See [50, Theorem B]. For the construction of $\log_{\varphi}$, see for example [297].   □

The following is a further conjecture from [50]. (In the terminology of *op. cit.*, the restriction on φ is that it is *overconvergent*.)

Conjecture 15.5.3. *Let φ be a K-linear substitution on $\mathcal{R}$ with $\varphi(t) \in \mathcal{E}^{\dagger}$. Then $(\operatorname{Frac}\mathcal{R})^{\varphi=1} = K$.*

Notes

Much of the existing literature makes the restriction that Frobenius lifts must be absolute. The generality we consider here is relevant for some applications (e.g., to families of Galois representations in *p*-adic Hodge theory), so it is prudent to allow it as much as possible.

On the other hand, we have not opted to consider cases where φ does not carry K into K. (In that level of generality, one might consider our definition to restrict to *scalar-preserving* Frobenius lifts.) We have also not allowed K to be nondiscrete; this avoids having to worry about whether to allow Frobenius lifts φ for which $\varphi(t) - t^q$ has 1-Gauss norm 1 but each of its coefficients has norm less than 1. (This cannot occur for K discrete.)

As in Chapter 14, a number of the results are based on [224], at least when restricted to an absolute Frobenius lift.

- Theorem 15.3.2 is a local formulation of a geometric result of Grothendieck. The proof given is from [224, Theorem 2.3.1].
- Theorem 15.3.3 is a variant of [224, Theorem 2.4.2].
- Theorem 15.3.4 is an adaptation of [224, Theorem 2.7.1].

For stronger global results along these lines (in the theory of *F-crystals*), see [125], [419], and [328].

Theorem 15.4.4 in the case of an absolute Frobenius lift is due to de Jong [123, Proposition 5.8]. Corollary 15.4.6 and its proof are a variant on [123, Proposition 8.1].

One can extend considerations about substitution maps to certain multivariate analogues of the Robba ring. See [265] for a development of this point of view.

Exercises

15.1 Verify the claims about Newton's method from the final paragraph of the proof of Lemma 15.1.3(c).

15.2 Suppose that κ_K is perfect and that $\varphi: \mathcal{E} \to \mathcal{E}$ is a continuous iso-metric endomorphism inducing the absolute q-power Frobenius on $\kappa_K((t))$. Prove that φ is an absolute Frobenius lift in the sense of Definition 15.2.3. (Hint: use Witt vector functoriality to show that φ carries K into K. Otherwise put, for $x \in K$, show that $\varphi(x^{p^n}) = \varphi(x)^{p^n}$ is p-adically close to an element of K.)

15.3 Prove Lemma 15.2.7. (Hint: show that the map

$$\lambda \mapsto \varphi_K^{-1}(\varphi(t)|_{t=\lambda})$$

on $\mathfrak{m}_K$ is contractive, where $\varphi(t)|_{t=\lambda}$ denotes the substitution $t \mapsto \lambda$.)

15.4 Let K be the completion of $\mathbb{Q}_p(x)$ for the 1-Gauss norm. Prove that the Frobenius lift φ on K defined by $\varphi(t) = t^p + px$ does not have a center.

15.5 Prove Lemma 15.4.2.

15.6 Check the omitted details in the proof of Lemma 15.4.5. (Hint: show that in fact $|x|_\alpha = \sup_i\{|\varphi^{-1}(x_i)|_\alpha\}$ by considering x in the dense subset $\bigcup_{m=0}^\infty \varphi^{-m}(\mathcal{E}^\dagger)$ of $\mathcal{E}_\varphi^\dagger$.)

15.7 Prove Corollary 15.4.6. (Hint: if the claim fails, we can choose a nonzero element $z = \sum_{i=1}^n x_i \otimes y_i$ of the kernel with n as small as possible. Now use Lemma 15.4.5 to construct an element with an even shorter presentation, as in the the proof of Lemma 1.3.11. A related argument in Galois theory is Artin's proof of Dedekind's independence theorem, as in [217, §4.14].)

16

Frobenius modules over the Robba ring

In Chapter 14, we discussed some structure theory for finite difference modules over a complete isometric nonarchimedean difference field. This theory can be applied to the field $\mathcal{E}$, which is the p-adic completion of the bounded Robba ring $\mathcal{E}^\dagger$; however, it yields somewhat limited information.

For the purposes of studying Frobenius structures on differential modules (see Part V), it would be useful to have a structure theory over $\mathcal{E}^\dagger$ itself. This is a bit too much to ask for; what we can provide is a structure theory that applies over the Robba ring $\mathcal{R}$, which is somewhat analogous to what we obtain over $\mathcal{E}$. In particular, with an appropriate definition of pure modules, we obtain a slope filtration theorem analogous to Theorem 14.4.15 but valid over $\mathcal{R}$.

Given a difference module over $\mathcal{E}^\dagger$, one gets slope filtrations and Newton polygons over both $\mathcal{E}$ and $\mathcal{R}$. For a module over $K[\![t]\!]_0$, these turn out to match the generic and special Newton polygons; in particular, they need not coincide. However, they do admit a specialization property analogous to Theorem 15.3.2.

As a proof of the slope filtration theorem over $\mathcal{R}$ would take us rather far afield, we do not include one here. Instead, we limit ourselves to a brief overview of the proof, and consign further discussion and references to the notes.

Hypothesis 16.0.1. Throughout this chapter, let φ be a Frobenius lift on the Robba ring $\mathcal{R}$. (For a possible relaxation of this hypothesis, see the notes.) All difference modules over $\mathcal{R}$ will be taken with respect to φ unless otherwise specified.

16.1 Frobenius modules on open discs

We start with the fact that over the ring $K\{t\}$, the classification of finite free difference modules reduces (mostly) to the classification over K.

Theorem 16.1.1. *Suppose that κ_K is strongly difference-closed (e.g., φ is an absolute Frobenius lift and κ_K is algebraically closed). Let M be a finite free dualizable difference module over $K\{t\}$ for a zero-centered Frobenius lift. Then there exists a noncanonical isomorphism of difference modules $M \cong (M/tM) \otimes_K K\{t\}$.*

Proof Let c be the reduction of $\varphi(t)/t$ modulo t, so that $c \in \mathfrak{m}_K$. Let $e_1, \ldots, e_n$ be a basis for M. Let A be the matrix of action of Φ on this basis, which is invertible because M is dualizable. Let A_0 be the constant term of A, which is an invertible matrix over K. We wish to find a matrix U over $K\{t\}$ with $U \equiv I_n$ (mod t) such that $U^{-1}A\varphi(U) = A_0$.

We first construct a sequence of matrices $U_0, U_1, U_2, \ldots$ over $K[\![t]\!]$ with $U_0 = I_n$ and $U_{i+1} = U_i(I_n + X_i t^i)$ for some matrix X_i over K, such that $U_i^{-1}A\varphi(U_i)A_0^{-1} \equiv I_n$ (mod t^{i+1}). Namely, given U_i, we find X_i by solving the equation

$$c^i A_0 \varphi(X_i) A_0^{-1} - X_i + t^{-i}(U_i^{-1}A\varphi(U_i)A_0^{-1} - I_n) \equiv 0 \quad (\text{mod } t), \quad (16.1.1.1)$$

which amounts to trivializing a class in $H^1(V)$ for some finite difference module V over K (namely, the space of $n \times n$ matrices with the action $X \mapsto c^i A_0 \phi(X) A_0^{-1}$). This is possible by Corollary 14.6.6, although not necessarily uniquely. However, let us choose $h \geq 0$ such that $|c^h| \|A_0\| |A_0^{-1}| < 1$. Then for $i \geq h$, for any choice of X_i, $|c^i A_0 \varphi(X_i) A_0^{-1}| < |X_i|$, so X_i is uniquely determined by (16.1.1.1). Moreover, X_i has the same norm as the reduction of $t^{-i}U_i^{-1}A\varphi(U_i)A_0^{-1} - I_n$ modulo t.

Choose α so that for $i = h$,

$$|U_i - I_n|_\alpha, |U_i^{-1} - I_n|_\alpha, |U_i^{-1}A\varphi(U_i)A_0^{-1} - I_n|_\alpha < 1. \quad (16.1.1.2)$$

By the previous paragraph, $|X_i t^i|_\alpha \leq |U_i^{-1}A\varphi(U_i)A_0^{-1} - I_n|_\alpha$. By our choice of h, $|A_0\varphi(X_i t^i)A_0^{-1}|_\alpha < |X_i t^i|_\alpha$; since

$$U_{i+1}^{-1}A\varphi(U_{i+1})A_0^{-1} = (I_n + X_i t^i)^{-1}U_i^{-1}A\varphi(U_i)A_0^{-1}(I_n + A_0\varphi(X_i t^i)A_0^{-1}),$$

we conclude that (16.1.1.2) holds also with i replaced by $i + 1$. By induction, we see that (16.1.1.2) holds for all $i \geq h$.

This means that for any $\beta < \alpha$, the matrices U and U^{-1} have entries in $K\langle t/\beta \rangle$. Using the relation $U = A\varphi(U)A_0^{-1}$, we may conclude that U and U^{-1} also have entries in $K\langle t/\beta' \rangle$, for β' equal to the minimum norm of a root of $\varphi(t) - c$ over all $c \in K^{\text{alg}}$ of norm β. Since φ is zero-centered, by considering the Newton polygon of $\varphi(t) - c$ we see that $\beta' \geq \min\{\beta^{1/q}, \beta/|\pi|\}$ for π a generator of $\mathfrak{m}_K$. Iterating the function $\beta \mapsto \min\{\beta^{1/q}, \beta/|\pi|\}$, we see that U and U^{-1} have entries in $K\langle t/\beta \rangle$ for all $\beta \in (0, 1)$, proving the claim. $\square$

Remark 16.1.2. We will see a similar result later in the presence of a differential structure, namely Dwork's trick (Corollary 17.2.2). The main differences are that in the setting of Dwork's trick, there will be a *canonical* isomorphism $M \cong (M/tM) \otimes_K K\{t\}$, and no restriction on K is needed.

Remark 16.1.3. Note that the function $\beta \mapsto \min\{\beta^{1/q}, \beta/|\pi|\}$ becomes exactly the function $\lambda \mapsto \min\{\lambda^{1/p}, p\lambda\}$ from Chapter 10 in case $q = p$ and K is absolutely unramified. See also Theorem 17.2.1.

One can apply Theorem 16.1.1 to coherent locally free modules on the open unit disc, by virtue of the following observation. (Compare Definition 8.4.3, remembering that throughout Part IV we are assuming that K is discretely valued.)

Proposition 16.1.4. *Any coherent locally free module on the open unit disc over K is freely generated by finitely many global sections.*

Proof Let M be such a module. Choose a sequence $0 < \beta_1 < \beta_2 < \cdots$ with limit 1, and put $M_i = M \otimes K\langle t/\beta_i \rangle$. Choose any basis B_1 of M_1. Given a basis B_i of M_i, there must exist $X_i \in \mathrm{GL}_n(K\langle t/\beta_i \rangle)$ such that changing basis from B_i via X_i produces a basis of M_{i+1}. By Lemma 8.3.4, we can factor $X_i = V_i W_i$ with $V_i \in \mathrm{GL}_n(K\langle t/\beta_i \rangle)$ such that $|V_i - I_n|_{\beta_i} < 1$ and $W_i \in \mathrm{GL}_n(K[t])$. Write $V_i = \sum_{j=0}^{\infty} V_{i,j} t^j$, let Y_i be the sum of $V_{i,j} t^j$ over only those j for which $|V_{i,j}|_1 \le 1$, and put $U_i = V_i Y_i^{-1}$. Now changing basis from B_i via U_i also produces a basis B_{i+1} of M_{i+1}.

Since K is discretely valued, for each positive integer j, for i sufficiently large (depending on j), $|V_{i,j} t^j|_1 \le 1$ if and only if $|V_{i,j} t^j|_{\beta_i} \le 1$. Consequently, the U_i converge t-adically to the identity matrix. Given any $\beta \in (0, 1)$, if we choose $\gamma \in (\beta, 1)$ and combine the previous observation with the fact that the U_i are bounded under $|\cdot|_\gamma$, we deduce that the U_i converge under $|\cdot|_\beta$ to the identity matrix. Hence the product $U_1 U_2 \cdots$ is the change-of-basis matrix from B_1 to a simultaneous basis of each M_i, as desired. □

16.2 More on the Robba ring

To discuss the classification of difference modules on annuli, we must recall some properties of the Robba ring beyond simply its definition.

Remark 16.2.1. Recall (Definition 15.1.4) that we have defined the Robba ring to be

$$\mathcal{R} = \bigcup_{\alpha \in (0,1)} K\langle \alpha/t, t\};$$

that is, $\mathcal{R}$ consists of formal sums $\sum_i c_i t^i$ which converge in some range $\alpha \le |t| < 1$, but need not have bounded coefficients. Unlike its subring $\mathcal{E}^\dagger$, $\mathcal{R}$ is not a field; for instance, the element

$$\log(1+t) = \sum_{i=1}^{\infty} \frac{(-1)^{i-1}}{i} t^i$$

is not invertible (because its Newton polygon has infinitely many slopes). More generally, we have the following fact.

Lemma 16.2.2. *We have* $\mathcal{R}^\times = (\mathcal{E}^\dagger)^\times = \mathcal{E}^\dagger \setminus \{0\}$.

Proof A unit in $K\langle \alpha/t, t \rangle$ must have empty Newton polygon, so it must belong to $\mathcal{E}^\dagger$. Since $\mathcal{E}^\dagger$ is a field by Lemma 15.1.3(a), this proves the claim. $\square$

Definition 16.2.3. Because $\mathcal{R}$ consists of series with possibly unbounded coefficients, it does not carry a natural p-adic norm or topology. The most useful topology on $\mathcal{R}$ is the *LF topology*, which is the direct limit of the Fréchet topology on each $K\langle \alpha/t, t \rangle$ defined by the $|\cdot|_\rho$ for $\rho \in [\alpha, 1)$. That is, a sequence converges if on one hand it is contained in some $K\langle \alpha/t, t \rangle$, and on the other hand it converges for the Fréchet topology on that ring. Note that it does not matter which α is chosen, since for $\gamma > \alpha$, the inclusion $K\langle \alpha/t, t \rangle \to K\langle \gamma/t, t \rangle$ is strict (i.e., it is a homeomorphism to its image equipped with the subspace topology).

In fact, the ring $\mathcal{R}$ is not even noetherian (by an argument similar to that for $K\{t\}$; see the exercises for Chapter 8), but it has the following useful properties. These follow from work of Lazard; see the chapter notes for a more precise attribution.

Proposition 16.2.4 (Lazard). *For an ideal I of $\mathcal{R}$, the following are equivalent.*

(a) *The ideal I is closed in the LF topology.*
(b) *The ideal I is finitely generated.*
(c) *The ideal I is principal.*

Remark 16.2.5. The equivalence of (b) and (c) in Proposition 16.2.4 implies that R is a *Bézout domain*, i.e., an integral domain in which every finitely generated ideal is principal. Such rings enjoy many properties analogous to principal ideal domains; see the exercises.

We also have the following analogue of Proposition 16.1.4.

Proposition 16.2.6. *Any coherent locally free module on the half-open annulus with closed inner radius α and open outer radius 1 is represented by a finite free module over $K\langle \alpha/t, t \rangle$, and so corresponds to a finite free module over $\mathcal{R}$.*

Proof Let M be such a module. Then $M \otimes K\langle \alpha/t, t/\alpha \rangle$ is finite free since $K\langle \alpha/t, t/\alpha \rangle$ is a principal ideal domain by Proposition 8.3.2. By choosing a basis of $M \otimes K\langle \alpha/t, t/\alpha \rangle$ and then invoking Lemma 8.3.6, we may extend M to a coherent locally free module on the whole unit disc. We may then deduce the claim from Proposition 16.1.4. $\square$

16.3 Pure difference modules

Over a complete nonarchimedean difference field such as $\mathcal{E}^\dagger$, we already have a notion of a pure difference module. Since $\mathcal{R}$ does not come with a Frobenius-invariant norm, we cannot use the same definition. The appropriate definition in this case is as follows.

Definition 16.3.1. A finite free difference module M over $\mathcal{R}$ is *pure of norm s* if there exists a finite free difference module $M^\dagger$ over $\mathcal{E}^\dagger$ which is pure of norm s (in the sense of Definition 14.4.6), such that $M \cong M^\dagger \otimes_{\mathcal{E}^\dagger} \mathcal{R}$. We will see shortly (Corollary 16.3.7) that the module $M^\dagger$ is uniquely determined by this requirement.

Lemma 16.3.2. *Let A be an $n \times n$ matrix over $\mathfrak{o}_{\mathcal{E}^\dagger}$. Then the map $v \mapsto v - A\varphi(v)$ induces a bijection on $(\mathcal{R}/\mathcal{E}^\dagger)^n$.*

Proof Exercise, or see [241, Proposition 1.2.6]. $\square$

Corollary 16.3.3. *Let M be a finite free difference module over $\mathcal{E}^\dagger$ such that $|\Phi|_{\mathrm{sp}, M \otimes \mathcal{E}} \leq 1$. Then the map $H^1(M) \to H^1(M \otimes_{\mathcal{E}^\dagger} \mathcal{R})$ is bijective.*

Proof This will follow from Lemma 16.3.2 once we produce a basis of M on which the matrix of action of Φ has entries in $\mathfrak{o}_{\mathcal{E}^\dagger}$. Such a basis can be seen to exist by arguing as in the proof of Proposition 14.5.9. $\square$

Corollary 16.3.4. *Let $M_1^\dagger, M_2^\dagger$ be two finite free dualizable difference modules over $\mathcal{E}^\dagger$. Suppose that every Newton slope of $M_1^\dagger \otimes_{\mathcal{E}^\dagger} \mathcal{E}$ is less than or equal to every Newton slope of $M_2^\dagger \otimes_{\mathcal{E}^\dagger} \mathcal{E}$. Then*

$$\mathrm{Hom}(M_1^\dagger, M_2^\dagger) = \mathrm{Hom}(M_1^\dagger \otimes_{\mathcal{E}^\dagger} \mathcal{R}, M_2^\dagger \otimes_{\mathcal{E}^\dagger} \mathcal{R}).$$

Proof Note that $(M_1^\dagger)^\vee \otimes M_2^\dagger$ has norm less than or equal to 1 (because its Newton slopes are all nonnegative). It thus suffices to check that for $M^\dagger$ of norm $s \leq 1$, $H^0(M^\dagger) = H^0(M^\dagger \otimes_{\mathcal{E}^\dagger} \mathcal{R})$. As in the previous proof, we can find a basis $e_1, \ldots, e_n$ of $M^\dagger$ such that the matrix of action A of Φ on this basis has entries in $\mathfrak{o}_{\mathcal{E}^\dagger}$. Any element of $H^0(M^\dagger \otimes_{\mathcal{E}^\dagger} \mathcal{R})$, when written in terms of the basis, corresponds to a column vector $v \in \mathcal{R}^n$ satisfying $A\varphi(v) = v$. By Lemma 16.3.2, this forces $v \in (\mathcal{E}^\dagger)^n$, so v corresponds to an element of M itself. $\qquad\square$

Corollary 16.3.5. *Let M_1, M_2 be finite free difference modules over $\mathcal{R}$ which are pure of norms s_1, s_2, respectively. If $s_1 > s_2 > 0$, then $\mathrm{Hom}(M_1, M_2) = 0$.*

Proof By Corollary 16.3.4, it suffices to check that if $M_1^\dagger, M_2^\dagger$ are finite free difference modules over $\mathcal{E}^\dagger$ which are pure of norms s_1, s_2, respectively, and $s_1 > s_2$, then $\mathrm{Hom}(M_1, M_2) = 0$. This holds because the image of any morphism $M_1^\dagger \to M_2^\dagger$, if nonzero, would have to be both pure of norm s_1 and pure of norm s_2, which is impossible. $\qquad\square$

Remark 16.3.6. In the proof of Corollary 16.3.5, we used the fact that when working over a difference field, purity of the middle term in a short exact sequence $0 \to M_1 \to M \to M_2 \to 0$ implies purity (of the same norm) at the extremes. This is not true over $\mathcal{R}$, which means that one cannot deduce Corollary 16.3.5 in case $s_1 < s_2$. For example, if φ is a Frobenius lift for which $\varphi(1 + t) = (1 + t)^p$, and $M = \mathcal{R}v$ with $\Phi(v) = p^{-1}v$, then

$$\log(1 + t)v \in H^0(M);$$

this constitutes a key example in p-adic Hodge theory (see Appendix C).

Corollary 16.3.7. *Let M be a finite free difference module over $\mathcal{R}$ which is pure of norm $s > 0$. Then there is a unique finite free difference module $M^\dagger$ over $\mathcal{E}^\dagger$ which is pure of norm s, such that $M \cong M^\dagger \otimes_{\mathcal{E}^\dagger} \mathcal{R}$.*

Remark 16.3.8. Even if M is pure of norm s, the uniqueness in Corollary 16.3.7 can fail if we do not require $M^\dagger$ to be pure.

Proposition 16.3.9. *Let $M_1^\dagger, M_2^\dagger$ be finite free difference modules over $\mathcal{E}^\dagger$ which are pure of the same norm $s > 0$. Then any finite free difference module M over $\mathcal{R}$ fitting into a short exact sequence of the form*

$$0 \to M_1^\dagger \otimes_{\mathcal{E}^\dagger} \mathcal{R} \to M \to M_2^\dagger \otimes_{\mathcal{E}^\dagger} \mathcal{R} \to 0$$

is also pure of norm s.

Proof It is equivalent to check that

$$\mathrm{Ext}(M_1^\dagger, M_2^\dagger) = \mathrm{Ext}(M_1^\dagger \otimes_{\mathcal{E}^\dagger} \mathcal{R}, M_2^\dagger \otimes_{\mathcal{E}^\dagger} \mathcal{R}).$$

If we set $M^\dagger = (M_1^\dagger)^\vee \otimes M_2^\dagger$, it is also equivalent to check that the natural map

$$H^1(M^\dagger) \to H^1(M^\dagger \otimes_{\mathcal{E}^\dagger} \mathcal{R})$$

is a bijection (see Definition 14.1.6); this holds by Corollary 16.3.3. $\qquad\square$

16.4 The slope filtration theorem

The fundamental theorem in the theory of difference modules over the Robba ring is the following result analogous to Theorem 14.4.15. We do not give a complete proof in this book, but we will sketch the argument in the remainder of this chapter. (As noted in Remark 14.4.7, this theorem is usually stated in additive rather than multiplicative terms; this explains the name "slope filtration theorem".)

Theorem 16.4.1 (Slope filtration theorem). *Let M be a finite free dualizable difference module over $\mathcal{R}$. Then there exists a unique filtration $0 = M_0 \subset \cdots \subset M_l = M$ by dualizable difference submodules with the following properties.*

(a) *Each successive quotient M_i/M_{i-1} is finite free, and is pure of some norm s_i (in the sense of Definition 16.3.1).*

(b) *We have $s_1 > \cdots > s_l$.*

In this book, the main application of the slope filtration will be to the p-adic local monodromy theorem for differential modules over $\mathcal{R}$ (Theorem 20.1.4). For the moment, let us record one or two additional corollaries. (For further applications, see the chapter notes.)

Definition 16.4.2. Let M be a finite free dualizable difference module over $\mathcal{R}$. Set notation as in Theorem 16.4.1. Define the *Newton polygon* of M to be the polygon associated to the multiset containing $-\log s_i$ with multiplicity $\text{rank}(M_i/M_{i-1})$.

By the compatibility of purity of difference modules over $\mathcal{E}^\dagger$ with tensor products (Corollary 14.4.9) and duals (Proposition 14.4.8), we get the usual behavior of Newton slopes under tensor products, exterior powers, and duals.

Lemma 16.4.3. *Let M, N be finite free dualizable difference modules over $\mathcal{R}$. Let $s_{M,1}, \ldots, s_{M,m}$ and $s_{N,1}, \ldots, s_{N,n}$ be the Newton slopes of M and N, respectively.*

(a) *The Newton slopes of $M \otimes N$ are $s_{M,i} + s_{N,j}$ for $i = 1, \ldots, m$ and $j = 1, \ldots, n$.*

(b) *For $j = 1, \ldots, \text{rank}(M)$, the Newton slopes of $\wedge^j M$ are $s_{M,i_1} + \cdots + s_{M,i_j}$ for $1 \leq i_1 < \cdots < i_j \leq m$.*

(c) *The Newton slopes of $M^\vee$ are $-s_{M,1}, \ldots, -s_{M,m}$.*

We have the following generalization of Corollary 16.3.5.

Proposition 16.4.4. *Let M_1, M_2 be finite free dualizable difference modules over $\mathcal{R}$. Suppose that every Newton slope of M_1 is less than every Newton slope of M_2. Then $\mathrm{Hom}(M_1, M_2) = 0$.*

Proof We induct on $\mathrm{rank}(M_1) + \mathrm{rank}(M_2)$. If M_1 and M_2 are both pure, then Corollary 16.3.5 yields the claim. If M_1 is not pure, by Theorem 16.4.1 there exists a proper nonzero difference submodule N of M_1 such that N and M_1/N are both finite free, and we deduce that $\mathrm{Hom}(M_1, M_2) = 0$ from the fact that $\mathrm{Hom}(N, M_2) = \mathrm{Hom}(M_1/N, M_2) = 0$. (Namely, any morphism from M_1 to M_2 must vanish on N, yielding a morphism from M_1/N to M_2 which also vanishes.) We argue similarly if M_2 fails to be pure. $\square$

We mention a result which extends the semicontinuity theorem for Newton polygons (Theorem 15.3.2). See Example 20.2.1 for an explicit example.

Definition 16.4.5. For M a finite free dualizable difference module over $\mathcal{E}^\dagger$, we can construct two different Newton polygons, namely the ones associated to $M \otimes_{\mathcal{E}^\dagger} \mathcal{E}$ and $M \otimes_{\mathcal{E}^\dagger} \mathcal{R}$. We call these the *generic Newton polygon* and the *special Newton polygon*, respectively. The terminology is justified by the fact that if κ_K is strongly difference-closed, φ is induced by a zero-centered Frobenius lift on $K[\![t]\!]_0$, and M is obtained by base extension from a finite free dualizable difference module M_0 over $K[\![t]\!]_0$, then by Theorem 16.1.1 the special Newton polygon of M coincides with the special Newton polygon of M_0.

Theorem 16.4.6. *Let M be a finite free dualizable difference module over $\mathcal{E}^\dagger$. Then the special Newton polygon lies on or above the generic Newton polygon, with the same endpoints.*

Proof It suffices to show that the least special slope of M is greater than or equal to the least generic slope, as then applying this inequality to the exterior powers of M yields the comparison of Newton polygons by Lemma 14.5.3 and Lemma 16.4.3.

Suppose on the contrary that the least special slope is less than the least generic slope. Let M_1 be the first step in the filtration of $M \otimes_{\mathcal{E}^\dagger} \mathcal{R}$ given by Theorem 14.4.15. Since M_1 is pure, it descends to a difference module N_1 over $\mathcal{E}^\dagger$ of the same norm. By hypothesis, the generic slopes of $N_1^\vee \otimes M$ are all positive. It follows that $H^0(N_1^\vee \otimes M) = 0$, and then by Corollary 16.3.4 that $H^0((N_1^\vee \otimes M) \otimes_{\mathcal{E}^\dagger} \mathcal{R}) = 0$. However, the latter contradicts the fact that N_1 is a submodule of $M \otimes_{\mathcal{E}^\dagger} \mathcal{R}$. We thus deduce the claim. $\square$

We can similarly use slope filtrations to finish the uniqueness part of Theorem 15.3.3.

Corollary 16.4.7. *With notation as in Theorem 15.3.3, the submodule N is always unique.*

Proof It suffices to check that $N \otimes_{K[\![t]\!]_0} \mathcal{R}$ is uniquely determined; this follows because this module appears as a step in the slope filtration of $M \otimes_{K[\![t]\!]_0} \mathcal{R}$. □

Remark 16.4.8. Note that in Corollary 16.4.7, we need only the uniqueness of the slope filtration in Theorem 16.4.1. The existence is guaranteed by Theorem 16.1.1 because we start with a difference module over $K[\![t]\!]_0$; by contrast, in the general case of Theorem 16.4.1, existence requires a much deeper argument.

16.5 Harder–Narasimhan filtrations

As a first approximation towards the existence statement of Theorem 16.4.1, we introduce a formal construction of filtrations.

Definition 16.5.1. Let M be a finite free dualizable difference module over $\mathcal{R}$ of rank $n > 0$. Define the *average norm* of M to be

$$\mu(M) = |\Phi|^{1/n}_{\mathrm{sp}, \wedge^n M};$$

this quantity is multiplicative in the sense that $\mu(M_1 \otimes M_2) = \mu(M_1)\mu(M_2)$. We say M is *semistable* if $\mu(M) \geq \mu(N)$ for any nonzero difference submodule N of M.

Lemma 16.5.2. *Let M be a nonzero finite free dualizable difference module over $\mathcal{R}$ which is pure of some norm. Then M is semistable.*

Proof Suppose on the contrary that there exists a difference submodule N of M with $\mu(N) > \mu(M)$. Put $e = \mathrm{rank}(N)$; then $\wedge^e N$ is of rank 1 and hence pure. Consequently, the inclusion $\wedge^e N \to \wedge^e M$ violates Corollary 16.3.5; this contradiction yields the claim. □

Lemma 16.5.3. *Let M be a nonzero finite free dualizable difference module over $\mathcal{R}$. There exists a unique filtration $0 = M_0 \subset \cdots \subset M_l = M$ by difference submodules, such that each quotient M_i/M_{i-1} is free and semistable, and*

$$\mu(M_1/M_0) > \cdots > \mu(M_l/M_{l-1}).$$

Proof First note that the claim holds for any module of rank 1 by Lemma 16.5.2. From this, the rest of the proof is formal; we leave it as an exercise. □

Definition 16.5.4. The filtration described in Lemma 16.5.3 is called the *Harder–Narasimhan filtration*, or *HN filtration*, of the module M. To prove Theorem 16.4.1, it suffices to check that semistable modules over $\mathcal{R}$ are pure.

Remark 16.5.5. The reader may wonder what content is carried by Theorem 16.4.1 given that the definition of the HN filtration is a formality. The answer is that the formal definition does not imply much about the norms occurring in the HN filtration; for instance, it is not formal that the tensor product of semistable modules is semistable, but this does follow from Theorem 16.4.1. A similar situation arises in the theory of vector bundles, on which the usage of the word "semistable" is modeled; see the discussion preceding [241, Theorem 1.7.1]. A comparable situation is Deligne's use of *determinantal weights* in his second proof of the Weil conjectures [128]; there one has matrices over a p-adic field whose eigenvalues are supposed to be algebraic numbers with certain archimedean norms, but one has direct control only over the determinants.

16.6 Extended Robba rings

To upgrade HN filtrations to slope filtrations, we need to introduce something like a residual difference closure of the ring $\mathcal{R}$. (This characterization is justified by Corollary 16.7.2.) This construction, based on Mal'cev–Neumann series (Example 1.5.8), is called the *extended Robba ring* in [241].

Definition 16.6.1. Let $\tilde{\mathcal{R}}$ be the ring of formal sums $x = \sum_{i \in \mathbb{Q}} x_i t^i$ with $x_i \in K$ that satisfy the following properties. (Here α is a number in $(0, 1)$ which may vary with x.)

(a) For $\rho \in (\alpha, 1)$, $\lim_{i \to \pm\infty} |x_i| \rho^i = 0$.
(b) For $\rho \in (\alpha, 1)$, for $c > 0$, the set of indices i for which $|x_i| \rho^i \geq c$ is a well-ordered subset of $\mathbb{R}$ (i.e., it contains no infinite decreasing subsequence).

We may also construct $\tilde{\mathcal{R}}$ as follows. Recall that the field $K((t^{\mathbb{Q}}))$ of Mal'cev–Neumann series over K consists of formal sums $\sum_{i \in \mathbb{Q}} x_i t^i$ for which $\{i \in \mathbb{Q} : x_i \neq 0\}$ is well-ordered. On the subring of $K((t^{\mathbb{Q}}))$ consisting of series with bounded coefficients, for $\rho \in (0, 1)$ define the Gauss norm

$$\left| \sum_{i \in \mathbb{Q}} x_i t^i \right|_\rho = \sup_i \{|x_i| \rho^i\}.$$

For each $\alpha \in (0, 1)$, take the Fréchet completion for the norms $|\cdot|_\rho$ for $\rho \in (\alpha, 1)$; the union of these completions over all α gives $\tilde{\mathcal{R}}$.

Like the usual Robba ring, the ring $\tilde{\mathcal{R}}$ turns out to be a Bézout domain (compare Proposition 16.2.4). Let $\tilde{\mathcal{E}}^\dagger$ be the subring of $\tilde{\mathcal{R}}$ consisting of formal sums with bounded coefficients. By imitating the proofs for $\mathcal{E}^\dagger$ (Lemma 15.1.3), it can be shown that $\tilde{\mathcal{E}}^\dagger$ is a henselian discretely valued field with residue field $\kappa_K((t^\mathbb{Q}))$. Various definitions (e.g., purity, semistability) carry over from $\mathcal{R}$ to $\tilde{\mathcal{R}}$; we will not write these out explicitly.

We view $\tilde{\mathcal{R}}$ as a difference ring with the Frobenius given by

$$\varphi\left(\sum_{i \in \mathbb{Q}} x_i t^i\right) = \sum_{i \in \mathbb{Q}} \varphi_K(x_i) t^{qi}.$$

We have the following analogue of Lemma 16.3.2.

Lemma 16.6.2. *Let A be an $n \times n$ matrix over $\mathfrak{o}_{\tilde{\mathcal{E}}^\dagger}$. Then the map $v \mapsto v - A\varphi(v)$ induces a bijection on $(\tilde{\mathcal{R}}/\tilde{\mathcal{E}}^\dagger)^n$.*

Proof   Exercise, or see [241, Proposition 2.2.8].   □

16.7 Proof of the slope filtration theorem

With the definition of extended Robba rings in hand, we sketch the remainder of the proof of Theorem 16.4.1. See the chapter notes for further discussion.

The following lemma is where most of the hard work is concentrated.

Lemma 16.7.1. *Suppose that κ_K is strongly difference-closed. Then every semistable dualizable finite difference module over $\tilde{\mathcal{R}}$ is pure of some norm.*

Sketch of proof   One first constructs (using an explicit calculation) a filtration in which the successive quotients are each pure of some norm, but these norms may not increase in the right direction as we proceed up the filtration. One then argues (using a second explicit calculation) that if there are two steps in the wrong order which cannot be switched (because the extension between them is not split), then one can change the filtration to move the norms of these steps closer to each other. (Note the similarity with Grothendieck's classification of vector bundles on the projective line; see the notes for Chapter 26.) See [241, Theorem 2.1.8] for full details.   □

Although we will not need this explicitly, we mention the following analogue of the generalized Dieudonné–Manin classification (Theorem 14.6.3). It is not stated explicitly in [241], but is a direct consequence of results from there.

Corollary 16.7.2. *Suppose that κ_K is strongly difference-closed. Then every dualizable finite difference module over $\tilde{\mathcal{R}}$ can be split (non-uniquely) as a direct sum of difference submodules, each of the form $V_{\lambda,d}$ for some λ, d (as in Definition 14.6.1).*

Proof By [241, Proposition 2.1.6], the categories of pure dualizable difference modules of a given norm over K and over $\tilde{\mathcal{R}}$ are equivalent. Thus by Lemma 16.7.1 and Theorem 14.6.3, it suffices to split any short exact sequence $0 \rightarrow M_1 \rightarrow M \rightarrow M_2 \rightarrow 0$ in which M_1, M_2 are base extensions of pure difference modules over K with $\mu(M_1) > \mu(M_2)$. This can be deduced from the proof of [241, Proposition 2.1.5] or by a simple direct calculation. □

It remains to descend our understanding of semistability and purity from $\tilde{\mathcal{R}}$ to $\mathcal{R}$. We first make the following observation.

Remark 16.7.3. Let L be a complete extension of K as a difference field, such that κ_L is strongly difference-closed. (Such an L always exists; see [241, Proposition 3.2.4] or the exercises for Chapter 14.) Let $\tilde{\mathcal{R}}_L$ denote the extended Robba ring constructed using L as the coefficient field. Then there is an embedding of difference rings $\psi : \mathcal{R} \rightarrow \tilde{\mathcal{R}}_L$ which is isometric; that is, there exists $\alpha \in (0, 1)$ such that for $\rho \in (\alpha, 1)$, $|\psi(x)|_\rho = |x|_\rho$ for all $x \in \mathcal{R}$ (in particular, the two quantities are either both finite or both infinite). This is clear if $\varphi(t) = t^q$, as we can then set $\psi(t) = t$; it is verified in general by [241, Proposition 2.2.6].

We then obtain the following results; the key tool used is descent for modules along a faithfully flat ring homomorphism. (One also makes heavy use of an analogue of the projection $\mathcal{E}_\varphi \rightarrow \mathcal{E}$ of Lemma 15.4.5.)

Lemma 16.7.4. *Let M be a semistable dualizable finite difference module over $\mathcal{R}$. Then $M \otimes_\mathcal{R} \tilde{\mathcal{R}}_L$ is semistable.*

Proof See [241, Theorem 3.1.2]. □

Lemma 16.7.5. *Let M be a dualizable finite difference module over $\mathcal{R}$ such that $M \otimes_\mathcal{R} \tilde{\mathcal{R}}_L$ is pure. Then M is pure.*

Proof See [241, Theorem 3.1.3]. □

Putting everything together, we prove Theorem 16.4.1 as follows.

Proof of Theorem 16.4.1 We wish to show that the HN filtration of M is pure. By Lemma 16.7.4, if we start with the HN filtration of M and tensor with $\tilde{\mathcal{R}}_L$, the resulting filtration still has semistable successive quotients. Consequently, it must be the HN filtration of $M \otimes_\mathcal{R} \tilde{\mathcal{R}}_L$. By Lemma 16.7.1, each successive quotient of the HN filtration of M becomes pure when tensored with $\tilde{\mathcal{R}}_L$. By Lemma 16.7.5, each successive quotient is itself pure. □

Notes

The restriction that φ must be a Frobenius lift is probably stronger than necessary. In fact, as suggested in [241], it should suffice for the action of φ on $\kappa_K((t))$ to satisfy $v(\varphi(x)) = mv(x)$ for some integer $m > 1$.

Our proof of Theorem 16.1.1 is essentially that of [235, Proposition 4.3], except that the latter only treats the absolute case and further assumes that M descends to $K[\![t_0]\!]$.

Proposition 16.2.4 is the essential content of a paper of Lazard [289]. Note that it depends on K being spherically complete, and is false otherwise; however, we have assumed throughout Part IV that K is discretely valued, so we are safe.

The treatment of pure modules over $\mathcal{R}$ is an abridged version of [241, §1].

Corollary 16.3.4 has appeared in several guises previously. It figures in the work of Cherbonnier–Colmez [84], which we will discuss in Appendix C (see Remark C.2.6); it also appears in work of Tsuzuki [392, Proposition 4.1.1].

The existence of slope filtration for Frobenius modules over the Robba ring was anticipated by Tsuzuki [395], who introduced the definition in the case of a Frobenius structure on a differential module [395, Definition 5.1.1]. For more on why he did this, see the notes for Chapter 20.

The slope filtration theorem itself (Theorem 16.4.1), in the case of an absolute Frobenius lift with K of characteristic 0, was originally proved by Kedlaya in [230, Theorem 6.10]. That proof was significantly more complicated than the one given here. A second proof in the absolute case (again with K of characteristic 0), which also gives an important generalization (see below), was given in [234]; this proof introduced the formalism of semistability, inspired by some parallel work of Hartl and Pink [193] (more on which below). Our sketch is modeled on a third proof, this time for an arbitrary Frobenius lift, given in [241, Theorem 1.7.1]; this proof adds several technical simplifications to the previous ones. One is the direct characterization of pure modules; the previous proofs used a characterization in terms of the appropriate analogue of the Dieudonné–Manin classification (Corollary 16.7.2). Another is the use of faithful flat descent, replacing a more complicated Galois descent argument used in the previous proofs. A third simplification is the use of the extended Robba ring $\tilde{\mathcal{R}}$ as described here. In the previous proofs, the role of $\tilde{\mathcal{R}}$ was played by a somewhat smaller ring: its bounded elements (plus 0) form a field whose residue field is the algebraic closure of $\kappa_K((t))$, rather than the much larger field of generalized power series.

The proof of Theorem 16.4.1 in [234] includes a nontrivial generalization of the original slope filtration theorem beyond [241]. Herein the ring $\mathcal{E}$ is replaced by a Cohen ring (or a ramified extension thereof) not for a field of power series, but for a more general complete nonarchimedean field of characteristic p; the rings $\mathcal{E}^\dagger$ and $\mathcal{R}$ are replaced by appropriate analogues. The resulting theorem plays an important role in the study of isocrystals; see the notes for Appendix B. It also has applications to Rapoport–Zink spaces; see below.

In the case of an absolute Frobenius lift, Theorem 16.4.6 becomes [234, Proposition 5.5.1]. (This reference is also the source of the terminology of special and generic Newton polygons over $\mathcal{E}^\dagger$.) The proof in [234] uses the reverse filtration of de Jong (Theorem 15.4.4).

At the level of generality treated in [234], yet another approach to the slope filtration theorem was introduced by Fargues and Fontaine [160]. In their work, difference modules over $\tilde{\mathcal{R}}$ correspond to vector bundles on a certain space (a "curve in p-adic Hodge theory", commonly known as a *Fargues–Fontaine curve*), and the property of semistability translates into the usual definition derived from geometric invariant theory [317]. See [258, Lecture 3] for further discussion in the context of the theory of *perfectoid spaces*.

Besides the p-adic local monodromy theorem (Theorem 20.1.4), several additional applications of the slope filtration theorem have been found. One of these is in p-adic Hodge theory; this is discussed further in the notes for Appendix C. Another one is a q-difference analogue of the p-adic local monodromy theorem due to André and Di Vizio [20]. A third application has been pursued by Hartl in the context of period morphisms for Rapoport–Zink spaces. These are moduli spaces for deformations of certain p-divisible groups into mixed characteristic; the main conjecture of Rapoport and Zink is to construct a period morphism between this space and a space of linear-algebraic objects via which the Galois representations corresponding to the p-divisible groups correspond to some coherent data. In equal positive characteristic, Hartl [190] has established the appropriate analogue of this conjecture, using a form of Theorem 16.4.1 in which K is of positive characteristic. (Hartl proves that case of Theorem 16.4.1 directly, building on work of Hartl and Pink [193] giving a form of Corollary 16.7.2 for K of characteristic p.) In the original mixed characteristic setting, some results were obtained by Hartl [191, 192] and Faltings [162] before the language of perfectoid spaces emerged as a natural framework for this question; see [258, Lecture 3] and [366] for further discussion.

Exercises

16.1   Show that if φ is a Frobenius lift such that $\varphi(t) \equiv 0$ (mod t^2), then the conclusion of Theorem 16.1.1 holds without any restriction on K.

16.2   Let R be a *Bézout domain* (an integral domain in which every finitely generated ideal is principal).

   (a) Prove that any $x_1, \ldots, x_n \in R$ which generate the unit ideal appear as the first row in some $n \times n$ invertible matrix over R.

   (b) Prove that every finitely generated torsion-free R-module is free.

16.3   Prove Lemma 16.3.2 and Lemma 16.6.2. (Hint: Reduce to the case where $|A|_\rho \leq 1$ for $\rho \in [\alpha, 1)$. For injectivity, given that $v - A\varphi(v) \in (\mathcal{E}^\dagger)^n$, show that $|v|_\rho$ is bounded for $\rho \in [\alpha, 1)$, by comparing $|v|_\rho$ with $|v|_{\rho^{1/q}}$ using Lemma 15.2.1. For surjectivity, to find the class of $w \in \mathcal{R}^n$ in the image of the map, separate off the positive terms w_+ of w, replace w with $A\varphi(w_+) + (w - w_+)$, then repeat.)

16.4   Prove that any $A \in \mathrm{GL}_n(\mathcal{R})$ can be factored as UV with $U \in \mathrm{GL}_n(K\{t\})$ and $V \in \mathrm{GL}_n(\mathcal{E}^\dagger)$. (Hint: imitate the proof of Proposition 8.3.5.)

16.5   Let M be a finite free difference module over $K\{t\}$ such that $M \otimes_{K\{t\}} \mathcal{R}$ is pure of some norm s. Let $M^\dagger$ be the pure module over $\mathcal{E}^\dagger$ satisfying $M \otimes_{K\{t\}} \mathcal{R} \cong M^\dagger \otimes_{\mathcal{E}^\dagger} \mathcal{R}$. Prove that there exists a basis $e_1, \ldots, e_n$ of M which is also a basis of $M^\dagger$. (Hint: imitate the proof of Lemma 8.3.6, using the previous exercise.)

16.6   Prove Lemma 16.5.3. (Hint: first check that $\mu(N) \leq \mu(M)$ whenever N is a difference submodule of M of full rank, by comparing the top exterior powers of M and N. See also [241, Proposition 1.4.15].)

16.7   Let M be a nonzero finite dualizable difference module over $\mathcal{R}$. Let $0 = M_0 \subset \cdots \subset M_l = M$ be any filtration of M by difference submodules, such that each quotient M_i/M_{i-1} is free and semistable. Form the convex polygon of length n associated (as in Definition 4.3.2) to the multiset of slopes given by $- \log \mu(M_i/M_{i-1})$ with multiplicity $\mathrm{rank}(M_i/M_{i-1})$ for $i = 1, \ldots, l$. Prove that this polygon lies on or below the corresponding polygon for the HN filtration, with the same endpoint. (Again, this is a purely formal consequence of the definitions of semistability and the HN filtration, so the proof is as in the theory of vector bundles.)

Part V

Frobenius Structures

17

Frobenius structures on differential modules

In this part of the book, we bring together the streams of differential algebra (from Part III) and difference algebra (from Part IV), realizing Dwork's fundamental insight that the study of differential modules on discs and annuli is greatly enhanced by the introduction of Frobenius structures.

This chapter sets the foundations for this study. We first introduce the notion of a Frobenius structure on a differential module, with some examples. We then consider the effect of Frobenius structures on the generic radius of convergence, and obtain the fact that a differential module on a disc has a full basis of horizontal sections ("Dwork's trick"). We also show that the existence of a Frobenius structure does not depend on the particular choice of a Frobenius lift; this independence plays an important role in rigid cohomology (Appendix B).

Throughout Part V, Hypothesis 14.0.1 remains in force unless explicitly contravened. In particular, K will by default be a *discretely valued* complete nonarchimedean field.

17.1 Frobenius structures

We start with the basic compatibility between differential and difference structures.

Definition 17.1.1. Let R be a ring as in Definition 15.2.3. For M a finite free differential module over R, a *Frobenius structure* on M with respect to a Frobenius lift φ on R is an isomorphism $\Phi \colon \varphi^* M \cong M$ of differential modules. In more explicit terms, we must equip M with the structure of a dualizable difference module over (R, φ), such that

$$D(\Phi(m)) = \frac{d\varphi(t)}{dt} \Phi(D(m)) \qquad (m \in M).$$

311

In even more explicit terms, if A, N are the matrices of action of Φ, D on some basis, then A must be invertible, and we must have the compatibility

$$NA + \frac{dA}{dt} = \frac{d\varphi(t)}{dt} A\varphi(N). \qquad (17.1.1.1)$$

Remark 17.1.2. We may also speak about Frobenius structures on finite free differential modules for the derivation $t\frac{d}{dt}$; the analogue of (17.1.1.1) is

$$NA + t\frac{dA}{dt} = \frac{t}{\varphi(t)}\frac{d\varphi(t)}{dt} A\varphi(N). \qquad (17.1.2.1)$$

However, if R is a subring of $K[\![t]\!]$, then (17.1.2.1) only makes sense if $\varphi(t) = t^q u$ for $u \in R^\times$, in which case taking constant terms in (17.1.2.1) yields $N_0 A_0 = qu_0 A_0 \varphi(N_0)$. This gives $|N_0|_{\mathrm{sp}} = |A_0^{-1} N_0 A_0|_{\mathrm{sp}} = q^{-1}|\varphi(N_0)|_{\mathrm{sp}} = q^{-1}|N_0|_{\mathrm{sp}}$, so N_0 must have spectral radius 0 and hence must be nilpotent.

We now describe some examples where Frobenius structures can be constructed explicitly.

Definition 17.1.3. Suppose $\pi \in K$ satisfies $\pi^{p-1} = -p$. Then the power series $E(t) = \exp(\pi t - \pi t^p)$ (sometimes called the *Dwork exponential series*) has radius of convergence $p^{(p-1)/p^2}$ (exercise), even though the series $\exp(\pi t)$ has radius of convergence 1.

Example 17.1.4 (Dwork). Assume that K contains an element π with $\pi^{p-1} = -p$. Pick any $f \in \mathfrak{o}_{\mathcal{E}^\dagger}$, and let M_f be the differential module over $\mathcal{E}^\dagger$ of rank 1 with a generator v satisfying $D(v) = \pi\frac{df}{dt}v$. (Note that this is the pullback along f of Example 9.3.5.) This module is typically nontrivial because the exponential series $\exp(-\pi f)$ does not represent an element of $\mathcal{E}^\dagger$ (e.g., if $f = t^{-1}$). However, the exponential series provides an important clue about how to associate a Frobenius structure to M_f; namely, for φ an absolute Frobenius lift, we would like to define

$$\Phi(v) = \exp(\pi f - \pi\varphi(f))v.$$

To verify that this expression is meaningful, note that it suffices to do so in the case $\varphi(t) = t^q$, since $\exp(\pi(f^q - \varphi(f)))$ is already well-defined in $\mathcal{E}^\dagger$. We can then write

$$\exp(\pi f - \pi f^{p^a}) = E(f)E(f^p)\cdots E(f^{p^{a-1}}),$$

where E is the Dwork exponential series from Definition 17.1.3.

Example 17.1.5. We can similarly construct a Frobenius structure for Example 9.9.3. (The case $h = 0$ will reproduce Example 17.1.4; see Remark 17.1.7.) For h a nonnegative integer and $\zeta \in K$ a primitive p^{h+1}-st root of unity, let M

be the differential module of rank 1 over $\mathcal{E}^\dagger$ with the action of D on a generator v given by

$$D(v) = -\sum_{i=0}^{h}(\zeta^{p^i} - 1)t^{-p^i-1}v.$$

(Note that we have pulled back the formula from Example 9.9.3 along the substitution $t \mapsto t^{-1}$.) Let $\varphi: \mathcal{E}^\dagger \to \mathcal{E}^\dagger$ be an absolute Frobenius lift fixing ζ. On the disc $|t^{-1}| < 1$, we have already constructed the horizontal section

$$\frac{E_p(t^{-1})}{E_p(\zeta t^{-1})}v = \exp\left(\sum_{i=0}^{h}\frac{1 - \zeta^{p^i}}{p^i}t^{-p^i}\right)v,$$

where

$$E_p(t) = \exp\left(\sum_{i=0}^{\infty}\frac{t^{p^i}}{p^i}\right)$$

is the Artin–Hasse exponential. This suggests the possibility of defining a Frobenius action fixing this horizontal section, which would then be given by the formula

$$\Phi(v) = \frac{E_p(t^{-1})E_p(\zeta t^{-p})}{E_p(t^{-p})E_p(\zeta t^{-1})}v. \tag{17.1.5.1}$$

In fact, this gives a Frobenius structure defined over $\mathcal{E}^\dagger$, because the coefficient of v in (17.1.5.1), as a power series in t^{-1}, has radius of convergence strictly greater than 1. We will not verify this here; it was shown for $p > 2$ by Matsuda [307] by an explicit calculation, and for all p by Pulita [343] as part of a much broader result. We do note here that the existence of this Frobenius structure implies the following strengthening of Theorem 12.7.2.

Theorem 17.1.6. *Let K be a complete nonarchimedean field (not necessarily discretely valued) containing the p^h-th roots of unity for all $h \leq \log_p b$ and having perfect residue field. Let M denote a finite differential module of rank 1 on a half-open annulus with open outer radius 1, which is solvable at 1 with differential slope b. Then there exist $c_1, \ldots, c_b \in \{0\} \cup \mathfrak{o}_K^\times$ and nonnegative integers $j_1, \ldots, j_b$ such that*

$$M \otimes M_{1,c_1} \otimes \cdots \otimes M_{b,c_b}$$

has differential slope 0, for M_{i,c_i} defined as in Theorem 12.7.2.

Proof Since κ_K is perfect, K contains a subfield K_0 isomorphic to the fraction field of the Witt vectors of κ_K. Let φ_0 be the Witt vector Frobenius on K_0. Let

ψ be the substitution $t \mapsto t^p$, and put $\varphi = \psi \circ \varphi_0$ as an endomorphism of $\mathcal{E}_0^\dagger = \bigcup_{\alpha \in (0,1)} K_0 \langle \alpha/t, t \rrbracket_0$.

On one hand, by Example 17.1.5, for any $c_b \in \mathfrak{o}_{K_0}$, $M_{b,c_b} \cong \varphi^*(M_{b,c_b}) = \psi^*(M_{b,\varphi_0(c_b)})$. On the other hand, for $c_b, d_b \in \mathfrak{o}_K$ with $|c_b - d_b| < 1$, $M_{b,c_b}^\vee \otimes M_{b,d_b}$ is trivial because we can use (9.9.3.1) to write down a horizontal section. Since κ_K is perfect, given $c_b \in \mathfrak{o}_K$, we can choose $c_b' \in \mathfrak{o}_{K_0}$ with $|(c_b')^p - c_b| < 1$, and so $|\varphi_0(c_b') - c_b| < 1$. We thus deduce that

$$M_{b,c_b'} \cong \psi^*(M_{b,c_b}).$$

With this fact, we may deduce the claim from Theorem 12.7.2. □

Remark 17.1.7. We will have special need later for the case of Theorem 17.1.6 in which $M^{\otimes p}$ has differential slope 0. In this case, we only deal with objects of the form M_{i,c_i} with i not divisible by p, thanks to Corollary 12.7.3.

This implies first that we only need K to contain the p-th roots of unity, not the p^h-th roots of unity for all $h \le \log_p b$. It also implies that we can use the Frobenius structure from Example 17.1.4 instead of that from Example 17.1.5. Namely, for ζ a primitive p-th root of unity, there exists a unique $\pi \in \mathbb{Q}_p(\zeta)$ with $\pi^{p-1} = -p$ and $|\pi - (\zeta_p - 1)| < p^{-1/(p-1)}$ (exercise), and for this choice, if $c_i \in \mathfrak{o}_K$ and $f = c_i t^{-i}$, then $M_f^\vee \otimes M_{i,c_i}$ has differential slope 0 by an explicit calculation (exercise).

Remark 17.1.8. In addition to the above examples, Dwork managed to construct explicit Frobenius structures in several classical examples, using explicit formal solutions of the corresponding differential equations; for instance, see Example 20.2.1. These examples are uniformly explained by the fact that Picard–Fuchs modules carry Frobenius structures in rather broad generality. See Appendix A for further discussion.

17.2 Frobenius structures and the generic radius of convergence

One of Dwork's early discoveries is that the presence of a Frobenius structure forces solvability at the boundary. (There is also a converse for modules of rank 1; see notes.)

Theorem 17.2.1. *Let M be a finite differential module on the half-open annulus with closed inner radius α and open outer radius 1, equipped with a Frobenius structure. Then*

$$\lim_{\rho \to 1^-} IR(M_\rho) = 1,$$

that is, M is solvable at 1. More precisely, for $\rho \in (0, 1)$ sufficiently close to 1,

$$IR(M_{\rho^{1/q}}) \geq IR(M_\rho)^{1/q}.$$

Proof By imitating the proof of Lemma 10.3.2 (using Lemma 15.2.1), we may show that for $\rho \in (0, 1)$ sufficiently close to 1,

$$IR(M_{\rho^{1/q}}) \geq \min\{IR(M_\rho)^{1/q}, qIR(M_\rho)\}.$$

The function $f(s) = \min\{s^{1/q}, qs\}$ on $(0, 1]$ is strictly increasing, and any sequence of the form $s, f(s), f(f(s)), \ldots$ converges to 1 (as in the proof of Theorem 16.1.1). This proves the first claim; for the second claim, note that $f(s) = s^{1/q}$ for s sufficiently close to 1. $\quad\square$

The following corollary is sometimes called *Dwork's trick*. It may be viewed as a true nonarchimedean analogue of the fundamental theorem of ordinary differential equations.

Corollary 17.2.2 (Dwork). *Let M be a finite differential module on the open unit disc, such that the restriction of M to some half-open annulus with open outer radius 1 admits a Frobenius structure. Then M admits a basis of horizontal sections.*

Proof By Theorem 17.2.1, for each $\lambda < 1$, there exists $\rho \in (\lambda, 1)$ such that $R(M_\rho) > \lambda$. By Dwork's transfer theorem (Theorem 9.6.1), $M \otimes K\langle t/\lambda\rangle$ admits a basis of horizontal sections. Taking λ arbitrarily close to 1 yields the claim. $\quad\square$

Remark 17.2.3. The proof of Corollary 17.2.2 admits the following geometric interpretation. By Proposition 9.3.3, the horizontal sections converge on some disc of positive radius ρ. Pulling back by Frobenius gives a new space of horizontal sections on the disc of radius $\min\{\rho^{1/q}, q\rho\}$, but this space must coincide with the original space. Repeating the construction, we eventually stretch the horizontal sections over the entire open unit disc.

One also has a nilpotent analogue of Dwork's trick, by using Theorem 13.7.1 in place of Theorem 9.6.1.

Corollary 17.2.4. *Let M be a finite differential module on the open unit disc for the derivation $t\frac{d}{dt}$ with a nilpotent singularity at $t = 0$, such that the restriction of M to some annulus with open outer radius 1 admits a Frobenius structure. Then M has radius of convergence 1; that is, for any $\beta < 1$, for any basis of $M \otimes K\langle t/\beta\rangle$, the fundamental solution matrix for that basis has entries in $K\langle t/\beta\rangle$. (Note that the nilpotency of the singularity is automatic if the Frobenius* *lift φ is of the form described in Remark 17.1.2 and the Frobenius structure is defined on the entire disc.)*

A nice application of Dwork's trick is the following.

Proposition 17.2.5. *Let M be a finite differential module over $K[[t]]_0$ for the derivation $t\frac{d}{dt}$ with $R(M) = 1$. (For instance, this holds if M admits a Frobenius structure, by Dwork's trick.) Then $H^0(M) = H^0(M \otimes_{K[[t]]_0} \mathcal{E}^\dagger)$.*

Proof By Theorem 9.6.1, there exists a horizontal basis $e_1, \ldots, e_n$ of $M \otimes_{K[[t]]_0} K\{t\}$. If $v \in H^0(M \otimes_{K[[t]]_0} \mathcal{E}^\dagger)$, then when we write $v = \sum_{i=1}^n v_i e_i$ with $v_i \in \mathcal{R}$, we must have $t\frac{d}{dt}(v_i) = 0$ for $i = 1, \ldots, n$. This forces $v_i \in K$ for $i = 1, \ldots, n$, so by Lemma 12.2.4,

$$v \in (M \otimes_{K[[t]]_0} \mathcal{E}^\dagger) \cap (M \otimes_{K[[t]]_0} K\{t\}) = M \otimes_{K[[t]]_0} (\mathcal{E}^\dagger \cap K\{t\}) = M. \quad \square$$

17.3 Independence from the Frobenius lift

Another key property of Frobenius structures is that their existence does not depend on the exact shape of the Frobenius lift.

Proposition 17.3.1. *Let φ_1, φ_2 be two Frobenius lifts on R which agree on K. Let M be a finite free differential module over R equipped with a Frobenius structure for φ_1. Then there is a functorial way to equip M with a Frobenius structure for φ_2.*

Proof The Frobenius structure for φ_2 is defined by the following Taylor series:

$$\Phi_2(m) = \sum_{i=0}^\infty \frac{(\varphi_2(t) - \varphi_1(t))^i}{i!} \Phi_1(D^i(m)). \tag{17.3.1.1}$$

By Theorem 17.2.1 and the fact that $|\varphi_2(t) - \varphi_1(t)|_1 < 1$, this series converges under $|\cdot|_\rho$ for $\rho \in (0, 1)$ sufficiently close to 1 (if this makes sense for R), and also under $|\cdot|_1$ (if this makes sense for R). $\quad \square$

Corollary 17.3.2. *Let φ_1, φ_2 be two Frobenius lifts on R which agree on K. Then there is a canonical equivalence between the categories of finite free differential modules over R equipped with Frobenius structures with respect to φ_i for $i = 1, 2$; this equivalence is the identity functor on the underlying difference modules.*

Definition 17.3.3. Corollary 17.3.2 allows us to switch from one Frobenius lift on R to another one which may be more convenient. One useful choice is what we call the *standard q-power Frobenius lift* for a given choice of φ_K, namely the Frobenius lift φ for which $\varphi(t) = t^q$.

We may also see that changing Frobenius lifts preserves purity.

Lemma 17.3.4. *Let M be a finite free differential module over $\mathcal{E}^\dagger$ equipped with a unit-root Frobenius structure. Let M_0 be a finite free dualizable difference module over $\mathfrak{o}_{\mathcal{E}^\dagger}$ such that M is isomorphic as a difference module to $M_0 \otimes_{\mathfrak{o}_{\mathcal{E}^\dagger}} \mathcal{E}^\dagger$. Then under any such isomorphism, M_0 is stable under $D^i/i!$ for each nonnegative integer i.*

Proof We proceed by induction on i. Note that $\mathfrak{o}_{\mathcal{E}^\dagger}$ is stable under $d^i/i!$ for each i; hence by the Leibniz rule, given the claim for all $j < i$, it suffices to exhibit a single basis of M_0 on which $D^i/i!$ acts via a basis over $\mathfrak{o}_{\mathcal{E}^\dagger}$.

Choose a matrix of M_0 and let A, N be the matrices of action of Φ, D on this basis. If we apply Φ to this basis, the matrices of action of Φ, D on the resulting basis are $\varphi(A)$, $\frac{d\varphi(t)}{dt}\varphi(N)$. By repeating this construction, for any $\epsilon < 0$, we can find a basis of M_0 on which the matrix of action of D has norm at most ϵ.

In particular, we deduce the base case $i = 1$. For $i > 1$, use the previous paragraph to choose a basis $e_1, \ldots, e_n$ of M_0 such that $(D/i)(e_j) \in M_0$ for $j = 1, \ldots, n$, then invoke the induction hypothesis to deduce that $(D^i/i!)(e_j) = (D^{i-1}/(i-1)!)((D/i)(e_j)) \in M_0$. $\qquad\square$

Proposition 17.3.5. *Suppose $R = \mathcal{E}^\dagger$ or $R = \mathcal{R}$. Let φ_1, φ_2 be two Frobenius lifts on R which agree on K. Let M be a finite free differential module over R equipped with a Frobenius structure for φ_1 which is pure of some norm. Then M is also pure of the same norm with respect to the induced Frobenius structure for φ_2.*

Proof It suffices to check the case $R = \mathcal{E}^\dagger$, by the following reasoning: if M is pure over $\mathcal{R}$, and $M^\dagger$ is the unique pure module over $\mathcal{E}^\dagger$ with $M \cong M^\dagger \otimes_{\mathcal{E}^\dagger} \mathcal{R}$, then Φ acts on M by Corollary 16.3.4. We may also reduce to the case where M is pure of norm 1, by replacing φ_1, φ_2 by powers of themselves and invoking Corollary 14.4.5.

By Proposition 14.4.16, we can write $M = M_0 \otimes_{\mathfrak{o}_{\mathcal{E}^\dagger}} \mathcal{E}^\dagger$ for some finite free $\mathfrak{o}_{\mathcal{E}^\dagger}$-module M_0 such that M_0 and $M_0^\vee$ are stable under the action of φ_1. By Lemma 17.3.4, M_0 is stable under $D^i/i!$ for each nonnegative integer i. By (17.3.1.1), the Frobenius structure with respect to φ_2 carries M_0 into itself, and similarly for $M_0^\vee$. This yields the claim. $\qquad\square$

17.4 Slope filtrations and differential structures

In order to apply the slope filtration theorem (Theorem 16.4.1) to differential modules with Frobenius structures, we must check some compatibilities between slope filtrations and differential modules.

Lemma 17.4.1. *Let M be a finite differential module over $\mathcal{R}$ equipped with a Frobenius structure. Then the steps of the filtration of Theorem 16.4.1 are differential modules, not just difference modules.*

Proof It suffices to check for M_1, as then we may quotient by M_1 and repeat the argument. Since M_1 is pure of some norm which is greater than every norm appearing in the slope filtration of M/M_1, $\mathrm{Hom}(M_1, M/M_1) = 0$ by Proposition 16.4.4. We may thus apply the criterion from Remark 5.1.7: the composition $M_1 \xrightarrow{D} M \to M/M_1$ is a morphism of difference modules over $\mathcal{R}$ and hence must be zero, implying that M_1 is a differential submodule of M. $\quad\square$

Lemma 17.4.2. *Let M be a finite unit-root difference module over $\mathcal{E}^\dagger$ such that $M \otimes_{\mathcal{E}^\dagger} \mathcal{R}$ admits a compatible differential structure. Then this structure is induced by a corresponding differential structure on M itself.*

Proof Let N, A be the matrices via which D, Φ act on a basis of M. Write the commutation relation (17.1.2.1) between N and A in the form

$$N - \frac{t}{\varphi(t)} \frac{d\varphi(t)}{dt} A\varphi(N)A^{-1} = \frac{d}{dt}(A)A^{-1}.$$

We deduce from Lemma 16.3.2 that N has entries in $\mathcal{E}^\dagger$. $\quad\square$

From Lemma 17.4.1 and 17.4.2, we deduce the following refinement of the slope filtration theorem in the presence of a differential structure.

Theorem 17.4.3. *Let M be a finite free differential module over $\mathcal{R}$ equipped with a Frobenius structure. Then there exists a unique filtration $0 = M_0 \subset \cdots \subset M_l = M$ by differential submodules preserved by the Frobenius structure, with the following properties.*

(a) *Each successive quotient M_i/M_{i-1} is finite free, and descends uniquely to a differential module over $\mathcal{E}^\dagger$ with an induced Frobenius structure that is pure of some norm s_i (in the sense of Definition 16.3.1).*

(b) *We have $s_1 > \cdots > s_l$.*

17.5 Extension of Frobenius structures

The following result allows us to extend certain Frobenius actions. An important application will be to Picard–Fuchs modules (Appendix A).

Proposition 17.5.1. *Let M be a finite differential module on the open unit disc for the derivation $t\frac{d}{dt}$ with a nilpotent singularity at $t = 0$. Assume either that:*

(a) the Frobenius lift φ on $K[\![t]\!]_0$ is arbitrary, and M has no singularity at $t = 0$; or

(b) the Frobenius lift φ on $K[\![t]\!]_0$ satisfies $\varphi(t) = t^q u$ for some $u \in \mathfrak{o}_K[\![t]\!]^\times$.

Then any Frobenius structure with respect to φ on the restriction of M to some annulus with open outer radius 1 is induced by a Frobenius structure on M itself.

Proof It suffices to check this for a standard Frobenius lift, as under either (a) or (b) we may switch back and forth between the given Frobenius lift and a standard one using Proposition 17.3.1. We thus assume hereafter that $\varphi(t) = t^q$.

Let Φ be a Frobenius structure with respect to φ on the restriction of M to some annulus with open outer radius 1. By Corollary 17.2.4, M admits a basis on which the matrix of action of D is a nilpotent matrix N over K. Let $A = \sum_{i \in \mathbb{Z}} A_i t^i$ be the matrix of action of Φ on the same basis; then the commutation relation (17.1.2.1) between Φ and D states that $NA + t\frac{dA}{dt} = qA\varphi_K(N)$. Consequently, for each $i \in \mathbb{Z}$, $NA_i + iA_i = qA_i\varphi_K(N)$. By Lemma 7.3.5, the operator $X \mapsto NX + iX - qX\varphi_K(N)$ on $n \times n$ matrices over K has all eigenvalues equal to i because N and $\varphi_K(N)$ are both nilpotent. Hence $A_i = 0$ for $i \neq 0$, so $A \in \mathrm{GL}_n(K)$. In particular, the Frobenius structure Φ can be defined on the entire open unit disc. $\qquad\square$

The following corollary is [270, Lemma 2.3].

Corollary 17.5.2. *Let φ be a standard q-power Frobenius lift. Let $N = \sum_{i=0}^{\infty} N_i t^i$ be an $n \times n$ matrix over $K\{t\}$ such that N_0 has eigenvalues $\alpha_1, \ldots, \alpha_n \in \mathbb{Q} \cap \mathbb{Z}_p$. Let A be an $n \times n$ matrix over $K\langle\alpha/t, t\rangle$ for some $\alpha \in (0, 1)$ satisfying (17.1.2.1). Then A has entries in $t^{-m}K\{t\}$ for*

$$m = \lfloor \max_j\{\alpha_j\} - p\min_j\{\alpha_j\}\rfloor.$$

Proof By adjoining $t^{1/k}$ for some k prime to p (which has the effect of multiplying each of $\alpha_1, \ldots, \alpha_n$ by k), we may reduce to the case where $\alpha_1, \ldots, \alpha_n \in \mathbb{Z}$. Set

$$a = \min_j\{\alpha_j\}, \qquad b = \max_j\{\alpha_j\}.$$

We now apply shearing transformations (Proposition 7.3.10) to construct a matrix W over $K(t)$ such that $t^b W$ and $t^{-a} W^{-1}$ have entries in $K\{t\}$, $N_W = W^{-1} N W + W^{-1} t \frac{dW}{dt}$ has entries in $K\{t\}$, and the constant term of N_W is nilpotent. As per Remark 14.1.3, the matrices N_W and $A_W = W^{-1} A \varphi(W)$ again satisfy (17.1.2.1), so we may apply Proposition 17.5.1 to deduce that A_W has entries in $K\{t\}$. Writing $A = W A_W \varphi(W^{-1})$, we see that the claim holds with $m = pa - b$. $\square$

17.6 Frobenius intertwiners

In some settings, it is natural to consider a notion of Frobenius structure that is more permissive than what we have considered so far.

Definition 17.6.1. Let R be a ring as in Definition 15.2.3. For two finite free differential modules M_1, M_2 over R, a *Frobenius intertwiner* from M_1 to M_2 with respect to a Frobenius lift φ on R is an isomorphism $\Phi \colon \varphi^* M_1 \cong M_2$ of differential modules.

Since a Frobenius intertwiner cannot be used for an iterative argument such as Dwork's trick, its value is somewhat diminished compared to a Frobenius structure. The reason to consider this definition is that Frobenius intertwiners can often be constructed in p-adic analytic families; this is illustrated by the following example.

Definition 17.6.2. The *generalized hypergeometric equation* with parameters in K given by

$$\underline{\alpha}; \underline{\beta} = \alpha_1, \ldots, \alpha_m; \beta_1, \ldots, \beta_n$$

is the linear differential equation of the form

$$P(\underline{\alpha}; \underline{\beta})(y) = 0, \quad P(\underline{\alpha}; \underline{\beta}) = t \prod_{i=1}^{m} \left(t\frac{d}{dt} + \alpha_i \right) - \prod_{j=1}^{n} \left(t\frac{d}{dt} + \beta_j - 1 \right).$$
(17.6.2.1)

This specializes to the hypergeometric equation (0.3.2.2) by taking

$$m = n = 2; \alpha = (a, b); \beta = (c, 1); t = z.$$

Let $M_{\underline{\alpha};\underline{\beta}}$ denote the differential module of rank $\max\{m, n\}$ over $K(t)$ corresponding to (17.6.2.1).

Theorem 17.6.3 (Dwork). *Assume that K contains an element π satisfying $\pi^{p-1} = -p$. Let $\underline{\alpha}; \underline{\beta}$ and $\underline{\alpha}', \underline{\beta}'$ be two sequences in $\mathbb{Z}_p$ such that $p\underline{\alpha}; p\underline{\beta}$*

are congruent modulo $\mathbb{Z}$ to some permutations of $\underline{\alpha}', \beta'$. Then there exists a Frobenius intertwiner from $M_{\underline{\alpha};\beta,\rho=1}$ to $M_{\underline{\alpha}';\beta',\rho=1}$ with respect to the Frobenius lift $\varphi(t) = t^p$.

Proof This result is originally due to Dwork [147]; sere [266, Theorem 4.1.2] for more precise references. $\square$

Notes

The statement of Theorem 17.1.6 (whose proof includes that of Theorem 12.7.2) is a slight weakening of [309, Corollaire 2.0–2], with a similar proof; the technique goes back to Robba [353]. A complete classification of rank 1 solvable modules on an open annulus with outer radius 1 (and unspecified inner radius) has been given by Pulita [343], and can be formulated in terms of any Lubin–Tate group; using the formal multiplicative group recovers Theorem 17.1.6.

Theorem 17.2.1 admits the following partial converse: if M is a finite free differential module over $\mathcal{R}$ of rank 1 which is solvable at 1, then there exists $\lambda \in \mathbb{Z}_p$ such that $M \otimes_{\mathcal{R}} V_\lambda$ admits a Frobenius structure for some absolute Frobenius lift. (Here V_λ is defined as in Example 9.5.2.) For M defined over F_1 (the field of analytic elements), this was shown by Chiarellotto and Christol [85]; their argument extends directly to $\mathcal{E}^\dagger$, but not to $\mathcal{R}$. The general case was shown by Pulita [342]; it constitutes a refinement of our Theorem 17.1.6.

We cannot resist viewing Dwork's trick (Corollary 17.2.2) as an instance of a general principle articulated beautifully by Coleman [106, §III]:

Rigid analysis was created to provide some coherence in an otherwise totally disconnected p-adic realm. Still, it is often left to Frobenius to quell the rebellious outer provinces.

A direct proof of Corollary 17.2.4, not employing the transfer theorem for a nilpotent regular singularity (Theorem 13.7.1), can be found in [239, §3.6].

The proof of Proposition 17.3.1 is taken from [395, Theorem 3.4.10].

Theorem 17.4.3, in the case of an absolute Frobenius lift, is the original form of the slope filtration theorem suggested, though not explicitly conjectured, by Tsuzuki in [395]. Its derivation from Theorem 16.4.1 is the same as the derivation of [230, Theorem 6.12] from [230, Theorem 6.10]. See also [234, §7.1].

Dwork's book [147] builds upon the study of generalized hypergeometric equations made by Dwork in [143, 145], incorporating what is now called the theory of *A-hypergeometric systems* attributed to Gelfand–Kapranov–Zelevinsky [175] (though Dwork's development in [147] was done independently).

Exercises

17.1 Verify the unproved assertion in Definition 17.1.3. (Hint: see [355, §7.2].)

17.2 Verify the unproved assertions in Remark 17.1.7.

17.3 Give an example of a differential module M of rank 2 over F_1 admitting a Frobenius structure and a short exact sequence

$$0 \to M_1 \to M \to M_2 \to 0$$

of difference modules which is not a sequence of differential modules. (Hint: use Remark 5.1.7 as in the proof of Lemma 17.4.1, to figure out how to choose M_1 and M_2.)

18

Effective convergence bounds

In this chapter, we discuss some effective bounds on the solutions of p-adic differential equations with nilpotent singularities. These come in two forms. We start by discussing bounds that make no reference to a Frobenius structure, due to Christol, Dwork, and Robba. These could have been presented earlier, and indeed one of these bounds was invoked in Chapter 13; we chose to postpone them until this point so that we can better contrast them against the bounds available in the presence of a Frobenius structure. The latter are original, though they are strongly inspired by some recent results of Chiarellotto and Tsuzuki.

These results carry both theoretical and practical interest. Besides their application in the study of p-adic exponents mentioned above (and in the proof of the unit-root p-adic local monodromy theorem to follow; see Theorem 19.3.1), another theoretical point of interest is their use in the study of logarithmic growth of horizontal sections at a boundary. We discuss some recent advances in this study due to André, Chiarellotto–Tsuzuki, and Ohkubo. (An area of application we will not discuss is the theory of G-functions, as in [149].)

A point of practical interest is that effective convergence bounds are useful for carrying out rigorous numerical calculations, e.g., in the machine computation of zeta functions of varieties over finite fields. See the notes for Appendix B for further discussion.

Hypothesis 18.0.1. In this chapter, we drop the running restriction that K is discretely valued, imposing it only when we discuss Frobenius structures. We retain the condition $p > 0$.

18.1 A first bound

We first go back to the p-adic Fuchs theorem for discs (Theorem 13.2.2) and extract an effective convergence bound in the case of a nilpotent singularity. One can also give effective bounds for more general regular singularities (see Proposition 18.5.1), but the nilpotent case is sufficient for many applications in algebraic geometry, and the bounds are much easier to describe in this case. The case of no singularity reproduces the p-adic Cauchy theorem (Proposition 9.3.3).

Proposition 18.1.1. *Let* $N = \sum_{i=0}^{\infty} N_i t^i$ *be an* $n \times n$ *matrix over* $K[\![t/\beta]\!]_0$ *corresponding to the differential system* $D(v) = Nv + d(v)$, *where* $d = t\frac{d}{dt}$. *Assume that* N_0 *is nilpotent with nilpotency index* m; *that is,* $N_0^m = 0$ *but* $N_0^{m-1} \neq 0$. *Assume also that* $|N|_\beta \leq 1$. *Then the fundamental solution matrix* $U = \sum_{i=0}^{\infty} U_i t^i$ *over* $K[\![t]\!]$ *(as in Proposition 7.3.6) satisfies*

$$|U_i|\beta^i \leq |i!|^{-2m+1} \qquad (i = 1, 2, \dots). \tag{18.1.1.1}$$

Consequently, U *has entries in* $K[\![t/(p^{-(2m-1)/(p-1)}\beta)]\!]_0$, *as does its inverse.*

Proof Recall that U is determined by the recursion (7.3.6.1):

$$N_0 U_i - U_i N_0 + iU_i = -\sum_{j=1}^{i} N_j U_{i-j} \qquad (i > 0).$$

By Lemma 7.3.5, the map $f(X) = N_0 X - X N_0$ on $n \times n$ matrices is nilpotent with nilpotency index $2m - 1$. Hence the map $X \mapsto iX + f(X)$ has inverse

$$X \mapsto \sum_{j=0}^{2m-2} (-1)^j i^{-j-1} f^j(X).$$

This gives the claim by induction on i. □

18.2 Effective bounds for solvable modules

We now give an improved version of Proposition 18.1.1 under the hypothesis that U has entries in $K\{t/\beta\}$. The hypothesis is only qualitative, in that it implies that $|U_i|\beta^i \to 0$ as $i \to \infty$, but does not give a specific bound on $|U_i|$ for any particular i. Somewhat surprisingly, this hypothesis plus an explicit bound on N together imply a rather strong explicit bound on $|U_i|$. (We continue to restrict to the case of nilpotent singularities; see Section 18.5 for the story when the exponents are allowed to range over $\mathbb{Z}_p$.)

Theorem 18.2.1. *Let* $N = \sum_{i=0}^{\infty} N_i t^i$, $U = \sum_{i=0}^{\infty} U_i t^i$ *be* $n \times n$ *matrices over* $K[\![t]\!]$ *satisfying the following conditions.*

(a) *The matrix* N *has entries in* $K[\![t/\beta]\!]_0$.

(b) *We have* $U_0 = I_n$.

(c) *We have* $U^{-1}NU + U^{-1}t\frac{d}{dt}(U) = N_0$.

(d) *The matrix* N_0 *is nilpotent.*

(e) *The matrices* U *and* U^{-1} *have entries in* $K\{t/\beta\}$.

Then for every nonnegative integer i,

$$|U_i|\beta^i \le p^{(n-1)\lfloor \log_p i \rfloor} \max\{1, |N|_\beta^{n-1}\}.$$

The first step in the proof of Theorem 18.2.1 is to change basis to reduce $|N|_\beta$, at the expense of enlarging K slightly and decreasing β slightly.

Lemma 18.2.2. *With notation as in Theorem 18.2.1, suppose that* K *has value group* $\mathbb{R}$. *Then for any* $\lambda < 1$, *there exists an invertible* $n \times n$ *matrix* X *over* $K[\![t/(\lambda\beta)]\!]_0$ *such that*

$$|X^{-1}NX + X^{-1}t\frac{d}{dt}(X)|_{\lambda\beta} \le 1$$
$$|X^{-1}|_{\lambda\beta} \le 1$$
$$|X|_{\lambda\beta} \le |N|_{\lambda\beta}^{n-1}.$$

Proof Let M be the differential module over $K[\![t/\beta]\!]_0$ for the operator $t\frac{d}{dt}$, with a basis on which D acts via N, and let $|\cdot|$ be the supremum norm defined by this basis. Over the closed disc of radius $\lambda\beta$, M becomes isomorphic to a successive extension of trivial differential modules. Consequently, the generic radius of convergence of $M \otimes_{K[\![t/\beta]\!]_0} F_{\lambda\beta}$ is equal to $\lambda\beta$. In particular,

$$p^{-1/(p-1)}\lambda\beta = |t^{-1}D|_{\mathrm{sp},M\otimes F_{\lambda\beta}} \le \left|\frac{d}{dt}\right|_{F_{\lambda\beta}} = \lambda\beta.$$

By Proposition 6.5.6 plus Lemma 8.6.1, we obtain the desired matrix X. $\quad\square$

Using Lemma 18.2.2, we wish to prove Theorem 18.2.1 by using Frobenius antecedents to reduce the index from i to $\lfloor i/p \rfloor$. One can improve upon this argument if one has a Frobenius structure on the differential module; see Lemma 18.3.2.

Lemma 18.2.3. *With notation as in Theorem 18.2.1, suppose that* $|N|_\beta \le 1$.

Then there exist $n \times n$ matrices N', U' over $K[\![t/\beta^p]\!]$ satisfying the hypotheses of Theorem 18.2.1, such that

$$|N'|_{\beta^p} \le p$$

$$\max\{|U_j|\beta^j : 0 \le j \le i\} \le \max\{|U'_j|\beta^{pj} : 0 \le j \le i/p\}.$$

Proof   Define the invertible $n \times n$ matrix $V = \sum_{i=0}^{\infty} V_i t^i$ over $K[\![t/\beta]\!]$ as follows. Start with $V_0 = I_n$. Given $V_0, \dots, V_{i-1}$, if $i \equiv 0 \pmod{p}$, put $V_i = 0$. Otherwise, put $W = \sum_{j=0}^{i-1} V_j t^j$ and $N_W = W^{-1} N W + W^{-1} t \frac{d}{dt}(W)$, and let V_i be the unique solution of the matrix equation

$$N_0 V_i - V_i N_0 + i V_i = -(N_W)_i.$$

By induction on i, $|V_i|\beta^i \le 1$ for all i, so V is invertible over $K[\![t/\beta]\!]_0$.

Let $\varphi: K[\![t]\!] \to K[\![t]\!]$ denote the substitution $t \mapsto t^p$. Put $N'' = V^{-1} N V + V^{-1} t \frac{d}{dt}(V)$; then N'' has entries in $K[\![t^p]\!]$, and $|\varphi^{-1}(N'')|_{\beta^p} \le 1$. Put $U'' = V^{-1} U$; then

$$(U'')^{-1} N'' U'' + (U'')^{-1} t \frac{d}{dt}(U'') = N''_0 = N_0,$$

which forces U'' also to have entries in $K[\![t^p]\!]$. We may then take $N' = p^{-1} \varphi^{-1}(N'')$ and $U' = \varphi^{-1}(U'')$.   □

We now put everything together.

Proof of Theorem 18.2.1   There is no harm in enlarging K, so we may assume that K has value group $\mathbb{R}$. We then prove the claim by induction on i, in three stages. First, if $i < p$ and $|N|_\beta \le 1$, then the desired estimate follows from Proposition 18.1.1. Second, for any given i, the desired estimate for general N follows from the estimate for the same i in the case $|N|_\beta \le 1$, by Lemma 18.2.2. (More precisely, for any $\lambda < 1$, replace the pair N, U by $X^{-1} N X + X^{-1} t \frac{d}{dt}(X), X^{-1} U X_0$; then take the limit as $\lambda \to 1$.) Third, if $|N|_\beta \le 1$, then the desired estimate for any given i follows from the corresponding estimate for general N with i replaced by $\lfloor i/p \rfloor$, by Lemma 18.2.3.   □

Example 18.2.4.   We show with an example that one cannot significantly improve the bound of Theorem 18.2.1 without extra hypotheses. (There is a tiny improvement possible; see the chapter notes.) The functions

$$f_i = \frac{1}{i!} (\log(1+t))^i \qquad (i = 0, \dots, n-1)$$

satisfy the differential system

$$\frac{d}{dt} f_0 = 0, \qquad \frac{d}{dt} f_i = \frac{1}{1+t} f_{i-1} \qquad (i = 1, \dots, n-1),$$

in which the coefficients have 1-Gauss norm at most 1.

One important special case of these results is that of a generic disc. This gives an alternate proof of Lemma 13.5.4.

Corollary 18.2.5. *Let V be a finite differential module over F_β (for the deriva-tion $\frac{d}{dt}$) for some $\beta > 0$, such that $IR(V) = 1$. Choose a basis of V, and for $i \geq 0$, let D_i be the matrix of action of D^i on this basis. Then*

$$|D_i/i!|_\beta \beta^i \leq p^{(n-1)\lfloor \log_p i \rfloor} \max\{1, |D_1|_\beta^{n-1}\} \qquad (i \geq 0).$$

Proof Let L be the completion of $K(t_\beta)$ for the β-Gauss norm, so that $t_\beta \in L$ is a generic point of norm β. As in Section 9.4, we may form the base change of V to the open disc of radius β with series parameter $t - t_\beta$. The fundamental solution matrix at t_β can be computed using Remark 5.8.4: it is

$$\sum_{i=0}^{\infty} \frac{(t_\beta - t)^i}{i!} D_i.$$

If we write this matrix as $\sum_{i=0}^{\infty} T_i (t - t_\beta)^i$ with T_i having entries in L, we obtain from Theorem 18.2.1 the bound

$$|T_i|\beta^i \leq p^{(n-1)\lfloor \log_p i \rfloor} \max\{1, |D_1|_\beta^{n-1}\}.$$

We now deduce the claim by induction on i. Write D_i as a power series $\sum_{j=0}^{\infty} D_{i,j}(t-t_\beta)^j$ whose coefficients have entries in L, so that $|D_{i,j}|_\beta \beta^j \leq |D_i|_\beta$ for $j \geq 0$ with equality for $j = 0$. (See Section 9.4 to recall where these inequalities come from.) We then have

$$T_i = \sum_{j=0}^{i} (-1)^j \frac{D_{j,i-j}}{j!}.$$

By the induction hypothesis, for $j < i$,

$$
\begin{aligned}
|D_{j,i-j}/j!|_\beta \beta^i &\leq |D_j/j!|_\beta \beta^j \\
&\leq p^{(n-1)\lfloor \log_p j \rfloor} \max\{1, |D_1|_\beta^{n-1}\} \\
&\leq p^{(n-1)\lfloor \log_p i \rfloor} \max\{1, |D_1|_\beta^{n-1}\}.
\end{aligned}
$$

Combined with the bound on T_i, this yields

$$|D_i/i!|_\beta \beta^i = |D_{i,0}/i!|_\beta \beta^i \leq p^{(n-1)\lfloor \log_p i \rfloor} \max\{1, |D_1|_\beta^{n-1}\},$$

as desired. □

18.3 Better bounds using Frobenius structures

Although Theorem 18.2.1 is close to optimal under its hypotheses, it can be improved if the differential module in question admits a Frobenius structure. For simplicity, we restrict to standard Frobenius structures.

Hypothesis 18.3.1. In this section, we restore the hypothesis that K is discretely valued. Fix a power q of p, and let φ be the standard q-th power Frobenius lift on $K[\![t]\!]_0$ with respect to some isometry $\varphi_K : K \to K$.

The key here is to imitate the proof of Theorem 18.2.1, but with the differential equation replaced by a certain difference equation.

Lemma 18.3.2. *Let $U = \sum_{i=0}^{\infty} U_i t^i$, $A = \sum_{i=0}^{\infty} A_i t^i$ be $n \times n$ matrices over $K[\![t]\!]$ satisfying the following conditions.*

(a) The matrix A has entries in $K[\![t]\!]_0$.
(b) We have $U_0 = I_n$, and the matrix A_0 is invertible.
(c) We have $U^{-1} A \varphi(U) = A_0$.

Then

$$\max\{|U_j| : 0 \le j \le i\} \le |A_0^{-1}||A|_1 \max\{|U_j| : 0 \le j \le \lfloor i/q \rfloor\}.$$

Consequently, for every nonnegative integer i,

$$|U_i| \le (|A_0^{-1}||A|_1)^{\lceil \log_q i \rceil}.$$

Proof Note that (c) can be rewritten as

$$U = A \varphi(U) A_0^{-1}.$$

This gives the first inequality. To deduce the second inequality, we proceed as in the proof of Theorem 18.2.1, except that we iterate $\lceil \log_q i \rceil$ times to reach the case $i = 0$ (rather than iterating $\lfloor \log_q i \rfloor$ times to get to the case $0 < i < p$). $\square$

Theorem 18.3.3. *Let $N = \sum_{i=0}^{\infty} N_i t^i$, $U = \sum_{i=0}^{\infty} U_i t^i$, $A = \sum_{i=0}^{\infty} A_i t^i$ be $n \times n$ matrices over $K[\![t]\!]$ satisfying the following conditions.*

(a) The matrix A has entries in $K[\![t]\!]_0$.
(b) We have $U_0 = I_n$, and the matrix A_0 is invertible.
(c) We have $U^{-1} N U + U^{-1} t \frac{d}{dt}(U) = N_0$.
(d) We have $N A + t \frac{d}{dt}(A) = q A \varphi(N)$.

Then $U^{-1} A \varphi(U) = A_0$, and for every nonnegative integer i,

$$|U_i| \le (|A_0^{-1}||A|_1)^{\lceil \log_q i \rceil}.$$

Proof As noted in Remark 17.1.2, the commutation relation (d) implies that $N_0 A_0 = q A_0 \varphi(N_0)$, which forces N_0 to be nilpotent. Put $B = U^{-1} A \varphi(U) = \sum_{i=0}^{\infty} B_i t^i$. Then $B_0 = A_0$, and $N_0 B + t \frac{d}{dt}(B) = q B \varphi(N_0)$. Hence

$$N_0 B_i + i B_i = q B_i \varphi(N_0) = B_i A_0^{-1} N_0 A_0,$$

or

$$N_0(B_i A_0^{-1}) + i(B_i A_0^{-1}) = (B_i A_0^{-1}) N_0. \tag{18.3.3.1}$$

By Lemma 7.3.5, the operator $X \mapsto N_0 X - X N_0 + iX$ on $n \times n$ matrices is invertible for $i \neq 0$, so (18.3.3.1) implies $B_i = 0$ for $i > 0$.

We conclude that indeed $U^{-1} A \varphi(U) = A_0$, so we may conclude by applying Lemma 18.3.2 to reduce to the case $i < q$, then applying Theorem 18.2.1. $\quad\square$

Remark 18.3.4. By combining Theorem 18.3.3 with Theorem 18.2.1 (applying the latter for $i < q$), we can obtain the bound

$$|U_i| \leq |N|_1^{n-1} p^{(n-1)\lfloor \log_p i - (\log_p q) \lfloor \log_q i \rfloor \rfloor} (|A_0^{-1}||A|_1)^{\lfloor \log_q i \rfloor}.$$

Remark 18.3.5. In applications to Picard–Fuchs modules, the difference between the bounds given by Theorem 18.2.1 and Theorem 18.3.3 can be quite significant. For instance, given a Picard–Fuchs module arising from a family of curves of genus g, the bound of Theorem 18.2.1 contains the factor $p^{(2g-1)\lfloor \log_p i \rfloor}$, but the bound of Theorem 18.3.3 replaces the factor of $2g - 1$ by 1. In general, it should be possible to use Theorem 18.3.3 (and perhaps also Theorem 18.3.6) to explain various instances in which a calculation of n terms of a power series involves a precision loss of $p^{O(\log n)}$, even though the accumulated factors of p by which one divides throughout the calculation amount to $p^{O(n)}$. (A typical example of this is [229, Lemma 3].)

We record also a sharper form of Theorem 18.3.3 for use in the discussion of logarithmic growth in the next section.

Theorem 18.3.6. *Let v be a column vector of length n over $K[\![t]\!]$, let $A = \sum_{i=0}^{\infty} A_i t^i$ be an $n \times n$ matrix over $K[\![t]\!]$, and let $\lambda \in K$ be chosen to satisfy the following conditions.*

(a) The matrix A has entries in $K[\![t]\!]_0$.
(b) The matrix A_0 is invertible.
(c) We have $A\varphi(v) = \lambda v$.

Then

$$\max\{|v_j| : 0 \leq j \leq i\} \leq |\lambda^{-1}||A|_1 \max\{|v_j| : 0 \leq j \leq \lfloor i/q \rfloor\}.$$

Consequently, for every nonnegative integer i,

$$|v_i| \le |v_0|(|\lambda^{-1}||A|_1)^{\lceil \log_q i \rceil}.$$

Proof Rewrite (c) as $v = \lambda^{-1}A\sigma(v)$ and proceed as in Lemma 18.3.2. □

We sketch an application of Theorem 18.3.3 to the complexity of Frobenius structures; this is relevant for some computational applications (as in Appendix A). See [270, Theorem 2.1] for a complete development.

Definition 18.3.7. A *wide open* subspace of the rigid analytic projective line over K is one which is the complement of a finite union of closed discs, each contained in an open residue disc.

In what follows, fix a finite collection S of residue discs, and let R be the union of the rings of rigid analytic functions on all wide open subspaces whose complements are contained in the union of the discs in S.

To describe elements of R more concretely, let us assume S contains the residue discs at 0 and ∞. Choose a polynomial $P \in K[t]$ with $|P|_1 = 1$ having at least one root in each of the other residue discs in S and no roots in any other residue discs. Then each element of R has the form

$$f_0 + \sum_{i=1}^{\infty} \frac{f_i}{P^i}$$

where $f_0 \in \bigcup_{\alpha<1<\beta} K\langle \alpha/t, t/\beta \rangle$, $f_i \in K[T]$, $\deg(f_i) < \deg(P)$, and there exists $\gamma > 1$ such that $|f_i|\gamma^i \to 0$ as $i \to \infty$. In particular, there is a natural map $R \to F_1$.

Remark 18.3.8. Equip R with a standard q-power Frobenius lift. Let M be a finite differential module over $K(t)$ with regular singularities with all exponents in $\mathbb{Q} \cap \mathbb{Z}_p$, such that 0 and ∞ are singular points and no two singular points lie in the same residue disc. In case $M \otimes_{K(t)} R$ admits a Frobenius structure, we would like to write this in terms of a basis, map from R to F_1, reduce modulo a power of p, and then approximate the result with an element of $K(t)$ with explicitly bounded zero and pole orders at all points.

This is done in [270, Theorem 2.1] by combining two ingredients. One is Corollary 17.5.2, which gives an explicit bound on the pole orders at 0 and ∞. The other is Proposition 17.3.1, which allows us to compare Frobenius structures defined using different choices of Frobenius lifts. For example, if 1 is a singular point of M, we may apply Corollary 17.5.2 to a Frobenius lift φ for which $\varphi(t-1) = (t-1)^p$, then change the Frobenius structure to a standard one.

The resulting estimates are in some sense best possible; see [270, §3].

18.4 Logarithmic growth

In general, the fundamental solution matrix of a differential system with a nilpotent regular singularity at 0 need not be bounded on a closed disc, even if the matrix defining the system is bounded and the solution matrix converges on the open disc. However, one gets a fairly mild growth condition at the boundary; better yet, one can extract some interesting information by distinguishing different solutions by their order of growth. This line of inquiry was initiated by Dwork [141, 142] (see also [349]); it is loosely inspired by archimedean considerations, for which see the chapter notes.

Definition 18.4.1. For $\delta \geq 0$, let $K[\![t]\!]_\delta$ be the subset of $K[\![t]\!]$ consisting of those $f = \sum_{i=0}^{\infty} f_i t^i$ for which

$$|f|_\delta = \sup_i \left\{ \frac{|f_i|}{(i+1)^\delta} \right\} < \infty;$$

note that $K[\![t]\!]_\delta$ is complete under the norm $|\cdot|_\delta$. For $\delta = 0$, we recover the ring $K[\![t]\!]_0$ of bounded power series. However, $K[\![t]\!]_\delta$ is not a ring for $\delta > 0$: we only have

$$K[\![t]\!]_{\delta_1} \cdot K[\![t]\!]_{\delta_2} \subset K[\![t]\!]_{\delta_1+\delta_2}.$$

Also, $K[\![t]\!]_\delta$ is stable under $\frac{d}{dt}$, but antidifferentiation carries it into $K[\![t]\!]_{\delta+1}$. Put

$$K[\![t]\!]_{\delta+} = \bigcap_{\delta' > \delta} K[\![t]\!]_{\delta'}.$$

For another useful characterization of $K[\![t]\!]_\delta$, see the exercises.

Definition 18.4.2. For $f \in K[\![t]\!]$, we say that f has *order of log-growth* δ if $f \in K[\![t]\!]_\delta$ but $f \notin K[\![t]\!]_{\delta'}$ for any $\delta' < \delta$. We say f has *order of log-growth* $\delta+$ if $f \notin K[\![t]\!]_\delta$ but $f \in K[\![t]\!]_{\delta'}$ for any $\delta' > \delta$. We have similar definitions for vectors or matrices over $K[\![t]\!]$, and for elements of $M \otimes_{K[\![t]\!]_0} K[\![t]\!]$ if M is a finite free module over $K[\![t]\!]_0$ (by computing in terms of a basis, the choice of which will not affect the answer).

We then deduce the following from Theorem 18.2.1.

Proposition 18.4.3. *Let M be a differential module of rank n over $K[\![t]\!]_0$ for the operator $t\frac{d}{dt}$, such that the action of D on M/tM is nilpotent and the fundamental solution matrix with respect to some basis is an invertible matrix over $K\{t\}$. Then any element of $H^0(M \otimes_{K[\![t]\!]_0} K[\![t]\!])$ has order of log-growth at most $n - 1$.*

Remark 18.4.4. One can make a slightly stronger assertion than Proposition 18.4.3 by observing that $M \otimes_{K[\![t]\!]_0} K[\![t]\!][\log t]$ admits a basis of horizontal sections, each of which has degree at most $n - 1$ in $\log t$ (exercise). If we write some $m \in H^0(M \otimes_{K[\![t]\!]_0} K[\![t]\!][\log t])$ as a formal sum $\sum_{i=0}^{n-1} m_i (\log t)^i$ with $m_i \in M \otimes_{K[\![t]\!]_0} K[\![t]\!]$, then Theorem 18.2.1 implies that for $i = 0, \ldots, n-1$, m_i has order of log-growth at most $n - 1$. However, we suspect that this can be improved to $n - 1 - i$.

If we drop the condition that M is solvable on the open unit disc, we can still hope to get some information. The following is [141, Conjecture 1]. (See [303] for a result in this direction.)

Conjecture 18.4.5 (Dwork). *Let M be a finite differential module over $K[\![t]\!]_0$ and set*

$$\delta = \dim_K H^0(M \otimes_{K[\![t]\!]_0} K\{t\}).$$

Then

$$H^0(M \otimes_{K[\![t]\!]_0} K[\![t]\!]_{\delta-1}) = H^0(M \otimes_{K[\![t]\!]_0} K\{t\}).$$

Theorem 18.4.6 (Ohkubo). *Conjecture 18.4.5 holds when the rank of M is 2.*

Proof See [325, Main Theorem]. □

In the presence of a Frobenius structure, one obtains a much sharper bound, due to Chiarellotto and Tsuzuki [87, Theorem 6.17]. (One can also formulate a refinement in the manner of Remark 18.4.4.)

Theorem 18.4.7. *Assume that K is discretely valued. Let M be a differential module of rank n over $K[\![t]\!]_0$ for the operator $t\frac{d}{dt}$, equipped with a Frobenius structure for a q-power Frobenius lift as in Remark 17.1.2. Then any element $v \in H^0(M \otimes_{K[\![t]\!]_0} K[\![t]\!])$ satisfying $\Phi(v) = \lambda v$ for some $\lambda \in K$ has order of log-growth at most $(-\log |\lambda| - s_0)/(\log q)$, where s_0 is the least generic Newton slope of M.*

Proof Thanks to Proposition 17.3.1, we may assume the Frobenius lift is standard. By then replacing the Frobenius lift by some power, we can reduce to the case where s_0 is a multiple of $-\log p$. We can then twist into the case $s_0 = 0$. By Proposition 14.5.9, we can choose a basis of M such that the least generic Hodge slope of M is also 0. Then the claim follows immediately from Theorem 18.3.6. □

Refining a conjecture of Dwork, Chiarellotto and Tsuzuki [87] have conjectured that if M is indecomposable, then Theorem 18.4.7 is optimal. This is now a theorem of Ohkubo.

Theorem 18.4.8 (Ohkubo). *In the notation of Theorem 18.4.7, suppose further that M is indecomposable. Then v has order of log-growth exactly $(-\log|\lambda| - s_0)/(\log q)$.*

Proof See [87, Theorem 7.1] for the case where M has rank 2 and [326, Theorem 0.1] for the general case. □

The following corollary of Theorem 18.4.8 was known previously (see [324, Main Theorem]); we formulate it so that we can point out its dependence on the Frobenius structure (see Remark 18.4.12).

Corollary 18.4.9 (Ohkubo). *With notation as in Theorem 18.4.8, the following statements hold.*

(a) For all $\delta \geq 0$,

$$H^0(M \otimes_{K[\![t]\!]_0} K[\![t]\!]_\delta) = H^0(M \otimes_{K[\![t]\!]_0} K[\![t]\!]_{\delta+}).$$

(b) For every $\delta > 0$ not in $\mathbb{Q}$, there exists $\delta' \in [0, \delta)$ with

$$H^0(M \otimes_{K[\![t]\!]_0} K[\![t]\!]_{\delta'}) = H^0(M \otimes_{K[\![t]\!]_0} K[\![t]\!]_\delta).$$

In other words, the order of log-growth of every horizontal section of M is rational.

Proof This is immediate from Theorem 18.4.8. □

We now introduce a related result of André which does not require a Frobenius structure.

Definition 18.4.10. Let M be a finite differential module of rank n over $K[\![t]\!]_0$. Define the nondecreasing sequence $\delta_1, \ldots, \delta_n$ of real numbers by setting δ_i to be the supremum of all $\delta \geq 0$ for which $\dim_K H^0(M \otimes_{K[\![t]\!]_0} K[\![t]\!]_\delta) < i$ (or 0 if no such δ exist). By Proposition 18.4.3, $\delta_n \leq n - 1$. We assemble the sequence $\delta_1, \ldots, \delta_n$ into the *log-growth polygon* of M, but with the convention that we fix the *right* endpoint at $(n, 0)$ rather than fixing the left endpoint at $(0, 0)$.

As in Theorem 11.9.2, we can obtain a second such polygon by performing base change to a generic disc and computing there. We refer to this as the *generic log-growth polygon* of M, and to the original definition as the *special log-growth polygon* of M.

In the case of a differential module with Frobenius structure, the following statement is a consequence of Theorem 18.4.8 on account of Theorem 15.3.2.

Theorem 18.4.11. *For any finite differential module over $K[\![t]\!]_0$, the special log-growth polygon lies on or above the generic log-growth polygon. (We do not make any claim about the left endpoint; see Remark 18.4.12.)*

Proof See [17, Theorem 4.1.1]. ◻

Remark 18.4.12. One might hope from Theorem 18.4.11 that additional properties of Frobenius Newton polygons are absorbed by the special and generic log-growth polygons in the absence of a Frobenius structure, but this hope turns out to be too optimistic. For example, without a Frobenius structure, there is no analogue of Corollary 18.4.9: the order of log-growth of a horizontal section need not be rational, and it can have the form $\delta+$ instead of δ [87, §5.2]. Moreover, the left endpoints of the special and generic log-growth polygons need not coincide [322].

18.5 Nonzero exponents

So far, we only have considered regular differential systems with all exponents equal to zero. We now allow for arbitrary exponents in $\mathbb{Z}_p$; for this, we must revisit Proposition 18.1.1.

Proposition 18.5.1. *Let $N = \sum_{i=0}^{\infty} N_i t^i$ be an $n \times n$ matrix over $K[\![t/\beta]\!]_0$ corresponding to the differential system $D(v) = Nv + d(v)$, where $d = t\frac{d}{dt}$. Assume that N_0 has prepared eigenvalues $\alpha_1, \ldots, \alpha_l \in \mathbb{Z}_p$ and that its minimal polynomial is $\prod_{j=1}^{l} (T - \alpha_j)^{m_j}$. Assume also that $|N|_\beta < 1$. Then the fundamental solution matrix $U = \sum_{i=0}^{\infty} U_i t^i$ over $K[\![t]\!]$ (as in Proposition 7.3.6) satisfies*

$$|U_i|\beta^i \leq \prod_{j,k=1}^{l} \prod_{h=1}^{i} |h - \alpha_j + \alpha_k|^{-(e_j + e_k - 1)} \qquad (i = 1, 2, \ldots). \qquad (18.5.1.1)$$

Proof By Lemma 7.3.5, the map $f(X) = N_0 X - X N_0$ on $n \times n$ matrices has minimal polynomial $P(T) = \prod_{j,k=1}^{l} (T - \alpha_j + \alpha_k)^{e_j + e_k - 1}$. We may thus invert $X \mapsto iX + f(X)$ using the formula

$$X \mapsto \sum_{j=1}^{\deg(P)} -\frac{P^{(j)}(i)}{j! P(i)} X^j.$$

Since $P^{(j)}(i)/j! \in \mathbb{Z}_p$, this yields the asserted bound. ◻

Theorem 18.5.2. *Let $N = \sum_{i=0}^{\infty} N_i t^i$, $U = \sum_{i=0}^{\infty} U_i t^i$ be $n \times n$ matrices over $K[\![t]\!]$ satisfying the following conditions for some $\beta > 0$.*

(a) *The matrix N has entries in $K[\![t/\beta]\!]_0$.*
(b) *We have $U_0 = I_n$.*
(c) *We have $U^{-1}NU + U^{-1}t\frac{d}{dt}(U) = N_0$.*

(d) The matrix N_0 has prepared eigenvalues $\alpha_1, \ldots, \alpha_l \in \mathbb{Z}_p$ and its minimal polynomial is $\prod_{j=1}^{l}(T - \alpha_j)^{m_j}$.

(e) The matrices U and U^{-1} have entries in $K\{t/\beta\}$.

Then for every positive integer i,

$$|U_i|\beta^i \leq c_i p^{n+(n-1)\lceil \log_p i \rceil} \max\{1, |N|_\beta^{n-1}\}$$

where

$$c_i = \max\left\{\prod_{1 \leq j < k \leq l} \left|\frac{\alpha_j - \alpha_j^{(s)} - \alpha_k + \alpha_k^{(s)}}{p^s}\right|^{-(m_j + m_k - 1)} : 0 \leq s \leq \lfloor \log_p i \rfloor\right\}.$$

Proof The statement of Lemma 18.2.2 carries over to this situation without change. However, we can only carry out the construction in the proof of Lemma 18.2.3 if the exponents are divisible by p. Otherwise, we must first enforce this condition by performing up to $p - 1$ shearing transformations, yielding the weaker bound

$$\max\{|U_j|\beta^j : 0 \leq j \leq i\} \leq \max\{|U_j'|\beta^{pj} : 0 \leq j \leq (i+p-1)/p\}.$$

If $i \leq p^{h+1}$ for some $h \geq 0$, then after h iterations of the map $i \mapsto \lfloor (i+p-1)/p \rfloor$ we end up with a quantity which is at most p. We may then obtain the claimed bound by applying 18.5.1. $\qquad\square$

This example limits the extent to which Theorem 18.5.2 can be improved.

Example 18.5.3. For $\alpha \in \mathbb{Z}_p \setminus \mathbb{Z}$ with type 1 in the sense of Definition 13.1.1, the matrices

$$N = \begin{pmatrix} 0 & t \\ 0 & \alpha + \sum_{n=1}^{\infty} t^n \end{pmatrix}, \qquad U = \begin{pmatrix} 1 & \sum_{n=1}^{\infty} \frac{t^n}{\alpha - n} \\ 0 & \sum_{n=0}^{\infty} t^n \end{pmatrix}$$

satisfy the hypotheses of Theorem 18.5.2. Such examples can have arbitrarily large log-growth: for k a positive integer and $\alpha = -\sum_{i=1}^{\infty} p^{k^i}$, U has order of log-growth k. In particular, the constant c_i in Theorem 18.5.2 cannot be chosen independently of the eigenvalues of N_0. (In fact, U need not exhibit logarithmic growth at all; see the exercises.)

Notes

In the case of no singularities ($N_0 = 0$), the effective bound of Theorem 18.2.1 is due to Dwork and Robba [151], with a slightly stronger bound. For the key argument, see Lemma 9.11.1 and also [149, Theorem IV.3.1].

The general case of Theorem 18.2.1 is due to Christol and Dwork [94], except that their bound is significantly weaker: it is roughly $p^{c(n-1)\lfloor \log_p i \rfloor}$ with $c = 2 + \frac{1}{p-1}$. The discrepancy comes from the fact that the role of Proposition 6.5.6 is played in [94] by an effective version of the cyclic vector theorem, which does not give optimal bounds. As usual, use of cyclic vectors also introduces singularities which must then be removed, leading to some technical difficulties. See also [149, Theorem V.2.1].

Theorems 18.3.3 and 18.3.6 are original, but they owe a great debt to the proof of [87, Theorem 7.2]. The main difference is that we prefer to argue in terms of matrices rather than cyclic vectors.

In the case of no singularities, Proposition 18.4.3 was first proved by Dwork; it appears in [141] and [142]. (See also [89].) The extension suggested by Remark 18.4.4 is original; as noted above, the effective bounds in [94] are not strong enough to imply this.

The study of logarithmic growth initiated by Chiarellotto and Tsuzuki in [87] is continued in [88]. A key role is played by a theorem of Christol that relates logarithmic growth to certain canonical filtrations of a differential module over $\mathcal{E}$; see [89, §4.3] and [87, §3.2].

The archimedean motivation for the study of logarithmic growth comes from Deligne's study of regular singularities [126]. He showed that a differential module over the ring of germs of meromorphic functions at a point has a regular singularity if and only if the horizontal sections have at worst logarithmic growth at the singular point.

Theorem 18.5.2 is derived from [149, Theorem V.9.1], which does not include an explicit bound. Example 18.5.3 was suggested by Takahiro Nakagawa.

Exercises

18.1 Prove that for $\delta \geq 0$,

$$K[\![t]\!]_\delta = \{f \in K[\![t]\!] : \limsup_{\rho \to 1^-} \frac{|f|_\rho}{(-\log \rho)^\delta} < \infty\}.$$

(Hint: the inequality

$$\sup_i \{(i+1)^\delta \rho^i\} \leq \rho^{-1} \left(\frac{\delta}{e}\right)^\delta (-\log \rho)^{-\delta}$$

may be helpful.)

18.2 Let M be a differential module of rank n over $K[\![t]\!]$ for the operator $t\frac{d}{dt}$, such that the action of D on M/tM is nilpotent. Show that if we extend the action of $t\frac{d}{dt}$ to $K[[t]][\log t]$ by setting $(t\frac{d}{dt})(\log t) = 1$, then $M \otimes_{K[[t]]} K[[t]][\log t]$ admit a basis of horizontal sections and each horizontal section has degree in $\log t$ bounded by $n - 1$.

18.3　In this exercise, set notation as in Example 18.5.3.

(a) Verify the assertion that for $\alpha = -\sum_{i=1}^{\infty} p^{k^i}$, U has order of log-growth k.

(b) Show that for $\alpha = -\sum_{i=1}^{\infty} p^{i!}$, U has entries in $K\{t\}$ but does not have order of log-growth δ for any $\delta > 0$.

18.4　In this exercise, we prove a lemma of Dwork [146]. Let M_1 and M_2 be differential modules over $K(t)$ satisfying the following conditions.

(a) We have $R(M_{1,\rho=1}) = 1$.

(b) The singularities of M_2 are all regular with entries in $\mathbb{Q} \cap \mathbb{Z}_p$, and no two of them lie in the same residue disc.

(c) The differential module M_2 is irreducible.

Let R be a ring as in Definition 18.3.7. Show that if there exists a Frobenius intertwiner from $M_1 \otimes_{K(t)} R$ to $M_2 \otimes_{K(t)} R$, then it is unique up to multiplication by a nonzero element of K. (Hint: let A_1, A_2 be the matrices of action of two intertwiners on a single pair of bases; then $A = A_2^{-1}A_1$ defines a horizontal morphism from M_2 to M_2. Use Theorem 13.7.1 to show that the entries of A are meromorphic in each residue disc.)

19

Galois representations and differential modules

In this chapter, we construct a class of examples of differential modules on open annuli which carry Frobenius structures, and hence are solvable at a boundary. These modules are derived from continuous linear representations of the absolute Galois group of a positive-characteristic local field.

We first construct a correspondence between Galois representations and differential modules over $\mathcal{E}$ carrying a unit-root Frobenius structure. The basic mechanism for producing these modules is to tensor with a large ring carrying a Galois action, then take Galois invariants. This mechanism will reappear when we turn to p-adic Hodge theory, at which point we will attempt to simulate this construction using the Galois group of a mixed-characteristic local field. See Appendix C.

We then refine the construction to compare Galois representations with finite image of inertia and differential modules over $\mathcal{E}^{\dagger}$ carrying a unit-root Frobenius structure; the main result here is an equivalence of categories due to Tsuzuki. It is also a special case of the absolute case of the p-adic local monodromy theorem (Theorem 20.1.4), and indeed can be used together with the slope filtration theorem (Theorem 17.4.3) to prove the monodromy theorem in the absolute case. This result also has an analogue in p-adic Hodge theory; see Theorem C.2.5.

We finally describe (without proof) a numerical relationship between wild ramification of a Galois representation and convergence of solutions of p-adic differential equations. Besides making explicit the analogy between wild ramification of Galois representations and irregularity of meromorphic differential systems, it also suggests an approach to higher-dimensional ramification theory. We reserve most discussion of the latter to the notes.

Remark 19.0.1. The reader encountering this material for the first time is strongly encouraged to assume that κ_K is perfect. For an explanation of what gets complicated otherwise, see Section 20.5.

Notation 19.0.2. In this chapter, we write d and φ instead of D and Φ for the actions on a differential or difference module. This is to avoid confusion with another standard usage of the letter D, which will first appear in Definition 19.1.2.

19.1 Representations and differential modules

We first describe a simple correspondence between Galois representations and differential modules, due to Fontaine. This will serve as a model later when we introduce (φ, Γ)-modules associated to Galois representations of local fields in mixed characteristic (Appendix C).

Definition 19.1.1. For L a finite separable extension of $\kappa_K((t))$, let $\mathcal{E}_L$ be the finite unramified extension of $\mathcal{E}$ with residue field L (see Corollary 3.2.4). Let $\tilde{\mathcal{E}}$ be the completion of the maximal unramified extension of $\mathcal{E}$, which in particular contains the completion $\tilde{K}$ of the maximal unramified extension of K. Then $G_{\kappa_K((t))}$ acts on $\tilde{\mathcal{E}}$ with fixed field $\mathcal{E}$.

Definition 19.1.2. Let V be a finite-dimensional vector space over K, and let $\tau \colon G_{\kappa_K((t))} \to \mathrm{GL}(V)$ be a continuous homomorphism for the p-adic topology on $\mathrm{GL}(V)$. Let us view $V \otimes_K \tilde{\mathcal{E}}$ as a left $G_{\kappa_K((t))}$-module with the action on the first factor coming from τ and the natural action on the second factor. Put

$$D(V) = (V \otimes_K \tilde{\mathcal{E}})^{G_{\kappa_K((t))}}.$$

Lemma 19.1.3. *The space $D(V)$ is an $\mathcal{E}$-vector space of dimension $\dim_K(V)$. Equivalently, the natural map $D(V) \otimes_{\mathcal{E}} \tilde{\mathcal{E}} \to V \otimes_K \tilde{\mathcal{E}}$ is an isomorphism.*

Proof We first check this in case τ has finite image, in which case τ factors through $G_{E/\kappa_K((t))}$ for some finite separable extension E of $\kappa_K((t))$. In this case, the claim is a consequence of Noether's nonabelian version of Hilbert's Theorem 90: for any finite Galois extension E/F of fields, the nonabelian cohomology set $H^1(G_{E/F}, \mathrm{GL}_n(E))$ is trivial.

In the general case, we must argue a bit more carefully. Since $G_{\kappa_K((t))}$ is compact as a topological group, its image under τ is also compact. This implies that we can find an $\mathfrak{o}_{\mathcal{E}}$-lattice T in V which is stable under the Galois action, e.g., by starting with any lattice and taking the span of its images under the Galois action. (The compactness ensures that the resulting $\mathfrak{o}_{\mathcal{E}}$-submodule of V is indeed finitely generated.) For any positive integer i, the induced topology on $T/\mathfrak{m}_K^i T$ is discrete, so the Galois action factors through the Galois group of a finite separable extension of $\kappa_K((t))$. We may thus argue for each i as above and then take the inverse limit. $\square$

Definition 19.1.4. Note that $\frac{d}{dt}$ extends uniquely to $\mathcal{E}_L$, and hence to $D(V)$ by taking the action on V to be trivial. Since the action of $\frac{d}{dt}$ commutes with the Galois action, we also obtain an action on $D(V)$. That is, $D(V)$ is a differential module over $\mathcal{E}$. By the same token, if we equip V with the structure of a difference module with respect to φ_K, then $D(V)$ inherits a Frobenius structure for any Frobenius lift φ on $\mathcal{E}$ acting on K via φ_K. Since we can always start with a unit-root Frobenius structure on V (e.g., by forcing a basis of V to be fixed), we deduce that $D(V)$ admits a unit-root Frobenius structure.

We obtain an instance of nonabelian Artin–Schreier theory. See the notes for Chapter 14 for some background.

Proposition 19.1.5. *Suppose that φ is an absolute q-power Frobenius lift, and that the fixed field K_0 of φ on K has residue field $\mathbb{F}_q$ and the same value group as K. Given a continuous representation of $G_{\kappa_K((t))}$ on a finite-dimensional K_0-vector space V_0, equip $V = V_0 \otimes_{K_0} K$ with the Frobenius action induced by the trivial action on V_0. Then $V_0 \mapsto D(V)$ is an equivalence of categories from the category of continuous representation of $G_{\kappa_K((t))}$ on finite-dimensional K_0-vector spaces to the category of finite differential modules over $\mathcal{E}$ equipped with unit-root Frobenius structures.*

Proof We will show that the reverse equivalence is provided by the functor V_0 on finite differential modules over $\mathcal{E}$ equipped with unit-root Frobenius structures defined by

$$V_0(D) = (D \otimes_{\mathcal{E}} \tilde{\mathcal{E}})^{\varphi=1}.$$

Since $D(V_0) \otimes_{\mathcal{E}} \tilde{\mathcal{E}} \to V_0 \otimes_{K_0} \tilde{\mathcal{E}}$ is an isomorphism by Lemma 19.1.3, we may naturally identify $V_0(D(V_0))$ with $(V \otimes_K \tilde{\mathcal{E}})^{\varphi=1} = V_0$. (Here the conditions on K_0 are needed to ensure that $\tilde{\mathcal{E}}^{\varphi=1} = K_0$. See the exercises.)

It remains to give a canonical isomorphism from $D(V_0(D))$ to D; by the same token, it is sufficient to check that the natural map $V_0(D) \otimes_{K_0} \tilde{\mathcal{E}} \to D \otimes_{\mathcal{E}} \tilde{\mathcal{E}}$ is a bijection. This is exactly the statement that $D \otimes_{\mathcal{E}} \tilde{\mathcal{E}}$ is a trivial difference module. This holds by the first part of the proof of Theorem 14.6.3, since the residue field of $\tilde{\mathcal{E}}$ is separably closed and hence weakly difference-closed for absolute Frobenius. $\square$

Corollary 19.1.6. *Suppose that φ is an absolute q-power Frobenius lift and that $K = \mathbb{Q}_q$ is the unramified extension of $\mathbb{Q}_p$ with residue field $\mathbb{F}_q$. Then the* *functor D, from continuous representations of $G_{\mathbb{F}_q((t))}$ on finite-dimensional $\mathbb{Q}_q$-vector spaces to finite differential modules over $\mathcal{E}$ equipped with unit-root Frobenius structure, is an equivalence of categories.*

Remark 19.1.7. In Proposition 19.1.5, the differential structure is not necessary; see the exercises. What we do use is a symmetry between two actions on $\tilde{\mathcal{E}}$: that of the Galois group and that of the monoid given by the nonnegative powers of φ. The proof works because $\tilde{\mathcal{E}}$ is large enough to trivialize both the Galois action (by the Hilbert–Noether–Speiser theorem) and the φ-action (by the Dieudonné–Manin classification). Starting from this point, Fontaine realized that one could set up an analogous situation when $G_{\mathbb{F}_q((t))}$ is replaced by the Galois group of a finite extension of $\mathbb{Q}_p$; the result is a central construction in p-adic Hodge theory. See Appendix C.

Remark 19.1.8. If φ is a Frobenius lift which is not absolute, then Definition 19.1.4 remains valid. This is because even though an individual finite separable extension of $\kappa_K((t))$ may not carry an action of φ, the separable closure of $\kappa_K((t))$ does carry such an action. However, Proposition 19.1.5 does not remain valid because the residue field of $\tilde{\mathcal{E}}$ need not be weakly difference-closed.

19.2 Finite representations and overconvergent differential modules

Following an idea of Crew, we give a refinement of the previous construction for some special representations.

Definition 19.2.1. Since $\mathcal{E}^\dagger$ is henselian (Lemma 15.1.3), for each finite separable extension L of $\kappa_K((t))$, there exists a unique finite unramified extension $\mathcal{E}_L^\dagger$ with residue field L. In fact,

$$\mathcal{E}_L \cong \mathcal{E} \otimes_{\mathcal{E}^\dagger} \mathcal{E}_L^\dagger,$$

and $\mathcal{E}_L^\dagger$ is the integral closure of $\mathcal{E}^\dagger$ in $\mathcal{E}_L$. In particular, $G_{\kappa_K((t))}$ acts on $\mathcal{E}_L^\dagger$ with fixed field $\mathcal{E}^\dagger$.

Definition 19.2.2. Let V be a finite-dimensional vector space over K, and let $\tau: G_{\kappa_K((t))} \to \mathrm{GL}(V)$ be a continuous homomorphism for the *discrete* topology on $\mathrm{GL}(V)$. That is, τ factors through $G_{L/\kappa_K((t))}$ for some finite separable extension L of $\kappa_K((t))$. Let us view $V \otimes_K \mathcal{E}_L^\dagger$ as a $G_{\kappa_K((t))}$-module with the action on the first factor coming from τ and the action on the second factor as above. Put

$$D^\dagger(V) = (V \otimes_K \mathcal{E}_L^\dagger)^{G_{\kappa_K((t))}}.$$

Lemma 19.2.3. *The space $D^\dagger(V)$ is an $\mathcal{E}^\dagger$-vector space of dimension $\dim_K(V)$. Equivalently, the natural map $D^\dagger(V) \otimes_{\mathcal{E}^\dagger} \mathcal{E}_L^\dagger \to V \otimes_K \mathcal{E}_L^\dagger$ is an isomorphism. (In particular, $D^\dagger(V)$ is canonically independent of the choice of L.)*

Proof This again follows from the Hilbert–Noether–Speiser theorem. □

Definition 19.2.4. As in Definition 19.1.4, $D^\dagger(V)$ is a differential module over $\mathcal{E}^\dagger$ admitting a unit-root Frobenius structure for any Frobenius lift φ on $\mathcal{E}^\dagger$. However, now that we have a module over $\mathcal{E}^\dagger$, there is a sense in which it makes sense to compute the subsidiary radii of $D^\dagger(V) \otimes_{\mathcal{E}^\dagger} F_\rho$ for $\rho \in (0, 1)$ sufficiently close to 1. Namely, realize $D^\dagger(V)$ as a differential module over $K\langle\alpha/t, t\rrbracket_0$ for some α and compute there. Beware that any two such realizations for a given α need only become isomorphic over $K\langle\beta/t, t\rrbracket_0$ for some $\beta \in [\alpha, 1)$. However, statements about the germ at 1 of the function $\rho \mapsto R(D^\dagger(V) \otimes_{\mathcal{E}^\dagger} F_\rho)$ are unambiguous.

Proposition 19.2.5. *The generic radius of convergence of $D(V)$ is equal to 1. Consequently (by the continuity of the generic radius of convergence, as in Theorem 11.3.2(a)), $D^\dagger(V)$ is solvable at 1.*

Proof This follows from the existence of a Frobenius structure on $D^\dagger(V)$, using Theorem 17.2.1. □

Example 19.2.6. Assume that K contains an element π with $\pi^{p-1} = -p$; then K contains a unique p-th root of unity ζ_p satisfying $1 - \zeta_p \equiv \pi \pmod{\pi^2}$ (see Remark 17.1.7). Let $L = \kappa_K((t))[z]/(z^p - z - \bar{f})$ be an Artin–Schreier extension and let V be the Galois representation corresponding to the character of $G_{L/\kappa_K((t))}$ taking the automorphism $z \mapsto z + 1$ to ζ_p. We can then explicitly describe $D^\dagger(V)$: it is the module M_f of Example 17.1.4 (exercise).

Similarly, one can realize the construction of Example 9.9.3 as $D^\dagger(V)$ for a certain explicit character of order p^h. This observation has been thoroughly generalized by Pulita [343].

Remark 19.2.7. Note that the kernel of d on $\mathcal{E}_L^\dagger$ is the integral closure K' of K in $\mathcal{E}_L^\dagger$ (exercise). Consequently, the space of horizontal sections of $D^\dagger(V) \otimes_{\mathcal{E}^\dagger} \mathcal{E}_L^\dagger$ is equal to $V \otimes_K K'$. This suggests that we cannot recover all of V from $D^\dagger(V)$ as long as we only use the differential structure; instead, we only recover the restriction of V to the inertia subgroup of $G_{\kappa_K((t))}$, which we can identify with $G_{\kappa_K^{\mathrm{sep}}((t))}$.

The previous remark suggests the following construction.

Definition 19.2.8. Let V be a finite-dimensional vector space over K, and let $\tau \colon G_{K_K((t))} \to \mathrm{GL}(V)$ be a continuous homomorphism for the p-adic topology. We say that τ has *finite local monodromy* if the image of the inertia subgroup of $G_{K_K((t))}$ is finite. (That inertia subgroup is isomorphic to $G_{K_K^{\mathrm{sep}}((t))}$.) In this case, let $\mathcal{E}^\dagger_{K_K^{\mathrm{sep}}((t))}$ be the ring defined in the same fashion as $\mathcal{E}^\dagger$ but using $\widehat{K^{\mathrm{unr}}}$ (the completion of the maximal unramified extension of K) for the coefficients; let $G_{K_K((t))}$ act on this ring via the quotient by its inertia subgroup. We can then define

$$D^\dagger(V) = (V \otimes_K (\mathcal{E}^\dagger_{K_K^{\mathrm{sep}}((t))})^{\mathrm{unr}})^{G_{K_K((t))}}$$

and this will be a differential module over $\mathcal{E}^\dagger$ of the right dimension, again admitting a unit-root Frobenius structure for any Frobenius lift.

19.3 The unit-root p-adic local monodromy theorem

We have the following refinement of Proposition 19.1.5. See the chapter notes for a detailed attribution. (For an analogous fact in p-adic Hodge theory, see Theorem C.2.5.)

Theorem 19.3.1 (Tsuzuki). *Suppose that φ is an absolute q-power Frobenius lift, and that the fixed field K_0 of φ on K has residue field $\mathbb{F}_q$ and the same value group as K. Given a continuous representation of $G_{K_K((t))}$ with finite local monodromy on a finite-dimensional K_0-vector space V_0, equip $V = V_0 \otimes_{K_0} K$ with the Frobenius action induced by the trivial action on V_0. Then $V_0 \mapsto D^\dagger(V)$ is an equivalence of categories with the category of finite differential modules over $\mathcal{E}^\dagger$ equipped with unit-root Frobenius structure.*

Although it is possible to deduce this theorem from the p-adic local monodromy theorem (see Remark 20.1.5), it is both instructive and historically accurate to give a proof using the tools we have available at this point. We proceed to this task now.

Definition 19.3.2. Let M be a finite differential module over $\mathcal{E}^\dagger$ equipped with a unit-root Frobenius structure. For $c \in (0, 1)$, we say that M is *c-constant* if there exists a basis of M on which Φ acts via a matrix A with $|A - I_n|_1 \le c$; we call such a basis a *c-constant basis*.

In case φ is absolute, we can see directly that the property of M being c-constant is invariant under changing the choice of the Frobenius lift (Corollary 17.3.2). Namely, by Proposition 19.1.5, M is c-constant if and only if it occurs as $D(V)$ for some representation $\tau \colon G_{K_K((t))} \to \mathrm{GL}(V)$ such that V

admits a supremum norm $|\cdot|$ for which $|\tau(g)(x) - x| \le c|x|$ for all $g \in G_{\kappa_K((t))}$ and $x \in V$.

For arbitrary φ, we may reach the same conclusion by imitating the proof of Proposition 17.3.5. See exercises.

Lemma 19.3.3. *Let M be a finite differential module over $\mathcal{E}^\dagger$ admitting a unit-root Frobenius structure for a standard (but not necessarily absolute) Frobenius lift. Then there exists a positive integer m coprime to p such that $M \otimes_{\mathcal{E}^\dagger} \mathcal{E}^\dagger[t^{1/m}]$ admits a basis on which the matrices A, N of actions of Φ, tD have entries in $\mathcal{E}^\dagger[t^{1/m}] \cap \mathfrak{o}_K[\![t^{-1/m}]\!]$, as does A^{-1}. Moreover, if M is c-constant for some $c < 1$, then we can ensure that $m = 1$ and $|A - I_n|_1 \le c$.*

Proof Since M admits a unit-root Frobenius structure, we can choose a basis on which the matrix of action A of Φ belongs to $\mathrm{GL}_n(\mathfrak{o}_{\mathcal{E}^\dagger})$. We first put A in the desired form modulo $\mathfrak{m}_K$. By reordering the basis vectors, we can ensure that the minimum t-adic valuation of the reduction $\overline{A}_{ij}$ of A_{ij} occurs for $i = j = 1$. By adjoining $t^{1/(q-1)}$ and then rescaling, we can force this minimum valuation to equal 0. We can then conjugate to ensure that $\overline{A}_{i1} = 0$ for $i = 2, \ldots, n$. Proceeding in this manner, we can force $\overline{A}$ to be upper triangular and invertible over $\kappa_K[\![t^{1/m}]\!]$.

We next put A in the desired form, by repeating the following operation. (If M is c-constant for some $c < 1$, we may start the argument here.) Write $A = \sum_i A_i t^{i/m}$, so that A_0 is upper triangular and invertible modulo $\mathfrak{m}_K$, and A_i vanishes modulo $\mathfrak{m}_K$ for $i < 0$. Then replace A by $U^{-1} A \varphi(U)$ with

$$U = I_n + A_0^{-1}\left(\sum_{i>0} A_i t^{i/m}\right).$$

The matrices U then converge to the identity under the $(t^{1/m}, \mathfrak{m}_K)$-adic topology on $\mathfrak{o}_K[\![t^{1/m}]\!]$.

Finally, we note that the compatibility $N - qA\varphi(N)A^{-1} = t\frac{d}{dt}(A)A^{-1}$ from (17.1.2.1) implies that having A in the desired form forces N to be in the desired form as well. Namely, the compatibility first implies that N modulo q only involves nonnegative powers of $t^{-1/m}$. We then derive the same conclusion modulo q^2, q^3, and so on. $\qquad\square$

The crux of the proof is the following lemma.

Lemma 19.3.4. *Let M be a finite differential module over $\mathcal{E}^\dagger$ admitting a unit-root Frobenius structure. Suppose that M is c-constant for some $c < p^{-1/(p-1)}$. Then M is trivial as a differential module.*

Proof We may assume that the Frobenius lift is standard. By Lemma 19.3.3, we may assume that there is a c-constant basis on which Φ, tD act via matrices A, N over $\mathcal{E}^\dagger \cap \mathfrak{o}_K[\![t^{-1}]\!]$. Then A and N together represent a finite differential module over $\bigcup_{\alpha \in (0,1)} K\langle \alpha/t \rangle$ equipped with a unit-root Frobenius structure. Moreover, the commutation relation $NA + t\frac{d}{dt}(A) = qA\varphi(N)$ from (17.1.2.1) and the fact that $|A_0| = |A_0^{-1}| = |A|_1 = |A^{-1}|_1 = 1$ force $N_0 = 0$ and $|N|_1 \leq c$; consequently, there exists $\alpha \in (0, 1)$ for which $|N|_\alpha \leq 1$.

We now proceed as in the proof of Lemma 18.2.3. As in that proof, we construct a matrix V with entries in $\mathcal{E}^\dagger \cap K[\![t^{-1}]\!]$ with $|V - I_n|_1 \leq c$ and $|V - I_n|_\alpha \leq 1$, for which

$$A' = V^{-1}A\varphi(V), \qquad N' = V^{-1}NV + V^{-1}t\frac{d}{dt}(V)$$

have entries in $\mathcal{E}^\dagger \cap K[\![t^{-p}]\!]$. Let $\psi : K[\![t^{-1}]\!] \to K[\![t^{-1}]\!]$ be the K-linear substitution $t^{-1} \mapsto t^{-p}$. Since φ is standard, φ commutes with ψ; consequently, $A'' = \psi^{-1}(A'), N'' = p^{-1}\psi^{-1}(N')$ again satisfy the commutation relation $N''A'' + t\frac{d}{dt}(A'') = qA''\varphi(N'')$. Since $|A'' - I_n|_1 = |A' - I_n|_1 \leq c$, we deduce $|N''|_1 \leq c$; we also have $|N''|_{\alpha^p} = p|N'|_\alpha \leq p$. To choose $u \in [0, 1]$ such that $p^u c^{1-u} = 1$, or in other words $u \log p + (1 - u) \log c = 0$, we take

$$u = \frac{\log c}{\log c - \log p}.$$

Using Proposition 8.2.3(b), we obtain $|N''|_\beta \leq 1$ for

$$\log \beta = up \log \alpha = \frac{p \log c}{\log c - \log p} \log \alpha.$$

Since $p \log c < \log c - \log p < 0$, $-\log \beta \geq (1 + \epsilon)(-\log \alpha)$ for some fixed $\epsilon > 0$.

Consequently, we can replace A, N by another pair for which it is equivalent to derive the desired result, with α arbitrarily small. In particular, we can force $\alpha < p^{-1/(p-1)}$. Now let M' be the differential module on the disc of radius α^{-1} in the coordinate t^{-1} with action of $t^{-1}\frac{d}{dt^{-1}} = -t\frac{d}{dt}$ given by $-N$. The fact that $|N|_\alpha \leq 1$ implies by Theorem 6.5.3 that the spectral radius of $\frac{d}{dt^{-1}}$ on M' is at most α. Hence the generic radius of convergence of M' at radius α^{-1} is at least $p^{-1/(p-1)}\alpha^{-1} > 1$, so Theorem 9.6.1 proves that the local horizontal sections at infinity converge on a disc of radius greater than 1 in the parameter t^{-1}. We may then restrict these to a basis of horizontal sections of M. $\square$

Proof of Theorem 19.3.1 It follows from Proposition 19.1.5 that $D^\dagger$ is fully faithful, so the key point is to show that $D^\dagger$ is essentially surjective. That is, for any finite differential module M over $\mathcal{E}^\dagger$ equipped with a unit-root Frobenius structure, there exists a finite separable extension L of $\kappa_K((t))$ such that

$M \otimes_{\mathcal{E}^\dagger} \mathcal{E}_L^\dagger$ is a trivial differential module (and so corresponds to an unramified representation of $G_{\kappa_K((t))}$). Using Proposition 19.1.5, we may choose L so that $M \otimes_{\mathcal{E}^\dagger} \mathcal{E}_L^\dagger$ is c-constant for some $c < p^{-1/(p-1)}$. Then Lemma 19.3.4 implies the desired result. $\qquad\square$

19.4 Ramification and differential slopes

In the equal-characteristic case, we can relate the upper numbering ramification filtration (Definition 3.4.3) of a Galois representation with finite local monodromy to the generic radius of convergence of an associated differential module, as follows. We will not give a proof here; see the chapter notes for attribution and references, plus some speculation about a mixed-characteristic analogue.

Theorem 19.4.1. *Assume that κ_K is perfect. Let V be a finite-dimensional vector space over K, and let $\tau \colon G_{\kappa_K((t))} \to \mathrm{GL}(V)$ be a continuous homomorphism for the p-adic topology on $\mathrm{GL}(V)$, with finite local monodromy. Then for $\rho \in (0, 1)$ sufficiently close to 1,*

$$R(D^\dagger(V)_\rho) = \rho^b, \qquad b = \max\{i \geq 1 \colon G^i_{\kappa_K((t))} \not\subseteq \ker(\tau)\}.$$

Corollary 19.4.2. *With notation as in Theorem 19.4.1, let $V_1, \ldots, V_m$ be the constituents of V, and let $\tau_j \colon G_{\kappa_K((t))} \to \mathrm{GL}(V_j)$ be the corresponding homomorphisms for $j = 1, \ldots, m$. For $\rho \in (0, 1)$ sufficiently close to 1, the multiset of subsidiary radii of $D^\dagger(V)_\rho$ consists of $\rho^{b_1}, \ldots, \rho^{b_n}$, where the multiset $\{b_1, \ldots, b_n\}$ consists of $\max\{i \geq 1 \colon G^i_{\kappa_K((t))} \not\subseteq \ker(\tau_j)\}$ with multiplicity $\dim(V_j)$ for $j = 1, \ldots, m$.*

Remark 19.4.3. One interpretation of Theorem 19.4.1 is that the decomposition of V by ramification numbers matches up with the Christol–Mebkhout decomposition of $D^\dagger(V) \otimes \mathcal{R}$ provided by Theorem 12.6.4. While the latter was inspired by analogues for meromorphic connections in the complex analytic setting, the analogy with wild ramification was anticipated somewhat before it was instantiated in Theorem 19.4.1.

Remark 19.4.4. Using the integrality properties of subsidiary radii (Theorem 11.3.2(b)), we may deduce that for $\rho \in (0, 1)$ sufficiently close to 1, the product of the subsidiary radii is an integral power of ρ; this amounts to verifying the *Hasse–Arf theorem* for V (integrality of the Artin conductor).

An interesting corollary of Theorem 19.4.1 is the following.

Proposition 19.4.5. *Let M be a finite free differential module on the open annulus with inner radius α and outer radius β, satisfying the Robba condition. Suppose that for some closed interval $[\gamma, \delta]$ with $\alpha < \gamma < \delta < \beta$, there exists a finite étale extension R of $K\langle\gamma/t, t/\delta\rangle$ such that the differential module $M \otimes_{K\langle\gamma/t,t/\delta\rangle} R$ is unipotent (i.e., a successive extension of trivial modules). Then there exists a positive integer m coprime to p such that the pullback of M along the map $t \mapsto t^m$ is unipotent.*

Proof Note that the conclusion may be checked after replacing K by a finite Galois extension K'; this follows from the fact that

$$H^0(M \otimes_K K')^{G_{K'/K}} = H^0(M).$$

We may thus enlarge K and then rescale to ensure $1 \in (\alpha, \beta)$. We may also assume that R is Galois over $K\langle\gamma/t, t/\delta\rangle$ and (possibly after enlarging K again) geometrically connected. That is, R is connected and remains so after any further finite enlargement of K.

Put $G = R \otimes_{K\langle\gamma/t,t/\delta\rangle} F_1$. Our hypotheses so far ensure that G is a field, so by Theorem 1.4.9, G carries a unique multiplicative norm extending $|\cdot|_1$. Since that norm is computed using Newton polygons, we can replace K by a finite extension to ensure that the multiplicative value group of G is the same as that of K.

By Lemma 8.6.1, the norm on G can be defined as the supremum norm for some basis of $R \otimes_{K\langle\gamma/t,t/\delta\rangle} K\langle t, t^{-1}\rangle$. After moving γ and δ closer to 1, we can define the same norm with a basis $e_1, \ldots, e_n$ of R itself.

Let A be the matrix of the trace pairing of R in terms of this basis, i.e., A_{ij} is the trace of multiplication by $e_i e_j$ as an endomorphism of R over $K\langle\gamma/t, t/\delta\rangle$. On one hand, because G has the same multiplicative value group as K, we must have $|\det(A)|_1 = 1$. On the other hand, since R is étale over $K\langle\gamma/t, t/\delta\rangle$, $\det(A)$ must be a unit in R. We conclude that $\det(A) = ct^m$ for some $c \in \mathfrak{o}_K^\times$ and some $m \in \mathbb{Z}$. In particular, R induces a finite étale extension $\overline{R}$ of $\kappa_K[t, t^{-1}]$ (see Definition 19.4.6).

Let $0 = M_0 \subset M_1 \subset \cdots \subset M_l = M$ be a filtration of M such that for $i = 1, \ldots, l$, $M_i/M_{i-1} \otimes_{K\langle\gamma/t,t/\delta\rangle} R$ is a trivial differential module. View the K-vector space

$$\bigoplus_i H^0((M_i/M_{i-1}) \otimes_{K\langle\gamma/t,t/\delta\rangle} R)$$

as a representation of $\mathrm{Aut}(\overline{R}/\kappa_K[t, t^{-1}])$. By restricting to the inertia groups at $t = 0$ and $t = \infty$, applying Theorem 19.4.1, and possibly changing the choice of $[\gamma, \delta]$ (still with 1 in its interior), we can construct a subextension R' of R which induces a tamely ramified extension of $\kappa_K[t, t^{-1}]$, such that $M \otimes_{K\langle\gamma/t,t/\delta\rangle} R'$

is unipotent (this is an instance of the Katz–Gabber construction; see Proposition 19.4.7). However, a finite étale extension of $\kappa_K[t, t^{-1}]$ which is tamely ramified at $t = 0$ and $t = \infty$ must be contained in an extension of the form $\kappa'_K[t^{1/m}, t^{-1/m}]$ for some positive integer m and some finite separable extension κ'_K of κ_K. (This is usually deduced as a consequence of Grothendieck's theory of the tame quotient of the étale fundamental group of a scheme; see [183, Exposé XIII].)

This proves the claim for the restriction of M to the open annulus with inner radius γ and outer radius δ. The claim for M itself follows by Corollary 13.6.4. □

The Katz–Gabber construction used above is the following.

Definition 19.4.6. By a *finite étale extension* of a ring R, we will mean an R-algebra S which is finite projective as an R-module and for which the trace pairing

$$S \times S \to R, \qquad (s_1, s_2) \mapsto \mathrm{Trace}_{S/R}(s_1 s_2)$$

(that is, view multiplication by $s_1 s_2$ as an R-linear map from S to S and take the trace) defines an isomorphism $S \to \mathrm{Hom}_R(S, R)$. This is equivalent to requiring S to be finite and étale over R, except that we have not defined the latter; see the chapter notes. For R a field, a finite étale extension is the same thing as a finite direct sum of finite separable field extensions.

Proposition 19.4.7 (Katz–Gabber). *Let F be a finite extension of $\kappa_K((t))$. Then there exists a finite étale extension R of $\kappa_K[t, t^{-1}]$ such that $R \otimes_{\kappa_K[t, t^{-1}]} \kappa_K((t)) \cong F$ and $R \otimes_{\kappa_K[t, t^{-1}]} \kappa_K((t^{-1}))$ is tamely ramified over $\kappa_K((t^{-1}))$. Moreover, R is uniquely determined by F up to unique isomorphism.*

Proof   See [225, Main Theorem 1.4.1].   □

Notes

A more detailed survey of most of the material in this chapter (excluding the unit-root p-adic local monodromy theorem) can be found in [233].

Proposition 19.1.5 was originally formulated by Fontaine [165, A1.2.6], in slightly less general form and with no reference to differential modules. Our presentation more closely follows [393, Theorem 4.1.3]; a related result in the language of F-isocrystals is the theorem of Crew [117, Theorem 2.1].

For κ_K perfect, the rank 1 case of Theorem 19.3.1 is due to Crew [117, Theorem 4.12], while the general case is due to Tsuzuki [393, Theorem 5.1.1];

however, both arguments can be extended to the general case. An alternate exposition was given by Christol [91] in the case of a standard Frobenius, although without discussion of the fact that standardness is not stable under replacing $\kappa_K((t))$ by a finite separable extension. (In our argument, this is treated by the invariance of the c-constant property under changing Frobenius lifts, as in Definition 19.3.2. The representation-theoretic approach therein is the one taken in [393].) Yet another exposition may be inferred from [240, Theorem 4.5.2], where a stronger result is proved. (The stronger result is used in the study of semistable reduction for isocrystals; see the notes for Appendix B.) All of these proofs are similar in form to the one given here; by contrast, one may infer a rather different proof by specializing Theorem 20.1.4 to the unit-root case (Remark 20.1.5).

The fact that one can give a direct quantitative relationship between wild ramification and spectral properties of differential modules fits nicely into a well-developed analogy between the structures of irregular formal meromorphic connections and of wildly ramified ℓ-adic étale sheaves. An early suggestion along these lines is given by Katz [226] and pursued further by Terasoma [385]; some further work in this direction is that of Beilinson, Bloch, and Esnault [44].

Theorem 19.4.1 was originally stated in its present form by Matsuda [307, Corollary 8.8]; a reformulation in the formalism of Tannakian categories was given by André [16, Complement 7.1.2] as part of his formulation and proof of the p-adic local monodromy theorem. However, thanks to the p-adic global index theorem of Christol and Mebkhout [99, Theorem 8.4–1], [100, Corollaire 5.0–12], this could have already been deduced from a Grothendieck–Ogg–Shafarevich formula for unit-root overconvergent F-isocrystals in rigid cohomology; such a formula was proved by Tsuzuki [394, Theorem 7.2.2] (by Brauer induction, as is possible in the ℓ-adic case) and Crew [119, Theorem 5.4] (using the Katz–Gabber construction, as also is possible in the ℓ-adic case). For a proof by direct computations and Brauer inductions (not using the Christol–Mebkhout index theory), see [233, Theorem 5.23].

In the case of an imperfect residue field, it was originally suggested by Matsuda [308] to formulate an analogue of Theorem 19.4.1 relating the Abbes–Saito conductor (discussed in the notes for Chapter 3) to a suitable differential analogue. That differential analogue was described by Kedlaya [238]; the comparison with the Abbes–Saito conductor has been established by Chiarellotto and Pulita [86] for one-dimensional representations, and by Xiao [411] in the general case. This has the side effect of establishing integrality of the Abbes–Saito conductor in equal characteristic, which is not evident from the original construction.

In mixed characteristic, the appropriate analogue of the functor $V \mapsto D(V)$ is provided by the theory of (φ, Γ)-modules; see Appendix C. It is distinctly less clear what sort of analogue of Theorem 19.4.1 should exist in mixed characteristic. Even in the case of a perfect residue field, where one is asking for a differential interpretation of the usual conductor, only a partial answer is known, by a result of Marmora [304]: one has a differential interpretation of the conductor once one passes to a suitably large cyclotomic extension. (See [323] for a generalization to the case where the residue field may be imperfect.) In general, one does at least have integrality of the Abbes–Saito conductor when the residue characteristic is odd, by work of Xiao [412].

The usual definition of an étale morphism can be found in [378, Tag 00U0]. If R is a field, then a ring homomorphism $R \to S$ is étale in this sense if and only if it is a finite product of finite separable field extensions [378, Tag 00U3]. To see that a morphism $R \to S$ that is finite and étale in the usual sense is finite étale in our sense, note that S is finitely presented and flat [378, Tag 00U2]; hence S is finite projective as an R-module and we may check pointwise (on the prime spectrum of R) that the trace morphism induces an isomorphism $S \to \mathrm{Hom}_R(S, R)$. To see that a finite étale morphism $R \to S$ in our sense is étale in the usual sense, we may use [378, Tag 00U6].

Exercises

19.1 Suppose that the fixed field K_0 of φ on K has residue field $\mathbb{F}_q$ and the same value group as K. Prove that $\tilde{\mathcal{E}}^{\varphi=1} = K_0$. (Hint: reduce to the corresponding equality of residue fields.)

19.2 Prove that any finite free unit-root difference module over $\mathcal{E}$ admits a unique compatible differential structure.

19.3 Prove that the kernel of $\frac{d}{dt}$ on $\tilde{\mathcal{E}}$ equals $\tilde{K}$. In particular, for any finite separable extension L of $\kappa_K((t))$, the kernel of $\frac{d}{dt}$ on $\mathcal{E}_L^\dagger$ is the integral closure of K in $\mathcal{E}_L^\dagger$.

19.4 Write out explicitly the isomorphism $D^\dagger(V) \cong M$ implied by Example 19.2.6. (Hint: this module is generated by $(1 + p\pi f)^{1/p}$ for any lift $f \in \mathcal{E}^\dagger$ of $\overline{f}$.)

19.5 Prove that for an arbitrary Frobenius lift, the property of a finite differential module over $\mathcal{E}$ or $\mathcal{E}^\dagger$ equipped with a unit-root Frobenius structure being c-constant is invariant under changing the Frobenius lift. (Hint: in Lemma 17.3.4, if the matrix of action of D/i has norm at most ϵ, then so does the matrix of action of $D^i/i!$. Apply this observation to (17.3.1.1).)

Part VI

The p-adic local monodromy theorem

20

The p-adic local monodromy theorem

We are now ready to state the capstone theorem of this book, the p-adic local monodromy theorem. This asserts that a finite differential module over an annulus carrying a Frobenius structure has "finite local monodromy", in the sense that it becomes unipotent after making a suitable finite étale cover of the annulus. In this chapter, we give the precise statement of the theorem, illustrate it with an example and a couple of basic applications, and discuss some technical points that arise if the field K has imperfect residue field. We postpone discussion of the proof(s) of the theorem to the next two chapters.

We will discuss two broad areas of applications of the p-adic local monodromy theorem in the appendices. One of these is in the subject of rigid cohomology, where the theorem plays a role analogous to the ℓ-adic local monodromy theorem of Grothendieck in the subject of étale cohomology (hence the name); see Appendix B. The other is in p-adic Hodge theory, where the theorem clarifies the structure of certain p-adic Galois representations; see Appendix C.

Hypothesis 20.0.1. Throughout this chapter and the next, let K be a discretely valued complete nonarchimedean field. Fix a homomorphism $\overline{\varphi} \colon \kappa_K((t)) \to \kappa_K((t))$ preserving (but not necessarily fixing) κ_K, and carrying t to t^q for some power q of p. We will assume that all Frobenius lifts considered are lifts of this particular $\overline{\varphi}$.

20.1 Statement of the theorem

Definition 20.1.1. Let L be a finite separable extension of $\kappa_K((t))$ to which $\overline{\varphi}$ extends. Put

$$\mathcal{R}_L = \mathcal{R} \otimes_{\mathcal{E}^\dagger} \mathcal{E}_L^\dagger.$$

353

We say a finite free differential module M over $\mathcal{R}$ is *quasiconstant* if there exists L such that $M \otimes_{\mathcal{R}} \mathcal{R}_L$ is trivial. We say M is *quasiunipotent* if it is a successive extension of quasiconstant modules; this holds if and only if $M \otimes_{\mathcal{R}} \mathcal{R}_L$ is unipotent for some L (exercise).

Remark 20.1.2. It is more typical to make Definition 20.1.1 without reference to $\overline{\varphi}$, so that L may be any finite separable extension of $\kappa_K((t))$. This corresponds precisely to the case where $\overline{\varphi}$ is an absolute Frobenius, as then $\overline{\varphi}$ extends to any finite separable extension of $\kappa_K((t))$.

The condition of quasiunipotence implies solvability.

Proposition 20.1.3. *Let M be a finite free quasiunipotent differential module over $\mathcal{R}$. Then M is solvable at 1.*

Proof We may reduce to the case where M is irreducible, and hence quasiconstant. We may thus pick L for which $M \otimes_{\mathcal{R}} \mathcal{R}_L$ is trivial. It is then clear that $M \otimes_{\mathcal{R}} \mathcal{R}_L$ is solvable at 1, and one can argue by direct calculation that this implies the same for M. Here, we will give a slightly different argument.

There is no harm in enlarging K, so we may assume K is integrally closed in $\mathcal{R}_L$. In this case, $H^0(M \otimes_{\mathcal{R}} \mathcal{R}_L)$ is a vector space over K equipped with a linear action of $G_{L/\kappa_K((t))}$; that is, we have a representation of a finite group on a finite-dimensional vector space over a field of characteristic 0. (Without the enlargement of K, we might only have a semilinear group action.) It is a basic fact of the representation theory of finite groups that this representation can also be defined over a subfield of K which is finite over $\mathbb{Q}$; in particular, we can pick a finite extension K_0 of $\mathbb{Q}_p$ contained in K and a K_0-lattice T in $H^0(M \otimes_{\mathcal{R}} \mathcal{R}_L)$ stable under the Galois action.

Since the desired result makes no reference to $\overline{\varphi}$, we can choose φ to be an absolute Frobenius lift fixing K_0. We then obtain a Frobenius structure on $M \otimes_{\mathcal{R}} \mathcal{R}_L$ by declaring T to be fixed under the Frobenius action. Since T is stable under $G_{L/\kappa_K((t))}$, this action descends to a Frobenius structure on M itself. We may thus deduce that M is solvable at 1 by applying Theorem 17.2.1. (The astute reader may recognize this construction from the proof of Theorem 19.3.1.) □

Conversely, the following important theorem asserts that many naturally occurring solvable differential modules, including Picard–Fuchs modules, are quasiunipotent. In the case of an absolute Frobenius lift, it is due independently to André [16], Kedlaya [230], and Mebkhout [309]; the generalization to non-absolute Frobenius lifts is original to this book. See Chapter 21 for proof and the chapter notes for further discussion.

Theorem 20.1.4 (*p*-adic local monodromy theorem). *Let M be a finite free* *differential module over $\mathcal{R}$ admitting a Frobenius structure for some Frobenius lift. Then M is quasiunipotent.*

Remark 20.1.5. In the special case where φ is absolute and $M = M^\dagger \otimes_{\mathcal{E}^\dagger} \mathcal{R}$ for $M^\dagger$ a finite differential module over $\mathcal{E}^\dagger$ equipped with a unit-root Frobenius structure, Theorem 20.1.4 almost recovers Tsuzuki's theorem (Theorem 19.3.1). The missing ingredients are as follows.

- We must show that M is quasiconstant, not just quasiunipotent. This follows from the fact that a unipotent differential module over $\mathcal{R}$ equipped with a unit-root Frobenius structure must be trivial (exercise).
- We must show that $M^\dagger \otimes_{\mathcal{E}^\dagger} \mathcal{E}_L^\dagger$ is a trivial differential module. This follows from the triviality of $M \otimes_{\mathcal{R}} \mathcal{R}_L$ using Corollary 16.3.4.

20.2 An example

It may be worth seeing what Theorem 20.1.4 says in an explicit example. This example, corresponding to a Bessel equation, was originally considered by Dwork [144]; the analysis given here is due to Tsuzuki [395, Example 6.2.6], and was cited in the introduction of [230].

Example 20.2.1. Let p be an odd prime, put $K = \mathbb{Q}_p(\pi)$ with $\pi^{p-1} = -p$, and take φ to be the absolute p-power Frobenius. Let M be the differential module of rank 2 over $\mathcal{R}$ with the action of D on a basis e_1, e_2 given by

$$N = \begin{pmatrix} 0 & t^{-1} \\ \pi^2 t^{-2} & 0 \end{pmatrix}.$$

Then M admits a Frobenius structure; this was shown by explicit calculation in [144], but can also be derived by consideration of a suitable Picard–Fuchs module. Define the tamely ramified extension L of $\kappa_K((t))$ and the corresponding extension $\mathcal{E}_L^\dagger$ of $\mathcal{E}^\dagger$ by adjoining to $\kappa_K((t))$ an element u such that $4u^2 = t$, and put

$$u_\pm = 1 + \sum_{n=1}^{\infty} (\pm 1)^n \frac{(2n)!^2}{(32\pi)^n n!^3} u^n \in K\{u\}.$$

Define the matrix

$$U = \begin{pmatrix} u_+ & u_- \\ u\frac{d}{du}(u_+) + (\frac{1}{2} - \pi u^{-1})u_+ & u\frac{d}{du}(u_-) + (\frac{1}{2} + \pi u^{-1})u_- \end{pmatrix}$$

and use it to change basis; then the action of $\frac{d}{du}$ on the new basis e_+, e_- of

$M \otimes_{\mathcal{R}} \mathcal{R}_L$ is via the matrix

$$\begin{pmatrix} -\frac{1}{2}u^{-1} + \pi u^{-2} & 0 \\ 0 & -\frac{1}{2}u^{-1} - \pi u^{-2}. \end{pmatrix}.$$

That is, $M \otimes_{\mathcal{R}} \mathcal{R}_L$ splits into two differential submodules of rank 1. To render these quasiconstant, we must adjoin to L a square root of u (to eliminate the term $-\frac{1}{2}u^{-1}$) and a root of the polynomial $z^p - z = u^{-1}$ (which by Example 19.2.6 eliminates the terms $\pm\pi u^{-2}$).

By further analysis (carried out in [395, Example 6.2.6]), one determines that in this example, the special Newton slopes are $\frac{1}{2}\log p, \frac{1}{2}\log p$. By contrast, the generic Newton slopes (obtained by writing $M = M^{\dagger} \otimes_{\mathcal{E}^{\dagger}} \mathcal{R}$ using the chosen basis, then extending scalars to $\mathcal{E}$) are $0, \log p$. This illustrates the conclusion of the semicontinuity theorem for Newton polygons (Theorem 16.4.6).

20.3 Descent of horizontal sections

The property of quasiunipotence is rather handy because it often allows for complicated-looking statements about nontrivial differential modules to be reduced to statements about trivial differential modules, which can be checked by direct calculation. Over a disc, this can be achieved using Dwork's trick (Corollary 17.2.2); over an annulus, Theorem 20.1.4 often serves as a usable replacement. Here is a typical example, which builds on techniques of de Jong; see the notes for further discussion.

We start with a calculation which is in some sense dual to Lemma 16.3.2. Recall that $\mathcal{E}_{\varphi}$ and $\mathcal{E}_{\varphi}^{\dagger}$ denote the φ-perfections of $\mathcal{E}$ and $\mathcal{E}^{\dagger}$, respectively (Definition 15.4.1).

Lemma 20.3.1. *Let A be an $n \times n$ matrix over $\mathfrak{o}_{\mathcal{E}_{\varphi}^{\dagger}}$, and suppose $v \in \mathcal{E}_{\varphi}^n$, $w \in (\mathcal{E}_{\varphi}^{\dagger})^n$ satisfy $Av - \varphi(v) = w$. Then $v \in (\mathcal{E}_{\varphi}^{\dagger})^n$.*

Proof Exercise. □

We next bring to bear de Jong's reverse filtration.

Lemma 20.3.2. *Let M be a finite dualizable difference module over $\mathcal{E}_{\varphi}^{\dagger}$. Let $\psi : M \to \mathcal{E}_{\varphi}$ be a nonzero φ-equivariant map. Then $\psi^{-1}(\mathcal{E}_{\varphi}^{\dagger})/\ker(\psi)$ has rank 1 and Newton slope 0, and $M/\psi^{-1}(\mathcal{E}_{\varphi}^{\dagger})$ has all Newton slopes less than 0.*

Proof We may assume ψ is injective. Let M_1 be the first step of the filtration on M given by Theorem 15.4.4. Then ψ induces a nonzero element of $H^0(M_1^{\vee} \otimes_{\mathcal{E}_{\varphi}^{\dagger}}$

$\mathcal{E}_\varphi$), which can only exist if M_1 is pure of norm 1. Moreover, by Lemma 20.3.1 (applied after using Proposition 14.4.16 to choose a suitable basis),

$$H^0(M_1^\vee) = H^0(M_1^\vee \otimes_{\mathcal{E}_\varphi^\dagger} \mathcal{E}_\varphi),$$

so ψ is induced by a nonzero morphism $M_1 \to \mathcal{E}_\varphi^\dagger$. We deduce that $\psi^{-1}(\mathcal{E}_\varphi^\dagger)$ is of rank 1; since we have assumed that ψ is injective, M_1 must also be of rank 1. This proves the desired results. □

To eliminate the perfection in the previous lemma, we use the projection $\mathcal{E}_\varphi \to \mathcal{E}$ constructed in Lemma 15.4.5.

Proposition 20.3.3. *Let M be a finite dualizable difference module over $\mathcal{E}^\dagger$. Let $\psi \colon M \to \mathcal{E}$ be a nonzero φ-equivariant map. Then $\psi^{-1}(\mathcal{E}^\dagger)/\ker(\psi)$ has rank 1 and Newton slope 0, and $M/\psi^{-1}(\mathcal{E}^\dagger)$ has all Newton slopes less than 0.*

Proof Put $M' = M \otimes_{\mathcal{E}^\dagger} \mathcal{E}_\varphi^\dagger$, and let $\psi' \colon M' \to \mathcal{E}_\varphi$ be the composition

$$M' \xrightarrow{\psi \otimes 1} \mathcal{E} \otimes_{\mathcal{E}^\dagger} \mathcal{E}_\varphi^\dagger \to \mathcal{E}_\varphi,$$

with the last map being multiplication in $\mathcal{E}_\varphi$. The map ψ' is nonzero because it restricts to ψ on M. (Note that the multiplication map is injective by Corollary 15.4.6, but we will use the proof technique of that result here rather than its statement.)

By Lemma 20.3.2, the desired results will follow from the assertion that the natural inclusions

$$\psi^{-1}(0) \otimes_{\mathcal{E}^\dagger} \mathcal{E}_\varphi^\dagger \to (\psi')^{-1}(0)$$
$$\psi^{-1}(\mathcal{E}^\dagger) \otimes_{\mathcal{E}^\dagger} \mathcal{E}_\varphi^\dagger \to (\psi')^{-1}(\mathcal{E}_\varphi^\dagger)$$

are surjective. To check this assertion, choose a basis $m_1, \ldots, m_n$ of M. Given v in $(\psi')^{-1}(0)$ (resp. $v \in (\psi')^{-1}(\mathcal{E}_\varphi^\dagger)$), write $v = \sum_{i=1}^n v_i m_i$ with $v \in \mathcal{E}_\varphi^\dagger$. We induct on the largest integer j for which $v_j \neq 0$, using the case where no such j exists as a vacuous base case.

Define the map $\lambda \colon \mathcal{E}_\varphi \to \mathcal{E}$ as in Lemma 15.4.5, then define $\lambda_M = \mathrm{id}_M \otimes \lambda \colon M' \to M$, so that $\psi \circ \lambda_M = \lambda \circ \psi'$. Then the quantity

$$\lambda_M(v/v_j) = \sum_{i=1}^j \lambda(v_i/v_j) m_i$$

belongs to $\psi^{-1}(0)$ (resp. to $\psi^{-1}(\mathcal{E}^\dagger)$). Moreover, the coefficient of m_j in $v/v_j - \lambda_M(v/v_j)$ vanishes, so by the induction hypothesis the latter belongs to $\psi^{-1}(0) \otimes_{\mathcal{E}^\dagger} \mathcal{E}_\varphi^\dagger$ (resp. to $\psi^{-1}(\mathcal{E}^\dagger) \otimes_{\mathcal{E}^\dagger} \mathcal{E}_\varphi^\dagger$). Hence v does also. □

We now add the p-adic local monodromy theorem in order to split some exact sequences.

Lemma 20.3.4. *Suppose that κ_K is perfect. Let $0 \to M_1 \to M \to M_2 \to 0$ be a short exact sequence of finite differential modules over $\mathcal{E}^\dagger$ equipped with Frobenius structures. Suppose that every generic Newton slope of M_1 is strictly greater than every generic Newton slope of M_2. Then the exact sequence splits.*

Proof As in Lemma 5.3.3, we may replace M_1, M_2 with $M_2^\vee \otimes M_1, \mathcal{E}^\dagger$; that is, we may reduce to the case where $M_2 = \mathcal{E}^\dagger$ and every generic Newton slope of M_1 is positive. By Theorem 16.4.6, every special Newton slope of M_1 is also positive. By Theorem 20.1.4, we can find a finite Galois extension L of $\kappa_K((t))$ such that $M_1 \otimes_{\mathcal{E}^\dagger} \mathcal{R}_L$ is unipotent. Since we have assumed that κ_K is perfect, $\mathcal{R}_L$ can itself be written as a Robba ring over some finite extension of K (see Remark 20.5.3). By a direct calculation (exercise), the exact sequence splits over $\mathcal{R}_L$. By Corollary 16.3.3, in the category of difference modules the map $H^1(M_1 \otimes_{\mathcal{E}^\dagger} \mathcal{E}_L^\dagger) \to H^1(M_1 \otimes_{\mathcal{E}^\dagger} \mathcal{R}_L)$ is injective, so the original exact sequence splits in the category of difference modules over $\mathcal{E}_L^\dagger$. That splitting is unique (as can be seen after base extension to $\mathcal{E}_L$), so it descends to a splitting of the original sequence in the category of difference modules over $\mathcal{E}^\dagger$.

We now check that this splitting is also a splitting of differential modules. We start with an injective homomorphism $\mathcal{E}^\dagger \to M$ of difference modules. By comparing special slopes, we see that the resulting copy of $\mathcal{R}$ in $M \otimes_{\mathcal{E}^\dagger} \mathcal{R}$ must be the first step in the slope filtration of $M \otimes_{\mathcal{E}^\dagger} \mathcal{R}$, in the sense of Theorem 17.4.3. In particular, the copy of $\mathcal{R}$ in $M \otimes_{\mathcal{E}^\dagger} \mathcal{R}$ is a differential submodule, as is its pure descent to $\mathcal{E}^\dagger$. The latter is just the image of $\mathcal{E}^\dagger$ in M, so the claim follows. $\square$

We now put the preceding lemmas together to get a theorem on the descent of horizontal sections.

Theorem 20.3.5. *Let M be a finite differential module over $R = K[[t]]_0$ or $R = \mathcal{E}^\dagger$ admitting a Frobenius structure. Then in the category of differential modules,*

$$H^0(M) = H^0(M \otimes_R \mathcal{E}).$$

Proof For the case $R = K[[t]]_0$, $H^0(M) = H^0(M \otimes_R \mathcal{E}^\dagger)$ by Proposition 17.2.5, so we may assume that $R = \mathcal{E}^\dagger$ hereafter. We may enlarge K to force κ_K to be perfect. We first check that any nonzero Φ-invariant horizontal section v of $M \otimes_{\mathcal{E}^\dagger} \mathcal{E}$ descends to M. The section v corresponds to a nonzero φ-equivariant map $\psi: M^\vee \to \mathcal{E}$. By Proposition 20.3.3, $\psi^{-1}(\mathcal{E}^\dagger) \neq 0$ and the generic slopes of $M^\vee / \psi^{-1}(\mathcal{E}^\dagger)$ are all negative. By Lemma 20.3.4, the exact sequence

$$0 \to \psi^{-1}(\mathcal{E}^\dagger)/\ker(\psi) \to M^\vee/\ker(\psi) \to M^\vee/\psi^{-1}(\mathcal{E}^\dagger) \to 0$$

must split. However, by Proposition 20.3.3 again, any complement of the difference submodule $\psi^{-1}(\mathcal{E}^\dagger)/\ker(\psi)$ of $M^\vee/\ker(\psi)$ must map to zero under ψ. We conclude that $\psi(M^\vee) = \mathcal{E}^\dagger$, forcing $v \in H^0(M)$.

To check the original claim (still for $R = \mathcal{E}^\dagger$), we may enlarge K to have algebraically closed residue field. In this case, Corollary 14.6.4 implies that $H^0(M \otimes_{\mathcal{E}^\dagger} \mathcal{E})$ is spanned by one-dimensional fixed subspaces for some power of the Frobenius action. The previous argument shows that any generator of one of these subspaces belongs to M, proving the claim. □

20.4 Local duality

Here is another useful property of quasiunipotent differential modules, which by Theorem 20.1.4 is also present whenever one has a Frobenius structure.

Lemma 20.4.1. *Put $R = \mathcal{E}^\dagger$ or $\mathcal{R}$. Let M be a finite free differential module over R. Then for $i = 0, 1$, the natural maps*

$$H^i(M) \to H^i(M \otimes_{\mathcal{E}^\dagger} \mathcal{E}_L^\dagger)^{G_{L/\kappa_K((t))}}$$

are bijections.

Proof For $i = 0$, we clearly have

$$H^0(M \otimes_{\mathcal{E}^\dagger} \mathcal{E}_L^\dagger)^{G_{L/\kappa_K((t))}} = H^0((M \otimes_{\mathcal{E}^\dagger} \mathcal{E}_L^\dagger)^{G_{L/\kappa_K((t))}}) = H^0(M).$$

For $i = 1$, we have injectivity because M is a direct summand of $M \otimes_{\mathcal{E}^\dagger} \mathcal{E}_L^\dagger$. We have surjectivity because if $x \in M \otimes_{\mathcal{E}^\dagger} \mathcal{E}_L^\dagger$ represents a Galois-invariant class in $H^1(M \otimes_{\mathcal{E}^\dagger} \mathcal{E}_L^\dagger)$, then the average of $g(x)$ over $g \in G_{L/\kappa_K((t))}$ is an element of M representing the same class. □

Proposition 20.4.2. *Suppose that κ_K is perfect. (This hypothesis can be removed; see Section 20.5.) Let M be a finite free quasiunipotent differential module over $\mathcal{R}$. Then the spaces $H^0(M), H^1(M)$ are finite-dimensional, and there is a perfect pairing*

$$H^0(M) \times H^1(M^\vee) \to H^1(M \otimes M^\vee) \to H^1(\mathcal{R}) \cong K\frac{dt}{t}.$$

In particular, by Theorem 20.1.4, this holds whenever M admits a Frobenius structure.

Proof This may be checked directly when M is unipotent (exercise). In general, we may choose a finite Galois extension L of $\kappa_K((t))$ such that $M \otimes_{\mathcal{R}} \mathcal{R}_L$ is unipotent. Since we have assumed that κ_K is perfect, $\mathcal{R}_L$ can itself be written as a Robba ring over some finite extension of K (see Remark 20.5.3), so the desired assertions hold for $M \otimes_R R_L$. We then deduce the desired results using Lemma 20.4.1. (See also [233, Proposition 4.26].)   □

This leads to the following result of Matsuda [307, Theorem 7.8].

Corollary 20.4.3 (Matsuda). *Every indecomposable finite free quasiunipotent differential module over R has the form $M \otimes N$ with M quasiconstant and N unipotent.*

Proof Exercise.   □

20.5 When the residue field is imperfect

We suggested earlier (Remark 19.0.1) that the reader may wish to assume that the residue field κ_K of K is perfect. We now discuss the possible confusion that may occur when this hypothesis is omitted.

Remark 20.5.1. The Cohen structure theorem asserts that if F is a complete discretely valued field of characteristic $p > 0$, then there exists an isomorphism $F \cong \kappa_F((t))$. However, this isomorphism is far from unique; it depends not only on the choice of the series parameter t, but also on the choice of an embedding of the residue field κ_F into F. This choice is unique when κ_F is perfect (because in that case one has multiplicative lifts; see the notes for Chapter 14), but not otherwise.

Example 20.5.2. Suppose $F = \mathbb{F}_p(x)((t))$. Then for any $y = \sum_{i=1}^{\infty} y_i t^i$, there is an embedding $\mathbb{F}_p(x) \hookrightarrow F$ given by $x \mapsto x + y$.

Remark 20.5.3. Suppose further that E is a finite separable extension of F. By the Cohen structure theorem, we can find copies of κ_E and κ_F inside E and F, respectively, and use these to present E and F as power series fields (possibly in different series parameters, if $|E^\times| \neq |F^\times|$). If κ_E is separable over κ_F, we can choose the copy of κ_E to be the integral closure in E of the chosen copy of κ_F, by Proposition 3.2.3 (or directly applying Hensel's lemma). If $F = \kappa_K((t))$ and $E = L$, we can then view R_L as a copy of the Robba ring with coefficients in the unramified extension of $\kappa_K = \kappa_F$ with residue field κ_E. However, this may not be possible if κ_E is not separable over κ_F, as in the following example.

Example 20.5.4. Suppose again that $F = \mathbb{F}_p(x)((t))$, and put

$$E = F[z]/(z^p - z - xt^{-p}).$$

Then $\kappa_E \cong \mathbb{F}_p(x^{1/p})$, but there is no p-th root of x inside E itself. To write $E \cong \kappa_E((t))$, we must make a different choice of the copy of κ_E inside E, e.g., $\mathbb{F}_p(zt)$.

Fortunately, in our setting, there is a convenient way to skirt this issue using the Frobenius map $\overline{\varphi}$.

Lemma 20.5.5. *Suppose that $\overline{\varphi}$ is bijective on κ_K. Then for any finite Galois extension E of $F = \kappa_\mathcal{E}$ equipped with an extension of $\overline{\varphi}$, κ_E is separable over κ_F.*

Proof Let U and T be the maximal unramified and tamely ramified subextensions, respectively, of E over F. Then the claim holds for the extensions U/F and T/U (the former by the definition of an unramified extension, the latter by Proposition 3.3.6). We may thus assume E is totally wildly ramified over F.

By Remark 3.3.10, the extension E/F can be written as a tower of Artin–Schreier extensions $F = E_0 \subset E_1 \subset \cdots \subset E_l = E$. It may happen that E_1 is not preserved by the action of $\overline{\varphi}$ on E; however, it must be carried to another $\mathbb{Z}/p\mathbb{Z}$-subextension of E over F, of which there are only finitely many. Consequently, E_1 is preserved by some power of $\overline{\varphi}$; similarly, we can choose a power of $\overline{\varphi}$ preserving $E_2, \ldots, E_l$.

Since the desired result is insensitive to replacing $\overline{\varphi}$ by a power, we may thus reduce to considering a single Artin–Schreier extension

$$E = F[z]/(z^p - z - P), \qquad P = \sum_i c_i t^i.$$

Since E admits an extension of $\overline{\varphi}$, there must exist $a \in \mathbb{F}_p^\times$ such that $P - a\overline{\varphi}(P) = y^p - y$ for some $y \in F$. By replacing $\overline{\varphi}$ by a suitable power, we may reduce to the case $a = 1$.

Remember that we do not change the extension by replacing P with $P + y^p - y$ for $y \in F$. We may thus choose P so that for any two indices $i, j < 0$ such that $c_i, c_j \neq 0$, the ratio i/j is not a power of p. Now let j be the smallest integer for which $c_j \neq 0$. We are done if either $j \geq 0$, in which case E is unramified, or $j < 0$ is not divisible by p, in which case $[|E^\times| : |F^\times|] > 1$ and so by Lemma 3.1.1, $\kappa_E = \kappa_F$.

Suppose to the contrary that j is divisible by p. By the choice of P and the fact that $P - \overline{\varphi}(P) = y^p - y$ for some $y \in F$, we must have $\overline{\varphi}(c_j t^j) = (c_j t^j)^q$. That is, $c_j^{1/q} = \overline{\varphi}^{-1}(c_j) \in \kappa_K$ since $\overline{\varphi}_K$ was assumed to be bijective. We can thus replace P by $P + (c_j t^j)^{1/p} - c_j t^j$ to increase j. This process can only repeat finitely many times, after which we may deduce the claim. □

20.6 Minimal slope quotients

As an aside, we introduce a result of Tsuzuki whose proof is closely related to the previous discussion. We use the rings $\tilde{\mathcal{R}}, \tilde{\mathcal{E}}^\dagger$ defined in Definition 16.6.1 using Mal'cev–Neumann series.

Theorem 20.6.1. *Suppose that $\overline{\varphi}$ is bijective on κ_K. Let V be a finite free dualizable difference module over $\tilde{\mathcal{E}}^\dagger$. Then there exists a unique filtration*

$$0 = V_0 \subset V_1 \subset \cdots \subset V_l = V$$

of difference modules, such that each successive quotient V_i/V_{i-1} is pure of some norm s_i, and $s_1 < \cdots < s_l$.

Proof The proof of Theorem 15.4.4 carries over. $\qquad\square$

Definition 20.6.2. In the following lemmas, let $\tilde{\mathcal{E}}$ denote the completion of $\tilde{\mathcal{E}}^\dagger$ with respect to its discrete valuation. Note that this overrides our previous use of this notation (Definition 19.1.1).

We have the following analogues of Lemma 15.4.5 and Corollary 15.4.6.

Lemma 20.6.3. *There exists a $\tilde{\mathcal{E}}$-linear map $\lambda: \tilde{\mathcal{E}} \to \mathcal{E}$ sending $\tilde{\mathcal{E}}^\dagger$ to $\mathcal{E}^\dagger$, such that $\lambda(x) = x$ for all $x \in \mathcal{E}$.*

Proof An element x of $\tilde{\mathcal{E}}$ can be expressed as a formal sum $\sum_{i \in \mathbb{Q}} x_i t^i$; we define the map λ by mapping x to $\sum_{i \in \mathbb{Z}} x_i t^i$. $\qquad\square$

Corollary 20.6.4. *The multiplication map $\mu: \mathcal{E} \otimes_{\mathcal{E}^\dagger} \tilde{\mathcal{E}}^\dagger \to \tilde{\mathcal{E}}$ is injective.*

Proof This follows from Lemma 20.6.3 in the same way that Corollary 15.4.6 follows from Lemma 15.4.5. $\qquad\square$

Lemma 20.6.5. *Let A be an $n \times n$ matrix over $\mathfrak{o}_{\tilde{\mathcal{E}}^\dagger}$, and suppose $v \in \tilde{\mathcal{E}}^n$, $w \in (\tilde{\mathcal{E}}^\dagger)^n$ satisfy $Av - \varphi(v) = w$. Then $v \in (\tilde{\mathcal{E}}^\dagger)^n$.*

Proof The proof of Lemma 20.3.1 carries over. $\qquad\square$

Lemma 20.6.6. *Suppose that $\overline{\varphi}$ is bijective on κ_K. Let M be a finite dualizable difference module over $\tilde{\mathcal{E}}^\dagger$. Let $\psi: M \to \tilde{\mathcal{E}}$ be a nonzero φ-equivariant map. Then $\psi^{-1}(\tilde{\mathcal{E}}^\dagger)/\ker(\psi)$ has rank 1 and Newton slope 0, and $M/\psi^{-1}(\tilde{\mathcal{E}}^\dagger)$ has all Newton slopes less than 0.*

Proof The proof of Lemma 20.3.2 carries over, replacing Theorem 15.4.4 with Theorem 20.6.1 and Lemma 20.3.1 with Lemma 20.6.5. $\qquad\square$

Theorem 20.6.7 (Tsuzuki). *Let M be a finite difference module over $\mathcal{E}$ having a single Newton slope. Let N be a finite difference module over $\mathcal{E}^{\dagger}$. Let $\eta\colon N \to M$ be a horizontal, $\mathcal{E}^{\dagger}$-linear, Frobenius-equivariant morphism for which the induced map $N \otimes_{\mathcal{E}^{\dagger}} \mathcal{E} \to M$ is surjective. Then the latter map identifies M with the final quotient in the slope filtration of $N \otimes_{\mathcal{E}^{\dagger}} \mathcal{E}$.*

Proof We may assume at once that η is injective. We may check the claim after enlarging K as per Remark 16.7.3, so we may assume that κ_K is strongly difference-closed. We can also replace φ with a power and twist, to reduce to the case where the Newton slope of M is 0 and the Newton slopes of N are all integers.

Define $\tilde{M} = M \otimes_{\mathcal{E}} \tilde{\mathcal{E}}$, $\tilde{N} = N \otimes_{\mathcal{E}^{\dagger}} \tilde{\mathcal{E}}^{\dagger}$ and let $\tilde{\eta}\colon \tilde{N} \otimes_{\tilde{\mathcal{E}}^{\dagger}} \tilde{\mathcal{E}} \to \tilde{M}$ be the map induced by η. By Corollary 20.6.4, $\tilde{\eta}$ is again injective, so it will suffice to check that the induced surjection $\tilde{N} \otimes_{\tilde{\mathcal{E}}^{\dagger}} \tilde{\mathcal{E}} \to \tilde{M}$ identifies $\tilde{M}$ with the quotient of $\tilde{N} \otimes_{\tilde{\mathcal{E}}^{\dagger}} \tilde{\mathcal{E}}$ by the final step of its slope filtration.

Consider the reverse filtration

$$0 = \tilde{N}_0 \subset \tilde{N}_1 \subset \cdots \subset \tilde{N}_l = \tilde{N}$$

of $\tilde{N}$ given by Theorem 20.6.1. By Lemma 16.6.2 and Corollary 16.7.2, each step of this filtration admits a basis of eigenvectors, as then does $\tilde{M}$ because $\tilde{N} \otimes_{\tilde{\mathcal{E}}^{\dagger}} \tilde{\mathcal{E}} \to \tilde{M}$ is surjective. In particular, we have a family of φ-equivariant morphisms $\tilde{M} \to \tilde{\mathcal{E}}$ whose joint kernel is trivial.

We now apply Lemma 20.6.6 to the compositions of η with these morphisms. This implies first that $\tilde{N}_1$ has Newton slope 0, and then that $\tilde{\eta}$ induces an isomorphism $\tilde{N}_1 \otimes_{\tilde{\mathcal{E}}^{\dagger}} \tilde{\mathcal{E}} \to \tilde{M}$. Since $\tilde{N}_1 \otimes_{\tilde{\mathcal{E}}^{\dagger}} \tilde{\mathcal{E}}$ is isomorphic to the final quotient in the slope filtration of $\tilde{N} \otimes_{\tilde{\mathcal{E}}^{\dagger}} \tilde{\mathcal{E}}$ (see Remark 15.4.7), this yields the desired result. $\square$

Notes

The p-adic local monodromy theorem (Theorem 20.1.4) can be viewed as an archimedean analogue of the following theorem of Borel [362, Theorem 6.1]. Given a vector bundle with connection over the punctured unit disc in the complex plane, equipped with a polarized variation of (rational) Hodge structures, Borel's theorem asserts that the monodromy transformation is forced to be quasiunipotent (i.e., its eigenvalues are roots of unity). It appears that the Frobenius structure in the p-adic setting plays the role of the variation of Hodge structures in the complex analytic realm.

The p-adic local monodromy theorem was originally formulated and proved only in the absolute case. In this case, it is often referred to in the literature as *Crew's conjecture*, because it emerged from the work of Crew [118] on finite dimensionality of rigid cohomology with coefficients in an overconvergent *F*-isocrystal. Crew's original conjecture was somewhat more limited still, as it only concerned modules such that the differential and Frobenius structures were both defined over $\mathcal{E}^{\dagger}$; this form was restated in a more geometric form by de Jong [124]. A closer analysis of Crew's conjecture was then given by Tsuzuki [395], who explained (using Theorem 19.3.1, which he had proved in [393]) how Theorem 20.1.4 in the absolute case would follow from a slope filtration theorem [395, Theorem 5.2.1].

The original proofs of Theorem 20.1.4 in the absolute case can be briefly described as follows. The proof of Kedlaya uses slope filtrations (Theorem 17.4.3) to reduce everything to the unit-root case (Theorem 19.3.1). The proofs of André and Mebkhout proceed by using properties of solvable differential modules to reduce everything to the *p*-adic Fuchs theorem of Christol and Mebkhout (Theorem 13.6.1). André's proof is phrased in the language of Tannakian categories, whereas Mebkhout's proof is more explicit; the proof in the relative case that we give in the next chapter is most closely modeled on Mebkhout's proof. (See also the treatment in [93].)

For applications to rigid cohomology, as far as we know only the absolute case of Theorem 20.1.4 is of any relevance. However, the nonabsolute case occurs in the context of relative *p*-adic Hodge theory. Namely, Berger and Colmez [51] use the full strength of Theorem 20.1.4 to prove an analogue of Fontaine's conjecture on potential semistability of de Rham representations (Corollary C.4.5) for a family of de Rham representations parametrized by an affinoid base space.

To our knowledge, no proof of the nonabsolute case of Theorem 20.1.4 has appeared prior to the one we are about to describe in Chapter 21. The proof by Kedlaya does not apply because it relies on Theorem 19.3.1, whose given proof does not extend to the relative case. The use of the nonabsolute case of Theorem 20.1.4 in [51] provided a strong motivation for working at the level of generality pursued in the present book.

Lemma 20.3.1 is a mild generalization of [392, Proposition 2.2.2]. A weaker result in the same spirit appears in the work of Cherbonnier and Colmez [84].

In the case of an absolute Frobenius lift, the case of Theorem 20.3.5 with $R = \mathcal{E}^{\dagger}$ was originally conjectured by Tsuzuki [396, Conjecture 2.3.3] and proved by Kedlaya [231, Theorem 1.1]. However, most of the ideas are already present in the subcase $R = K[\![t]\!]_0$, which was established by de Jong [123, Theorem 9.1]; the main difference is that de Jong can use Dwork's trick (Corollary 17.2.2) where we have to use the *p*-adic local monodromy theorem (Theorem 20.1.4).

Since a weak form of Dwork's trick applies even without a differential structure (Theorem 16.1.1), it should be possible to extend the case $R = K[\![t]\!]_0$ of Theorem 20.3.5 to difference modules; this is carried out in the absolute case in [235], but we expect the general case to follow similarly.

The proof of Theorem 20.3.5 given here is in substance the same as the proof in the absolute case given in [231]. In particular, Proposition 20.3.3 is essentially [231, Proposition 4.2], which in turn is close to de Jong's [123, Corollary 8.2].

The original application of Theorem 20.3.5 in the case $R = K[\![t]\!]_0$ is de Jong's proof of the equal-characteristic analogue of Tate's extension theorem on p-divisible groups (Barsotti–Tate groups). Tate proved [382] that for R a complete discrete valuation ring of characteristic 0 and residual characteristic $p > 0$, given any two p-divisible groups over R, any morphism between their generic fibers extends to a full morphism. Tate's proof introduced the seeds that grew into the subject of p-adic Hodge theory (see Appendix C); de Jong recognized that an appropriate analogue of Tate's method for R of characteristic p would go through crystalline Dieudonné theory. In this manner, de Jong [123, Theorem 1.1] reduced the analogue of Tate's theorem to an instance of Theorem 20.3.5.

The case $R = \mathcal{E}^\dagger$ of Theorem 20.3.5 has applications in the theory of overconvergent F-isocrystals (for more on which see Appendix B). Namely, it implies (with some work) that on a smooth variety over a field of characteristic $p > 0$, the restriction functor from overconvergent F-isocrystals to convergent F-isocrystals is fully faithful [242, Theorem 4.2.1].

The local duality for quasiunipotent differential modules (Proposition 20.4.2) is due to Matsuda [307]. See also the treatment in [233, §4].

Theorem 20.6.7 is stated in [398, Theorem 2.14] for differential modules with Frobenius structure; however, the proof is the one we have given here, which does not in fact use the differential structure. This result is the local version of a global theorem of Tsuzuki on overconvergent F-isocrystals on curves [398, Proposition 5.8]; it resolves a question raised in [262].

One can promote the p-adic local monodromy theorem to a statement about G-linear objects for G a reductive algebraic group over K, with the case $G = \mathrm{GL}_n$ recovering the original statement for objects of rank n; this is relevant to applications in p-adic Hodge theory concerning Galois representations valued in algebraic groups. See [414].

Exercises

20.1   Prove that a finite free differential module M over $\mathcal{R}$ is quasiunipotent if and only if $M \otimes_{\mathcal{R}} \mathcal{R}_L$ is unipotent for some L. (Hint: produce a nonzero quasiconstant submodule of M, e.g., by Galois descent.)

20.2   Let $0 \rightarrow M_1 \rightarrow M \rightarrow M_2 \rightarrow 0$ be a nonsplit extension of differential modules over $\mathcal{R}$, where M_1 and M_2 are both trivial of rank 1. Prove that M cannot admit a Frobenius structure which induces unit-root Frobenius structures on both M_1 and M_2. (Hint: use the fact that $H^1(\mathcal{R}) = K \, dt/t$.)

20.3   Prove Lemma 20.3.1 and Lemma 20.6.5. (Hint: reduce to the case where $|A|_\rho \leq 1$ for some $\rho \in (0, 1)$ for which $|w|_\rho < \infty$. Then use $|w|_\rho$ to bound the terms of v of norm greater than some $c > 0$.)

20.4   Complete the proof of Lemma 20.3.4 by proving the following assertion. Let M be a finite unipotent differential module over $\mathcal{R}$, equipped with a Frobenius structure whose Newton slopes are all positive. Prove that any exact sequence $0 \rightarrow M \rightarrow N \rightarrow \mathcal{R} \rightarrow 0$ in the category of differential modules with Frobenius structures must split. (Hint: apply Lemma 9.2.3 to N, then see how the resulting basis behaves under a standard Frobenius lift.)

20.5   In the notation of Theorem 20.3.5, suppose that $\varphi(t) = t^q$. Prove that the equality $H^0(M) = H^0(M \otimes_{\mathcal{R}} \mathcal{E})$ also holds in the category of difference modules. (Hint: if $v \notin H^0(M)$, then $t^{-1}D(v) \subset H^0(M)$ also.)

20.6   Prove Proposition 20.4.2 in case M is unipotent. (Hint: use Lemma 9.2.3.)

20.7   Prove Corollary 20.4.3 using Proposition 20.4.2. (Hint: let P be an indecomposable finite free quasiunipotent differential module over $\mathcal{R}$. Prove first that P is a successive extension of copies of a single irreducible differential module M. Then construct an isomorphism $P \cong M \otimes N$ with N unipotent by induction on the rank of P.)

21

The p-adic local monodromy theorem: proof

In this chapter, we give a proof of the p-adic local monodromy theorem, at the full level of generality at which we have stated it (Theorem 20.1.4). After some initial reductions, we start with the case of a module of differential slope 0, i.e., one satisfying the Robba condition. We describe how this case can be treated using either the p-adic Fuchs theorem for annuli (Theorem 13.6.1), or the slope filtration theorem (Theorem 16.4.1). We then treat the rank-1 case using the classification of rank-1 solvable modules from Chapter 12. We then show that any module of rank greater than 1 and prime to p can be made reducible, by comparing the module with its top exterior power and using properties of refined differential modules. We finally handle the case of a module M of rank divisible by p by considering $M^\vee \otimes M$ instead.

The reader may notice some similarities with the proof of the Turrittin–Levelt–Hukuhara decomposition theorem (Theorem 7.5.1). In fact, this theorem is also known as the *p-adic Turrittin theorem* for this reason.

Besides the running hypothesis for this part (Hypothesis 14.0.1) and the one from the previous chapter (Hypothesis 20.0.1), it will be convenient to set several more hypotheses. We consign these to Section 21.1.

21.1 Running hypotheses

We are going to make a number of calculations under similar hypotheses. Rather than repeat the hypotheses each time, we enunciate them once and for all here.

We first explain how to deal with the case where κ_K is imperfect.

Remark 21.1.1. We first point out that it suffices to prove quasiunipotence after replacing K by a complete extension K' to which φ_K extends, either by

Proposition 6.9.1 or a more elementary argument (see [230, Proposition 6.11]). In particular, we may take K' to be the φ-perfection of K, on which φ is bijective.

Here is a variant of the previous remark.

Remark 21.1.2. One can also proceed without φ-perfecting K initially, but replacing K by its inverse image under φ at any time when needed. We end up with a finite extension L of $\varphi^{-m}(\kappa_K)((t))$ for some positive integer m, such that $M \otimes_{\mathcal{R}} \mathcal{R}_L$ is quasiunipotent. In particular, $H^0(M \otimes_{\mathcal{R}} \mathcal{R}_L)$ is a nonzero space on which φ acts bijectively. Applying φ^m gives a nonzero element of $H^0(M \otimes_{\mathcal{R}} \mathcal{R}_{L'})$ for a finite separable extension L' of $\kappa_K((t))$. We may deduce that M has a nonzero Φ-stable differential submodule which is quasiunipotent; repeating this argument will recover Theorem 20.1.4 in full.

We are now ready to introduce the new running hypotheses.

Hypothesis 21.1.3. For the rest of this chapter, assume that φ_K is bijective on K; by the preceding remarks, this is harmless for the purpose of proving Theorem 20.1.4. Let F be a finite Galois (but not necessarily unramified) extension of K. Put $\mathcal{R}_F = \mathcal{R} \otimes_K F$; we will not attempt to extend φ to $\mathcal{R}_F$. Let M be a finite differential module over $\mathcal{R}$ equipped with a Frobenius structure.

Hypothesis 21.1.4. Within each lemma in this chapter, set notation as follows. Let N be a nonzero differential submodule (not a subquotient) of $M \otimes_{\mathcal{R}} \mathcal{R}_F$ for some specified F. We will use L to indicate an initially unspecified finite separable extension of $\kappa_K((t))$ to which $\overline{\varphi}$ extends; since φ_K is bijective on K, Lemma 20.5.5 implies that $\mathcal{R}_L$ may be viewed as a Robba ring over a finite unramified extension of K. We will use F' to indicate an initially unspecified finite extension of F which is Galois over the integral closure K' of K in $\mathcal{R}_L$. (This may require identifying a subfield of F larger than K with an isomorphic subfield of $\mathcal{R}_L$.) Write $\mathcal{R}_{L,F'} = \mathcal{R}_L \otimes_{K'} F'$.

Notation 21.1.5. We write N_0 to refer to the component of N of differential slope 0, as provided by Theorem 12.6.4.

21.2 Modules of differential slope 0

We start by describing objects of differential slope 0. This requires the use of either of two pieces of heavy machinery: the theory of p-adic exponents (Chapter 13) or the theory of slope filtrations for difference modules over the Robba ring (Theorem 17.4.3).

Lemma 21.2.1. *Suppose that N has differential slope 0. Then we may choose L such that $N \otimes_{R_F} R_{L,F}$ is unipotent.*

Before giving either proof, we insert a reduction common to both.

Remark 21.2.2. In Lemma 21.2.1, the existence of N forces M to have a nontrivial summand M_0 of differential slope 0, such that N appears in $M \otimes_R R_F$. It thus suffices to prove that if M is of differential slope 0, then we may choose L such that $M \otimes_R R_L$ is unipotent.

We first give the proof using p-adic exponents.

First proof of Lemma 21.2.1 As in Remark 21.2.2, we may assume that M itself is of differential slope 0. By Corollary 17.3.2, we may change to a standard Frobenius lift. We may then apply Corollary 13.6.2 to deduce that $M \otimes_R R[t^{1/m}]$ is unipotent for some positive integer m coprime to p. $\square$

We next give the proof using slope filtrations.

Second proof of Lemma 21.2.1 Again as in Remark 21.2.2, we may assume that M itself is of differential slope 0. By Theorem 17.4.3 (and changing Frobenius, as in Corollary 17.3.2), we may reduce to the case where M is pure of norm 1 as a difference module.

In other words, Lemma 21.2.1 is reduced to the following claim. Let M be a finite differential module over $\mathcal{E}^{\dagger}$ admitting a unit-root Frobenius structure for a standard Frobenius lift. Suppose that $IR(M_\rho) = 1$ for $\rho \in (0, 1)$ sufficiently close to 1. Then there exists a positive integer m coprime to p such that $M \otimes_R R[t^{1/m}]$ is unipotent.

To prove this claim, apply Lemma 19.3.3 to construct a basis of $M \otimes_R \mathcal{E}^{\dagger}[t^{1/m}]$ for some m, on which Φ, tD act via matrices A, N such that A, A^{-1}, N have entries in $\mathcal{E}^{\dagger} \cap \mathfrak{o}_K[\![t^{-1/m}]\!]$. As in the proof of Lemma 19.3.4, this forces $N_0 = 0$.

For notational simplicity, assume hereafter $m = 1$. Let $f: K[\![t^{-1}]\!] \to K[\![t]\!]$ denote the substitution $t^{-1} \mapsto t$. We may then view $f(N)$ as defining a differential module M' over $K\langle t/\beta \rangle$ for the operator $\frac{d}{dt}$ for some $\beta > 1$, such that $IR(M'_\beta) = 1$. By Theorem 9.6.1, this module is trivial on the open disc of radius β; this implies that the original module M is trivial. In particular, M is unipotent. $\square$

Remark 21.2.3. It would be interesting to know whether one can prove Lemma 21.2.1 without using either p-adic exponents or slope filtrations, but instead simply using the fact that the hypothesis forces M to extend across the entire punctured open unit disc (by pasting together Frobenius antecedents).

21.3 Modules of rank 1

We next consider the rank-1 case, using the classification of solvable rank-1 differential modules (Theorem 12.7.2). This proof is the only point in the course of the proof of Theorem 20.1.4 at which we will introduce any specific wildly ramified extensions of $\kappa_K((t))$. (We made some tamely ramified extensions in Lemma 21.2.1 and will make some unramified extensions in Lemma 21.4.1.)

Lemma 21.3.1. *Suppose that* $\mathrm{rank}(N) = 1$. *Then we may choose L such that* $N \otimes_{\mathcal{R}_F} \mathcal{R}_{L,F}$ *is trivial.*

Proof We may assume F contains all p-th roots of 1. By Theorem 17.2.1, M is solvable at 1, as then is $M \otimes_{\mathcal{R}} \mathcal{R}_F$, as then is N. By Theorem 12.7.2, there exists a nonnegative integer h such that $N^{\otimes p^h}$ has differential slope 0. If $h = 0$, we may deduce the claim from Lemma 21.2.1. It suffices to check the case $h = 1$, as we may repeatedly apply this case to deduce the general case.

In case $h = 1$, by Theorem 17.1.6 (or the special case thereof discussed in Remark 17.1.7), there exist $c_1, \ldots, c_b \in \{0\} \cup \mathfrak{o}_F^{\times}$ with $c_i = 0$ whenever i is divisible by p, such that the rank 1 differential module $N_1 = M_{1,c_1} \otimes \cdots \otimes M_{b,c_b}$ over $\mathcal{R}_F$ has the property that $N_1^{\vee} \otimes N$ has differential slope 0. Since $N_1 \otimes_{\mathcal{R}_F} \mathcal{R}_{L_0,F}$ is unipotent for L_0 equal to the Artin–Schreier extension L_0 of $\kappa_F((t))$ defined by the parameter $\overline{c_1}t^{-1} + \cdots + \overline{c_b}t^{-b}$, so is $N \otimes_{\mathcal{R}_F} \mathcal{R}_{L_0,F}$.

Unfortunately, the construction does not guarantee that L_0 admits an action of $\overline{\varphi}$. However, suppose we run over all possible choices of N for the given M and F, defining each L_0 that occurs using the unique possible Artin–Schreier parameter in the additive group

$$\bigoplus_{i<0,\, p\nmid i} \kappa_K^{\mathrm{alg}} t^{-i}.$$

The resulting parameters are then limited to a finite set S.

Pick an integer i with $i < 0$, $p \nmid i$. Let S_i be the set of coefficients of t^{-i} appearing in the elements of S. Let $\overline{P}_i(x) \in \kappa_K[x]$ be the product of the distinct minimal polynomials of the elements of S_i over κ_K. Since M admits a Frobenius structure, the roots of $\overline{\varphi}(\overline{P}_i(x))$, when multiplied with t^{-qi}, must define Artin–Schreier parameters equivalent to the roots of $\overline{P}_i(x)$, when multiplied with t^{-i}. In other words,

$$\overline{\varphi}(\overline{P}_i(x)) = \overline{P}_i(x^{1/q})^q.$$

We may thus extend $\overline{\varphi}$ to the splitting field of $\overline{P}_i$ in such a way that for each root a of $\overline{P}_i$, $\overline{\varphi}(a) = b^q$ for some root b of $\overline{P}_i$.

This allows us to define a compositum of Artin–Schreier extensions L to which $\overline{\varphi}$ extends, including all of the choices of L_0 made above. For this L,

$N \otimes_{\mathcal{R}_F} \mathcal{R}_{L,F}$ has differential slope 0, so we can make it trivial by Lemma 21.2.1. This gives the desired result. □

21.4 Modules of rank prime to p

We next pass from rank 1 to rank prime to p, using what we know about refined differential modules over F_ρ.

Lemma 21.4.1. *Suppose that* rank(N) *is coprime to p. Then we may choose* L, F' *such that* $N \otimes_{\mathcal{R}_F} \mathcal{R}_{L,F'}$ *is either unipotent or reducible.*

Proof Put $n = $ rank(N). The case $n = 1$ is covered by Lemma 21.3.1, so we may assume n is greater than 1 and coprime to p. Suppose by way of contradiction that $N \otimes_{\mathcal{R}_F} \mathcal{R}_{L,F'}$ is irreducible and nontrivial for all L, F'. By Lemma 21.2.1 and Lemma 21.3.1, we may reduce to the case where $(M^\vee \otimes M)_0$ is unipotent and $\wedge^n N$ is trivial.

Let us realize M, M_0, N as differential modules on some annulus $\alpha < |t| < 1$. For $\rho \in (\alpha, 1)$, let $F_\rho = F[\![\rho/t, t/\rho]\!]_{\mathrm{an}}, F'_\rho = F'[\![\rho/t, t/\rho]\!]_{\mathrm{an}}$ denote the fields of analytic elements on the circle of radius ρ over the respective base fields F, F'. We claim that $N_\rho = N \otimes F_\rho$ must be refined for every $\rho \in (\alpha, 1)$. Otherwise, by Theorem 10.6.7, for a suitable choice of F' and some positive integer m coprime to p, we could split N nontrivially over $F'_\rho(t^{1/m})$. Each projector for this splitting would define a horizontal section of $(N^\vee \otimes N) \otimes F'_\rho(t^{1/m})$. Since $(M^\vee \otimes M)_0$ is unipotent, so is its subquotient $(N^\vee \otimes N)_0$; by expanding elements of $(N^\vee \otimes N)_0$ in terms of a basis as in Lemma 9.2.3, this would yield

$$H^0((N^\vee \otimes N) \otimes F'_\rho(t^{1/m})) = H^0(N^\vee \otimes N).$$

We would thus get a nontrivial splitting of N itself, contrary to hypothesis.

We now have that N_ρ is refined for every $\rho \in (\alpha, 1)$. Choose $\rho \in (\alpha, 1)$ so that $p^{-p^{-h+1}/(p-1)} < IR(N_\rho) < p^{-p^{-h}/(p-1)}$ for some nonnegative integer h. Apply Theorem 10.4.2 to form the h-fold Frobenius antecedent P of N_ρ, which is still refined. We may then apply Proposition 6.8.4(a) to deduce that $P^{\otimes n}$ has the same spectral radius as P, and Proposition 6.8.4(b) to deduce that $(\wedge^n P^\vee) \otimes P^{\otimes n}$ has strictly greater spectral radius than P. However, these two contradict each other because $\wedge^n P^\vee$ is the h-fold Frobenius antecedent of the trivial module $(\wedge^n N^\vee)_\rho$. This contradiction yields the desired result. □

21.5 The general case

We now make the step from rank prime to p to arbitrary rank. The trick used here is one familiar from elementary group theory; for example, it occurs in the proof of the Sylow theorems. See the exercises for another example of its use.

Lemma 21.5.1. *For arbitrary N, we may choose L, F' such that $N \otimes_{\mathcal{R}_F} \mathcal{R}_{L,F'}$ is either unipotent or reducible.*

Proof By Lemma 21.4.1, it suffices to consider N of rank n divisible by p. Then the trace-zero component of $N^\vee \otimes N$ has rank $n^2 - 1$, which is not divisible by p. By repeated application of Lemma 21.4.1, we can force the trace-zero component of $(N^\vee \otimes N) \otimes_{\mathcal{R}_F} \mathcal{R}_{L,F'}$ to acquire a unipotent component. In particular, the space $V = H^0((N^\vee \otimes N) \otimes_{\mathcal{R}_F} \mathcal{R}_{L,F'})$ has F'-dimension greater than 1.

We may view V as a finite-dimensional not necessarily commutative F'-algebra. A standard fact about such algebras (exercise) is that for some finite extension F'' of F', $V \otimes_{F'} F''$ fails to be a division algebra. Thus for suitable F'', we can find a pair of nonzero horizontal endomorphisms of $N \otimes_{\mathcal{R}_F} \mathcal{R}_{L,F''}$ which compose to zero. This forces N to be reducible. $\square$

Proof of Theorem 20.1.4 It suffices to consider the case where M is irreducible. By Lemma 21.5.1, we may choose L, F' such that $M \otimes_{\mathcal{R}} \mathcal{R}_{L,F'}$ is either unipotent or reducible. In the former case we are done, as this implies that $M \otimes_{\mathcal{R}} \mathcal{R}_L$ is also unipotent. Otherwise, $M \otimes_{\mathcal{R}} \mathcal{R}_{L,F'}$ contains a proper nonzero differential submodule N. By applying Lemma 21.5.1 repeatedly, we may keep replacing N by a submodule (after changing L and F') until N becomes unipotent.

To $(M \otimes_{\mathcal{R}} \mathcal{R}_L) \otimes_{K'} F'$, apply Φ on the left and $G_{F'/K'}$ on the right. The images of N fill out a submodule of the form $N_0 \otimes_{K'} F'$, where N_0 is a nonzero differential submodule of M stable under Φ. The module N_0 is unipotent because $N_0 \otimes_{K'} F'$ is unipotent. Consequently, $M \otimes_{\mathcal{R}} \mathcal{R}_L$ has a nonzero unipotent submodule N_0 stable under Φ, so we may apply the induction hypothesis to $(M \otimes_{\mathcal{R}} \mathcal{R}_L)/N_0$ to conclude. $\square$

Notes

The approach to Theorem 20.1.4 presented here is modeled on that given by Mebkhout [309] in the absolute case, except that Mebkhout only describes the first of our two proofs of Lemma 21.2.1. The approach of André [16] is substantively similar but formally different, as it is phrased in the language of

Tannakian categories; it is thus more similar to the argument we present in the next chapter.

In the course of proving Lemma 21.3.1, we noticed that if $\overline{\varphi}$ is not an absolute Frobenius, then it is a highly nontrivial restriction on a finite separable extension L of $\kappa_K((t))$ to require that it must support an extension of $\overline{\varphi}$. This phenomenon was noted in the context of p-adic Hodge theory by Berger and Colmez [51, Proposition 6.2.2].

Exercises

21.1 Let G be a finite p-group and let $\tau\colon G \to \mathrm{GL}(V)$ be a complex linear representation of G. Prove that if τ is nontrivial, then either τ or $\tau^\vee \otimes \tau$ has a nontrivial one-dimensional subrepresentation. (Hint: consider the possible dimensions of irreducible representations of G.)

21.2 Let V be a finite-dimensional not necessarily commutative F-algebra. Prove that there exists a finite extension F' of F such that $V \otimes_F F$ is not a division algebra. (Hint: we may assume that V itself is a division algebra. Pick any nonzero $x \in V$, view x as a linear transformation from V to V via left multiplication, and subtract an eigenvalue of this transformation.)

22

p-adic monodromy without Frobenius structures

In this chapter, we introduce an alternate approach to the *p*-adic local monodromy theorem (taken from [255], incorporating corrections from [269]), in which we first prove a corresponding statement for solvable differential modules *without* a Frobenius structure. The argument thus makes minimal use of either *p*-adic exponents or slope filtrations over Robba rings, at the expense of requiring an application of the basic formalism of Tannakian categories.

Throughout this chapter, assume that K is a nonarchimedean field of residue characteristic $p > 0$; we add additional hypotheses as needed.

As in Part III, we write ω to mean 1 if $p = 0$ and $p^{-1/(p-1)}$ if $p > 0$.

22.1 The Robba ring revisited

Definition 22.1.1. As in Definition 15.1.4 over K, we define the *Robba ring* over K as

$$\mathcal{R} = \bigcup_{\alpha \in (0,1)} K\langle \alpha/t, t\}.$$

As before, we may define the subring

$$\mathcal{E}^\dagger = \bigcup_{\alpha \in (0,1)} K\langle \alpha/t, t]\!]_0;$$

however, if K is not discretely valued then $\mathcal{E}^\dagger$ need not be a field. Nonetheless, we can recover part of Lemma 15.1.3.

Lemma 22.1.2. *The ring $\mathcal{E}^\dagger$ has the following properties.*

(a) *Under the 1-Gauss norm $|\cdot|_1$, the valuation ring $\mathfrak{o}_{\mathcal{E}^\dagger}$ is a local ring with residue field $\kappa_K((t))$.*

(b) The local ring $\mathfrak{o}_{\mathcal{E}^\dagger}$ is henselian in the sense of Remark 3.0.2.

Proof To prove (a), it suffices to check that any $x = \sum_{i \in \mathbb{Z}} x_i t^i \in \mathfrak{o}_{\mathcal{E}^\dagger}$ for which $|x_0 - 1| < 1$ and $|x_i| < 1$ for all $i \neq 0$ is a unit. In this case, the indices $i > 0$ do not contribute to the Newton polygon of x, so we can choose α so that $x \in K\langle \alpha/t, t]\!]_0$ has no slopes between 0 and $-\log \alpha$. We may then continue as in Lemma 15.1.3. Given (a), we may deduce (b) as in the proof of Lemma 15.1.3. $\qquad\square$

Definition 22.1.3. Let L be a finite separable field extension of $\kappa_K((t))$. By Lemma 22.1.2, there exists a unique (up to unique isomorphism) finite étale extension $\mathfrak{o}_{\mathcal{E}^\dagger_L}$ of $\mathfrak{o}_{\mathcal{E}^\dagger}$ with residue field L; we define

$$\mathcal{E}^\dagger_L = \mathcal{E}^\dagger \otimes_{\mathfrak{o}_{\mathcal{E}^\dagger}} \mathfrak{o}_{\mathcal{E}^\dagger_L}, \qquad \mathcal{R}_L = \mathcal{R} \otimes_{\mathfrak{o}_{\mathcal{E}^\dagger}} \mathfrak{o}_{\mathcal{E}^\dagger_L}.$$

If κ_K is perfect, then we can view L as a Laurent series field over the integral closure of κ_K in L (see Remark 20.5.1) and so we can view $\mathcal{R}_L$ as the Robba ring over a finite extension of K.

Let C denote the category whose objects are finite differential modules on an (unspecified) open annulus of the form $\alpha < |t| < 1$ over K which are solvable at 1, with morphisms being morphisms of differential modules over a (possibly smaller) such annulus. We think of these as "finite solvable differential modules over $\mathcal{R}$" although they are not guaranteed to be finitely generated (unless K is spherically complete; see the chapter notes). In particular, we can define *differential slopes* for objects of C; these remain unchanged after tensoring over $\mathcal{R}$ with the Robba ring over some extension of K. This means that we can also define differential slopes even if κ_K is not perfect, by first replacing K with a sufficiently large extension (e.g., a completed algebraic closure). Let $b(M)$ denote the largest differential slope of $M \in C$.

We can now state the theorem whose proof will occupy much of this chapter.

Theorem 22.1.4. *For $M \in C$, there exists a finite separable extension L of $\kappa_K((t))$ such that $b(M \otimes_{\mathcal{R}} \mathcal{R}_L) = 0$.*

22.2 Modules of cyclic type

One key issue in the absence of a Frobenius structure is that we cannot count on being able to apply the theory of p-adic exponents without running into problems with p-adic Liouville numbers; we thus cannot force arbitrary modules over $\mathcal{R}$ of differential slope 0 to become successive extensions of modules of rank 1. For this reason, we must consider certain modules which look like the tensor product of a module of rank 1 with a module of differential slope 0, and then show that in the case of solvable modules over $\mathcal{R}$ they actually do have this form (Theorem 22.2.4).

Definition 22.2.1. For M a finite differential module on a closed, open, or half-open annulus over K, we say that M is of *cyclic type* if $M^\vee \otimes M$ satisfies the Robba condition. Similarly, for $M \in C$, we say that M is of cyclic type if $b(M^\vee \otimes M) = 0$.

We record some direct consequences of this definition.

- If M satisfies the Robba condition, then M is of cyclic type.
- If M is of rank 1, then M is of cyclic type.
- If M is of cyclic type, then so is $M^\vee$.
- If M_1, M_2 are of cyclic type, then so is $M_1 \otimes M_2$.
- If $0 \to M_1 \to M \to M_2 \to 0$ is a short exact sequence of differential modules, then M is of cyclic type if and only if M_1 and M_2 are of cyclic type and $M_1^\vee \otimes M_2$ satisfies the Robba condition.

In light of these facts, we may define an equivalence relation on differential modules of cyclic type by declaring M_1, M_2 to be equivalent if $M_1^\vee \otimes M_2$ satisfies the Robba condition. The set of equivalence classes then forms a group under tensor product, with the identity element being the class of modules which themselves satisfy the Robba condition.

Lemma 22.2.2. *Let M be a finite differential module on the annulus $|t| \in I$ which is of cyclic type. For each $\lambda \in K$ with $|\lambda - 1| < 1$, let λ^* denote the pullback operation on differential modules along the substitution $t \mapsto \lambda t$. Use the Taylor series construction to produce an isomorphism $T_\lambda \colon \lambda^*(M^\vee \otimes M) \cong M^\vee \otimes M$, then reinterpret this map as a horizontal section of*

$$\lambda^*(M^\vee \otimes M)^\vee \otimes (M^\vee \otimes M) \cong \lambda^* M \otimes \lambda^* M^\vee \otimes M^\vee \otimes M$$

$$\cong \lambda^* M \otimes M^\vee \otimes \lambda^* M^\vee \otimes M$$

$$\cong (\lambda^* M^\vee \otimes M)^\vee \otimes (\lambda^* M^\vee \otimes M).$$

Then reinterpreting again yields a horizontal morphism $h_\lambda \colon \lambda^* M^\vee \otimes M \to \lambda^* M^\vee \otimes M$ which is a projector of rank 1.

Proof From the expression of T_λ in terms of the Taylor series, it is apparent that the diagram

$$\begin{array}{ccc}
\lambda^*(M^\vee \otimes M) \otimes \lambda^*(M^\vee \otimes M) & \xrightarrow{\;-\circ-\;} & \lambda^*(M^\vee \otimes M) \\
\downarrow{\scriptstyle T_\lambda \otimes T_\lambda} & & \downarrow{\scriptstyle T_\lambda} \\
(M^\vee \otimes M) \otimes (M^\vee \otimes M) & \xrightarrow{\;-\circ-\;} & M^\vee \otimes M
\end{array}$$

commutes. We may thus apply Remark 5.3.5. □

For modules of cyclic type, we have the following improvement of Theorem 11.3.2(b) to the effect that these modules behave like modules of rank 1.

Theorem 22.2.3. *Let M be a finite differential module on the annulus $|t| \in I$ which is of cyclic type. Then the slopes of the function $r \mapsto f_1(M,r)$ on $-\log I$ belong to $\mathbb{Z}$.*

Proof It suffices to check the slopes at a fixed point $r_0 \in -\log I$ at which $f_1(M,r_0) \neq r_0 - p^{-j} \log \omega$ for any nonnegative integer j. By repeatedly forming Frobenius antecedents (Theorem 10.4.2), we may reduce to the case where $f_1(M,r_0) > r_0 - \log \omega$. Choose a basis of M as in the proof of Lemma 11.5.1 (possibly after shrinking I) and equip $M^\vee$ with the dual basis.

For any $\lambda \in K$ with $|\lambda - 1| < 1$, equip $\lambda^* M^\vee$ with the pullback of the chosen basis of $M^\vee$ and equip $\lambda^* M^\vee \otimes M$ with the product basis; we may then apply Theorem 6.7.4 to deduce that there exists $a > 0$ such that for r in some neighborhood of r_0, for $|\lambda - 1|$ sufficiently close to 1,

$$f_1(\lambda^* M^\vee \otimes M, r) = f_1(M,r) + a \log |\lambda - 1|.$$

(The value of a can be read off from a certain Newton polygon, as in the proof of Lemma 12.8.5, but it is irrelevant to what follows.)

Now note that $\lambda^* M^\vee \otimes M$ is of cyclic type, but by Lemma 22.2.2 it contains a direct summand Q of rank 1. Therefore for r, λ as above,

$$f_1(Q,r) = f_1(\lambda^* M^\vee \otimes M, r) = f_1(M,r) + a \log |\lambda - 1|$$

and so $f_1(Q,r)$ and $f_1(M,r)$ have the same slopes at r_0. Since $f_1(Q,r)$ has slopes in $\mathbb{Z}$ by Theorem 11.3.2(b), this proves the claim. □

Using Theorem 22.2.3, we can prove an extension of Theorem 17.1.6 to modules of cyclic type.

Theorem 22.2.4. *Suppose that K is algebraically closed. Let M denote a finite differential module of cyclic type on a half-open annulus with open outer* *radius* 1, *which is solvable at* 1 *with differential slope* b. *Then there exist* $c_1, \ldots, c_b \in \{0\} \cup \mathfrak{o}_K^\times$ *and nonnegative integers* $j_1, \ldots, j_b$ *such that*

$$M \otimes M_{1,c_1} \otimes \cdots \otimes M_{b,c_b}$$

has all differential slopes equal to 0, *for* M_{i,c_i} *defined as in Theorem* 12.7.2.

Proof   As in the proof of Theorem 17.1.6, it suffices to prove the claim with M_{i,c_i} replaced by some Frobenius pullback; we follow the proof of Theorem 12.7.2. By Theorem 22.2.3, $b \in \mathbb{Z}$; we may thus proceed by induction on b, with trivial case case $b = 0$. Pick $0 < \alpha < \beta < 1$ such that for some nonnegative integer j,

$$\omega^{p^{-j+1}} < IR(M_\alpha) = \alpha^b < IR(M_\beta) = \beta^b < \omega^{p^{-j}}.$$

By Theorem 10.4.4, M admits a j-fold Frobenius antecedent N over the ring $K\langle \alpha^{p^j}/t^{p^j}, t^{p^j}/\beta^{p^j} \rangle$. Put $F = \mathrm{Frac}(K\langle \alpha^{p^j}/t^{p^j}, t^{p^j}/\beta^{p^j} \rangle)$ and choose a cyclic vector (by Theorem 5.4.2) to obtain an isomorphism $M \otimes F \cong F\{T\}/F\{T\}P$ for some monic twisted polynomial P over F. Choose a value $\rho_0 \in (\alpha, \beta)$; by Corollary 6.5.4 and the hypothesis that M is of cyclic type, any two roots $\lambda_1, \lambda_2 \in F^{\mathrm{alg}}$ of P satisfy $|\lambda_1 - \lambda_2|_{\rho_0^{p^j}} < |\lambda_1|_{\rho_0^{p^j}}$. By Corollary 6.5.4 again and the hypothesis that K is algebraically closed, we can choose $c \in K$ such that $IR((N \otimes M_{b,c})_{\rho^{p^j}}) > \rho^{p^j b}$ for ρ in some neighborhood of ρ_0. We may use the log-concavity of the intrinsic radius (Theorem 11.3.2(e)) to deduce that the differential slope of $N \otimes M_{b,c}$ is strictly less than b. Thus the induction hypothesis gives the desired result.   □

22.3 A Tannakian construction

At this point, our desired strategy for proving Theorem 22.1.4 is to proceed by induction as follows: given $M \in C$, we find some related module N which is of cyclic type and apply Theorem 22.2.4, then argue as in Chapter 21 to show that the resulting module of rank 1 becomes a module of differential slope 0 after tensoring with $\mathcal{R}_L$ for a suitable choice of L. In fact, this is what we did in Chapter 21, but there we were lucky enough to be able to find N in a rather controlled manner. Here we are not so lucky, so we need some alternate construction to provide an upper bound on the search space for the module N, as well as to show that resolving the problem for N constitutes measurable progress towards solving the problem for M.

Based on the model of Chapter 19, we may eventually hope to show that the structure of M is controlled by some finite linear group. This is achieved by the formalism of Tannakian categories except that the resulting group is only algebraic, not necessarily finite. To limit the technical background needed to read this section, we have taken some shortcuts around the general definitions associated with the Tannakian formalism; see the chapter notes for appropriate context.

Definition 22.3.1. For $M \in C$, let $[M]$ be the smallest K-linear subcategory of C which:

- contains M;
- is closed under isomorphisms;
- is closed under the operations $\oplus, \otimes, \vee$; and
- is closed under formation of subquotients.

By Proposition 9.1.2, this is a K-linear abelian category; it is sometimes called the *category generated by M*.

Definition 22.3.2. Let F be an algebraic closure of $\mathrm{Frac}(\mathcal{R})$ (viewed as a field with no extra structure). For $M \in C$, write M_F as shorthand for $M \otimes_{\mathcal{R}} F$, and let $[M] \otimes_K F$ be the F-linear category obtained from $[M]$ by preserving the objects, but tensoring all sets of morphisms from K to F.

Let $G(M)$ be the group of natural transformations of $[M] \otimes_K F$ (as an F-linear abelian category) that commute with the functor $N \mapsto N_F$. A more concrete way to make $G(M)$ is to take the subgroup of $g \in \mathrm{GL}(M_F)$ satisfying the following condition.

(a) For every positive integer m, for every differential submodule N of $(M \oplus M^\vee)^{\otimes m}$, the image of g in

$$\mathrm{GL}((M \oplus M^\vee)^{\otimes m}_F) \cong \mathrm{GL}((M_F \oplus M_F^\vee)^{\otimes m})$$

carries N_F into itself.

From this description, we see that $G(M)$ is a linear algebraic group over F.

For $r > 0$ (resp. $r \geq 0$), let $G^r(M)$ (resp. $G^{r+}(M)$) be the subgroup of $g \in G(M)$ satisfying the following additional condition.

(b) For every positive integer m, for every subquotient N of $(M \oplus M^\vee)^{\otimes m}$ with $b(N) < r$ (resp. $b(N) \leq r$), the action of g on N (which is well-defined by (a)) fixes N_F.

Note that for $r < s$, $G^s(M) \subseteq G^{r+}(M) \subseteq G^r(M)$. Moreover, $G^s(M)$ is normal in $G(M)$ and hence in $G^r(M)$: conjugating an element satisfying (b) by an element satisfying (a) yields another element satisfying (b).

Before studying this definition in general, let us reconcile it with the work of Chapter 19.

Remark 22.3.3. Suppose that K is discretely valued and $M = D^\dagger(V) \otimes_{\mathcal{E}^\dagger} \mathcal{R}$ for some representation $\tau \colon G_{\kappa_K((t))} \to \mathrm{GL}(V)$ as in Definition 19.2.2. Let G be the image of τ, and let $I \subseteq G$ be the image of the inertia subgroup of $G_{\kappa_K((t))}$ under τ. Then $(M \oplus M^\vee)^{\otimes m}$ decomposes as a direct sum of irreducible differential modules according to the decomposition of $(V \oplus V^\vee)^{\otimes m}$ into irreducible representations of G, with the G-fixed subspace corresponding to the maximal trivial submodule. By Theorem 22.3.5 below, this will imply that $G(M) = G$. Moreover, since every irreducible representation of G/I, viewed as an irreducible representation of G, appears in $(V \oplus V^\vee)^{\otimes m}$ for some m (see Lemma 22.3.4 below), and there are only finitely many isomorphism classes of such representations, we deduce that $G^{0+}(M) = G^s(M) = I$ for $s > 0$ sufficiently small.

Lemma 22.3.4. *Let F be an algebraically field of characteristic 0. Let G be a finite subgroup of $\mathrm{GL}(V)$ for some finite-dimensional F-vector space V. Then every irreducible finite-dimensional F-linear representation of G can be found as a subquotient of $V^{\otimes m}$ for some positive integer m.*

Proof Exercise.   □

Theorem 22.3.5 (Tannaka reconstruction). *Let F be a field of characteristic 0. Let V be a finite-dimensional vector space over F. Let G be an algebraic subgroup of $\mathrm{GL}(V)$. Then G coincides with the set of $g \in \mathrm{GL}(V)$ satisfying the following conditions.*

(a) *For every positive integer m, every G-stable subspace of $(V \oplus V^\vee)^{\otimes m}$ is also g-stable.*

(b) *For every positive integer m, every G-invariant subquotient of $(V \oplus V^\vee)^{\otimes m}$ is also g-invariant. (The fact that g acts on such a subquotient is guaranteed by (a).)*

Proof We follow the proof of the Tannaka reconstruction theorem given in [377, Theorem 2.5.3]. Let H be the group defined by (a) and (b); there is an evident injection $G \to H$. Let $v_1, \ldots, v_n$ be a basis of V and let $v_1^\vee, \ldots, v_n^\vee$ be the dual basis of $V^\vee$; we can then view the coordinate ring A of G as a quotient of the polynomial ring in F on the generators $v_i^\vee \otimes v_j$ of the vector space $V^\vee \otimes V$. We can rephrase this in a coordinate-free way by saying that A is a quotient of the symmetric algebra of $V^\vee \otimes V$ over F. From this point of view, we see that the action of G on A by right translation (viewing A as the ring of regular functions on G; here we need characteristic 0 to ensure that G is reduced) extends to an action of H which is equivariant for both the multiplication map $\mu: A \otimes A \to A$ and the comultiplication $\Delta: A \to A \otimes A$ corresponding to the multiplication map on G.

For any $h \in H$, the previous paragraph implies that the action of g on $V^\vee \otimes V$ carries G into itself. In particular, it carries the identity matrix to some element g of G; using the H-equivariance of Δ, we see that g and h act the same way on A, and so must actually coincide. $\square$

Remark 22.3.6. A ring equipped with a comultiplication structure, in addition to its usual multiplication structure, is called a *Hopf algebra*. More precisely, such a Hopf algebra is commutative (as a ring) but not necessarily *cocommutative* because the corresponding algebraic group need not be commutative.

For general M, we have the following result which allows us to interpret the category $[M] \otimes_K F$ in group-theoretic terms.

Definition 22.3.7. From the definition of $G(M)$, we obtain a functor from $[M] \otimes_K F$ to the category $\mathrm{Rep}_F(G(M))$ of finite-dimensional F-linear (algebraic) representations of $G(M)$.

Theorem 22.3.8. *In Definition 22.3.7, the functor* $[M] \otimes_K F \to \mathrm{Rep}_F(G(M))$ *is an equivalence of categories.*

The subtle point here is to ensure that $G(M)$ does not fix *more* elements of $(M \oplus M^\vee)_F^{\otimes m}$ than the definition requires.

Proof The argument is parallel to the proof of Theorem 22.3.5. Define the ring A as the quotient of the symmetric algebra of $M_F^\vee \otimes M_F$ over F by the ideal generated by the elements of $H^0((M^\vee \otimes M)^{\otimes m})$ for all m. One can then equip A with a multiplication and comultiplication map and show that the resulting group variety G has the property that $\mathrm{Rep}_F(G)$ is equivalent to $[M] \otimes_K F$. Using Theorem 22.3.5, we may then deduce that $G \cong G(M)$. As the construction is rather involved, we defer to [130, Theorem 2.11] for further details. $\square$

Remark 22.3.9. With notation as in Definition 22.3.2, for any $N \in [M]$, we have a natural homomorphism $G(M) \to G(N)$; by Theorem 22.3.8, this map is surjective because $[N] \otimes_K F$ is a full subcategory of $[M] \otimes_K F$. By similar logic, the image of $G^r(M)$ (resp. $G^{r+}(M)$) in $G(N)$ is equal to $G^r(N)$ (resp. $G^{r+}(N)$).

Remark 22.3.10. Motivated by the previous discussion, we would like to show that the groups $G^{0+}(M)$ are finite for any finite solvable differential module M over $\mathcal{R}$. While this will turn out to be true (Theorem 22.5.7), the proof will require a fairly involved analysis. We assemble the key group-theoretic ingredients in the next section. By contrast, we make no attempt to control $G(M)$; this is related to the discussion at the start of Section 22.2.

22.4 Interlude on finite linear groups

We now take a deep dive into finite group theory to assemble the ingredients needed to prove Theorem 22.5.7. Throughout this section, let F be a field of characteristic 0, let n be a positive integer, and let V be a vector space of dimension n over F.

Remark 22.4.1. We will use frequently, and without comment, the fact that every nontrivial finite p-group admits a surjective homomorphism onto $\mathbb{Z}/p\mathbb{Z}$. See the exercises for Chapter 3.

Lemma 22.4.2 (Jordan's theorem). *There exists a constant $f(n)$ depending only on n, such that every finite subgroup G of $\mathrm{GL}(V)$ contains an abelian normal subgroup H with $[G : H] \leq f(n)$.*

Proof See [121, Chapter 36]. □

Remark 22.4.3. An important example in Lemma 22.4.2 is the semidirect product $H \rtimes S_n \subset \mathrm{GL}_n(\mathbb{C})$ where H is the group of diagonal matrices with entries in μ_n and S_n is the group of permutation matrices.

A slight variant is $H_1 \rtimes S_{n+1} \subset \mathrm{GL}_n(\mathbb{C})$ where $H_1 = H \cap \mathrm{SL}_n(\mathbb{C})$. This example is in a sense best possible: using the classification of finite simple groups, Collins [109] showed that Lemma 22.4.2 holds with $f(n) = (n+1)!$ for $n \geq 71$. See also [188] for more context.

Lemma 22.4.4. *Let $G_0 \subseteq G_1 \subseteq \cdots$ be an increasing sequence of finite subgroups of $\mathrm{GL}(V)$ such that G_i is normal in G_j whenever $i \leq j$.*

(a) The union $G = \bigcup_{i=0}^{\infty} G_i$ contains an abelian normal subgroup of finite index bounded above by the constant $f(n)$ in Lemma 22.4.2.

(b) There exists an index i such that G/G_i is isomorphic to a subgroup of $(\mathbb{Q}/\mathbb{Z})^n$ (and in particular is abelian).

Proof Let S_i be the set of abelian normal subgroups of G_i of index at most $f(n)$; by Lemma 22.4.2, S_i is nonempty. For $i \leq j$ and $H_j \in S_j$, the map $G_i/(G_i \cap H_j) \to G_j/H_j$ is injective, so $G_i \cap H_j \in S_i$. We may thus view the S_i as an inverse system, whose inverse limit is nonempty by Tikhonov's theorem (or an elementary compactness argument); for any sequence $H_0, H_1, \ldots$ in the inverse limit, $H = \bigcup_{i=0}^{\infty} H_i$ is an abelian normal subgroup of G of index at most $f(n)$. This yields (a).

Given (a), let F^{alg} be an algebraic closure of F. Since the matrices of H all commute, we may simultaneously diagonalize them over F^{alg}; this yields an embedding of H into $((F^{\text{alg}})^{\times})^n$. Since the torsion subgroup of $(F^{\text{alg}})^{\times}$ is isomorphic to $\mathbb{Q}/\mathbb{Z}$, we see that H is isomorphic to a subgroup of $(\mathbb{Q}/\mathbb{Z})^n$. Since G/H is finite and is the union of its subgroups $G_i/(G_i \cap H)$, there must be an index i for which $G_i/(G_i \cap H) \cong G/H$. For such i, G/G_i is isomorphic to $H/(G_i \cap H)$, which is turn isomorphic to a subgroup of $(\mathbb{Q}/\mathbb{Z})^n/(G_i \cap H)$; the latter is itself isomorphic to a subgroup of $(\mathbb{Q}/\mathbb{Z})^m$ for some $m \le n$. This yields (b). $\quad\square$

Lemma 22.4.5. *For $r > 0$, let G^r be an algebraic subgroup of $\mathrm{GL}(V)$; for $r \ge 0$, put $G^{r+} = \bigcup_{s>r} G^s$. Suppose that the following conditions hold.*

(a) *For $0 < r < s$, G^s is a normal subgroup of G^r.*

(b) *For some $r > 0$, G^r is the trivial group.*

(c) *For every $s > 0$, if G^s is finite, then there exists $r \in (0, s)$ such that G^r/G^s is a finite p-group.*

(d) *For every $r > 0$, if G^{r+} is finite, then G^r/G^{r+} is a finite p-group.*

(e) *For $r > 0$, there exists $s_0 \in (r, \infty)$ such that for every $s \in [r, s_0)$, if G^t is finite and G^t/G^{s+} is abelian for all $t \in (r, s]$, then G^{r+}/G^{s+} is finite.*

Then G^r is a finite p-group for all $r > 0$.

Proof Let S be the set of $r > 0$ for which G^r is a finite p-group. By (a)–(c), S is up-closed, nonempty, and does not contain its infimum; hence $S = (r, \infty)$ for some $r \ge 0$. Suppose by way of of contradiction that $r > 0$. Define s_0 as in (e). By Lemma 22.4.4, G^{r+}/G^{s+} is abelian for some $s > r$, and hence for some $s \in [r, s_0)$. By (e), G^{r+}/G^{s+} is finite; by (d), G^r/G^{r+} is finite. But now G^r is finite, yielding the desired contradiction. $\quad\square$

To control G^{0+}, we need some additional conditions which involve the ambient group G.

Lemma 22.4.6. *In Lemma 22.4.5, assume that F is algebraically closed and that there exists an algebraic subgroup G of $\mathrm{GL}(V)$ containing G^{0+} and satisfying the following additional conditions for every positive integer m.*

(f) *For every $r > 0$, the G^{r+}-invariant subspace of $(V \oplus V^{\vee})^{\otimes m}$ admits a direct sum decomposition into G-stable subspaces, each of which restricts to an isotypical representation of G^r/G^{r+}.*

(g) If W is a one-dimensional G-stable subspace of $(V \oplus V^\vee)^{\otimes m}$ such that G^{0+} fixes W and G fixes some power of W, then G fixes W itself.

(h) If W is a one-dimensional G-stable subspace of $(V \oplus V^\vee)^{\otimes m}$, then G^{0+} acts on W with finite image.

Then G^{0+} is also finite and G centralizes G^{0+}.

Proof   We first check that G centralizes G^{0+}. Let G_0 be the identity connected component of G; then for each $r > 0$, the action of G on G^r by conjugation must restrict trivially to G_0 (because G^r is finite by Lemma 22.4.5). Consequently, the action of G on G^{0+} factors through the finite group G/G_0. By (f) plus Theorem 22.3.8, G acts trivially on G^r/G^{r+} for every $r > 0$; hence the action of G/G_0 factors through a finite p-group. Suppose by way of contradiction that this group is nontrivial; then by Remark 3.3.10 it admits a quotient of order p. By Theorem 22.3.8 again, we can find some m and some one-dimensional G-stable subspace W of $(V \oplus V^\vee)^{\otimes m}$ on which G^{0+} acts trivially while G acts via a nontrivial scalar action of $\mathbb{Z}/p\mathbb{Z}$. This constitutes a violation of (g).

We now know that G centralizes G^{0+}, so the G^{0+}-isotypical decomposition of $(V \oplus V^\vee)^{\otimes m}$ is also G-stable. If G^{0+} were not finite, then by Lemma 22.4.4(b) it would have a quotient isomorphic to $\mathbb{Q}_p/\mathbb{Z}_p$; hence for some m, $(V \oplus V^\vee)^{\otimes m}$ would admit a G-stable subspace W on which G^{0+} acts via a nontrivial scalar action of $\mathbb{Q}_p/\mathbb{Z}_p$. By taking a suitable exterior power, we could achieve the same thing with W being one-dimensional; this would contradict (h). Hence G^{0+} is finite, as desired.   □

22.5 Back to the Tannakian construction

Throughout this section, fix $M \in C$. We apply the previous discussion to analyze the groups $G^r(M)$.

Remark 22.5.1. Our first goal is to apply Lemma 22.4.5 to the family of groups $G^r(M)$. To begin with, note that condition (a) was discussed already in Definition 22.3.2, while condition (b) holds for any $r > b(M)$.

Lemma 22.5.2. *Suppose that $b(M) = r$. Then $G^r(M)/G^{r+}(M)$ is a finite elementary abelian p-group.*

Proof   At this point, we are free to enlarge K or to adjoin $t^{1/m}$ for some positive integer coprime to p (as this will rescale all differential slopes by m); hence by Theorem 10.6.7, we may reduce to the case where M is refined. In this case, Proposition 6.8.4(c) implies that $G^r(M)/G^{r+}(M)$ is a cyclic group of order p.   □

Corollary 22.5.3. *For any $r > 0$, if $G^{r+}(M)$ is finite, then $G^r(M)/G^{r+}(M)$ is a finite elementary abelian p-group. Consequently, the groups $G^r = G^r(M)$ satisfy condition (d) of Lemma 22.4.5.*

Proof Since $G^{r+}(M)$ is finite, it has finitely many isomorphism classes of finite-dimensional F-linear representations; by Lemma 22.3.4, each of these occurs in $(M \oplus M^\vee)_F^{\otimes m}$ for some positive integer m. By Theorem 12.8.4, we can decompose $(M \oplus M^\vee)^{\otimes m}$ as a direct sum $N_1 \oplus N_2$ so that $N_{1,F}$ is the G^{r+}-invariant subspace of $(M \oplus M^\vee)_F^{\otimes m}$. By Lemma 22.5.2, $G^r(N_1)/G^{r+}(N_1)$ is a finite elementary abelian p-group; by Remark 22.3.9, this is the same as taking the image of $G^r(M)$ in $\mathrm{GL}(N_{1,F})$ modulo the image of $G^{r+}(M)$. Combining this information over the isomorphism classes of representations of $G^{r+}(M)$ yields the claim. $\qquad\square$

Lemma 22.5.4. *Suppose that* $\mathrm{rank}(M) = n$ *and that the differential slopes of M are all equal to r. Then there are at most p^n distinct values in the range $(r/p, r]$ that can occur as differential slopes for subquotients of $(M \oplus M^\vee)^{\otimes m}$ for all positive integers m.*

Proof It suffices to check that for any positive integer m_0 and any $s \in (r/p, r)$, there are at most p^n distinct values in the range $[s, r]$ that occur as differential slopes for subquotients of $(M \oplus M^\vee)^{\otimes m}$ for all positive integers $m \leq m_0$. Choose $\alpha \in (0, 1)$ for which for $m = 1, \dots, m_0$, M arises as the base extension of a differential module M_0 over the annulus $\alpha < |t| < 1$ such that $(M_0 \oplus M_0^\vee)^{\otimes m}$ is clean for $m = 1, \dots, m_0$. We can then choose a value $\rho \in (\alpha, 1)$ such that for some nonnegative integer h, $\omega^{p^{-h+1}} < \rho^r \leq \rho^s < \omega^{p^{-h}}$. By forming the h-th Frobenius antecedent (Theorem 10.4.4), we may reduce to the case $h = 0$.

Apply Theorem 5.4.2 to write $M_{0,\rho}$ as $F_\rho\{T\}/F_\rho\{T\}P$ for some $P \in F_\rho\{T\}$. Let $Q \in F_\rho[T]$ be the untwisted polynomial with the same coefficients as P and let $\lambda_1, \dots, \lambda_n \in F_\rho^{\mathrm{alg}}$ be the roots of Q. By Theorem 6.7.4 (as applied in the proof of Lemma 6.8.1), for $m = 1, \dots, m_0$, we may read off those intrinsic subsidiary radii of $(M_0 \oplus M_0^\vee)^{\otimes m}$ in the range $[\rho^r, \rho^s]$ from the valuations of the quantities

$$\pm\lambda_{i_1} \pm \cdots \pm \lambda_{i_m} \qquad (i_1, \dots, i_m \in \{1, \dots, n\}).$$

However, this gives the same set as taking the valuations of the quantities

$$e_1\lambda_1 + \cdots + e_n\lambda_n \qquad (e_1, \dots, e_n \in \{0, \dots, p - 1\}),$$

yielding the desired upper bound. $\qquad\square$

Corollary 22.5.5. *For any $r > 0$, if $G^{r+}(M)$ is finite, then $G^{(r/p)+}(M)/G^{r+}(M)$*

is finite. Consequently, the groups $G^r = G^r(M)$ satisfy conditions (c) and (e) of Lemma 22.4.5.

Proof By Lemma 22.5.2 and Lemma 22.5.4, this holds when M is irreducible and $b(M) = r$. We may reduce the general case to this one as in the proof of Corollary 22.5.3.   □

Corollary 22.5.6. *The group $G^r(M)$ is a finite p-group for every $r > 0$.*

Proof The conditions of Lemma 22.4.5 are satisfied on account of Remark 22.5.1, Corollary 22.5.3, and Corollary 22.5.5.   □

Theorem 22.5.7. *The group $G^{0+}(M)$ is a finite p-group.*

Proof Keeping in mind Corollary 22.5.6, we will check that after adjoining $t^{1/m}$ for some m coprime to p, the groups $G = G(M)$ and $G^r = G^r(M)$ satisfy the conditions of Lemma 22.4.6. (Note that adjoining $t^{1/m}$ might make $G(M)$ smaller, but it will not change $G^{0+}(M)$.)

In this context, condition (f) follows from the existence of refined decompositions (Theorem 10.6.7). To check (g), it will suffice to check that if K is algebraically closed, M is of rank 1, and $M^{\otimes r}$ is trivial for some positive integer r, then M becomes trivial after adjoining $t^{1/m}$ for suitable m; this follows at once from Remark 13.6.5. To check (h), it suffices to note that if M is of rank 1, then its differential slope is confined to the discrete set $\mathbb{Z}$ by Theorem 11.3.2(b).   ⊔

Remark 22.5.8. By Lemma 22.4.6, after adjoining $t^{1/k}$ for a suitable positive integer k (as in the proof of Theorem 22.5.7), $G(M)$ centralizes $G^{0+}(M)$. By this plus Theorem 22.3.8, we deduce that (after adjoining $t^{1/k}$ for suitable k) every quotient of $G^{0+}(M)$ occurs as $G^{0+}(N)$ for some $N \in [M]$.

22.6 Proof of the theorem

We are finally ready to approach Theorem 22.1.4. Now that we have the finiteness of $G^{0+}(M)$ (Theorem 22.5.7), we may use it to power an induction argument.

Proof of Theorem 22.1.4 We proceed by induction on the order of the finite (by Theorem 22.5.7) p-group $G^{0+}(M)$. In the base case where this group is trivial, $b(M) = 0$ and there is nothing to check. (Note that we are free to replace K by a finite extension at any point.)

Assuming that $G^{0+}(M)$ is a nontrivial finite p-group, it admits a nontrivial irreducible representation via a quotient onto $\mathbb{Z}/p\mathbb{Z}$. By Lemma 22.3.4, this occurs within $(M \oplus M^\vee)^{\otimes m}$ for some positive integer m. By Remark 22.5.8, after adjoining $t^{1/k}$ for a suitable positive integer k, we can find a subobject N of $(M \oplus M^\vee)^{\otimes m}$ for which $G^{0+}(N)$ is a quotient of $G^{0+}(M)$ of order p. In particular, N is of cyclic type, so we may apply Theorem 22.2.4 to find $P \in C$ of rank 1 such that $b(P^\vee \otimes N) = 0$. Arguing as in Lemma 21.3.1, we may choose L so that $b(P \otimes_{\mathcal{R}} \mathcal{R}_L) = 0$, and so $b(N \otimes_{\mathcal{R}} \mathcal{R}_L) = 0$. Hence $G^{0+}(M \otimes_{\mathcal{R}} \mathcal{R}_L)$ is a proper subgroup of $G^{0+}(M)$, and so we may apply the induction hypothesis to conclude. $\qquad\square$

22.7 Relation to Frobenius structures

In the case of an absolute Frobenius lift, Theorem 20.1.4 can be recovered from Theorem 22.1.4 using Lemma 21.2.1. Note that again this involves using either p-adic exponents or slope filtrations, but these hypotheses are less deeply embedded into the argument.

In the general case, it is slightly more subtle to recover Theorem 20.1.4 from Theorem 22.1.4, as we need to ensure that the Frobenius lift on $\mathcal{R}$ extends to $\mathcal{R}_L$. For this, we return to the well of Chapter 19.

Lemma 22.7.1. *Suppose that κ_K is perfect. For L a finite separable extension of $\kappa_K((t))$, let N be the restriction of scalars of the trivial differential module from $\mathcal{R}_L$ to $\mathcal{R}$. Then there is a natural isomorphism of $G^{0+}(N)$ with the wild inertia subgroup of $G_{L/\kappa_K((t))}$.*

Proof This is included in Remark 22.3.3. $\qquad\square$

Lemma 22.7.2. *With notation as in Lemma 22.7.1, for $M \in C$, the map $G^{0+}(M \otimes_{\mathcal{R}} \mathcal{R}_L) \to G^{0+}(M)$ is injective and induces an isomorphism*

$$G^{0+}(M \otimes_{\mathcal{R}} \mathcal{R}_L) \times G^{0+}(N) \to G^{0+}(M \oplus N).$$

Proof The injectivity of $G^{0+}(M \otimes_{\mathcal{R}} \mathcal{R}_L) \to G^{0+}(M)$ is apparent from the explicit description of $G(M)$. To check that the map

$$G^{0+}(M \otimes_{\mathcal{R}} \mathcal{R}_L) \times G^{0+}(N) \to G^{0+}(M \oplus N)$$

is surjective, we argue by contradiction. Suppose the contrary; then by Remark 3.3.10 and Remark 22.5.8, after adjoining $t^{1/k}$ for a suitable positive integer k, we can find $P \in [M \oplus N]$ such that $G^{0+}(P)$ is cyclic of order p and the composition

$$G^{0+}(M \otimes_{\mathcal{R}} \mathcal{R}_L) \times G^{0+}(N) \to G^{0+}(M \oplus N) \to G^{0+}(P)$$

is the trivial map. We may even take P to be irreducible, in which case we must have $P \in [M]$ or $P \in [N]$ (or both). The latter cannot hold because $G^{0+}(N) \to G^{0+}(P)$ is surjective, so we must have $P \in [M]$ and $b(P \otimes_{\mathcal{R}} \mathcal{R}_L) = 0$. By Lemma 21.3.1 and Theorem 22.2.4, $P = P_1 \otimes P_2$ where P_1 is quasiconstant of rank 1 and $b(P_2) = 0$, and hence $b(P_1 \otimes_{\mathcal{R}} \mathcal{R}_L) = 0$; however, this implies $P_1 \in [N]$ and so again $G^{0+}(N) \to G^{0+}(P)$ is surjective, a contradiction.

To check that the map

$$G^{0+}(M \otimes_{\mathcal{R}} \mathcal{R}_L) \times G^{0+}(N) \to G^{0+}(M \oplus N)$$

is injective, we proceed by induction on the degree $[L : \kappa_K((t))]$. After adjoining $t^{1/k}$ for a suitable positive integer k, we can find a subextension $L_1/\kappa_K((t))$ of L of degree p (as per Remark 3.3.10 again). We then have a commutative diagram

$$
\begin{array}{ccc}
G^{0+}(M \otimes_{\mathcal{R}} \mathcal{R}_L) \times G^{0+}(N \otimes_{\mathcal{R}} \mathcal{R}_{L_1}) & \longrightarrow & G^{0+}((M \oplus N) \otimes_{\mathcal{R}} \mathcal{R}_{L_1}) \\
\downarrow & & \downarrow \\
G^{0+}(M \otimes_{\mathcal{R}} \mathcal{R}_L) \times G^{0+}(N) & \longrightarrow & G^{0+}(M \oplus N)
\end{array}
$$

in which the top horizontal arrow is an isomorphism (by the induction hypothesis), the bottom horizontal arrow is surjective, the left vertical arrow is injective with cokernel of order p, and the right vertical arrow is also injective. Consequently, the only way for the desired injectivity to fail is for the right vertical arrow to be an isomorphism. This would imply that every $P \in [M \oplus N]$ satisfies $G^{0+}(P \otimes_{\mathcal{R}} \mathcal{R}_L) \cong G^{0+}(P)$, but this fails for $P \in [N]$ by Lemma 22.7.1. □

Corollary 22.7.3. *For $M \in C$, there is a natural surjective homomorphism from the wild inertia subgroup of $G_{\kappa_K((t))}$ to $G^{0+}(M)$, which is compatible with base extension from $\kappa_K((t))$ to a finite separable extension.*

Proof This follows by combining Lemma 22.7.1 with Lemma 22.7.2. □

Corollary 22.7.4. *Suppose that κ_K is algebraically closed. Then for any $M \in C$, there exists a finite Galois extension L of $\kappa_K((t))$ such that for any other finite (not necessarily Galois) extension L' of $\kappa_K((t))$, $b(M \otimes_{\mathcal{R}} \mathcal{R}_{L'}) = 0$ if and only if L' contains a copy of L.*

Proof By Corollary 22.7.3, $G^{0+}(M)$ corresponds to a finite-index subgroup W of the wild inertia subgroup of $G_{\kappa_K((t))}$. By Proposition 3.3.6 and the hypothesis that κ_K is algebraically closed, there is a least positive integer m for which $\kappa_K((t^{1/m}))$ admits a finite separable, totally wildly ramified extension L for which the wild inertia subgroup of G_L equals W. This extension has the desired form. □

Returning to the general case of Theorem 20.1.4, by Remark 21.1.1 we may assume that κ_K is algebraically closed. By Corollary 22.7.4, we have a unique minimal finite separable extension L of $\kappa_K((t))$ for which $b(M \otimes_R \mathcal{R}_L) = 0$. In particular, this uniqueness implies that the Frobenius lift on $\mathcal{R}$ extends to $\mathcal{R}_L$ (because otherwise the pullback of L by the Frobenius on $\kappa_K((t))$ would give a different extension with the same characterizing property).

Notes

This chapter is drawn mostly from [255]. That article required extensive post-publication corrections in collaboration with Atsushi Shiho, as recorded in [269]; we have accounted for these here.

If K is spherically complete, then the same result of Lazard [289] that was cited in connection with Proposition 16.2.4 shows that any object of the category C does in fact correspond to a finite differential module over $\mathcal{R}$. The same paper also shows how to construct counterexamples (at the level of underlying modules) when K is not spherically complete.

The notion of a module of cyclic type is taken from [255, §3.2]. Lemma 22.2.2 is taken from [255, Lemma 3.2.6]. Theorem 22.2.3 is taken from [255, Theorem 3.7.13(b)]. Theorem 22.2.4 is taken from [255, Corollary 3.8.10].

Definition 22.3.2 is the usual definition of the automorphism group of a fiber functor of a Tannakian category, incorporating André's definition of a tensor-bounded filtration [16, 18]. The key properties that ensure that Tannakian formalism applies to $[M] \otimes_K F$ (and in particular implies Theorem 22.3.8) are the following.

- It is an F-linear abelian category (by Proposition 9.1.2) and for M the trivial module, $\text{Hom}(M, M) = F$.
- It is a *symmetric monoidal category* with respect to the tensor product, with identity object given by R viewed as a differential module.
- It is a *rigid category*: the dual functor $M \mapsto M^\vee$ exists and is an involution, and moreover for any two objects M and N, the object $M^\vee \otimes N$ has the property that for any third object P, the map

$$\text{Hom}(P \otimes M, N) \cong \text{Hom}(P, M^\vee \otimes N)$$

taking $f \colon P \otimes M \to N$ to the composition

$$P \cong P \otimes R \to P \otimes M^\vee \otimes M \cong M^\vee \otimes P \otimes M \overset{1_{M^\vee} \otimes f}{\to} M^\vee \otimes N$$

is a bijection.

- The ring $\mathcal{R}$ is an integral domain, so there is a homomorphism $\mathcal{R} \to F$ for $F = \operatorname{Frac} \mathcal{R}$. We thus obtain a *neutral* fiber functor $M \mapsto M \otimes_{\mathcal{R}} F$. By contrast, the category $[M]$ does not admit a fiber functor to K-vector spaces.

The "Tannaka reconstruction theorem" is a slight misnomer, as the actual result of Tannaka is the corresponding statement in the context of compact Lie groups [381]. In that context, this result was upgraded by Krein [281] by identifying which categories occur as the representation category of a compact Lie group; this is known as *Tannaka–Krein duality*. These results were later transported to the setting of algebraic groups as part of the development of the theory of Tannakian categories, for which standard references are [127, 130]; these clarify and correct the original presentation in [358].

Lemma 22.4.4 is taken from [255, Lemma 1.1.1]. Lemmas 22.4.5 and 22.4.6 together are taken from the corrected version of [255, Proposition 1.1.2] given in [269].

The proof of Theorem 22.5.7 is modeled on [255, Theorem 3.8.16], again as corrected in [269].

Exercises

22.1 Let M_1, M_2 be finite differential modules on an annulus such that $M_1^\vee \otimes M_2$ satisfies the Robba condition. Show that both M and N are of cyclic type.

22.2 Prove Lemma 22.3.4. (Hint: let χ be the character of G acting on V and suppose that χ' is the character of an irreducible representation not appearing in $V^{\otimes n}$ for any n. Apply the orthogonality of characters to χ^n and χ', use it to deduce that $\chi(g)\chi'(g) = 0$ for every $g \in G$, and then obtain a contradiction by taking $g = e$.)

22.3 For $M \in C$, Theorem 22.3.8 implies that if we form the minimal decomposition $\bigoplus_{i \in I} V_i$ of M_F into $G^{0+}(M)$-isotypical representations, then $M^\vee \otimes M$ splits (as a differential module) as a direct sum $N_1 \oplus N_2$ where $N_{1,F}$ is the direct sum of the trace components of $V_i^\vee \otimes V_i$ over $i \in I$. Give an alternate proof of this using the strategy of Remark 5.3.5.

22.4 Let V be a finite-dimensional vector space over a field F of characteristic 0. Let G be a subgroup of $\operatorname{GL}(V)$ containing a normal abelian torsion subgroup N for which G/N is finite, and fix a maximal such subgroup. Show that the action of G/N on V^N is faithful. (Hint: this is similar to the proof of Lemma 22.4.6.)

Part VII

Global theory

23

Banach rings and their spectra

In this final part of the book, we return to the study of convergence of solutions of differential equations from Part III, but this time taking a more global viewpoint.

In this chapter, we introduce the key concept of the *Gelfand spectrum* associated to a (commutative) Banach ring. A running theme will be the analogy with the prime (Zariski) spectrum associated to a commutative ring.

23.1 Banach rings

Definition 23.1.1. By a *(commutative) Banach ring*, we will mean a ring R equipped with a submultiplicative norm $|\cdot|$ with respect to which it is complete. We denote by $|\cdot|_{sp}$ the associated spectral seminorm (see Definition 6.1.3).

For R a Banach ring and I a closed ideal, R/I is again a Banach ring for the quotient norm.

Lemma 23.1.2. *Let R be a Banach ring. Then the closure of any nonunit ideal I of R is again a nonunit ideal.*

Proof Suppose to the contrary that the closure of I contains 1; then there must exist $r \in I$ with $|1 - r| < 1$. But then the geometric series $\sum_{n=0}^{\infty}(1 - r)^n$ converges to an inverse of $1 - (1 - r)$ in I, so I is already the unit ideal. $\quad\square$

Corollary 23.1.3. *Let R be a Banach ring. Then any maximal ideal of R is closed.*

Remark 23.1.4. By Corollary 23.1.3, the quotient of a Banach ring by a maximal ideal is a Banach ring which is also a field. However, such an object need not be a nonarchimedean field; see Remark 1.1.10.

23.2 The spectrum of a Banach ring

We next define the spectrum of a ring equipped with a submultiplicative norm.

Definition 23.2.1. Let R be a ring equipped with a submultiplicative norm $|\cdot|$. Define the *Gelfand spectrum* of R, denoted $\mathcal{M}(R)$, as the set of multiplicative seminorms on R bounded above by $|\cdot|$. Note that this does not change upon replacing R with its completion, which is a Banach ring.

Remark 23.2.2. Recall that the prime spectrum of a ring R admits both an internal definition, as the set of prime ideals, and also a characterization in terms of homomorphisms of rings. To wit, the prime spectrum is in natural bijection with the set of equivalence classes of homomorphisms $R \to K$ with K being any field, where two maps $R \to K_1, R \to K_2$ are considered to be equivalent if there exist field embeddings $K_1 \to K, K_2 \to K$ into a common target for which the diagram

commutes.

We may similarly characterize the Gelfand spectrum of a Banach ring R as the set of equivalence classes of bounded homomorphisms $R \to K$ with K being any nonarchimedean field.

Definition 23.2.3. Let R, S be two rings equipped with submultiplicative norm $|\cdot|_R, |\cdot|_S$. Let $\varphi \colon R \to S$ be a bounded homomorphism; then the restriction along φ of any multiplicative seminorm on S bounded by $|\cdot|_S$ is a multiplicative seminorm on R bounded by some fixed constant multiple of $|\cdot|_R$, and hence by $|\cdot|_R$ itself (exercise). We thus obtain a map $\varphi^* \colon \mathcal{M}(S) \to \mathcal{M}(R)$.

23.3 Topological properties

Definition 23.3.1. For R a Banach ring, we equip $\mathcal{M}(R)$ with the *evaluation topology*, i.e., the coarsest topology such that for each $r \in R$, the map $\mathcal{M}(R) \to \mathbb{R}$ taking x to $x(r)$ is continuous.

With notation as in Definition 23.2.3, the map φ^* is continuous for the evaluation topologies. That is because the evaluation map defined by $r \in R$ pulls back to the evaluation map defined by $\varphi(r) \in S$, so the evaluation topology on $\mathcal{M}(S)$ is at least as fine as the pullback topology.

Lemma 23.3.2. *Let R be a Banach ring. Then the topological space* $\mathcal{M}(R)$ *is Hausdorff and compact.*

Proof The topology on $\mathcal{M}(R)$ is the subspace topology induced via the inclusion

$$\mathcal{M}(R) \to \prod_{r \in R} [0, |r|], \qquad x \mapsto (x(r))_{r \in R}$$

for the product topology on the target. Since the target is a product of compact topological spaces, it is itself compact by Tikhonov's theorem [318, Chapter 5]; since $\mathcal{M}(R)$ is a closed subspace of the product, it is also compact. □

Lemma 23.3.3. *Let R be a nonzero Banach ring. Let S be the set of nonzero bounded submultiplicative seminorms on R.*

(a) The set S contains at least one minimal element.

(b) Any minimal element of S is multiplicative.

Proof By Corollary 23.1.3, any maximal ideal of R is closed, so we may assume at once that R is a field. To prove (a), note that S is nonempty because the original norm on R belongs to S, then apply Zorn's lemma.

To prove (b), we may as well assume (to simplify notation) that the original norm on R is itself minimal. We first show that $|r||r^{-1}|_{sp} \leq 1$ for every $r \in R^{\times}$. Suppose by way of contradiction that this fails for some r, then choose $\rho \in (|r^{-1}|_{sp}^{-1}, |r|)$. Let R_1 be the completion of $R[t]$ for the ρ-Gauss norm, viewed as a subring of $R[\![t]\!]$. In the larger ring we have $(t - r)^{-1} = -\sum_{n=0}^{\infty} r^{-n-1} t^n$, which does not belong to R_1 by our choice of ρ; consequently, $t - r$ is not a unit in R_1. By Lemma 23.1.2, the quotient of R_1 by the closure of the ideal $(t - r)$ is a nonzero Banach ring; if we restrict the quotient norm back to R, the norm of r drops from $|r|$ to ρ. This contradicts our minimality assumption on the norm on R, yielding the claim.

We now know that $|r||r^{-1}|_{sp} \leq 1$ for every $r \in R^{\times}$. Since $|r^n||r^{-n}| \geq 1$ for all positive integers n, using Fekete's lemma (Lemma 6.1.4) we may take limits to obtain $|r|_{sp}|r^{-1}|_{sp} \geq 1$, and so $|r|_{sp} \geq |r|$. As the reverse inequality holds for trivial reasons, we must have $|r| = |r|_{sp}$. Since the same holds for r^{-1}, we also have $|r||r^{-1}| \leq 1$ and hence $|r^{-1}| = |r|$. Finally, for $r, s \in R$,

$$|s| \leq |rs||r^{-1}| = |rs||r|^{-1},$$

so $|r||s| \leq |rs|$ and hence $|\cdot|$ is multiplicative. □

In analogy with the fact that the prime spectrum of a nonzero ring is nonempty, we have the following result.

Theorem 23.3.4. *For any nonzero Banach ring R, $\mathcal{M}(R)$ is nonempty.*

Proof This is immediate from Lemma 23.3.3.   □

Corollary 23.3.5. *Let R be a Banach ring. Then an ideal I of R is the unit ideal if and only if for every $x \in M(R)$, there exists $r \in I$ with $x(r) \neq 0$. In particular, an element $r \in R$ is a unit if and only if $x(r) \neq 0$ for all $x \in M(R)$.*

Proof By Lemma 23.1.2, we may reduce to the case where I is closed, in which case we may apply Theorem 23.3.4 to R/I to conclude.   □

We may promote Theorem 23.3.4 to a quantitative statement as follows.

Theorem 23.3.6. *Let R be a Banach ring. Then for $r \in R$,*

$$|r|_{\mathrm{sp}} = \sup\{x(r) : x \in M(R)\},$$

interpreting the supremum to be 0 if $M(R) = \emptyset$. (Note that by Lemma 23.3.2, the supremum is actually a maximum.)

Proof Since any $x \in M(R)$ is multiplicative, $x(r) \leq |r|_{\mathrm{sp}}$ for all $x \in M(R)$, so $\sup\{x(r) : x \in M(R)\} \leq |r|_{\mathrm{sp}}$. To prove the reverse implication, it suffices to show that any $\rho > \sup\{x(r) : x \in M(R)\}$ is greater than or equal to $|r|_{\mathrm{sp}}$. For any $x \in R\langle t/\rho^{-1}\rangle$, $x(rt) < 1$ and so $x(1 - rt) = 1$. By Corollary 23.3.5, $1 - rt$ is a unit in $R\langle t/\rho^{-1}\rangle$; by computing in the larger ring $R[\![t]\!]$, we see that $\sum_{n=0}^{\infty} r^n t^n \in R\langle t/\rho^{-1}\rangle$, implying that $|r|_{\mathrm{sp}} \leq \rho$.   □

When studying Gelfand spectra, it will be useful to keep in mind some basic facts from point-set topology. (See [318, §26].)

Remark 23.3.7. Let $f : X \to Y$ be a continuous map between compact Hausdorff topological spaces.

(a) The map f is closed (that is, the image of any closed subset is closed).
(b) If f is a bijection, then it is a homeomorphism.
(c) More generally, if f is surjective, then f is a quotient map; that is, a map $g : Y \to Z$ is continuous if and only if $g \circ f : X \to Z$ is continuous. (By (a), f is always a quotient map onto its image.)

23.4 Complete residue fields

Continuing the analogy with the prime spectrum of a ring, we introduce an analogue of the residue field of a point.

Definition 23.4.1. Let R be a Banach ring and choose $x \in M(R)$. Let I be the kernel of x; then I is a prime ideal and x induces a multiplicative norm on

R/I. Let $\mathcal{H}(x)$ be the completion of the fraction field of R/I with respect to the multiplicative extension of this norm. We call $\mathcal{H}(x)$ the *complete residue field* of x; it comes equipped with a distinguished homomorphism $R \to \mathcal{H}(x)$.

Remark 23.4.2. Let R be a Banach ring over a nonarchimedean field K, choose $x \in \mathcal{M}(R)$, and let R' be the completion of $R \otimes_K \mathcal{H}(x)$ for the product seminorm (Definition 1.3.10). Then R' admits a bounded homomorphism to $\mathcal{H}(x)$ which restricts to the distinguished homomorphism $R \to \mathcal{H}(x)$ on the first factor and to the identity map on the second factor; this corresponds (via Remark 23.2.2) to a point $x' \in \mathcal{M}(R')$. We will make systematic use of this construction in subsequent chapters.

Notes

The definition of the spectrum of a Banach ring is due to Guennebaud [187] in the case of a Banach algebra over a nonarchimedean field. The terminology *Gelfand spectrum* is due to Berkovich [52]; it is motivated by the analogy with the spectrum of a commutative C^*-algebra [21, §1.1]. Our development generally follows [52, Chapter 1].

Before the spectrum was introduced, many arguments in nonarchimedean analysis had to be made directly in terms of subsets of K, leading to some complications. For example, for D a bounded closed subset of K, Krasner defined the Banach ring of analytic elements on D [280], and Escassut showed that this ring is connected if and only if D is *infraconnected*, meaning that for each $z \in D$ the image of the map $D \to \mathbb{R}$ taking x to $|x - z|$ has connected closure. In the context of Gelfand spectra, one can instead formulate a simpler general result analogous to a familiar fact about prime spectra of rings: for any Banach ring R, R is connected if and only if $\mathcal{M}(R)$ is connected [52, Corollary 7.4.2].

In general, the structure of the Gelfand spectrum is somewhat complicated. See the notes for Chapter 24 for further discussion.

Exercises

23.1 Let R be a ring equipped with a submultiplicative norm $|\cdot|$. Let x be a multiplicative seminorm on R bounded by $c\,|\cdot|$ for some $c > 0$. Prove that $x \in \mathcal{M}(R)$.

23.2 Let K be a nonarchimedean field and let L be a finite extension of K. Use

Theorem 23.3.4 to give an alternate proof of the existence of a nonzero multiplicative extension of the norm on K to L (Theorem 1.4.9). (Hint: construct a submultiplicative norm on L compatible with the norm on K, by expressing L in terms of a basis over K.)

23.3 Let R be a Banach ring.

(a) Suppose that R is a nonarchimedean field. Prove that $\mathcal{M}(R)$ consists of a single point.

(b) Suppose that $\mathcal{M}(R)$ consists of a single point. Prove that R is a local ring whose residue field is a nonarchimedean field. (For example, it could have the form $K[t]/(t^2)$ where K is a nonarchimedean field.)

24

The Berkovich projective line

In this chapter, we make the definition of the Gelfand spectrum explicit in the case of the rings $K\langle t/\rho \rangle$ and $K\langle \alpha/t, t/\beta \rangle$. This will lead us to the construction of the projective line over K in the Berkovich approach to nonarchimedean algebraic geometry.

To avoid some technical complications, we assume for the remainder of Part VII that K is an *algebraically closed* nonarchimedean field. For further discussion of the general case, see the chapter notes.

24.1 Points

We first define the affine and projective lines as topological spaces.

Definition 24.1.1. Let $\mathbb{A}_K$ be the set of multiplicative seminorms on $K[t]$, with no boundedness conditions. We call $\mathbb{A}_K$ the *Berkovich affine line* over K. We write $x \leq y$ if $x(P) \leq y(P)$ for all $P \in K[t]$, and say that x is *dominated* by y.

Let $\mathbb{P}_K$ be the union of $\mathbb{A}_K$ with a single element ∞, identified with the multiplicative seminorm on $K[t^{-1}]$ given by evaluation at $t^{-1} = 0$. We extend the partial order on $\mathbb{A}_K$ to $\mathbb{P}_K$ by specifying that $x < \infty$ for all $x \in \mathbb{A}_K$.

We equip $\mathbb{P}_K$ with the coarsest topology such that for each $r \in K(t) \cup \{\infty\}$, the evaluation map $x \mapsto x(r)$ is a continuous map from x to $[0, \infty]$. Note that $\mathbb{P}_K$ is compact for this topology; this will follow from identifying it as the union of two copies of $\mathcal{M}(K\langle t \rangle)$ (see Corollary 24.1.4) and applying Lemma 23.3.2, but can also be seen directly using Tikhonov's theorem.

We next identify some "accessible" points in $\mathbb{A}_K$.

Definition 24.1.2. For $z \in K$ and $\rho > 0$, let $x_{z,\rho} \in \mathbb{A}_K$ denote the ρ-Gauss norm on $K[t - z]$. We extend this to the case $\rho = 0$ by interpreting the 0-Gauss norm as the norm of the constant coefficient (or equivalently, of the evaluation at $t = z$). Let $\mathbb{A}_K^\circ$ be the set of points of $\mathbb{A}_K$ of the form $x_{z,\rho}$ for some z, ρ.

Lemma 24.1.3. *For $x \in \mathbb{A}_K$ and $z \in K$, $x \leq x_{z,\rho}$ for $\rho = x(t - z)$.*

Proof For every $P \in K[t - z]$, by writing P as a sum of monomials and applying the triangle inequality, we see that $x(P) \leq x_{z,\rho}(P)$. □

Corollary 24.1.4. *The set $\mathbb{A}_K$ is the union of its subsets $\mathcal{M}(K\langle t/\beta\rangle)$ over all $\beta > 0$.*

Remark 24.1.5. Note that the inclusion $\mathcal{M}(K\langle t/\beta\rangle) \to \mathbb{A}_K$ is not just an injective map of sets, but also a homeomorphism of $\mathcal{M}(K\langle t/\beta\rangle)$ with its image (for the subspace topology). Similarly, for $0 < \alpha \leq \beta$, the map $\mathcal{M}(K\langle \alpha/t, t/\beta\rangle) \to \mathbb{A}_K$ is an injective map which defines a homeomorphism onto its image.

Lemma 24.1.6. *For $0 < \alpha < \beta$, the following statements hold.*

(a) *The restriction map $\mathcal{M}(K[\![t/\beta]\!]_{\mathrm{an}}) \to \mathcal{M}(K\langle t/\beta\rangle)$ is injective and homeomorphic onto its image. The image consists of all points x for which $x(t) < \beta$, together with the single point $x_{0,\beta}$.*
(b) *The restriction map $\mathcal{M}(K\langle \alpha/t, t/\beta\rangle_{\mathrm{an}}) \to \mathcal{M}(K\langle \alpha/t, t/\beta\rangle)$ is injective and homeomorphic onto its image. The image consists of all points x for which $\alpha \leq x(t) < \beta$, together with the single point $x_{0,\beta}$.*
(c) *The restriction map $\mathcal{M}(K[\![\alpha/t, t/\beta]\!]_{\mathrm{an}}) \to \mathcal{M}(K\langle \alpha/t, t/\beta\rangle)$ is injective and homeomorphic onto its image. The image consists of all points x for which $\alpha < x(t) < \beta$, together with the two points $x_{0,\alpha}, x_{0,\beta}$.*

In particular, $\mathcal{M}(K[\![t/\beta]\!]_{\mathrm{an}})$, $\mathcal{M}(K\langle \alpha/t, t/\beta\rangle_{\mathrm{an}})$, and $\mathcal{M}(K[\![\alpha/t, t/\beta]\!]_{\mathrm{an}})$ are subspaces of $\mathbb{A}_K$.

Proof We treat (a) in detail, as (b) and (c) are similar. By definition, $K(t)$ is dense in $K[\![t/\beta]\!]_{\mathrm{an}}$, so any bounded multiplicative seminorm on $K[\![t/\beta]\!]_{\mathrm{an}}$ is uniquely determined by its restriction to $K(t)$. This implies that the map $\mathcal{M}(K[\![t/\beta]\!]_{\mathrm{an}}) \to \mathcal{M}(K\langle t/\beta\rangle)$ is injective. The image is a closed subspace of $\mathcal{M}(K\langle t/\beta\rangle)$: it consists of those x for which $x(t - z) = 1$ for all $z \in K$ with $|z| = 1$. By Lemma 23.3.2 and Remark 23.3.7, the restriction map is a homeomorphism onto its image. □

We next recall the geometric interpretation of Gauss norms.

Lemma 24.1.7. *For $z \in K$ and $\rho \geq 0$, $x_{z,\rho}$ is the supremum of $x_{z',0}$ over all $z' \in K$ with $|z - z'| \leq \rho$.*

Proof We again assume $z = 0$. By Lemma 24.1.3, we need only check that the supremum of the $x_{z',0}$ is greater than or equal to $x_{z,\rho}$. Choose a nonzero element $r = \sum_{n=0}^{\infty} r_n t^n \in K[t]$. Let n_0 be the smallest index n for which $|r_n|\rho^n$ achieves its maximal value. Since K is algebraically closed, $|K^{\times}|$ is divisible; hence for any $\epsilon > 0$, we can find $z \in K$ with $|z| \in (\rho - \epsilon, \rho)$. For ϵ sufficiently small, n_0 is the unique index for which $|r_n||z|^n$ achieves its maximal value, so $x_{z,0}(r) = |r_n||z|^n$. This proves the claim. □

Remark 24.1.8. Lemma 24.1.7 can be interpreted as saying that $x_{z,\rho}$ is the supremum norm over the closed disc $|t - z| \leq \rho$. In particular, if $z' \in K$ satisfies $|z - z'| = \rho$, then $x_{z,\rho} = x_{z',\rho}$. The converse is also true; see Lemma 24.3.1.

24.2 Classification of points

We next introduce Berkovich's classification of points of $\mathbb{A}_K$.

Definition 24.2.1. We sort the points of $\mathbb{A}_K$ into *types* as follows.

1. The points $x_{z,0}$ for $z \in K$. Such a point has complete residue field isomorphic to K.
2. The points $x_{z,\rho}$ for $z \in K$ and $\rho \in |K^{\times}|$. Such a point has complete residue field isomorphic to F_ρ, but with the variable $t - z$ in place of t.
3. The points $x_{z,\rho}$ for $z \in K$ and $\rho \in (0, \infty) \setminus |K^{\times}|$. Again, such a point has complete residue field isomorphic to F_ρ, but with the variable $t - z$ in place of t.
4. None of the above (i.e., the points of $\mathbb{A}_K \setminus \mathbb{A}_K^\circ$). See Remark 24.2.4 and Remark 24.3.6.

We extend the classification to $\mathbb{P}_K$ by declaring that ∞ has type 1.

Lemma 24.2.2. *For $x \in \mathbb{A}_K$ and $z \in K$, put $\rho = x(t - z)$. If $\rho \in (0, \infty) \setminus |K^{\times}|$, then $x = x_{z,\rho}$.*

Proof This follows from the fact that in this case $K\langle \rho/(t - z), (t - z)/\rho \rangle$ is a nonarchimedean field (by Remark 8.2.7). □

Corollary 24.2.3. *For $x \in \mathbb{A}_K$ not of type 3, $|\mathcal{H}(x)^{\times}| = |K^{\times}|$.*

Proof Since $\mathcal{H}(x)$ is obtained by taking a quotient of $K[t]$ and then completing, it suffices to check that the norm of any element of $K[t]$ (if nonzero) is in $|K^{\times}|$. This further reduces to checking elements of the form $t - z$; if any such element has norm not in $\{0\} \cup |X^{\times}|$, then x must be of type 3 by Lemma 24.2.2. □

Remark 24.2.4. For K' an algebraically closed nonarchimedean field containing K, there is a natural restriction map $\mathbb{A}_{K'} \to \mathbb{A}_K$. By Remark 23.4.2 and Corollary 24.1.4, every point of $\mathbb{A}_K$ can be written as the restriction of a point of $\mathbb{A}_{K'}$ of type 1 for some K' (depending on the original point). See Remark 24.3.6 for a more precise statement.

We may extend this map to a continuous map $\mathbb{P}_{K'} \to \mathbb{P}_K$. By Remark 23.3.7, this is a quotient map.

Lemma 24.2.5. *Let K' be an algebraically closed nonarchimedean field containing K. For $z \in K$ and $\rho \geq 0$, the point $x_{z,\rho} \in \mathbb{A}_{K'}$ is the supremum of all points whose restrictions to $\mathbb{A}_K$ are dominated by $x_{z,\rho}$.*

Proof If $y \in \mathbb{A}_{K'}$ has restriction to $\mathbb{A}_K$ dominated by $x_{z,\rho}$, then $y(t-z) \leq \rho$. We may thus deduce the claim from Lemma 24.1.3. $\square$

24.3 The domination relation

We next clarify the nature of the domination relation.

Lemma 24.3.1. *For $z, z' \in K$ and $\rho, \rho' \geq 0$, $x_{z,\rho} \leq x_{z',\rho'}$ if and only if $\rho \leq \rho'$ and $|z - z'| \leq \rho'$. In particular, this implies $x_{z,\rho'} = x_{z',\rho'}$.*

Proof If $x_{z,\rho} \leq x_{z',\rho'}$, then

$$\rho' = x_{z',\rho'}(t - z') \leq x_{z,\rho}((t - z) + (z - z')) = \max\{\rho, |z - z'|\}.$$

Conversely, if $\rho \leq \rho'$ and $|z - z'| \leq \rho'$, then $x_{z,\rho} \leq x_{z,\rho'} = x_{z',\rho'}$. $\square$

Lemma 24.3.2. *For $x \in \mathbb{A}_K$ and $y_1, y_2 \in \mathbb{A}_K^\circ$ with $x \leq y_1, x \leq y_2$, either $y_1 \leq y_2$ or $y_1 \leq y_2$.*

Proof By hypothesis, we can write $y_1 = x_{z_1,\rho_1}, y_2 = x_{z_2,\rho_2}$ for some $z, z_1, z_2 \in K$ and some $\rho_1, \rho_2 \geq 0$. By Remark 24.2.4, we may choose an algebraically closed nonarchimedean field K' containing K such that x is the restriction of $x_{z,0} \in \mathbb{A}_{K'}$ for some $z \in K'$; by Lemma 24.2.5, $x_{z,0} \leq x_{z_1,\rho_1}, x_{z,0} \leq x_{z_2,\rho_2}$.

Assume without loss of generality that $\rho_1 \leq \rho_2$. By Lemma 24.3.1, $|z_1 - z| \leq \rho_1$ and $|z_2 - z| \leq \rho_2$; by the triangle inequality, $|z_1 - z_2| \leq \rho_2$ and so

$$y_1 = x_{z_1,\rho_1} \leq x_{z_1,\rho_2} = x_{z_2,\rho_2} = y_2$$

as claimed. $\square$

This allows us to construct an analogue of the interpretation of real numbers as Dedekind cuts (the set-theoretic elucidation of Eudoxus's theory of proportionality).

Lemma 24.3.3. *For $x \in \mathbb{A}_K$, let S_x be the set of elements of $\mathbb{A}_K^\circ$ which dominate x. Then $x \mapsto S_x$ gives a bijection from $\mathbb{A}_K$ to the set of subsets of $\mathbb{A}_K^\circ$ which are totally ordered, up-closed, and contain their infima (when these infima exist in $\mathbb{A}_K^\circ$).*

Proof For $x \in \mathbb{A}_K$, the set S_x is up-closed and totally ordered by Lemma 24.3.2. By Lemma 24.1.3, x is the infimum of S_x, so S_x also contains its infimum when it exists in $\mathbb{A}_K^\circ$. We thus have a well-defined map. Since we can recover x from S_x, the map is injective.

To see that the map is surjective, suppose S is such a subset; we claim that $S = S_x$ for $x = \inf S$. It is obvious that $S \subseteq S_x$, so the issue is to prove that if $y \in S_x$, then $y \in S$. For any $y' \in S$, we may apply Lemma 24.3.2 to deduce that either $y \leq y'$ or $y' \leq y$. If $y' \leq y$ for some $y' \in S$, then $y \in S$ because S is up-closed. Otherwise, $y \leq y'$ for all $y' \in S$, so $y \leq \inf S = x$ and so $y = x$. Since S contains its infimum, $y \in S$. $\qquad\square$

Corollary 24.3.4. *The points of $\mathbb{A}_K$ of type 4 that are minimal under domination are precisely those of types 1 and 4.*

Proof It is clear that points of type 1 are minimal, whereas points of types 2 and 3 are not. The issue is thus to prove that points of type 4 are minimal.

Suppose by way of contradiction that $x < y$ with y of type 4. Define the sets S_x and S_y as in Lemma 24.3.3; then $S_y \subseteq S_x$ with equality if and only if $y = x$. Since $x \neq y$, we can find a point $y' \in S_x \setminus S_y$. By Lemma 24.3.2, y' is dominated by every element of S_y, and hence by y itself. Write $y' = x_{z,\rho}$ for some $z \in K$ and some $\rho > 0$, and put $\rho' = y(t - z)$.

By Lemma 24.1.3, $y \leq x_{z,\rho'}$. On the other hand, for $P \in K[t]$, we have $y(P) \geq y'(P) \geq |P(z)|$. For $P = \sum_{i=0}^{\infty} P_i(t - z)^i$, we prove that $y(P) = x_{z,\rho'}(P)$ by induction on $\deg(P)$, with the vacuous base case $P = 0$. Put $Q = \sum_{i=0}^{\infty} P_{i+1}(t - z)^i$. If $|P_0| = x_{z,\rho'}(P)$, then $y(P) \geq |P_0| = x_{z,\rho'}(P)$. Otherwise, $P = P_0 + (t - z)Q$ and (by the induction hypothesis)

$$y(P - P_0) = \rho' y(Q) = \rho' x_{z,\rho'}(Q) = x_{z,\rho'}(P - P_0) > |P_0|,$$

so $x_{z,\rho'}(P) = y(P)$. By continuity, $y = x_{z,\rho'} \in \mathbb{A}_K^\circ$, a contradiction. $\qquad\square$

This allows us to formally upgrade Lemma 24.3.2.

Corollary 24.3.5. *For $x \in \mathbb{A}_K$ and $y_1, y_2 \in \mathbb{A}_K$ with $x \leq y_1, x \leq y_2$, either $y_1 \leq y_2$ or $y_1 \leq y_2$.*

Proof If y_1 or y_2 is of type 4, then Corollary 24.3.4 implies that it is equal to x and there is nothing to check. Otherwise, $y_1, y_2 \in \mathbb{A}_K^\circ$ and so Lemma 24.3.2 applies. $\qquad\square$

We can now clarify the nature of the points of type 4 by relating them to the spherical completeness of K.

Remark 24.3.6. We may reformulate Lemma 24.3.3 as follows: every element of $\mathbb{A}_K$ of type 4 is the intersection of a maximal totally ordered sequence of closed discs in K with positive limiting radius and empty intersection. (The limiting radius must be positive because otherwise the completeness of K guarantees that the intersection is nonempty.) Consequently, $\mathbb{A}_K = \mathbb{A}_K^\circ$ if and only if K is spherically complete.

24.4 The tree structure

We now obtain a description of $\mathbb{P}_K$ as an "infinitely branched tree".

Definition 24.4.1. For $x \in \mathbb{A}_K^\circ$, Lemma 24.3.1 implies that there is a unique value of ρ for which $x = x_{z,\rho}$ for some $z \in K$. We write this value as $\rho(x)$ and call it the *radius* of the point x.

For $x \in \mathbb{A}_K \setminus \mathbb{A}_K^\circ$, we define $\rho(x)$ as the infimum of $\rho(y)$ over all $y \in \mathbb{A}_K^\circ$ dominating x. We thus obtain a function $\rho: \mathbb{A}_K \to [0, \infty)$. We extend this to a function $\rho: \mathbb{P}_K \to [0, \infty]$ by setting $\rho(\infty) = \infty$. This function is not continuous; see the exercises.

Lemma 24.4.2. *For $x \in \mathbb{A}_K$, for each $\rho \in [\rho(x), \infty)$ there exists a unique $y \in \mathbb{A}_K$ with $x \leq y$ and $\rho(y) = \rho$.*

Proof By Corollary 24.3.4, the set of $y \in \mathbb{A}_K$ dominating x consists of the set S_x of Lemma 24.3.3, plus x itself if it is of type 4. On S_x, the function ρ is strictly increasing by Lemma 24.3.1 If x is of type 4, then $\{\rho(y): y \in S_x\}$ does not contain its infimum, which equals $\rho(x)$. This proves the claim. $\square$

Definition 24.4.3. For $x \in \mathbb{P}_K$ and $\rho \in [0, \infty]$, define $H(x, \rho) \in \mathbb{P}_K$ to be x if $\rho(x) \geq \rho$, or else the unique $y \in \mathbb{P}_K$ dominating x with $\rho(y) = \rho$ (given by Lemma 24.4.2). This construction commutes with base extension in the following sense: with notation as in Remark 24.2.4, if $x' \in \mathbb{A}_{K'}$ restricts to $x \in \mathbb{A}_K$, then the restriction of $H(x', \rho)$ to $\mathbb{A}_K$ equals $H(x, \rho)$.

Theorem 24.4.4. *The map $H: \mathbb{P}_K \times [0, \infty] \to \mathbb{P}_K$ is continuous.*

Proof By Remark 24.2.4, we may check the claim after enlarging K. We may thus apply Theorem 1.5.3 to replace K with a spherically complete (and still algebraically closed) field, which by Remark 24.3.6 means that $\mathbb{A}_K$ contains no points of type 4.

By definition, a neighborhood basis for the topology of $\mathbb{P}_K$ is given by open subsets of the form

$$U = \{x \in \mathbb{P}_K : x(P_1) \in I_1, \ldots, x(P_n) \in I_n\}$$

where $P_1, \ldots, P_n \in K[t]$ are some polynomials and $I_1, \ldots, I_n$ are some open subintervals of $[0, \infty]$ (which could have left endpoint 0 or right endpoint ∞). To simplify the exposition, we consider only neighborhoods in $\mathbb{A}_K$, leaving the remainder of the argument as an exercise. For this, it suffices to check that for any such U as above with $\infty \notin I_1, \ldots, I_n$ and any point $(x, \rho) \in H^{-1}(U)$, there exist a neighborhood V of x in $\mathbb{P}_K$ and an open subinterval of $[0, \infty)$ containing ρ (which could have left endpoint 0) contained in $H^{-1}(U)$. In fact, it suffices to do this for $n = 1$, as we can then intersect the resulting neighborhoods and intervals; we relabel P_1, I_1 as P, I.

Since we assumed $\mathbb{A}_K$ contains no points of type 4, $x = x_{z,\rho_0}$ for some $z \in K$ and some $\rho_0 \in [0, \infty)$. We may assume P is monic; it then factors as a product of linear polynomials $(t-z_1) \cdots (t-z_m)$. By hypothesis, $\prod_{i=1}^{m} \max\{\rho, |z-z_i|\} \in I$. We may rewrite this condition as $\prod_{i=1}^{m} \max\{\rho, x(t - z_i)\} \in I$, in which form it evidently defines an open subset of $\mathbb{P}_K \times [0, \infty]$. □

Corollary 24.4.5. *The map H is a strong deformation retract of $\mathbb{P}_K$ onto the singleton set $\{\infty\}$. Consequently, the space $\mathbb{P}_K$ is contractible (and in particular connected and path-connected).*

Remark 24.4.6. For two elements x_1, x_2 in a partially ordered set, a *join* of x_1 and x_2 is an element y which is minimal for the property that $x_1 \leq y$ and $x_2 \leq y$. Such an element is of course unique if it exists; it is commonly denoted $x_1 \wedge x_2$.

In $\mathbb{P}_K$, the join of $x_1, x_2 \in \mathbb{P}_K$ always exists. Namely, there is a unique minimal value $\rho \in [0, \infty)$ such that $H(x_1, \rho) = H(x_2, \rho)$, and this common value is the join.

24.5 Skeleta

Definition 24.5.1. For each $x \in \mathbb{P}_K$, the restriction of H to $\{x\} \times [0, \infty]$ maps to a subset of $\mathbb{P}_K$ homeomorphic to a (possibly trivial) closed interval. We call this subset the *root path* from x. Using these paths, we may form a picture of $\mathbb{P}_K$ as an "infinite tree" rooted at ∞, with every point being joined to the root by a unique path; see Figure 24.1 for a few of the branches drawn in. (Note that consistent with the custom in both mathematics and computer science, we draw this "tree" with the root at the *top* of the diagram.)

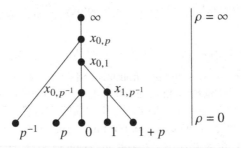

Figure 24.1 A (partial) illustration of $\mathbb{P}_K$ for $K = \mathbb{C}_p$

We can promote the level of mathematical rigor in this picture as follows.

Definition 24.5.2. Define a *skeleton* (resp. a *strict skeleton*) of $\mathbb{P}_K$ to be the union of a nonempty finite set of root paths from points of type 1 or 4 (resp. of type 1 only).

Lemma 24.5.3. *Let S be a skeleton of $\mathbb{P}_K$.*

(a) *For each $x \in \mathbb{P}_K$, there is a least value of $\rho \in [0, \infty]$ for which $H(x, \rho) \in S$.*
(b) *Define the map $H_S \colon \mathbb{P}_K \to S$ taking x to $H(x, \rho)$ for ρ as in (a). Then H_S is continuous, and thus gives a retraction of $\mathbb{P}_K$ onto S. (This can be upgraded to a strong deformation retraction; see exercises.)*

Proof Part (a) is immediate from the continuity of H (Theorem 24.4.4). To prove (b), note that the map from the disjoint union of the root paths making up S to S itself is a quotient map (again by Remark 23.3.7); it thus suffices to check the claim when S consists of a single such path. By applying Remark 24.2.4, we may further reduce to the case where S is the root path from a single type-1 point, which without loss of generality we take to be 0. Then the map H_S is simply given by the evaluation map $x \mapsto x(t)$, which by definition is continuous. □

Theorem 24.5.4. *The space $\mathbb{P}_K$ is the inverse limit of its strict skeleta for the inverse system associating to an inclusion $S \subseteq S'$ the restriction of the map $S' \to S$ induced by H_S.*

Proof Each strict skeleton is a finite union of closed intervals, and hence is compact. By Tikhonov's theorem, the inverse limit X is also compact (being a closed subspace of the product over all strict skeleta).

The family of maps $H_S \colon \mathbb{P}_K \to S$, being compatible with the inverse system, defines a continuous map $\mathbb{P}_K \to X$. If $x, y \in \mathbb{P}_K$ have the same image in X, then for each $z \in K$ we have $x(t - z) = y(t - z)$ (as these values are determined by H_S where S is the root path from z); since K is algebraically closed, this implies that x and y agree on $K[t]$ and hence must be identical. Hence the map $\mathbb{P}_K \to X$ is injective.

Since both $\mathbb{P}_K$ and X are compact, the image of $\mathbb{P}_K$ in X is closed (Remark 23.3.7). We may see that it is also dense as follows. One may form a neighborhood basis of X by taking an arbitrary strict skeleton S and an arbitrary open subset U of S, then taking the inverse limit in X. For any such S and U, we can choose a type-2 point $x_{z,\rho} \in U$, and then $x_{z,\rho} \in \mathbb{P}_K$ maps to the inverse image of U in X.

We now have that $\mathbb{P}_K \to X$ is a continuous bijection; by Remark 23.3.7 again, it is a homeomorphism. □

Definition 24.5.5. We define the *branches* of $\mathbb{P}_K$ at x to be the connected components of $\mathbb{P}_K \setminus \{x\}$ for the subspace topology. We can make these explicit for the various types of points as follows.

1. Since x is minimal under domination, $\mathbb{P}_K \setminus \{x\}$ is connected, so we get a single branch. (Similarly, if $x = \infty$, then $\mathbb{P}_K \setminus \{x\} = \mathbb{A}_K$ is connected.)
2. Let S be the root path from x. The branch containing ∞ equals $H_S^{-1}(S \setminus \{x\})$. The other branches can be identified with the affine line over the residue field of K.
3. Let S be the root path from x. Again, the branch containing ∞ equals $H_S^{-1}(S \setminus \{x\})$. However, in this case there is only one other branch.
4. Since x is minimal under domination, $\mathbb{P}_K \setminus \{x\}$ is connected, so we get a single branch.

Definition 24.5.6. We define a *skeleton* (resp. a *strict skeleton*) in $\mathcal{M}(K\langle t/\beta \rangle)$ or $\mathcal{M}(K\langle \alpha/t, t/\beta \rangle)$ as the intersection of said space with a skeleton (resp. a strict skeleton) of $\mathbb{P}_K$. By Theorem 24.5.4, each of $\mathcal{M}(K\langle t/\beta \rangle)$ and $\mathcal{M}(K\langle \alpha/t, t/\beta \rangle)$ is again the inverse limit of its strict skeleta.

Remark 24.5.7. As discussed in the notes for Chapter 0, there is a close relationship between Berkovich's theory of nonarchimedean analytic spaces and Huber's theory of adic spaces. At the level of the affine and projective lines, this manifests in the classification of points: the analogous spaces in Huber's framework contain a fifth type of point, which corresponds to the choice of a branch at a Berkovich point of type 2.

In Huber's theory, the value group $|K^\times|$ plays a privileged role, with the effect that there are no points associated to the branches at a Berkovich point of type 3. This changes if one replaces Huber's construction with the related construction of *reified adic spaces* [254].

24.6 Harmonic and subharmonic functions

Using skeleta, we can define harmonic and subharmonic functions on $\mathbb{P}_K$.

Definition 24.6.1. A function $f\colon \mathbb{P}_K \to [-\infty, \infty]$ is *piecewise affine* (resp. *strictly piecewise affine*) if it satisfies the following conditions.

(a) It factors through the retraction onto some skeleton (resp. strict skeleton) S.

(b) For each $x \in S$, the function $r \mapsto f(H(x, e^{-r}))\colon \mathbb{R} \to [-\infty, \infty]$ takes values in $\mathbb{R}$ and is piecewise affine.

For such a function, we may define its *slope* at any branch at any point of type 2. By convention, we measure slopes away from the original point.

Definition 24.6.2. For f a piecewise affine function, define the *Laplacian* of f, denoted Δf, as the function on $\mathbb{A}_K^{\circ}$ mapping x to the sum of the slopes of f at the branches of $\mathbb{P}_K$ at x; this makes sense because only finitely many of these slopes are nonzero. We say that f is *harmonic* (resp. *subharmonic, superharmonic*) if Δf is everywhere zero (resp. everywhere nonnegative, everywhere nonpositive).

Example 24.6.3. For any $r \in K(t)$, the function $x \mapsto \log |r(x)|$ is harmonic.

Notes

A definitive treatment of the Berkovich projective line, with an eye towards applications in algebraic and arithmetic dynamics, is [31]. See also [163] for an alternate perspective, motivated by considerations from complex function theory, that ultimately leads to a similar picture. For an overview, we also suggest [27].

The description of $\mathbb{A}_K$ that we have given can be extended, with some modifications, to the case where K is not algebraically closed; the key point is that for $\mathbb{C}$ a completed algebraic closure of K, $\mathbb{A}_K$ equals the quotient of $\mathbb{A}_{\mathbb{C}}$ by the action of the group of continuous automorphisms of $\mathbb{C}$ over K [52, Corollary 1.3.6]. See [249, §2.2] or [253, §2].

Our description of the deformation retract H loosely follows [253, Theorem 2.5]. See also [52, Theorem 6.1.5] for a similar assertion for projective spaces in any dimension.

One can give a similar explicit description of any one-dimensional Berkovich analytic space over a nonarchimedean field, including the analytification of a curve over K [28, 29]; this records information about semistable reduction [383] which is of practical value in computational number theory when considering bad-reduction Euler factors of arithmetic L-functions [134, 135]. See [137] for a definitive treatment.

Using the same analogy in the opposite direction, one can model various concepts from the algebraic geometry of curves in the language of finite graphs, notably the Riemann–Roch theorem. See for example [32].

Figure 24.1 is modeled on [27, Figure 1, p. 127], although many similar illustrations exist in the literature (including the cover art of [163]). An atypical one is [210, Figure 1], which shows that for $K = \mathbb{C}_p$, the entirety of $\mathbb{P}_K$ admits a homeomorphic embedding into $\mathbb{R}^2$; this turns out to be a special case of a general result about Berkovich spaces of any dimension. See also [365, Example 2.20], which includes the points of type 5 that occur in the context of Huber's adic spaces.

While the situation is somewhat more complicated in higher dimensions, one can still view a general Berkovich analytic space as an inverse limit of some polyhedral approximations (again called *skeleta*). Such approximations admit strong links to toroidal and tropical geometry [220, 333, 390] and to semistable models [185, 186]. These links can often be expressed in terms of homotopy types; for example, smooth analytic spaces over a field are locally contractible [54, 55]. The corresponding global statements make full use of the machinery of modern birational geometry; see [320] for an overview.

A more robust theory of harmonic and subharmonic functions on Berkovich analytic curves than the one considered here has been developed by Thuillier [389] with a view towards applications in Arakelov theory (see also [30]). A related development is the work of Favre and Jonsson concerning potential theory on the valuative tree, with applications to the theory of plurisubharmonic singularities on complex surfaces. See [163] and [164], and also a more recent paper by Boucksom, Favre, and Jonsson [72] giving some higher-dimensional generalizations.

We conclude this discussion by reiterating that we have focused on the structure of the underlying spaces in Berkovich analytic geometry, at the expense of saying anything about additional structures that play a crucial role (particularly sheaves). We refer back to the notes for Chapter 0 for broader context.

Exercises

24.1 Complete the proof of Theorem 24.4.4 by showing that the inverse image of any neighborhood of ∞ in $\mathbb{P}_K$ is an open subset of $\mathbb{P}_K \times [0, \infty]$. (Hint:

as in the proof of Theorem 24.4.4, it suffices to consider a neighborhood of the form $\{x \in \mathbb{P}_K : x(P) > c\}$ for some $P \in K[t]$ and some $c > 0$.)

24.2   Show that the points of type 1 are dense in $\mathbb{P}_K$, then use this to show that the radius map $\rho : \mathbb{P}_K \to [0, \infty]$ is not continuous.

24.3   Extend the map H_S from Lemma 24.5.3 to a strong deformation retract of $\mathbb{P}_K$ onto S. (Hint: modify the map H so that for any fixed x, $H(x, \rho)$ increases until it reaches S and then stops.)

24.4   Show that the following properties of a function $f : \mathbb{P}_K \to \mathbb{R}$ are preserved under pullback by a linear fractional transformation:

(a) piecewise affine;
(b) piecewise affine with integral slopes;
(c) harmonic;
(d) subharmonic.

24.5   Let K' be an algebraically closed nonarchimedean field containing K.

(a) Show that the restriction map $\mathbb{P}_{K'} \to \mathbb{P}_K$ admits a unique continuous section taking $x_{z,\rho}$ to $x_{z,\rho}$ for any $z \in K$ and any $\rho > 0$.

(b) Show that the map in (a) is a strong deformation retract.

25

Convergence polygons

In this chapter, we reinterpret the discussion of variation of subsidiary radii of a p-adic differential equation in a manner that makes fundamental use of the Berkovich analytification of the projective line.

25.1 The normalized radius of convergence

We start by reinterpreting the radius of convergence in the manner of Section 9.7. We start with the case of a closed disc.

Definition 25.1.1. Let M be a finite differential module over $K\langle t/\beta\rangle$. For $z \in K$ with $|z| \leq \beta$, define the *radius of convergence* of M at z by forming the base extension along the substitution $t \mapsto t - z$ and then applying Definition 9.3.1. By Proposition 6.9.1, this definition is stable under extension of K; using Remark 24.2.4, we may thus view the radius of convergence as a function on $\mathcal{M}(K\langle t/\beta\rangle)$.

Define the *normalized radius of convergence* of M at a point of $\mathcal{M}(K\langle t/\beta\rangle)$ as the radius of convergence divided by β. By Proposition 9.3.3, it takes values in $(0, 1]$.

We may similarly define the normalized radius of convergence for a differential module M over $K[\![t/\beta]\!]_{an}$, as a function on $\mathcal{M}(K[\![t/\beta]\!]_{an})$.

Remark 25.1.2. Let M be a finite differential module over $K\langle t/\beta\rangle$. For $x = x_{0,\beta}$, the radius of convergence of M is precisely the generic radius of convergence of M_β, while the normalized radius of convergence is the intrinsic radius of M_β.

For $x = x_{0,\rho}$ with $\rho \in (0, \beta)$, the corresponding statement is that the intrinsic radius of M_ρ is the normalized radius of convergence multiplied by β/ρ, except if this gives a result greater than 1. In the latter case, we truncate down to 1;

411

that is, we lose track of what happens to horizontal sections beyond the disc bounded by x.

We next consider a closed annulus, for which we need a slightly different definition.

Definition 25.1.3. Let M be a finite differential module over one of the rings $K\langle \alpha/t, t/\beta \rangle$. For $z \in K$ with $\alpha \leq |z| \leq \beta$, define the *radius of convergence* of M at z by forming the base extension along the map $K\langle \alpha/t, t/\beta \rangle \to K[\![(t-z)/|z|]\!]_{\mathrm{an}}$ and then applying Definition 9.3.1. By Proposition 6.9.1 again, this definition is stable under base extension, and so induces a function on $\mathcal{M}(K\langle \alpha/t, t/\beta \rangle)$.

For $x \in \mathcal{M}(K\langle \alpha/t, t/\beta \rangle)$, define the *normalized radius of convergence* of M at x as the radius of convergence divided by $x(t)$. By Proposition 9.3.3, it takes values in $(0, 1]$.

We may similarly define the normalized radius of convergence for a differential module M over one of the rings $R = K\langle \alpha/t, t/\beta \rangle\!]_{\mathrm{an}}, K[\![\alpha/t, t/\beta]\!]_{\mathrm{an}}$, again as a function on $\mathcal{M}(R)$.

Remark 25.1.4. Let M be a finite differential module over $K\langle \alpha/t, t/\beta \rangle$. For $x = x_{0,\rho}$ with $\rho \in [\alpha, \beta]$, the radius of convergence of M is precisely the generic radius of convergence of M_ρ, while the normalized radius of convergence is the intrinsic radius of M_ρ.

For $x = x_{z,\rho}$ with $\alpha \leq |z| \leq \beta$ and $\rho \subset (0, |z|)$, let F be the completion of $K(t - z)$ for the ρ-Gauss norm. Then the corresponding statement is that the intrinsic radius of $M \otimes_{K\langle \alpha/t, t/\beta \rangle} F$ is the minimum of 1 and the normalized radius of convergence.

25.2 Normalized subsidiary radii and the convergence polygon

We next extend the reinterpretation to subsidiary radii.

Definition 25.2.1. Let M be a finite differential module over one of the rings

$$R = K\langle t/\beta \rangle, K[\![t/\beta]\!]_{\mathrm{an}}, K\langle \alpha/t, t/\beta \rangle, K\langle \alpha/t, t/\beta \rangle\!]_{\mathrm{an}}, K[\![\alpha/t, t/\beta]\!]_{\mathrm{an}}.$$

For $z \in K$ such that $x_{z,0} \in \mathcal{M}(R)$, define the *normalized subsidiary radii of convergence* of M at z by forming the base extension as in Definition 25.1.1 (resp. as in Definition 25.1.3), computing the radii of optimal convergence (Definition 11.9.1), and rescaling as in the definition of the normalized radius of convergence. Again by Proposition 6.9.1, this definition is stable under base extension, and so induces a function on $\mathcal{M}(R)$.

Remark 25.2.2. Let M be a finite differential module over $K[\![\alpha/t, t/\beta]\!]_{an}$. For $\rho \in [\alpha, \beta]$, by Theorem 11.9.2, the normalized subsidiary radii of M at $x_{0,\rho}$ coincide with the intrinsic subsidiary radii of M_ρ.

Now let M be a finite differential module over $K[\![t/\beta]\!]_{an}$. For $\rho \in (0, \beta]$, by Theorem 11.9.2, we may recover the intrinsic subsidiary radii of M_ρ by applying the function $\lambda \mapsto \min\{1, \lambda\beta/\rho\}$ to the normalized subsidiary radii of M at $x_{0,\rho}$.

Definition 25.2.3. With notation as in Definition 25.2.1, given $x \in \mathcal{M}(R)$, let $\rho_1 \le \cdots \le \rho_n$ be the normalized subsidiary radii of convergence of M at x. For $i = 0, \ldots, n$, define

$$h_i(M, x) = -\log \rho_i, \qquad H_i(M, x) = -\log \rho_1 - \cdots - \log \rho_i$$

and view h_i and H_i as functions from $\mathcal{M}(R)$ to $\mathbb{R}$.

The *convergence polygon* of M at x is the Newton polygon defined by the points

$$\{(-i, -H_i(M, x)): i = 0, \ldots, n\}.$$

25.3 A constancy criterion for convergence polygons

We next establish a criterion for the convergence polygon to be constant on a disc. (This strategy was used previously in the proof of Theorem 12.8.4.)

Lemma 25.3.1. *For M a finite differential module of rank n over $K[\![t/\beta]\!]_{an}$, let $\rho_1 \le \cdots \le \rho_n$ be the radii of optimal convergence of M at 0. Then for $i = 1, \ldots, n$,*

$$h_i(M, x_{0,e^{-r}}) = \begin{cases} f_i(M, r) & r \le -\log \rho_i \\ -\log \rho_i & r > -\log \rho_i \end{cases}$$

where $f_i(M, r)$ is the function defined in Notation 11.3.1.

Proof By Remark 25.2.2,

$$f_i(M, r) = \max\{r, h_i(M, x_{0,e^{-r}})\}.$$

It is evident that $f_i(M, r) = -\log \rho_i < r$ for $r > -\log \rho_i$. It thus suffices to obtain a contradiction assuming that $f_i(M, r) < r$ for some $r \le -\log \rho_i$. Namely, this would imply that M would have $n - i + 1$ linearly independent horizontal sections on an open disc of radius strictly greater than $e^{-r} > \rho_i$ centered at a point z with $|z| = r$; however, this disc would also contain 0 and thus contradict the definition of ρ_i. $\square$

Corollary 25.3.2. *Let M be a finite differential module of rank n over $K[\![\alpha/t, t/\beta]\!]_{\mathrm{an}}$ (where $\alpha = 0$ is allowed).*

(a) *For $i = 1, \ldots, n$, the function $r \mapsto h_i(M, x_{0,\mathrm{e}^{-r}})$ on $[-\log \beta, -\log \alpha]$ is continuous and piecewise affine, with only finitely many slopes even if $\alpha = 0$.*

(b) *The slopes of $r \mapsto h_i(M, x_{0,\mathrm{e}^{-r}})$ belong to $\frac{1}{1}\mathbb{Z} \cup \cdots \cup \frac{1}{n}\mathbb{Z}$. Moreover, the slopes of $r \mapsto H_n(M, x_{0,\mathrm{e}^{-r}})$ are in $\mathbb{Z}$.*

Proof For $\alpha > 0$, both claims follow immediately from Theorem 11.3.2(a)–(b). For $\alpha = 0$, we deduce (a) by splicing together the functions $f_i(M, r)$ and $-\log \rho_i$ as per Lemma 25.3.1 and applying Theorem 11.3.2(a) to the former. We then deduce (b) from Theorem 11.3.2(b).   □

Remark 25.3.3. Lemma 25.3.1 implies that the equality $h_i(M, x_{0,\mathrm{e}^{-r}}) = r$ holds for $r = -\log \rho_i$ and not for any larger value. However, it does not preclude the equality from holding for smaller values of r; for example, this can occur if $\mathcal{E}$ arises from an overconvergent isocrystal (Appendix B).

Lemma 25.3.4. *Let M be a finite differential module of rank n over $K[\![t/\beta]\!]_{\mathrm{an}}$. Suppose that for some $r_0 \in [-\log \beta, \infty)$, for r in some right neighborhood of r_0, for $i = 1, \ldots, n$, $f_i(M, r)$ is either constant or identically equal to r. Then the convergence polygon is constant on $\mathcal{M}(K[\![t/\mathrm{e}^{-r_0}]\!]_{\mathrm{an}})$.*

Proof By enlarging K, we can ensure that $r_0 = 0$ and (by Theorem 1.5.3 and Remark 24.3.6) that $\mathbb{P}_K$ contains no points of type 4. If any of the optimal radii of convergence of M at 0 are strictly greater than 1, then these radii are evidently constant on $\mathcal{M}(K[\![t]\!]_{\mathrm{an}})$ (as they correspond to convergence on an open disc containing this entire subspace). It is thus sufficient to check the constancy of the convergence polygon after renormalizing to the unit disc; that is, we need only check that the convergence polygon of $M \otimes_{K[\![t/\beta]\!]_{\mathrm{an}}} K[\![t]\!]_{\mathrm{an}}$ is constant on $\mathcal{M}(K[\![t]\!]_{\mathrm{an}})$. To simplify notation, we may assume that $\beta = 1$.

By Theorem 12.4.1, we may decompose M into summands M_j such that for each j, for r in some right neighborhood of 0, the functions $f_i(M_j, r)$ for $i = 1, \ldots, \mathrm{rank}(M_j)$ are all equal to a single function (which is itself either constant or identically equal to r). The convergence polygon of M is made by taking the union of the slopes of the convergence polygons of the M_j, so we may assume hereafter that only a single summand occurs.

Put $r_1 = f_1(M, 0)$. By Theorem 11.3.2(d) and induction on i, we deduce that

$$f_i(M, r) = \max\{r_1, r\} \qquad (i = 1, \ldots, n; r \geq 0).$$

By Lemma 25.3.1, the radii of optimal convergence of M are all at most

e^{-r_1}; while the reverse inequality is not included in Lemma 25.3.1 (see Remark 25.3.3), here it is guaranteed by the transfer theorem (Theorem 9.6.1). Using Lemma 25.3.1 again, we deduce that $h_i(M, x_{0,e^{-r}})$ is constant for $r \geq 0$.

By the same token, for any $z \in K$ with $z < 1$, $h_i(M, x_{z,e^{-r}})$ is constant for $r \geq 0$. It follows that the convergence polygon is constant on $\mathcal{M}(K[\![t]\!]_{an})$. □

25.4 Finiteness of the convergence polygon

We now establish a crucial finiteness property of the convergence polygon.

Lemma 25.4.1. *Let M be a finite differential module of rank n over $K[\![t/\beta]\!]_{an}$. For $z \in K$ with $|z| < \beta$, there exists a neighborhood U of $x_{z,0}$ on which the convergence polygon of M is constant.*

Proof Let ρ be the radius of convergence of M at z; recall that $\rho > 0$ by the p-adic Cauchy theorem (Proposition 9.3.3). Let U be the disc $|t - z| < \rho$ in $\mathbb{P}_K$. For any $x \in U$, the local horizontal sections of M at z converge on a disc containing x, and so the radii of optimal convergence at the two points coincide. □

Lemma 25.4.2. *Let M be a finite differential module of rank n over $K[\![t/\beta]\!]_{an}$. For $\rho \in (0, \beta)$, there exists a neighborhood U of $x_{0,\rho}$ such that the convergence polygon of M, as a function on U, factors through the retraction onto some skeleton S. Moreover, the restriction to $U \cap S$ is continuous and piecewise affine.*

Proof By enlarging K, we can ensure that $\rho = 1$; we now set notation as in Theorem 11.3.2(c). By Remark 11.3.5, for all but finitely many values $\overline{\mu} \in \kappa_K^\times$, $s_{\overline{\mu},i}(M) = 0$ for each i such that $f_i(M, 0) > 0$. (Remember that we are assuming that K is algebraically closed, so κ_K must be also.) Choose values $z_1, \ldots, z_m \in K$ representing the remaining residue classes in κ_K. By Theorem 11.3.2(a)–(b), we can choose $r_1 < 0, r_2 > 0$ such that for $i = 1, \ldots, n$, the function $f_i(M, r)$ is piecewise affine on $[r_1, 0]$ and $[0, r_2]$ and the functions $\max\{r, h_i(M, x_{z_j,e^{-r}})\}$ are affine on $[0, r_2]$ for $j = 1, \ldots, m$.

Let U be the set of $x \in \mathcal{M}(K[\![t/\beta]\!]_{an})$ such that $e^{-r_2} \leq x(t) \leq e^{-r_1}$ and $x(t - z_j) \geq e^{-r_2}$ for $j = 1, \ldots, m$. Let S be the union of the root paths from $0, z_1, \ldots, z_m$. The complement of S in U is (set-theoretically) a disjoint union of open subspaces, each of which is an open disc bounded by a point of $S \cap U$. On each such subspace, we may apply Lemma 25.3.4 to see that the convergence polygon is identically equal to its value at the bounding point. (For points which retract onto a point of $S \cap U$ other than $x_{0,1}$, we must apply Theorem 11.3.2(c)–(d) to see that the hypothesis of Lemma 25.3.4 is satisfied.) □

Lemma 25.4.3. *Let M be a finite differential module of rank n over $K[\![\alpha/t, t/\beta]\!]_{\mathrm{an}}$. For $\rho \in (\alpha, \beta)$, there exists a neighborhood U of $x_{0,\rho}$ such that the convergence polygon of M, as a function on U, factors through the retraction onto some skeleton. Moreover, the restriction to $U \cap S$ is continuous and piecewise affine.*

Proof The argument is similar to that of Lemma 25.4.2, so we omit the details. □

Theorem 25.4.4. *Let M be a finite differential module over one of the rings*

$$R = K\langle t/\beta \rangle, K[\![t/\beta]\!]_{\mathrm{an}}, K\langle \alpha/t, t/\beta \rangle, K\langle \alpha/t, t/\beta]\!]_{\mathrm{an}}, K[\![\alpha/t, t/\beta]\!]_{\mathrm{an}}.$$

Then the convergence polygon of M, as a function on $\mathcal{M}(R)$, factors through the retraction onto some skeleton. Moreover, the restriction to said skeleton is continuous and piecewise affine.

Proof We may check the claim after enlarging K; by Theorem 1.5.3, we may thus assume that K is spherically complete and so $\mathbb{P}_K$ has no points of type 4 (Remark 24.3.6).

Suppose first that $R = K[\![t/\beta]\!]_{\mathrm{an}}, K[\![\alpha/t, t/\beta]\!]_{\mathrm{an}}$. For $x \in \mathcal{M}(R)$ of type 1, 2, or 3, we may apply Lemma 25.4.1, Lemma 25.4.2, or Lemma 25.4.3 respectively to see that there exists a neighborhood of x on which the convergence polygon of M factors through the retraction onto some skeleton. Since $\mathcal{M}(R)$ is compact by Lemma 23.3.2, we can choose a finite set of these neighborhoods which cover $\mathcal{M}(R)$; by taking the union of the associated skeleta, we obtain a single skeleton of the desired form.

Suppose next that $R = K\langle t/\beta \rangle$. By Remark 11.3.5, the criterion from Lemma 25.3.4 applies to all but finitely many branches of $\mathbb{P}_K$ at $x_{0,\beta}$; we may then apply the previous argument to the remaining branches. Similar arguments can be used to reduce the cases $R = K\langle \alpha/t, t/\beta]\!]_{\mathrm{an}}, K[\![\alpha/t, t/\beta]\!]_{\mathrm{an}}$ to the previously enumerated cases. □

Corollary 25.4.5. *Let M be a finite differential module over one of the rings $R = K\langle t/\beta \rangle, K[\![t/\beta]\!]_{\mathrm{an}}, K\langle \alpha/t, t/\beta \rangle, K\langle \alpha/t, t/\beta]\!]_{\mathrm{an}}, K[\![\alpha/t, t/\beta]\!]_{\mathrm{an}}$. Then the convergence polygon of M, as a function on $\mathcal{M}(R)$, is continuous.*

Proof This is immediate from Theorem 25.4.4. □

Remark 25.4.6. Corollary 25.4.5 implies that the normalized radius of convergence is also continuous as a function on $\mathcal{M}(R)$. This was originally established by Baldassarri [41] using a different method; see the chapter notes.

Remark 25.4.7. The proof of Theorem 25.4.4 provides no mechanism for showing that the convergence polygon factors through a retraction onto a *strict* skeleton, as by design the proof retains no special knowledge about points of type 4. Overcoming this requires a separate proof with additional ideas, notably the use of the p-adic local monodromy theorem without Frobenius structures (Theorem 22.1.4). See Chapter 27.

25.5 Effect of singularities

We next consider differential modules with singularities. In this context, we make contact with the concept of irregularity as studied in Chapter 7.

Definition 25.5.1. Let M be a finite differential module over $K\langle t/\beta\rangle[t^{-1}]$. We define the *convergence polygon* of M as a function on $\mathcal{M}(K\langle t/\beta\rangle) \setminus \{x_{0,0}\}$ by viewing the latter as the union of $K\langle \alpha/t, t/\beta\rangle$ over all $\alpha > 0$.

Lemma 25.5.2. *Let M be a finite differential module of rank n over $K\langle t/\beta\rangle[t^{-1}]$. Then there exists $r_0 \geq -\log\beta$ such that the functions*

$$f_1(M,r) - r, \ldots, f_n(M,r) - r$$

are affine for $r \geq r_0$. Moreover, in the multiset of limiting slopes, for each $s > 1$ the multiplicity of $\log s$ is the dimension of V_s in the decomposition of $V = M \otimes_{K\langle t/\beta\rangle[t^{-1}]} K((t))$ given by Theorem 6.6.1 (for the absolute value on $K((t))$ with $|t| = e^{-1}$). In particular, the limiting slope of $f_1(M,r)-r+\cdots+f_n(M,r)-r$ equals the irregularity of V.

Proof We prove that $f_i(M,r)$ is eventually affine by induction on i. Given the claim for all smaller values of i, it is equivalent to prove that $F_i(M,r) = f_1(M,r) + \cdots + f_i(M,r)$ is affine. By Theorem 11.3.2(a) and (e), $F_i(M,r)$ is continuous, piecewise affine, and convex, as then is $f_i(M,r)$ by the induction hypothesis. Consequently, for all sufficiently large r, either $f_i(M,r) = r$ or $f_i(M,r) - r$ is bounded below by an affine function of r with positive slope. In the former case, there is nothing left to prove. In the latter case, $f_i(M,r) - r$ diverges to ∞, so for r sufficiently large we may read off $f_i(M,r)$ using a cyclic vector as in the proof of Lemma 11.5.1. $\quad\square$

Theorem 25.5.3. *Let M be a finite differential module of rank n over $K\langle t/\beta\rangle[t^{-1}]$.*

(a) *The convergence polygon of M, as a function on $\mathcal{M}(K\langle t/\beta\rangle) \setminus \{x_{0,0}\}$, factors through the retraction onto some skeleton. Moreover, the restriction to said skeleton is continuous and piecewise affine.*

(b) *This function extends continuously over $x_{0,0}$ if and only if M is regular at $t = 0$.*

Proof By Lemma 25.5.2, for some $r_0 \geq -\log\beta$, the functions $h_i(M, x_{0,e^{-r}})$ are affine for $r \geq r_0$. By Theorem 11.3.2(c)–(d) and Lemma 25.3.4, for $x \in \mathbb{P}_K$ with $x(t) = \rho \in (0, e^{-r_0})$ the convergence polygon of M takes the same value at x as at $x_{0,\rho}$. Combining this with Theorem 25.4.4 proves (a); we read off (b) from Lemma 25.5.2. □

25.6 Affinoid subspaces

We extend the previous discussion to more general affinoid subspaces of $\mathbb{P}_K$.

Definition 25.6.1. By an *affinoid* subset of $\mathbb{P}_K$ (or more precisely a *connected affinoid* subset), we will mean any nonempty closed set U of the form

$$|t - z_1| \geq \alpha_1, \ldots, |t - z_m| \geq \alpha_m \qquad (25.6.1.1)$$

for some $m > 0$, some $\alpha_1, \ldots, \alpha_m > 0$, and some $z_1, \ldots, z_m \in K$, or

$$|t| \leq \beta, |t - z_1| \geq \alpha_1, \ldots, |t - z_m| \geq \alpha_m \qquad (25.6.1.2)$$

for some $\beta > 0$, some $m \geq 0$, some $\alpha_1, \ldots, \alpha_m \in (0, \beta]$, and some $z_1, \ldots, z_m \in K$. Conceptually, such a subset is the complement of a finite union of open discs in $\mathbb{P}_K$, one of which may be an "open disc around ∞". To canonicalize the presentation, we may ensure that the discs

$$|t - z_1| < \alpha_1, \ldots, |t - z_m| < \alpha_m$$

are pairwise disjoint.

Definition 25.6.2. Let U be an affinoid subset of $\mathbb{P}_K$ as in (25.6.1.1) or (25.6.1.2). Define the ring R_U as the completion of $K[(t-z_1)^{-1}, \ldots, (t-z_m)^{-1}]$ or $K[t, (t - z_1)^{-1}, \ldots, (t - z_m)^{-1}]$ for the norm

$$\max\{x_{z_1,\gamma_1}, \ldots, x_{z_m,\gamma_m}\} \text{ or } \max\{x_{0,\delta}, x_{z_1,\gamma_1}, \ldots, x_{z_m,\gamma_m}\};$$

any element of R_U can be written as a sum consisting of one element of $K\langle t/\beta \rangle$ (in the latter case) and one element of $K\langle \alpha_j/(t - z_j) \rangle$ for $j = 1, \ldots, m$. If we assume the presentation of U is minimal (see Definition 25.6.1), then this series presentation is unique except for shifting elements of K between summands.

Lemma 25.6.3. *For U an affinoid subset of $\mathbb{P}_K$ as in (25.6.1.1) or (25.6.1.2), we have a homeomorphism $\mathcal{M}(R_U) \cong U$.*

Proof The map $\mathcal{M}(R_U)$ is given by taking an element of $\mathcal{M}(R_U)$ and restricting it to $R_U \cap K(t)$. From the decomposition of elements of R_U given in

Definition 25.6.2, we see that the map $\mathcal{M}(R_U) \to U$ is bijective. We may then conclude using Lemma 23.3.2 and Remark 23.3.7. □

In order to discuss convergence polygons, we restrict hereafter to affinoid subspaces not containing ∞.

Definition 25.6.4. Let U be an affinoid subset of $\mathbb{P}_K$ as in (25.6.1.2). Let $Z = \{z'_1, \ldots, z'_l\}$ be a finite set of type-1 points of U. By a *finite differential module* over $U \setminus Z$, we will mean a finite differential module over the ring $R_U[(t - z'_1)^{-1}, \ldots, (t - z'_l)^{-1}]$.

Let M be a finite differential module over $U \setminus Z$. We may then define the convergence polygon of M as a function on $U \setminus Z$, normalizing radii of convergence at $x \in U \setminus Z$ with respect to the maximal open disc within U containing x and not meeting Z. In the case where U is a closed disc and $Z = \{0\}$, this agrees with the definition given in Definition 25.5.1.

Theorem 25.6.5. *Let U be an affinoid subset of $\mathbb{P}_K$ as in (25.6.1.2). Let Z be a finite set of type-1 points of U. Let M be a finite differential module over $U \setminus Z$. Then the convergence polygon of M, as a function of $U \setminus Z$, factors through the retraction onto some skeleton. Moreover, the restriction to said skeleton is continuous and piecewise affine.*

Proof We follow the now-familiar approach. Draw the root paths from the points of Z and the points x_{z_i, α_i}. By Theorem 25.4.4 and Theorem 25.5.3, these paths decompose into finitely many segments on each of which the convergence polygon is affine. At each point in the interior of one of these segments, Lemma 25.3.4 applies to show that the convergence polygon is constant along every branch not meeting the root paths. At each of the finitely many remaining points, an argument as in Remark 11.3.5 shows that Lemma 25.3.4 applies to all but finitely many branches, and we may apply Theorem 25.4.4 to take care of the others. □

25.7 Meromorphic differential equations

On a related note, we consider meromorphic differential equations.

Definition 25.7.1. Let $Z = \{z_1, \ldots, z_m\}$ be a *nonempty* finite subset of K. By a *finite differential module* over $\mathbb{P}_K \setminus Z$, we will mean a finite differential module over the ring $K[t, (t - z_1)^{-1}, \ldots, (t - z_m)^{-1}]$. We may then define the convergence polygon of M as a function on $\mathbb{P}_K \setminus Z$, normalizing radii of convergence at $x \in \mathbb{P}_K \setminus Z$ with respect to the maximal open disc containing

x and not meeting Z. (We need Z to be nonempty so that such a maximal disc exists.)

Theorem 25.7.2. *Let Z be a nonempty finite subset of K. Let M be a finite differential module over $\mathbb{P}_K \setminus Z$. Then the convergence polygon of M, as a function of $\mathbb{P}_K \setminus Z$, factors through the retraction onto some skeleton. Moreover, the restriction to said skeleton is continuous and piecewise affine.*

Proof The proof of Theorem 25.6.5 carries over.   □

25.8 Open discs and annuli

In preparation for the discussion of index formulas in the next chapter, we also need to consider open discs and annuli, where we encounter an extra subtlety. We take the opportunity to extend the treatment to affinoid subsets.

Definition 25.8.1. By an *unbounded affinoid subset* of $\mathbb{P}_K$, we will mean any nonempty open set U of the form

$$|t - z_1| > \alpha_1, \ldots, |t - z_m| > \alpha_m \tag{25.8.1.1}$$

for some $m > 0$, some $\alpha_1, \ldots, \alpha_m > 0$, and some $z_1, \ldots, z_m \in K$, or

$$|t| < \beta, |t - z_1| > \alpha_1, \ldots, |t - z_m| > \alpha_m \tag{25.8.1.2}$$

for some $\beta > 0$, some $m \geq 0$, some $\alpha_1, \ldots, \alpha_m \in (0, \beta]$, and some $z_1, \ldots, z_m \in K$. We define the *boundary* of U as the set $\delta U = \{x_{z_1,\alpha_1}, \ldots, x_{z_m,\alpha_m}\}$ or $\delta U = \{x_{0,\beta}, x_{z_1,\alpha_1}, \ldots, x_{z_m,\alpha_m}\}$.

Definition 25.8.2. Let U be an unbounded affinoid subset of $\mathbb{P}_K$ as in (25.8.1.2). Let Z be a finite set of type-1 points of U. By a *finite differential module M over $U \setminus Z$*, we will mean a coherent sequence of finite differential modules over $V \setminus (V \cap Z)$ where V varies over affinoid subsets of $\mathbb{P}_K$ contained in U. We may again define the convergence polygon of M as a function on $U \setminus Z$, normalizing radii of convergence at $x \in U \setminus Z$ with respect to the maximal open disc within U containing x and not meeting Z.

For $x \in \delta_U$, we say that M is *tame at x* if the convergence polygon of M restricts to an affine function on some segment of a root path with endpoint x. For example, if M is solvable at x, then this holds by Lemma 12.6.2.

Remark 25.8.3. In Definition 25.8.2, one way for M to fail to be tame at x is for the convergence polygon to diverge to ∞ at x. However, it is also possible for M to fail to be tame even if this does not occur, as in the following example.

Example 25.8.4. We give an example of a differential module over an open annulus over $\mathbb{C}_p$ which is not tame at the outer boundary. (One can make similar examples over a nonarchimedean field K of residue characteristic 0.) Choose $\alpha \in (0, 1)$ and let M be the differential module on the open annulus $\alpha < |t| < 1$ which is free on a single generator v satisfying

$$D(v) = fv, \qquad f = \sum_{i=1}^{\infty} p^{-2-1/i} t^i.$$

By Theorem 6.5.3, $f_1(M, r) \to 2 - \frac{1}{p-1}$ as $r \to 0^+$; however, $f_1(M, r)$ changes slope infinitely many times as $r \to 0^+$, at values corresponding to the slopes of the Newton polygon of f. Hence M is not tame at $x_{0,1}$.

Theorem 25.8.5. *Let U be an unbounded affinoid subset of $\mathbb{P}_K$ as in (25.8.1.2). Let Z be a finite set of type-1 points of U. Let M be a finite differential module over $U \setminus Z$ which is tame at every point of δU. Then the convergence polygon of M, as a function of $U \setminus Z$, factors through the retraction onto some skeleton. Moreover, the restriction to said skeleton is continuous and piecewise affine.*

Proof In light of the tame hypothesis, the proof of Theorem 25.6.5 carries over. $\qquad\square$

Notes

The normalized radius of convergence, as a function on the Berkovich projective line, was introduced by Baldassarri [41], building on previous work of Baldassarri and Di Vizio [42]; Baldassarri also established its continuity using an argument in the style of Proposition 6.3.1. As this work appeared only after the first edition of this book was mostly written, it received only a passing mention therein:

The approach of Baldassarri and Di Vizio gives a more refined invariant than ours, since ours can be recovered by truncating. This is likely to have certain advantages in applications. One serious disadvantage is that at the moment, there is no good theory of subsidiary generic radii of convergence in the Baldassarri–Di Vizio framework. The correct definition is presumably the one suggested by Theorem 11.9.2, but it is not clear how to prove any good properties for it.

The desired theory of normalized subsidiary radii was provided (using the results of Part III) by Poineau and Pulita in a series of papers [335, 336, 337, 338, 345]. An independent but similar treatment, with some distinct results (see in particular Theorem 27.0.1), was given in [255]. See also [259] for a survey.

The previous references all treat $\mathbb{P}_K$ within the broader framework of Berkovich analytic curves. In this case, there is an additional subtlety: there are finitely many points of type 2 at which the complete residue field has residue field isomorphic not to a rational function field over a field of characteristic p, but to some function field of higher genus. One can deal with such points by viewing this function field as a finite extension of a rational function field; in fact one can even ensure that this corresponds to a covering of curves which is everywhere tamely ramified. (In characteristic 2, this is subtle and depends on the base field; see [380] for the case of an algebraically closed base field and [267] for the case of a finite base field.)

Although it is a bit orthogonal to our discussion here, we do wish to point out a link between convergence polygons and ramification in morphisms between Berkovich curves (somewhat in the spirit of Chapter 19). We limit our discussion here to an unstructured (and probably incomplete) list of appropriate references: [11, 12, 13, 65, 66, 70, 104, 158, 159, 213, 384].

26

Index theorems

We next relate convergence polygons to a topic introduced in Chapter 7: the *index* of a differential module. This discussion represents but the tip of a large iceberg; see the chapter notes.

As in Part III, we write ω to mean 1 if $p = 0$ and $p^{-1/(p-1)}$ if $p > 0$.

26.1 The index of a differential module

Definition 26.1.1. Let R be a differential ring whose constant subring K is a field. As in Definition 7.4.1, we define the *index* of a finite differential module (M, D) over R to be

$$\chi(M) = \dim_K H^0(M) - \dim_K H^1(M) = \dim_K \ker(D, M) - \dim_K \operatorname{coker}(D, M),$$

provided that both dimensions are finite (a state of affairs we characterize by saying that *the index of M exists* or that *M has an index*). Note that by Lemma 5.1.5, if R is an integral domain then $\dim_K H^0(M) \leq \operatorname{rank}_R M < \infty$, so the existence of the index is equivalent to the finite-dimensionality of $H^1(M)$.

By Proposition 6.9.1, the index of M remains unchanged after base extension from K to an extension K'. In particular, if the new module has an index, then so does M.

Remark 26.1.2. When computing $H^1(M)$, we will frequently use its interpretation as the Yoneda extension group (Lemma 5.3.3).

Lemma 26.1.3. *Let*

$$0 \to M_1 \to M \to M_2 \to 0$$

be a short exact sequence of differential modules over a differential ring R

whose constant subring K is a field. If any two of M, M_1, M_2 have an index, then so does the third and

$$\chi(M) = \chi(M_1) \oplus \chi(M_2).$$

Proof By the snake lemma, we have an exact sequence

$$0 \to H^0(M_1) \to H^0(M) \to H^0(M_2) \to H^1(M_1) \to H^1(M) \to H^1(M_2) \to 0.$$

From this, we may deduce both claims. □

26.2 More on affinoid subspaces of $\mathbb{P}_K$

Before continuing, we fill in some details on the rings and modules associated to affinoid subspaces. We start with the following analogue of Weierstrass preparation (Proposition 8.3.2); for the proof, we take an approach rooted in our explicit description of $\mathbb{P}_K$.

Proposition 26.2.1. *For U an affinoid subset of $\mathbb{P}_K$ and $f \in R_U$ nonzero, there exists a polynomial $P \in K[t]$ and a unit $g \in R_U^\times$ such that $f = Pg$. In particular, R_U is a principal ideal domain.*

Proof We first check that any $x \in \mathcal{M}(R_U)$ not of type 1 is a norm. By Remark 24.2.4, we may assume that x is of type 2, say $x = x_{z,\gamma}$. If we assume the presentation of U is minimal, then from the explicit description of elements of R_U from Definition 25.6.2, we may read off a formula for $x_{z,\gamma}$ which gives a nonzero answer on any nonzero input.

We next observe that for any $x \in x_{z,0}$ of type 1 in $\mathcal{M}(R_U)$, R_U injects into the ring $K[\![t-z]\!]$. Consequently, for any $f \in R_U$ nonzero, there exists a nonnegative integer e such that $(t - z)^e$ divides f in e and $x((t - z)^{-e} f) \neq 0$; in particular, there is a punctured neighborhood of x in $\mathcal{M}(R_U)$ on which f does not vanish.

By compactness (Lemma 23.3.2), we can find a polynomial $P(t) \in K[t]$ such that $P(t)$ divides f in R_U and $f/P(t)$ vanishes at no point of $\mathcal{M}(R_U)$ of type 1. By the first paragraph, $f/P(t)$ vanishes nowhere on $\mathcal{M}(R_U)$; by Corollary 23.3.5, $f/P(t) \in R_U^\times$ as desired. □

Lemma 26.2.2. *Let U, U_1, U_2 be affinoid subsets of $\mathbb{P}_K$ such that $U = U_1 \cup U_2$, and put $U_{12} = U_1 \cap U_2$. Let M be a finite free module on U and put*

$$M_1 = M \otimes_{R_U} R_{U_1}, M_2 = M \otimes_{R_U} R_{U_2}, M_{12} = M \otimes_{R_U} R_{U_{12}}.$$

Then the sequence

$$0 \to M \to M_1 \oplus M_2 \to M_{12} \to 0,$$

in which the last map is $(m_1, m_2) \mapsto m_1 - m_2$, *is exact.*

Proof We may assume at once that $M = R_U$. In this case, the claim follows directly from the explicit description of R_U from Definition 25.6.2. □

We have the following analogue of Lemma 8.3.6.

Lemma 26.2.3. *Let* U, U_1, U_2 *be affinoid subsets of* $\mathbb{P}_K$ *such that* $U = U_1 \cup U_2$, *and put* $U_{12} = U_1 \cap U_2$. *Let* M_1, M_2, M_{12} *be finite free modules over* $R_{U_1}, R_{U_2}, R_{U_{12}}$ *and fix isomorphisms*

$$\psi_1 : M_1 \otimes_{R_{U_1}} R_{12} \cong M_{12}, \qquad \psi_2 : M_2 \otimes_{R_{U_2}} R_{12} \cong M_{12}.$$

Then there exist a finite free R_U-*module* M *and isomorphisms* $M \otimes_{R_U} R_{U_1} \cong M_1, M \otimes_{R_U} R_{U_2} \cong M_2$ *inducing* ψ_1 *and* ψ_2. *Moreover,* M *is determined by this requirement up to unique isomorphism.*

Proof Using the surjectivity of $R_{U_1} \oplus R_{U_2} \to R_{U_{12}}$ from Lemma 26.2.2, we can establish a factorization statement analogous to Proposition 8.3.5 and deduce the claim as in the proof of Lemma 8.3.6. □

Lemma 26.2.4. *Let* U, U_1, U_2 *be unbounded affinoid subsets of* $\mathbb{P}_K$ *such that* $U = U_1 \cup U_2$ *and put* $U_{12} = U_1 \cap U_2$. *Let* Z *be a finite set of type-1 points of* U. *Let* M *be a finite differential module on* $U \setminus Z$ *and let* M_1, M_2, M_{12} *be the restrictions of* M *to* U_1, U_2, U_{12}. *If any three of* M, M_1, M_2, M_{12} *have indices, then so does the fourth and*

$$\chi(M) = \chi(M_1) + \chi(M_2) - \chi(M_{12}).$$

Proof By applying Lemma 26.2.2 to various affinoid subsets of U, U_1, U_2, we obtain an exact sequence

$$0 \to M \to M_1 \oplus M_2 \to M_2 \to 0.$$

We may then apply the snake lemma, as in the proof of Lemma 26.1.3, to conclude. □

26.3 The Laplacian of the convergence polygon

We will relate the index of a differential module to properties of the convergence polygon that follow from (and eventually improve upon) our earlier study of subsidiary radii.

Definition 26.3.1. Let U be an affinoid subset of $\mathbb{P}_K$. Let Z be a finite set of type-1 points of U. Let M be a finite differential module over $U \setminus Z$. By Theorem 25.6.5, the convergence polygon of M factors through the retraction onto some skeleton, on which it is continuous and piecewise affine.

We form a finite graph Γ_M, called the *controlling graph* of M, as follows. Let S be the set of $x \in U$ with the property that there is no open disc in U containing x on which the convergence polygon is constant. This set can be written as a finite union of line segments such that the following conditions hold.

(a) Each segment lies within a root path.
(b) On each segment, the convergence polygon is affine.
(c) No point of one segment lies in the interior of another.
(d) Subject to the preceding conditions, the number of segments is minimized. (This rules out the possibility of breaking a single segment in two.)

Take the vertices of Γ_M to be the endpoints of the segments, and take the edges of Γ_M to correspond to the segments themselves.

We may similarly define the controlling graph of a finite differential module over $\mathbb{P}_K \setminus Z$, or of a finite differential module over $U \setminus Z$ when U is an unbounded affinoid subset of $\mathbb{P}_K$ provided that the module is tame at every point of δU.

Definition 26.3.2. With notation as in Definition 26.3.1, put $n = \mathrm{rank}(M)$. We may define the Laplacian $\Delta H_n(M)$ as a function on $(U \setminus Z) \cap \mathbb{P}_K^\circ$. We extend the definition to the other points of $\mathbb{P}_K$ by declaring the function to be zero at those points.

We further extend the definition to $x \in Z \cap \delta(U)$ by taking $\Delta H_n(M)(x)$ to be the slope of the height of the convergence polygon on a segment leading *away* from x. For example, for $x \in Z$, this means we take x to be $-\mathrm{irr}_z(M)$, the negation of the irregularity in the sense of Definition 7.4.4.

We suggestively write $\int_{U \setminus Z} \Delta H_n(M)$, $\int_{Z \cup \delta U} \Delta H_n(M)$, or $\int_{U \cup \delta U} \Delta H_n(M)$ to mean the sum of $\Delta H_n(M)(x)$ as x runs over the set in question. Equivalently, we identify $\Delta H_n(M)$ with a discrete measure on the topological space $U \cup \delta U$ and integrate the indicator function of the set in question against this measure.

Lemma 26.3.3. *With notation as in Definition 26.3.1, we have*

$$\int_{U \cup \delta U} \Delta H_n(M) = 0.$$

Consequently,

$$\int_{U \setminus Z} \Delta H_n(M) = -\int_{Z \cup \delta U} \Delta H_n(M).$$

Proof We may rewrite $\int_{U \cup \delta U} \Delta H_n(M)$ as a sum over the edges of $\Gamma(M)$, each summand consisting of two equal and opposite contributions coming from the two endpoints of that edge (these contributions measure the slope of the same affine function, but with the domain read in opposite directions). We thus get cancellation in the sum. □

Remark 26.3.4. Based on Theorem 11.3.2(c), one might expect that in fact $\Delta H_n(M)(x) \geq 0$ for all $x \in U$. Unfortunately, it is unclear whether one can prove a general result of this form; see [337, Remark 2.4.10] for a discussion of the essential difficulty.

26.4 An index formula for algebraic differential equations

Using the index formula for differential equations on $\mathbb{P}^1_{\mathbb{C}}$ from Chapter 7, we can formulate a relationship between index of differential modules and the Laplacian of the height of the convergence polygon. As a first step, we do this for algebraic differential modules.

Definition 26.4.1. Let Z be a finite subset of $\mathbb{P}^1(K) = K \cup \{\infty\}$, identified with the corresponding set of type-1 points in $\mathbb{P}_K$. We define the *Euler characteristic*

$$\chi(\mathbb{P}_K \setminus Z) = 2 - 2\#Z.$$

Theorem 26.4.2. *Let Z be a finite subset of $\mathbb{P}^1(K)$ containing ∞. Let M be a finite differential module of rank n over $\mathbb{P}_K \setminus Z$. Then M has an index equal to*

$$\chi(M) = n\chi(\mathbb{P}_K \setminus Z) - \int_{\mathbb{P}_K \setminus Z} \Delta H_n(M).$$

Proof By Theorem 7.4.8,

$$\chi(M) = n\chi(\mathbb{P}_K \setminus Z) - \sum_{z \in Z} \mathrm{irr}_z(M).$$

Meanwhile, by Lemma 26.3.3,

$$\int_{\mathbb{P}_K \setminus Z} \Delta H_n(M) = \sum_{z \in Z} \mathrm{irr}_z(M). \tag{26.4.2.1}$$

This proves the claim. □

Note that one can interpret Theorem 26.4.2 as a formula about rigid-analytic differential equations via the analogue of the GAGA principle.

Theorem 26.4.3. *The following statements hold.*

(a) *Every vector bundle on $\mathbb{P}_K$ is the pullback of a unique vector bundle on the scheme $\mathbb{P}^1_K$.*

(b) *The global sections of a vector bundle on $\mathbb{P}^1_K$ coincide with those of its pullback to $\mathbb{P}_K$.*

Proof This can be proved either by adapting the proof of [181] to rigid analytic geometry, or by directly appealing to the general GAGA theorem for rigid analytic spaces; see the chapter notes. □

This suggests the possibility of generalizing Theorem 26.4.2 to differential modules on unbounded affinoid subsets; we will do this a bit later (Theorem 26.10.2). In the meantime, let us consider explicitly what Theorem 26.4.2 says in some examples.

Example 26.4.4. Let M be the differential module of rank 1 over $\mathbb{P}_K \setminus \{0, \infty\}$ associated to the differential equation $y' = \lambda^{-1} y$ for some $\lambda \in \mathbb{Z}_p$. Since M is regular at both 0 and ∞, Theorem 7.4.8 asserts that $\chi(M) = 0$. By Example 9.5.2,

$$h_1(M, x) = 0 \qquad (x \in \mathbb{P}_K \setminus \{0, \infty\}).$$

Similarly, for $\lambda \notin \mathbb{Z}_p$, $h_1(M, x)$ is equal to a nonzero constant value that depends on the distance from λ to the nearest element of $\mathbb{Z}_p$ (see the exercises for Chapter 9). We conclude that Γ_M consists solely of the segment from 0 to ∞. (Compare [259, Example 7.19].)

Example 26.4.5. Let M be the differential module of rank 2 over $\mathbb{P}_K \setminus \{0, 1, \infty\}$ associated to the hypergeometric differential equation (0.3.2.2) for some parameters $a, b, c \in \mathbb{Z}_p$ such that neither $a - b$ nor c is a p-adic Liouville number. Since M is regular at $0, 1, \infty$, Theorem 7.4.8 asserts that $\chi(M) = -2$. By Theorem 9.6.1, Example 9.6.2, and Theorem 13.7.1,

$$h_1(M, x) = h_2(M, x) = 0 \qquad (x \in \mathbb{P}_K \setminus \{0, 1, \infty\}).$$

We conclude that Γ_M consists of the paths from $x_{0,1}$ to $0, 1, \infty$. Similar logic applies to other Picard–Fuchs equations; see Appendix A.

Example 26.4.6. Assume $p > 2$. Let M be the differential module of rank 2 over $\mathbb{P}_K \setminus \{0, \infty\}$ associated to the Bessel differential equation (Example 7.1.5). This system is regular at 0 and has irregularity 1 at ∞, so $\chi(M) = -1$.

By Example 20.2.1, this equation admits a Frobenius structure, and hence $h_1(M, x_{0,1}) = h_2(M, x_{0,1}) = 0$. By Theorem 13.7.1, it follows that $h_1(M, x) = h_2(M, x) = 0$ for all $x \in \mathbb{P}_K \setminus \{0\}$ with $x(t) \le 1$.

For $r < 0$, the functions $h_1(M, x_{0,e^{-r}})$ and $h_2(M, x_{0,e^{-r}})$ are piecewise affine and their sum is convex. By Lemma 25.5.2, both functions have limiting slope -1 as $r \to -\infty$. On the other hand, by Theorem 19.4.1 and Example 20.2.1, both functions also have limiting slope -1 as $r \to 0^-$. It follows that

$$h_1(M, x) = h_2(M, x) = \log x(t) \qquad (x(t) > 1).$$

We find that Γ_M consists of the path from 0 to ∞, but broken into two segments at $x = x_{0,1}$ where we have $\Delta H_2(M)(x) = 1$.

26.5 Local analysis on a disc

As a special case of the general index formula on an unbounded affinoid subset, let us examine a special case where we can make everything quite explicit.

Lemma 26.5.1. *Let M be a finite differential module of rank n on the open disc $|t| < \beta$. Suppose that the convergence polygon of M is constant. Then $\chi(M)$ equals the multiplicity of 0 as a slope of the convergence polygon.*

Proof By Theorem 12.5.1, M decomposes into summands M_i on each of which the convergence polygon is not only constant, but also consists of a single slope. If this slope is nonzero, then $H^i(M_i) = 0$ by the Yoneda interpretation: any exact sequence

$$0 \to M_i \to N \to P \to 0$$

with P trivial is split by Theorem 12.5.1 again.

If this slope is 0, then M_i is a trivial differential module. Since differentiation on $K\{t/\beta\}$ is surjective with kernel K, $\chi(M_i) = \operatorname{rank} M_i$. $\square$

Let us break down what this argument says in the simplest of examples; it is already a nontrivial statement of p-adic analysis in this case.

Example 26.5.2. Assume $p > 0$ and let M be the differential module of rank 1 over the open unit disc corresponding to the differential equation $y' = y$. By Example 9.5.1,

$$h_1(M, x) = \frac{1}{p-1} \log p \qquad (x \in \mathcal{M}(K\langle t/\beta \rangle)).$$

Lemma 26.5.1 asserts that $\chi(M) = 0$. Since evidently $H^0(M) = 0$, we can corroborate the assertion by computing that $H^1(M) = 0$ as follows.

Write $y = \sum_{i=0}^{\infty} y_i t^i$ and $w = \sum_{i=0}^{\infty} w_i t^i$; the equation $y' - y = w$ implies

$$(i + 1)y_i - y_i = w_i \qquad (i = 0, 1, \dots).$$

In $K[\![t]\!]$, for each choice of y_0 we obtain a unique solution characterized by

$$i! y_i = y_0 + \sum_{j=0}^{i-1} j! w_j.$$

If $y \in K\{t\}$, then $i! y_i \to 0$ as $i \to \infty$, so we must have

$$y_0 = -\sum_{j=0}^{\infty} j! w_j, \qquad y_i = -\sum_{j=i}^{\infty} \frac{j!}{i!} w_j \qquad (i = 1, 2, \dots).$$

Moreover, for any $\rho \in (\omega, 1)$,

$$|y_i| \rho^i \le \sup_{j \ge i}\{|(j-i)!| \|w_j| \rho^i\} \le \sup_{j \ge i}\{\omega^{j-i} |w_j| \rho^i\} \le \sup_{j \ge i}\{\rho^j |w_j|\}$$

and this tends to 0 as $i \to \infty$. Hence $y \in K\{t\}$. (Compare [259, Example 7.9].)

Remark 26.5.3. The statement of Lemma 26.5.1 points to a technical complication in the formulation of a general index formula. Suppose that $\beta = 1$ and that the normalized radii of convergence of M at 0 are $\rho_1 \le \cdots \le \rho_n$. Now suppose we remove the point 0. One might expect that the value of $\Delta H_n(M)$ at $x_{0,1}$ changes from 0 to $-n$, but in fact it becomes $\dim_K H_0(M) - n$. Correspondingly, for any ρ which occurs among $\rho_1, \dots, \rho_n$, the value of $\Delta H_n(M)$ at $x_{0,\nu}$ changes to the number of occurrences of ρ. Again we get cancellation in the overall sum.

We mention a related result which on one hand can be viewed as a generalization of Example 26.5.2, but on the other hand will be a special case of the calculation of local indices.

Lemma 26.5.4. *Let M be a finite differential module of rank n over $K\langle t/\beta\rangle$ such that $f_i(M, -\log\beta) > 0$ for $i = 1, \dots, n$. Then M has an index equal to n plus the right slope of $F_n(M, r)$ at $r = -\log\beta$.*

Proof Using off-centered Frobenius descendants (Theorem 10.8.3), we may reduce to the case where $f_i(M, -\log\beta) > -\log\omega$ for all i. Choose a basis B_1 of M as in the proof of Lemma 11.5.1 and let N_1 be the matrix of action of D on B_1. By Theorem 6.7.4, for r in some right neighborhood of $-\log\beta$, the values $f_i(M, r)$ are measured by the eigenvalues of N_1. Consequently, if we rewrite the action of D on M as

$$\frac{d}{dt} + N_1 = (1 + \frac{d}{dt} N_1^{-1}) N_1$$

we see that $\left|\frac{d}{dt} N_1^{-1}\right|_{F_\beta^n} < 1$ and so $1 + \frac{d}{dt} N_1^{-1}$ is invertible on F_β^n. We cannot make the same conclusion directly on $K\langle t/\beta \rangle^n$ because N_1 is not invertible. However, we may define the projection map

$$\mathrm{pr}\colon N_1^{-1} K\langle t/\beta \rangle^n \to K\langle t/\beta \rangle^n$$

by discarding principal parts, and then on $K\langle t/\beta \rangle^n$ we do have

$$\frac{d}{dt} + N_1 = \left(1 + \frac{d}{dt}(\mathrm{pr} \circ N_1^{-1})\right) N_1.$$

Hence D has the same index as multiplication by N_1, which counts the number of zeroes of $\det(N_1)$. From the Newton polygon of the latter, we obtain the desired inequality. $\qquad\square$

26.6 Local analysis on an annulus

We next study the index on an annulus on which the convergence polygon is affine. This requires close attention to modules satisfying the Robba condition; since we have not yet considered these for $p = 0$, we must fill in the structure theory for such modules here.

Lemma 26.6.1. *Suppose that $p = 0$. Let M be a finite differential module of rank n over $K\langle \alpha/t, t/\beta \rangle$ for the derivation $t\frac{d}{dt}$. Suppose that M admits a basis on which D acts via a matrix $N = \sum_{i \in \mathbb{Z}} N_i T^i$ with $|N_0| \le 1$ and $|N - N_0|_\rho < 1$ for $\rho \in [\alpha, \beta]$. Then there exists a basis of M on which D acts via a matrix over K with norm at most 1.*

Proof Using shearing transformations (Proposition 7.3.10), we may change basis to ensure that the eigenvalues of N_0 have images in κ_K which are prepared. By Lemma 7.3.5, for any $n \times n$ matrix X over K and any nonzero $i \in \mathbb{Z}$,

$$|N_0 X - X N_0 + iX| = |X|.$$

Define the sequence of matrices $U_0, U_1, \ldots$ over $K\langle \alpha/t, t/\beta \rangle$ as follows. Take $U_0 = I_n$. Given U_l, write

$$N_l = U_l^{-1} N U_l + U_l^{-1} D(U_l) = \sum_{i \in \mathbb{Z}} N_{l,i} t^i.$$

By the previous paragraph, there exists a matrix X_l with $|X_l|_\rho = |N_l - N_{l,0}|_\rho$ for $\rho \in [\alpha, \beta]$ and

$$X_l N_0 - N_0 X_l + D(X_l) = N_{l,0} - N_l;$$

put $V_l = I_n - X_l$ and $U_{l+1} = U_l V_l$. For $\epsilon = \max\{|N - N_0|_\rho : \rho \in [\alpha, \beta]\} < 1$, by induction on l, $|N_l - N_{l,0}|_\rho \leq \epsilon^{l+1}$ for $\rho \in [\alpha, \beta]$. Hence the matrices U_l converge to a matrix U, changing basis by which has the desired effect. □

Lemma 26.6.2. *Suppose that $p = 0$. Fix a section s of the homomorphism $K \to K/\mathbb{Z}$. Let M be a finite differential module of rank n for the derivation $t\frac{d}{dt}$ on the open annulus $\alpha < |t| < \beta$ satisfying the Robba condition. Then M admits a basis on which D acts via a matrix over K with eigenvalues in the image of s and of norm at most 1; moreover, the K-span of this basis is uniquely determined by M.*

Proof We first check the uniqueness aspect. Say we are given two bases on which D acts via the matrices N_1 and N_2 with entries in K and eigenvalues in the image of s. The change-of-basis matrix $U = \sum_{i \in \mathbb{Z}} U_i t^i$ from one basis to the other then satisfies $U^{-1} N_1 U + U^{-1} D(U) = N_2$, or

$$\sum_{i \in \mathbb{Z}} (N_1 U_i + U_i N_2 + i U_i) t^i = 0.$$

By Lemma 7.3.5, this forces $U_i = 0$ for $i \neq 0$, proving the claim.

We next establish the existence aspect locally around $\rho_0 \in (\alpha, \beta)$. As in the proof of Lemma 11.5.1, for some γ, δ with $\alpha < \gamma < \rho_0 < \delta < \beta$, we may choose a basis of $M \otimes K\langle \gamma/t, t/\delta \rangle$ on which D acts via a matrix $N = \sum_{i \in \mathbb{Z}} N_i t^i$ for which $|N|_\rho \leq 1$ for $\gamma \leq \rho \leq \delta$. In particular, $|N_0| \leq 1$ and $|N - N_0|_\rho < 1$ for $\gamma \leq \rho \leq \delta$. By Lemma 26.6.1, we can find another basis of $M \otimes K\langle \gamma/t, t/\delta \rangle$ on which D acts via a matrix over K of norm at most 1; using shearing transformations (Proposition 7.3.10), we may further ensure that the eigenvalues of this matrix are in the image of s.

Using the previous paragraph, we may cover the interval (α, β) with the interiors of closed subintervals, on each of which we have a basis of the desired form. By the uniqueness aspect, we can glue these intervals to get larger intervals with the same property; by compactness, we can cover any given closed subinterval of (α, β) in this fashion. This proves the desired result. □

Corollary 26.6.3. *With notation as in Lemma 26.6.2, the following statements hold.*

(a) *The module M is a successive extension of differential modules of rank 1, each of which admits a generator v for which $D(v) = \lambda v$ for some $\lambda \in K$ with $|\lambda| \leq 1$.*

(b) *We have $\chi(M) = 0$.*

Proof Since K is algebraically closed, (a) follows at once from Lemma 26.6.2. Given (a), we may apply Lemma 26.1.3 to reduce (b) to the case where M is free of a single generator $D(v) = \lambda v$ for some $\lambda \in K$ with $|\lambda| \leq 1$. If $\lambda \in \mathbb{Z}$, then we may reduce to the case $\lambda = 0$, and then $H^0(M) = K$ generated by constants while $H^1(M) = K$ via the residue map

$$\sum_i c_i t^i \mapsto c_{-1}.$$

Otherwise, $H^0(M) = 0$ evidently, while $H^1(M) = 0$ because for $x = \sum_i x_i t^i \in K\{\alpha/t, t/\beta\}$, we have $xv = D(yv)$ for

$$y = \sum_i \frac{x_i}{\lambda - i} t^i \in K\{\alpha/t, t/\beta\}.$$

This proves the desired result. □

Lemma 26.6.4. *Let M be a finite differential module of rank n on the open annulus $\alpha < |t| < \beta$. Suppose that the convergence polygon of M is constant. If $p > 0$, suppose in addition that the summand of M satisfying the Robba condition admits an exponent consisting of integers and p-adic non-Liouville numbers. Then $\chi(M) = 0$.*

Proof As in the proof of Lemma 26.5.1, we may decompose M into summands M_i corresponding to distinct slopes of the convergence polygon, and then see from Theorem 12.3.1 that $\chi(M_i) = 0$ whenever M_i correspond to a nonzero slope. The summand M_i corresponding to the slope 0 satisfies the Robba condition; by Corollary 26.6.3, it has index 0 if $p = 0$. If $p > 0$, then by Theorem 13.8.6 and our original hypothesis, we may split M_i as $N_1 \oplus N_2$ where N_1 admits an exponent identically equal to 0 while N_2 admits an exponent containing no integers or p-adic Liouville numbers. By Theorem 13.8.6 again, $\chi(N_2) = 0$. Meanwhile, by Theorem 13.6.1, N_1 is unipotent; using Lemma 26.1.3, we may further reduce to showing that $\chi(N) = 0$ when N is the trivial differential module on the annulus. In this case, we compute that $H^0(N) = H^1(N) = K$ as in the proof of Corollary 26.6.3. □

Remark 26.6.5. Without the condition on the exponent in Lemma 26.6.4, it is not guaranteed that M has an index. This already occurs for M as in Example 9.5.2: if λ has type less than 1, then the cokernel of D is infinite-dimensional over K.

26.7 Some nonarchimedean functional analysis

In order to proceed further, we need a few basic ideas from nonarchimedean functional analysis. As a thorough development would lead us quite far afield, we instead give a summary of some results that can be found in [363] (and elsewhere).

Definition 26.7.1. Throughout this section, let V be a Fréchet space over K. The key examples will be finite free modules over $K\{\alpha/t, t/\beta\}$ (allowing $\alpha = 0$).

Lemma 26.7.2. *Let $f, g : V \to V$ be continuous K-linear endomorphisms. If both f and g are Fredholm (that is, each has an index), then so is $f \circ g$ and*

$$\chi(f \circ g, V) = \chi(f, V) + \chi(g, V).$$

Proof This is a consequence of the exact sequences

$$0 \to \ker(g) \to \ker(f \circ g) \xrightarrow{g} \ker(f) \cap \mathrm{image}(g) \to 0$$

$$0 \to \frac{\ker(f)}{\ker(f) \cap \mathrm{image}(g)} \to \mathrm{coker}(g) \xrightarrow{f} \frac{\mathrm{image}(f)}{\mathrm{image}(f \circ g)} \to 0.$$

(Compare [363, Lemma 22.1].) □

The following definition creates a temporary terminological conflict, and so will only be used in this section; see Remark 26.7.10.

Definition 26.7.3. A *lattice* in V is an $\mathfrak{o}_K$-submodule L of V with the property that every element of V has a nonzero K-multiple contained in L. Equivalently, the natural map $L \otimes_{\mathfrak{o}_K} K \to M$ is surjective, and hence bijective (exercise).

A subset B of V is *bounded* if for any open lattice L, there exists $\lambda \in K$ such that $B \subseteq \lambda L$.

A $\mathfrak{o}_K$-submodule B of V is *c-compact* if for any decreasing filtered family L_i of open lattices of V, the natural map

$$B \to \varprojlim_i B/(B \cap L_i)$$

is surjective.

Remark 26.7.4. Let M be a finite free module over $K\{t/\beta\}$. For any positive integer n, we may obtain an open lattice in M by choosing a positive integer m and taking the preimage in M of the unit ball for some norm on the finite-dimensional K-vector space $M/t^m M$. In particular, if B is a c-compact $\mathfrak{o}_K$-submodule of M, then

$$B \to \varprojlim_m \mathrm{image}(B \to M/t^m M)$$

is surjective.

Remark 26.7.5. Let M be a finite free module over $K\{t/\beta\}$. Fix a basis $e_1, \ldots, e_n$ of M, choose $\beta' \geq \beta$, and let B be the unit ball in M for the supremum norm on the basis $e_1, \ldots, e_n$ with respect to $|\cdot|_{\beta'}$. Then B is a c-compact $\mathfrak{o}_K$-submodule of M.

Definition 26.7.6. Let $g : V_1 \to V_2$ be a continuous K-linear transformation of Fréchet spaces. We say g is *compact* if there is an open lattice L in V_1 such that the closure of the image $g(L)$ is bounded and c-compact in V_2.

Remark 26.7.7. Using similar formalism, one can define the notion of a compact linear transformation between locally convex topological K-vector spaces. In the context of Banach spaces over K, by [363, Remark 18.10] a continuous linear transformation is compact if and only if it is *completely continuous*, meaning that it is in the closure (for the operator norm) of the space of finite-rank endomorphisms. For example, for $0 < \beta' < \beta$, the inclusion

$$K\langle t/\beta \rangle \to K\langle t/\beta' \rangle$$

is completely continuous because it is the limit of its compositions with the projection maps

$$\sum_{i=0}^{\infty} c_i t^i \mapsto \sum_{i=0}^{n} c_i t^i \qquad (n = 0, 1, \ldots).$$

Proposition 26.7.8. *Within the noncommutative ring of continuous K-linear endomorphisms of V, the compact endomorphisms form a two-sided ideal. That is, the sum of two compact endomorphisms is compact, and the composition of a compact endomorphism with a continuous endomorphism (on either side) is again compact.*

Proof See [363, Remark 16.7.i]. $\qquad\qquad\qquad\qquad\qquad\qquad\qquad\qquad$ □

Lemma 26.7.9. *Let $f : V \to V$ be a continuous K-linear endomorphism.*

(a) *The map f is Fredholm (i.e., f has an index) if and only if f has an inverse modulo the ideal (Proposition 26.7.8) of compact endomorphisms.*
(b) *If f is Fredholm, then for any compact endomorphism $g : V \to V$, we have*
$$\chi(f + g, V) = \chi(f, V).$$

Proof For (a), see [363, Corollary 22.11]. For (b), we first note that if f is the identity map, then the equality $\chi(f + g, V) = 0$ is a consequence of [363, Corollary 22.9]. For general f, apply (a) to construct a continuous K-linear endomorphism $h : V \to V$ such that $\mathrm{id}_V - f \circ h$ and $\mathrm{id}_V - h \circ f$ are both compact. By Lemma 26.7.2,

$$\chi(f+g,V) = \chi(h \circ f + h \circ g, V) - \chi(h,V)$$
$$= \chi(\mathrm{id}_V + (f \circ h - \mathrm{id}_V + h \circ g), V) + \chi(f, V)$$

which equals $\chi(f,V)$ because $f \circ h - \mathrm{id}_V + h \circ g$ is compact.   □

Remark 26.7.10. The notion of a lattice in a Fréchet space is incompatible with our earlier (and probably more familiar) notation of a lattice in a finite-dimensional K-vector space V, which is a *finitely generated* $\mathfrak{o}_K$-submodule L such that the natural map $L \otimes_{\mathfrak{o}_K} K \to V$ is an isomorphism. If we instead view V as a Fréchet space, then the definition of a lattice no longer imposes the finite generation condition on L, so for instance V becomes a lattice in itself. Fortunately, at this point we will make no further reference to the notion of a lattice in a Fréchet space.

26.8 Plus and minus indices

We now consider finite free modules over an open annulus more carefully, and separate the index into two separate contributions that can be studied individually.

Definition 26.8.1. Fix $\alpha < \beta$ and let M be a finite free module over $K\{\alpha/t, t/\beta\}$. Choose a basis $e_1, \ldots, e_n$ of M and define the K-subspaces

$$M_- = K\{t/\beta\}e_1 \oplus \cdots \oplus K\{t/\beta\}e_n$$
$$M_+ = t^{-1}K\{\alpha/t\}e_1 \oplus \cdots \oplus t^{-1}K\{\alpha/t\}e_n,$$

so that $M = M_- \oplus M_+$. Define also the inclusion maps

$$\iota_-: M_- \to M, \qquad \iota_+: M_+ \to M$$

and the projection maps

$$\pi_-: M \to M_-, \qquad \pi_+: M \to M_+.$$

We consider the action of $\frac{d}{dt}$ which is coefficientwise with respect to the chosen basis, that is,

$$f_1 e_1 + \cdots + f_n e_n \mapsto \frac{df_1}{dt} e_1 + \cdots + \frac{df_n}{dt} e_n.$$

Lemma 26.8.2. *Let f be a $K\{\alpha/t, t/\beta\}$-linear endomorphism of M. Then the maps $\pi_+ \circ f \circ \iota_-: M_- \to M_+$ and $\pi_- \circ f \circ \iota_+: M_+ \to M_-$ are compact.*

Proof See [99, Proposition 8.2-2].   □

Remark 26.8.3. On a related note, the maps

$$\pi_+ \circ \frac{d}{dt} \circ \iota_-, \pi_- \circ \frac{d}{dt} \circ \iota_+ \colon K\{\alpha/t, t/\beta\} \to K\{\alpha/t, t/\beta\}$$

are of finite rank, and hence compact.

Definition 26.8.4. Let $f \colon M \to M$ be a continuous K-linear endomorphism. We define the *minus index* and the *plus index* of f to be respectively

$$\chi_-(f, M) = \chi(\pi_- \circ f \circ \iota_-, M_-), \qquad \chi_+(f, M) = \chi(\pi_+ \circ f \circ \iota_+, M_+).$$

Lemma 26.8.5. *Let* $f, g \colon M \to M$ *be* $K\{\alpha/t, t/\beta\}$-*linear endomorphisms of* M. *Then*

$$\chi_-(f \circ g, M) = \chi_-(f, M) + \chi_-(g, M),$$

$$\chi_+(f \circ g, M) = \chi_+(f, M) + \chi_+(g, M),$$

$$\chi_-(f \circ (\frac{d}{dt} + g), M) = \chi_-(f, M) + \chi_-((\frac{d}{dt} + g), M),$$

$$\chi_+(f \circ (\frac{d}{dt} + g), M) = \chi_+(f, M) + \chi_+((\frac{d}{dt} + g), M),$$

$$\chi_-((\frac{d}{dt} + f) \circ g, M) = \chi_-((\frac{d}{dt} + f), M) + \chi_-(g, M),$$

$$\chi_+((\frac{d}{dt} + f) \circ g, M) = \chi_+((\frac{d}{dt} + f), M) + \chi_+(g, M).$$

Proof Write

$$\pi_- \circ f \circ g \circ \iota_- = \pi_- \circ f \circ (\iota_- \circ \pi_- + \iota_+ \circ \pi_+) \circ g \circ \iota_-$$

and apply Lemma 26.7.9, Lemma 26.8.2, and Remark 26.8.3. $\square$

Although we will not use this fact, we mention for completeness the following closely related fact.

Lemma 26.8.6. *Let* f *be a* $K\{\alpha/t, t/\beta\}$-*linear endomorphism of* M. *Then*

$$\chi(f, M) = \chi_-(f, M) + \chi_+(f, M),$$

$$\chi(\frac{d}{dt} + f, M) = \chi_-(\frac{d}{dt} + f, M) + \chi_+(\frac{d}{dt} + f, M).$$

Proof Write

$$f = (\iota_- \circ \pi_- + \iota_+ \circ \pi_+) \circ f \circ (\iota_- \circ \pi_- + \iota_+ \circ \pi_+)$$

and apply Lemma 26.7.9, Lemma 26.8.2, and Remark 26.8.3. $\square$

Lemma 26.8.7. *Let f be a $K\{\alpha/t, t/\beta\}$-linear endomorphism of M. Then the plus and minus indices of f and $\frac{d}{dt} + f$ are independent of the choice of the basis $e_1, \ldots, e_n$.*

Proof Let $v_1, \ldots, v_n$ be a second basis and let U be the change-of-basis matrix from $e_1, \ldots, e_n$ to $v_1, \ldots, v_n$. We may then apply Lemma 26.8.5 to see that the plus and minus indices of f and $U^{-1} \circ f \circ U$ coincide, and that the plus and minus indices of $\frac{d}{dt} + f$ and $U^{-1} \circ (\frac{d}{dt} + f) \circ U$ coincide. □

26.9 Global analysis on a disc

We now obtain a general index theorem on an open disc.

Definition 26.9.1. Let M be a finite differential module of rank n on the open annulus $\alpha < |t| < 1$ for some $\alpha \in (0, 1)$, which is tame at $x_{0,1}$. By Theorem 12.3.1, for some $\gamma \in (\alpha, 1)$, the restriction of M to the annulus $\gamma < |t| < 1$ admits a direct sum decomposition $M_1 \oplus M_2$ where

$$\lim_{r \to 0^+} f_i(M_1, r) = 0 \qquad (i = 1, \ldots, \operatorname{rank}(M_1))$$

$$\lim_{r \to 0^+} f_i(M_2, r) > 0 \qquad (i = 1, \ldots, \operatorname{rank}(M_2)).$$

In particular, M_1 is solvable at 1, so (for some γ) it further decomposes according to its differential slopes (Theorem 12.6.4). The component of M_1 of differential slope 0 will also be called the *Robba component*, as it satisfies the Robba condition.

Theorem 26.9.2. *Let U be the open disc $|t| < \beta$. Let Z be a finite nonempty set of type-1 points of U. Let M be a finite differential module of rank n on U. Suppose that M is tame at $x_{0,1}$ and that (if $p > 0$) for some $\alpha \in (0, 1)$, the Robba component of the restriction of M to the annulus $\alpha < |t| < 1$ admits an exponent with p-adic non-Liouville differences. Then M has an index given by*

$$\chi(M) = n - \int_{U \setminus Z} \Delta H_n(M).$$

See Remark 26.5.3 for the reason why we insist that Z be nonempty.

Proof We begin with a reduction step. Let V be an annulus of the form $\alpha < |t| < \beta$ for some α, and let U' be a disc of the form $|t| < \beta'$ with $\alpha < \beta' < \beta$. By taking α sufficiently large and applying Lemma 26.6.4 and Lemma 26.2.4 to the covering of U by U' and V, we see that both sides of the desired equality remain unchanged upon replacing U with U' (here we have used the hypothesis that M is tame at $x_{0,1}$). In particular, we may assume that M admits a basis and that for $i = 1, \ldots, n$, either $f_i(M, r) = r$ for r in some right neighborhood of $-\log \beta$, or $f_i(M, r) - r$ has a positive limit as $r \to (-\log \beta)^+$.

Let N be the matrix of action of D on some basis of M. By Lemma 26.8.7, for any $\alpha \in (0, \beta)$,

$$\chi(M) = \chi_-(\frac{d}{dt} + N, M \otimes K\{\alpha/t, t/\beta\}).$$

In particular, choose α so that $M \otimes K\{\alpha/t, t/\beta\}$ admits a spectral decomposition. By Lemma 26.8.7, we may compute $\chi(M)$ by separately accounting for the different terms in the spectral decomposition. The Robba component evidently contributes 0; the other components may be handled by imitating the proof of Lemma 26.5.4. $\square$

As a corollary, we can formulate an improvement of Theorem 11.3.2(d).

Theorem 26.9.3. *In Theorem 11.3.2(d), suppose that if $p > 0$, then for every α', β' with $\alpha < \alpha' < \beta' < \beta$ such that the restriction of M to the annulus $\alpha' < |t| < \beta'$ admits a spectral decomposition, the Robba component admits an exponent with p-adic non-Liouville differences. Then $H_n(M, r)$ is everywhere nondecreasing.*

Proof It suffices to compute the slope on an annulus $\alpha' < |t| < \beta'$ on which M admits a spectral decomposition. In light of our hypothesis, we may apply Theorem 26.9.2 to the disc $|t| < \beta'$, taking $Z = \{0\}$. We may assume that M has no horizontal sections on this disc (otherwise, quotienting by the submodule generated by these sections does not change anything); the claim then follows from the fact that for the restriction of M to this disc, the dimension of H^1 is nonnegative. $\square$

26.10 A global index formula

In a similar vein, we obtain an index formula on a general unbounded affinoid subset.

Definition 26.10.1. For U an unbounded affinoid subspace of $\mathbb{P}_K$ and Z a finite subset of type-1 points of U, define

$$\chi(U \setminus Z) = 2 - \#(\delta U \cup Z).$$

Note that if U_1, U_2 are unbounded affinoid subspaces of $\mathbb{P}_K$ with union U and

$U_{12} = U_1 \cap U_2$, then

$$\chi(U \setminus Z) = \chi(U_1 \setminus Z) + \chi(U_2 \setminus Z) - \chi(U_{12} \setminus Z) \qquad (26.10.1.1)$$

(exercise).

Theorem 26.10.2 (Poineau–Pulita). *Let U be an unbounded affinoid subset of* $\mathbb{P}_K$. *Let Z be a finite set of type-1 points of U, which is nonempty in case U is a disc. Let M be a finite differential module of rank n over* $U \setminus Z$. *Suppose that for each* $x \in \delta U$, *there is an annulus in U bounded by x on which M is tame and (if* $p > 0$) *its Robba component admits an exponent with p-adic non-Liouville differences. Then M has an index equal to*

$$\chi(M) = n\chi(U \setminus Z) - \int_{U \setminus Z} \Delta H_n(M).$$

Proof The proof is similar to Theorem 26.9.2, but in lieu of giving the details we refer the interested reader to [337, Theorem 3.5.2]. □

Notes

The study of the index of a differential module is one of the basic questions motivating the study of p-adic differential equations, so unsurprisingly there is a long literature devoted to this question. Some older references are [9, 39, 349, 351, 352, 353, 415].

The study of local indices on annuli, in the case of a differential module which is solvable at one boundary, is the primary purpose of the series of articles [97, 98, 99, 100]. Some results in the nonsolvable case were given by Pons [339].

Our presentation, while drawing much from earlier references, is mostly modeled on the presentation in [259], which is in turn based on the papers of Poineau–Pulita [345, 335, 336, 337, 338]. In particular, Theorem 26.4.2 is [259, Theorem 7.7].

The description of the GAGA principle given in SGA 1 (see the notes for Chapter 7) makes it feasible to prove a corresponding result relating algebraic varieties and rigid analytic varieties over a nonarchimedean field. This was first done in the PhD thesis of Köpf [276]; a modern reference is [112, Appendix A].

Somewhat surprisingly, we are unaware of a "classical" reference for the analysis of Section 26.6. The source we follow is [255, §3.3].

The concepts from functional analysis used here (especially the study of compact operators) were largely systematized by Grothendieck, well before his more famous work in algebraic geometry; see [180]. Our application of these concepts to studying indices of differential modules is closely modeled on [99, §8]. In the context of rigid analytic geometry, similar ideas appear in the proof of Kiehl's theorem that the higher direct images of a coherent sheaf along a proper morphism of rigid analytic spaces are coherent [272].

The formulation of [337, Theorem 3.5.2] is slightly weaker than Theorem 26.10.2 in case $p > 0$: therein the condition on p-adic exponents is imposed not just near the points of δU, but along every edge of the controlling graph.

Exercises

26.1   Let V be a Fréchet space over a nonarchimedean field K. Prove that for any $\mathfrak{o}_K$-submodule L of V, the natural map $L \otimes_{\mathfrak{o}_K} K \to V$ is injective.

26.2   Verify (26.10.1.1).

27

Local constancy at type-4 points

We conclude this discussion by establishing the following refinement of Theorem 25.4.4, taken from [255, Theorem 4.5.15].

Theorem 27.0.1. *Let M be a finite differential module of rank n over $K\langle t/\beta \rangle$. For any $x \in M(K\langle t/\beta \rangle)$ of type 4, the convergence polygon of M is constant on some neighborhood of x.*

Corollary 27.0.2. *With notation as in Theorem 25.4.4, the convergence polygon of M factors through the retraction onto some strict skeleton.*

The proof requires an application of the p-adic local monodromy theorem without Frobenius structures (Chapter 22).

27.1 Geometry around a point of type 4

We start with some reminders about the geometry of $\mathbb{P}_K$ in the neighborhood of a point of type 4.

Remark 27.1.1. Fix $x \in \mathbb{P}_K$ of type 4. Then the radius $\rho(x)$ is positive, and the root path from x consists of the points $H(x, \rho)$ for $\rho \in [\rho(x), \infty]$.

Lemma 27.1.2. *For $x \in \mathbb{P}_K$ of type 4 and $P \in K\langle t/\beta \rangle$ for some $\beta > \rho(x)$, there exists $\rho_0 \in (\rho(x), \infty)$ such that $H(x, \rho)(P)$ is constant for $\rho \in [\rho(x), \rho_0]$.*

Proof Since K is algebraically closed, it suffices to check the claim for P of the form $t - z$ with $z \in K$. Since x is of type 4, by Corollary 24.3.4 it does not lie on the root path from $x_{0,z}$; consequently, there is a least value ρ_0 of ρ for which $H(x, \rho) = H(x_{0,z}, \rho)$. This value has the desired property. $\quad\square$

Corollary 27.1.3. *For $x \in \mathbb{P}_K$ of type 4, the discs containing x form a neighborhood basis of $\mathbb{P}_K$ at x.*

Proof Let U be a neighborhood of x in $\mathbb{P}_K$ of the form

$$U = \{x \in \mathbb{P}_K : x(P_1) \in I_1, \dots, x(P_n) \in I_n\}$$

for some polynomials $P_1, \dots, P_n \in K[t]$ and some open intervals $I_1, \dots, I_n$. By Lemma 27.1.2, there exists $\rho_0 \in (\rho(x), \infty)$ such that for $i = 1, \dots, n$, $H(x, \rho)(P_i)$ is constant for $\rho \in [\rho(x), \rho_0]$. Hence U contains the open disc of radius ρ_0 containing x. □

Definition 27.1.4. As per Remark 24.2.4, we can find an algebraically closed nonarchimedean field K_1 containing K such that x is the restriction of a type-1 point $x_{z,0} \in \mathbb{P}_{K_1}$. By Corollary 24.2.3, we may even assume that $|K_1^\times| = |K^\times|$ and $\kappa_{K_1} = \kappa_K$.

27.2 Local constancy in the visible range

Lemma 27.2.1. *With notation as in Theorem 27.0.1, for r in some left neighborhood of $-\log \rho(x)$, the function*

$$r \mapsto \max\{r - \log \omega, h_i(M, H(x, e^{-r}))\}$$

is either constant or identically equal to $r - \log \omega$.

Proof We deduce this by choosing a cyclic vector of $M \otimes_{K\langle t/\beta \rangle} \mathrm{Frac}(K\langle t/\beta \rangle)$ (Theorem 5.4.2), directly measuring the visible spectrum from the resulting twisted polynomial (Corollary 6.5.4), and applying Lemma 27.1.2 to each coefficient of that polynomial. □

Corollary 27.2.2. *Theorem 27.0.1 holds if $p = 0$.*

Proof This follows immediately from Lemma 27.2.1. □

Remark 27.2.3. Let us now consider the situation of Theorem 27.0.1 when $p > 0$. We proceed by induction on $\mathrm{rank}(M)$.

Suppose first that $h_1(M, x) > -\log \rho(x)$. By applying Lemma 27.2.1 to a suitable off-center Frobenius descendant (Theorem 10.8.3), we may deduce that $h_1(M, H(x, e^{-r}))$ is constant for r in some left neighborhood of $-\log \rho(x)$. We may then separate off a direct summand of M using Theorem 12.2.2, for which Lemma 25.3.4 implies constancy of the convergence polygon in a neighborhood of x, and then continue using the induction hypothesis.

Suppose next that $h_1(M, x) < -\log \rho(x)$. In this case, the convergence polygon is evidently constant on a neighborhood of x, as in the proof of Lemma 25.3.4.

We are thus left with the case where $h_1(M, x) = -\log \rho(x)$. By Corollary 11.8.2 applied at z, this case can only occur if $\rho(x) \in |K_1^\times| = |K^\times|$. In this case, we cannot hope to apply arguments about the visible spectrum directly, so we must instead look carefully at how things behave as we approach x. We take up this point of view in the next section.

27.3 Local monodromy at a point of type 4

Motivated by the previous discussion, we bring the Robba ring into the picture.

Definition 27.3.1. With notation as in Remark 27.2.3, define the rings

$$\mathcal{E}^\dagger_{x, K_1} = \bigcup_{\beta > \rho(x)} K[\![\rho(x)/(t - z), (t - z)/\beta]\!]_0,$$

$$\mathcal{R}_{x, K_1} = \bigcup_{\beta > \rho(x)} K\{\rho(x)/(t - z), (t - z)/\beta\}.$$

That is, $\mathcal{R}_{x, K_1}$ is an "inside-out" copy of the Robba ring over K_1: it consists of functions analytic on an open annulus with fixed *inner* boundary. If $\rho(x) \in |K^\times|$, we may choose $\lambda \in K$ with $|\lambda| = \rho(x)$ and then view $\mathfrak{o}_{\mathcal{E}^\dagger_{x, K_1}}$ as a local ring with residue field $\kappa_K((u^{-1}))$ where u is the reduction of $(t - z)/\lambda$.

Definition 27.3.2. Set notation as in Remark 27.2.3 and assume that $h_1(M, x) = -\log \rho(x)$. Then $M \otimes_{K\langle t/\beta\rangle} \mathcal{R}_{x, K_1}$ is solvable, so we may apply Corollary 22.7.4 to construct a unique minimal extension L of $\kappa_K((u^{-1}))$ corresponding to a finite extension of $\mathcal{R}_{x, K_1}$ over which M becomes a module of differential slope 0.

In order to make real use of this information, we need to eliminate the effect of the base extension from K to K_1. For $c \in \mathfrak{o}_K$, the map $t \mapsto t + c\lambda$ on $\mathbb{P}_K$ fixes $H(x, \rho)$ for any $\rho > \rho(x)$, so by continuity it also fixes x. By the uniqueness of L, we obtain a homomorphism from the additive group of κ_K to the group of continuous κ_K-linear automorphisms of L sending $\kappa_K((u^{-1}))$ into itself, where the action of $c \in \kappa_K$ sends u to $u + c$.

To study the effect of the existence of this group action on the ramification breaks of L, we compute in a somewhat more general situation. This allows to run an inductive argument.

Lemma 27.3.3. *Let k be an algebraically closed field of characteristic p. Let L be a finite Galois extension of $k((t))$. Suppose that for some positive integer m, there exists a homomorphism ψ from the additive group of k to the group of* *continuous k-linear automorphisms of L carrying $k((t))$ into itself such that for all $c \in k$,*

$$\psi(c)(t) = t + \sum_{j=m+1}^{\infty} P_j(c)t^j$$

for some polynomials $P_j(x) \in k[x]$ such that P_{m+1} is not a p-th power. Then the jumps in the upper numbering filtration of $G_{L/k((t))}$ are all at most m; that is, $G^i_{L/k((t))} = \{e\}$ for $i > m$.

Proof We first check the claim in the case where $L/k((t))$ is totally ramified and $G_{L/k((t))} \cong \mathbb{Z}/p\mathbb{Z}$, so that $L/k((t))$ is an Artin–Schreier extension (see Remark 3.3.10). Let e be the unique ramification break of $L/k((t))$; then e is not divisible by p and $L = k((t))[z]/(z^p - z - x)$ where $x = \sum_{j \geq -e} a_j t^j$ with $a = a_{-e}$ nonzero (see the exercises for Chapter 3).

Suppose by way of contradiction that $e > m$; we can then write

$$\psi(c)(x) - x \equiv \sum_{j=m-e}^{-1} Q_j(c)t^j \quad (\mathrm{mod}\ k[[t]])$$

for some polynomials $Q_j(x) \in k[x]$, where $Q_{m-e} = -eaP_{m+1}$. Since $\psi(c)$ extends to L, the elements x and $\psi(c)(x)$ define the same Artin–Schreier extension of $k((t))$, and so $\psi(c)(x) - x$ must have the form $y^p - y$ for some $y \in k((t))$. The elements of this shape form an additive subgroup, modulo which $\psi(c)(x) - x$ is congruent to

$$\sum_{\substack{h>0 \\ p \nmid h}} \left(\sum_{i=0}^{\infty} Q_{-hp^i}(c)^{p^{-i}} \right) t^{-h};$$

in this expression, for each h the coefficient of t^{-h} must vanish identically as a function of c. In particular, if we write $e - m = hp^j$ with h not divisible by p, then

$$\sum_{i=0}^{j} Q_{-hp^i}(x)^{p^{j-i}}$$

is a polynomial which vanishes for every $x \in c$, and hence is the zero polynomial (because k is an infinite field). However, $Q_{-hp^j} = -eaP_{m+1}$ is not a p-th power and so this is a contradiction. We deduce that $e \leq m$, as claimed.

We next reduce the general case to this one. Since we assumed k is algebraically closed, $L/k((t))$ is totally ramified. By replacing $k((t))$ with the maximal tamely subextension of $L/k((t))$ and multiplying m by the degree of that extension, we may reduce to the case where $L/k((t))$ is totally wildly ramified. Let e be the first jump in the upper numbering filtration of $G_{L/k((t))}$; the corresponding subextension F_e of $L/k((t))$ has Galois group which is abelian of exponent p, and the action of $\psi(c)$ must preserve F_e. Moreover, the resulting action of k on $G_{F_e/k((t))}$ must be trivial because the additive group has no nontrivial finite quotient. This shows that if we pick a $\mathbb{Z}/p\mathbb{Z}$-subextension F of F_e, then $\psi(c)$ also acts on F. By the previous paragraph, this forces $e \le m$. Moreover, ψ again satisfies the hypotheses of the lemma for the extension L/F_e and the integer $m' = (m - e)p + e$, so the jumps in the upper numbering filtration of G_{L/F_e} are at most m'. By converting back and forth between upper and lower numbering and applying Proposition 3.4.4, we deduce that the remaining jumps in the upper numbering filtration of $G_{L/k((t))}$ are also at most m. □

Corollary 27.3.4. *With notation as in Definition 27.3.2, the jumps in the upper numbering filtration of $L/k((t))$ are all in the range $[0, 1]$.*

Proof Apply Lemma 27.3.3 to the homomorphism constructed in Definition 27.3.2: the action of $c \in \kappa_K$ on $\kappa_K((u^{-1}))$ sends u^{-1} to $(u + c)^{-1} = u^{-1} - cu^{-2} + \cdots$, so the lemma applies with $m = 1$. □

27.4 End of the proof

To conclude, we combine our knowledge about the ramification of L with some information from the side of differential modules.

Proof of Theorem 27.0.1 From the discussion in Remark 27.2.3, we are reduced to treating the case where $\rho(x) \in |K^\times|$ and $h_1(M, x) = -\log \rho(x)$. By Theorem 19.4.1 and Corollary 27.3.4, for r in some left neighborhood of $-\log \rho(x)$, the functions $h_i(M, H(x, e^{-r}))$ are affine with slopes in $[0, 1]$. However, the sum of these slopes is nonpositive by Theorem 11.3.2(d), so the individual slopes must all be 0. We may thus apply Lemma 25.3.4 to conclude. □

Notes

Lemma 27.3.3 is taken from [255, Lemma 1.2.5, Proposition 1.2.6]. The remainder of the proof of Theorem 27.0.1 follows [255, Lemma 4.4.4, Lemma 4.4.5, Lemma 4.5.13]; these arguments are in turn adapted from [249, §5].

APPENDICES AND BACK MATTER

Appendix A

Picard–Fuchs modules

In these appendices, we touch briefly on some areas of application of the theory of p-adic differential equations. These are intended more to inspire than to inform, with statements that are more illustrative than definitive.

In this appendix, we revisit the territory of Chapter 0, briefly discussing how Picard–Fuchs modules give rise to differential equations with Frobenius structures, and what this has to do with zeta functions.

A.1 Picard–Fuchs modules

The original source for the p-adic differential equations that inspired the general theory is the following construction.

Definition A.1.1. Let K be a field of characteristic 0. Let t be a coordinate on the projective line $\mathbb{P}^1_K$. Let $f: X \to \mathbb{P}^1_K$ be a proper, flat, generically smooth morphism of algebraic varieties. Let $S \subset \mathbb{P}^1_K$ be a zero-dimensional subscheme containing ∞ (for convenience) and all points over which f is not smooth. The *Picard–Fuchs modules* on $\mathbb{P}^1_K \setminus S$ associated to f are certain finite locally free differential modules M_i for $i = 0, \ldots, 2\dim(f)$ over $\Gamma(\mathbb{P}^1_K \setminus S, O)$ with respect to the derivation $\frac{d}{dt}$; they also have regular singularities with rational exponents at each point of S. For $\lambda \notin S$, the fiber of M_i at λ can be canonically identified with the i-th de Rham cohomology of the fiber $f^{-1}(\lambda)$.

Although the classical construction of the Picard–Fuchs module is analytic (it involves viewing f as an analytically locally trivial fibration and integrating differentials against moving homology classes), there is an algebraic construction due to Katz and Oda [228], involving a Leray spectral sequence for the algebraic de Rham cohomology of the total space.

Example A.1.2. Perhaps the most fundamental example of a Picard–Fuchs module comes from the Legendre family of elliptic curves

$$y^2 = x(x - 1)(x - t)$$

with $S = \{0, 1, \infty\}$. In this case, the Picard–Fuchs module is precisely the differential module derived from the hypergeometric differential equation

$$y'' + \frac{(c - (a + b + 1)z)}{z(1 - z)} y' - \frac{ab}{z(1 - z)} y = 0$$

with parameters $(a, b, c) = (1/2, 1/2, 1)$, as considered in considered in Chapter 0. See [402, §7] for an algebraic derivation of this fact, in the manner of Katz and Oda.

A.2 Frobenius structures on Picard–Fuchs modules

The example studied in Chapter 0 was just one of a number of explicit examples studied by Dwork and others, in which there seemed to be some strong relationship between the Picard–Fuchs equations derived in characteristic 0 and the zeta functions observed over finite fields. Dwork was able to give a somewhat systematic explanation in some cases in terms of Frobenius structures; nowadays, the technology of p-adic cohomology (more about which in Appendix B) can be used to give a fairly general explanation.

We first give an explicit statement to the effect that Picard–Fuchs modules should always carry Frobenius structures. See notes for detailed discussion.

Theorem A.2.1. *With notation as above, assume that the field K is complete for a discrete valuation. Also suppose that f extends to a proper morphism $\mathfrak{X} \to \mathbb{P}^1_{\mathfrak{o}_K}$ such that the intersection of $\mathbb{P}^1_k$ with the nonsmooth locus is contained in the intersection of $\mathbb{P}^1_k$ with the Zariski closure of S (i.e., the morphism is smooth over all points of $\mathbb{P}^1_k$ that are not the reductions of points in S). Let M_i be the i-th Picard–Fuchs module for f, and let $\varphi: \mathbb{P}^1_{\mathfrak{o}_K} \to \mathbb{P}^1_{\mathfrak{o}_K}$ be a Frobenius lift (e.g., $t \mapsto t^p$) that acts on $\mathfrak{o}_K$ as a lift of the absolute Frobenius. Then for some $\alpha \in (0, 1)$, there exists an isomorphism $\varphi^*(M_i) \cong M_i$ over a ring R which is the Fréchet completion of $\Gamma(\mathbb{P}^1_K \setminus S, O)$ for the ρ^{-1}-Gauss norm and the Gauss norms $|t - \lambda| = \rho$ for $\rho \in [\alpha, 1)$ and $\lambda \in S$.*

Remark A.2.2. Geometrically, the Frobenius structure is defined on the complement in $\mathbb{P}^1_K$ of a union of discs around the points of S, each of radius less than 1. (This includes a disc of radius less than 1 around ∞, by which we mean the complement of a disc of radius greater than 1 around 0.) In particular, by working in a unit disc not containing any points of S, we obtain a differential module with Frobenius structure over $K[\![t]\!]_0$. In a unit disc containing one or more points of S, we only obtain a differential module with Frobenius structure over $\bigcup_{\alpha>0} K\langle \alpha/t, t]\!]_0$. If the disc contains exactly one point of S and the exponents at that point are all 0, we can also get a differential module with Frobenius structure over $K[\![t]\!]_0$ for the derivation $t\frac{d}{dt}$, provided that φ fixes that point.

Example A.2.3. In Example A.1.2, Theorem A.2.1 applies directly except when $p = 2$. In that case, the reduction modulo p fails to be generically smooth; one must change the defining equation to get a usable description mod 2.

Remark A.2.4. As in Example 26.4.5, by combining Theorem A.2.1 with Theorem 13.7.1 (and the fact that rational numbers are not p-adic Liouville numbers), one can deduce that the convergence polygon of a Picard–Fuchs module is identically zero.

A.3 Relationship with zeta functions

We next give an explicit statement to the effect that the Frobenius structure on a Picard–Fuchs module can be used to compute zeta functions. (The condition on λ allows for a unique choice in each residue disc, by Lemma 15.2.7.)

Theorem A.3.1. *Retain notation as in Theorem A.2.1, and assume now that* $\kappa_K = \mathbb{F}_q$ *with* $q = p^a$, *and that* φ *is a q-power Frobenius lift on* $\mathbb{P}^1_{\mathfrak{o}_K}$. *Suppose that* $\lambda \in \mathfrak{o}_K$ *satisfies* $\varphi(t - \lambda) \equiv 0 \pmod{t - \lambda}$, *and suppose that* f *extends smoothly over the residue disc containing* λ. *Then*

$$\zeta(f^{-1}(\overline{\lambda}), T) = \prod_{i=0}^{2\dim(f)} \det(1 - T\Phi, (M_i)_\lambda)^{(-1)^{i+1}}.$$

This suggests an interesting strategy for computing zeta functions, described by Alan Lauder.

Remark A.3.2. Suppose we have in hand the differential module M_i, plus the action of Φ on some individual $(M_i)_\lambda$. (This data would ordinarily be specified by a basis of M_i, the matrix of action of D, and the matrix of action of Φ modulo $t - \lambda$.) View the equation

$$NA + \frac{dA}{dt} = \frac{d\varphi(t)}{dt} A\varphi(N)$$

as a differential equation with initial condition provided by $(M_i)_\lambda$; we may then solve for A and then evaluate at another λ.

More explicitly, let us suppose for simplicity that $\lambda = 0$ is the starting value and $\lambda = 1$ is the target value. In the open unit disc around 0, we can compute U such that

$$U^{-1}NU + U^{-1}\frac{dU}{dt} = 0$$

and then write down

$$A = UA_0\varphi(U^{-1}).$$

This only gives a power series representation around $t = 0$ with radius of convergence 1, which does not give any way to specialize to $\lambda = 1$.

However, Theorem A.2.1 implies that the entries of A can be written as uniform limits of rational functions with limited denominators. For the purposes of computing the zeta function, we can limit how much p-adic accuracy is needed in the computations by giving some bounds on the degrees of the polynomials which appear and the sizes of the coefficients, say by using the Weil conjectures (Theorem 0.2.5). To obtain this much accuracy, we thus must compute A modulo some particular power of p. This means we must determine some rational function whose degree we can (in principle) control, so it suffices to determine suitably many terms in the power series expansion around 0. We can then reconstruct a sufficiently good rational function approximation to A, and then evaluate at $\lambda = 1$.

Remark A.3.3. One can recover from Theorem A.3.1 the example of Dwork discussed in Chapter 0. In that example, one is separating the Picard–Fuchs module, which has rank 2, into a unit-root component and a component of slope $\log p$. For this to be possible, one must be in the situation of Theorem 15.3.4; this fails precisely at the residue discs at which the Igusa polynomial vanishes, which is why one must invert the Igusa polynomial in the course of the computation.

Notes

The notion of algebraic de Rham cohomology was introduced by Grothendieck in [182], where he gave an algebraic construction of the topological de Rham cohomology of a complex algebraic variety. The construction involves commingling the cohomology of the de Rham complex (made out of the module of Kähler differentials) with the cohomology of coherent algebraic sheaves. The subject was further developed by Hartshorne [194, 195]. A thorough treatment of Grothendieck's theorem can be found in [179].

In the final footnote to [182], Grothendieck gives what we believe is the first suggestion to consider generally what we are calling Picard–Fuchs modules. His suggestion is based on Manin's work on the Mordell conjecture over function fields; partly for this reason, Picard–Fuchs modules are also commonly known as *Gauss–Manin connections*. This terminology is apparently due to Grothendieck. The name-check of Gauss presumably refers to Gauss's study of hypergeometric differential equations, which incorporated the original Picard–Fuchs equation governing periods of elliptic curves (Example A.1.2). A good collection of onward references can be found in [199].

The fact that a Picard–Fuchs module has regular singularities with rational exponents can be proved in several ways, both geometric (Griffiths, Landman) and arithmetic (Katz). One can also formulate a more abstract version, concerning polarized variations of Hodge structures; this was proved by Borel, and extended by Schmid (see the notes for Chapter 20 for further discussion of this last point). Again, see [199] for references.

The construction of Frobenius structures on Picard–Fuchs modules is a consequence of general results in the theory of p-adic cohomology, which we discuss further in Appendix B. For the moment, we point out that Theorem A.2.1 in its stated form can be found in [397, Theorem 3.3.1].

Lauder's strategy for computing zeta functions (also called the *deformation method*) was introduced in [286]; it has been worked out in detail for hyperelliptic curves by Hubrechts [214, 215]. A version for hypersurfaces has been described by Gerkmann [177], with some improvements by Pancratz and Tuitman [332] and Tuitman [401]. See also [243] and [252] for expository treatments.

From a computational point of view, some particularly accessible examples are the Frobenius structures (and intertwiners) on hypergeometric differential equations described in Theorem 17.6.3. See [22] for an algorithmic approach in the case of the Gaussian hypergeometric equation and [266] for some considerations in the general case. (See also [115] for a distinct but related point of view based on trace formulas.)

In another direction, one can combine the preceding considerations with ideas from the archimedean case [371] to compute deformations of cycle classes [116].

Appendix B
Rigid cohomology

It has been suggested several times in this book that the study of p-adic differential equations is deeply connected to a theory of *p-adic cohomology* for varieties over finite fields. In particular, the Frobenius structures arising on Picard–Fuchs modules, as discussed in Appendix A, appear within this theory.

In this appendix, we introduce a tiny bit of the theory of rigid p-adic cohomology, as developed by Berthelot and others. In particular, we illustrate the role played by the p-adic local monodromy theorem in a fundamental finiteness problem in the theory.

B.1 Isocrystals on the affine line

We start with a concrete description of p-adic cohomology in a very special case, namely the cohomology of "locally constant" coefficient systems on the affine line over a finite field. This is due to Crew [118].

Definition B.1.1. Let k be a perfect (for simplicity) field of characteristic $p > 0$. Let K be a complete discrete (again for simplicity) nonarchimedean field of characteristic 0 with $\kappa_K = k$. An *overconvergent F-isocrystal* on the affine line over k (with coefficients in K) is a finite differential module with Frobenius structure on the ring $\mathcal{A} = \bigcup_{\beta > 1} K \langle t/\beta \rangle$, for some absolute Frobenius lift φ; as in Proposition 17.3.1, the resulting category is independent of the choice of the Frobenius lift.

Definition B.1.2. Let M be an overconvergent F-isocrystal on the affine line over k. Let $\mathcal{R}$ be a copy of the Robba ring with series parameter t^{-1}, so that we can identify $\mathcal{A}$ as a subring of $\mathcal{R}$. (We can explicitly write $\mathcal{R} =$

454

$\bigcup_{\beta>1} K\langle \beta^{-1}/t^{-1}, t^{-1}\rangle$.) Define

$$H^0(\mathbb{A}_k^1, M) = \ker(D, M)$$
$$H^1(\mathbb{A}_k^1, M) = \operatorname{coker}(D, M)$$
$$H^0_{\mathrm{loc}}(\mathbb{A}_k^1, M) = \ker(D, M \otimes_{\mathcal{A}} \mathcal{R})$$
$$H^1_{\mathrm{loc}}(\mathbb{A}_k^1, M) = \operatorname{coker}(D, M \otimes_{\mathcal{A}} \mathcal{R})$$
$$H^1_c(\mathbb{A}_K^1, M) = \ker(D, M \otimes_{\mathcal{A}} (\mathcal{R}/\mathcal{A}))$$
$$H^2_c(\mathbb{A}_K^1, M) = \operatorname{coker}(D, M \otimes_{\mathcal{A}} (\mathcal{R}/\mathcal{A})).$$

By taking kernels and cokernels in the short exact sequence

$$0 \to M \to M \otimes_{\mathcal{A}} \mathcal{R} \to M \otimes_{\mathcal{A}} (\mathcal{R}/\mathcal{A}) \to 0$$

and applying the snake lemma, we get an exact sequence

$$0 \to H^0(\mathbb{A}_k^1, M) \to H^0_{\mathrm{loc}}(\mathbb{A}_k^1, M) \to H^1_c(\mathbb{A}_k^1, M) \to$$
$$\to H^1(\mathbb{A}_k^1, M) \to H^1_{\mathrm{loc}}(\mathbb{A}_k^1, M) \to H^2_c(\mathbb{A}_k^1, M) \to 0.$$

Remark B.1.3. Crew shows [118] that in this construction, H^i computes the *rigid cohomology* of M, H^i_c computes the *rigid cohomology with compact supports*, and H^i_{loc} computes some sort of local cohomology at ∞.

Crew's main result in this setting is the following.

Theorem B.1.4 (Crew). *The spaces $H^i(\mathbb{A}_k^1, M)$, $H^i_c(\mathbb{A}_k^1, M)$, $H^i_{\mathrm{loc}}(\mathbb{A}_K^1, M)$ are all finite-dimensional over K. Moreover, the Poincaré pairings*

$$H^i(\mathbb{A}_k^1, M) \times H^{2-i}_c(\mathbb{A}_k^1, M^\vee) \to H^2_c(\mathbb{A}_k^1, \mathcal{A}) \cong K$$
$$H^i_{\mathrm{loc}}(\mathbb{A}_k^1, M) \times H^{1-i}_{\mathrm{loc}}(\mathbb{A}_k^1, M^\vee) \to H^1_{\mathrm{loc}}(\mathbb{A}_K^1, \mathcal{A}) \cong K$$

are perfect.

The key ingredient is the fact that by the *p*-adic local monodromy theorem (Theorem 20.1.4), $M \otimes_{\mathcal{A}} \mathcal{R}$ is quasiunipotent; this implies finiteness of $H^i_{\mathrm{loc}}(\mathbb{A}_k^1, M)$. This implies the finite dimensionality of all of the terms except for $H^1_c(\mathbb{A}_k^1, M)$ and $H^1(\mathbb{A}_k^1, M)$; however, these are related by a map with finite-dimensional kernel and cokernel. Moreover, they carry incompatible topologies: the former is a Fréchet space, while the latter is dual to a Fréchet space. This incompatibility can only be resolved by both spaces being finite-dimensional.

B.2 Crystalline and rigid cohomology

We next briefly discuss how the previous example fits into a broader theory of p-adic cohomology, deferring most references to the chapter notes.

Motivated by the work of Dwork, and related work of Monsky and Washnitzer, Grothendieck proposed a method of constructing an analogue of algebraic de Rham cohomology in positive characteristic. This was developed by Berthelot and Ogus into the theory of *crystalline cohomology*. An important example of this is the fact that if X is a smooth proper scheme over $\mathbb{Z}_p$, then the algebraic de Rham cohomology of X carries a Frobenius action which computes the zeta function of the special fiber. (This generalizes to explain the results on Frobenius structures and zeta functions discussed in Appendix A.)

One defect of crystalline cohomology is that it only gives a coherent cohomology theory for schemes over a finite field which are smooth and proper. By contrast, the work of Monsky and Washnitzer had given a good theory for smooth *affine* schemes. To fuse these, Berthelot introduced the theory of *rigid cohomology*, as well as a theory of locally constant coefficient objects, the *overconvergent F-isocrystals*. The example of the previous paragraph demonstrates the computation of cohomology of coefficients on the affine line; there is also a theory of cohomology with compact supports, and a local cohomology theory, that are also demonstrated by Crew's construction.

For trivial coefficients, it was shown by Berthelot that rigid cohomology has all of the desired properties of a Weil cohomology: finite dimensionality, Poincaré duality, Künneth formula, cycle class maps, Lefschetz trace formula for Frobenius, etc. These were extended to nonconstant coefficients by Kedlaya, using a relative version of Theorem B.1.4.

Berthelot also suggested a theory of constructible coefficients, based on ideas from the theory of algebraic $\mathcal{D}$-modules. (These are modules over a ring of differential operators; they are the natural coefficient objects in algebraic de Rham cohomology.) Recent work of Abe, Caro, Kedlaya, and Tsuzuki has shown that these form a good theory of coefficients, enjoying the same formal properties as their ℓ-adic étale counterparts. For example, one can execute a proof of the Weil conjectures entirely using p-adic cohomology [237]. One can also adapt the methods of Drinfeld [136] and Lafforgue [284] to establish the Langlands correspondence (for GL_n over a global function field) for p-adic coefficients [4]; this can in turn be used to study compatible systems (including both étale and crystalline coefficient objects) on higher-dimensional varieties [263, 264].

B.3 Machine computations

In recent years, interest has emerged in explicitly computing the zeta functions of algebraic varieties defined over finite fields. Some of this interest has come from cryptography, particularly the use of Jacobians of elliptic (and later hyperelliptic) curves over finite fields as "black box abelian groups" for certain public-key cryptography schemes (Diffie–Hellman, ElGamal).

For elliptic curves, a good method for computing zeta functions was proposed by Schoof [364]. It amounts to computing the trace of Frobenius on the ℓ-torsion points, otherwise known as the étale cohomology with $\mathbb{F}_\ell$-coefficients, for enough small values of ℓ to determine uniquely the one unknown coefficient of the zeta function within the range prescribed by the Hasse–Weil bound.

It turns out to be somewhat more difficult to execute Schoof's scheme for curves of higher genus, as discovered by Pila [334]. One is forced to work with higher division polynomials in order to compute torsion of the Jacobian of the curve; the interpretation in terms of étale cohomology is of little value because the definition of étale cohomology is not intrinsically computable. (It is straightforward to write down cohomology classes, but difficult to test two such classes for equality.) A simplification of Pila's construction has recently been given by Abelard [5, 6], but the basic computational difficulties remain; beyond elliptic curves, the strategy has only been successfully implemented for curves of genus 2 [174] and some curves of genus 3 with extra endomorphisms [7].

It was noticed by several authors that rigid cohomology is intrinsically more computable, and so lends itself better to this sort of task. Specifically, Kedlaya [229] proposed an algorithm using rigid cohomology for computing the zeta function of a hyperelliptic curve over a finite field of small odd characteristic. The limitation to odd characteristic was lifted by Denef and Vercauteren [132]; the limitation to small characteristic was somewhat remedied by Harvey [197], who improved the dependence on the characteristic p from $O(p)$ to $O(p^{1/2+\epsilon})$. A comprehensive (but still practical) generalization to arbitrary curves has been given by Tuitman [399, 400] (see also [83]); this has proven important in computing p-adic path integrals for diophantine applications (see the chapter notes).

More recently, interest has emerged in considering also higher-dimensional varieties, partly coming from potential applications in the study of mirror symmetry for Calabi–Yau varieties. In this case, étale cohomology is of no help at all, since there is no geometric interpretation of H^i_{et} for $i > 1$ analogous to the interpretation for $i = 1$ in terms of the Jacobian. By contrast, computing the Frobenius action on rigid cohomology turns out to be practical at least up to dimension 4 [3, 114, 155].

Notes

Crew's work, and subsequent work which builds on it (e.g., [236]), makes essential use of nonarchimedean functional analysis, as is evident in the discussion of Theorem B.1.4. We recommend Schneider's book [363] as a user-friendly introduction to this topic.

For general surveys of the subject of p-adic cohomology, we recommend [216] (for crystalline cohomology only) and [246] (broader but more advanced). Also useful at this level of generality is Berthelot's original article outlining the theory of rigid cohomology [57].

For the basics of crystalline cohomology, see [62] and references within (largely to Berthelot's thesis [56]). For the work of Monsky and Washnitzer which motivated it, in addition to the original papers [315, 313, 314], there is also a useful survey by van der Put [402].

Until very recently, no comprehensive introductory text on rigid cohomology was available. That state of affairs has been remedied by the appearance of the book of Le Stum [291]. Berthelot's own attempt at a foundational treatise remains incomplete, but what does exist [58] may also be helpful. See also Berthelot's articles concerning finite dimensionality [59] and Poincaré duality [60].

For coefficients in an overconvergent F-isocrystal, the basic properties (finite dimensionality, Poincaré duality, Künneth formula) are proved in [236] using techniques similar to those of Crew in [118]. However, these techniques do not appear to suffice for the construction of a full coefficient theory. For this, one needs a higher-dimensional analogue of Theorem 20.1.4, called the *semistable reduction theorem* for overconvergent F-isocrystals. This asserts that an overconvergent F-isocrystal on a noncomplete variety can always be extended to a logarithmic isocrystal on some compactification after pulling back along a generically finite cover; such a result allows reductions to arguments about (logarithmic) crystalline cohomology. (Except for the introduction of a generically finite cover, this is analogous to Deligne's theory of canonical extensions in [126].) See [239, 242, 245, 249] for the proof; the original results from this book feature prominently in [249]. (See also [255, Remark 4.6.7] for a correction to [249].) See [262] for a retrospective overview on these developments.

Berthelot wrote a useful survey about his theory of arithmetic $\mathcal{D}$-modules [61]. It is this theory that has been developed by Caro (partly jointly with

Tsuzuki) into a full theory of p-adic coefficients. This work is quite expansive, and requires a summary more detailed than we can provide here; see [77, 78] for a representative sample.

On the subject of machine calculations, as a companion to our original paper on hyperelliptic curves [229], we recommend Edixhoven's course notes [152]. Some discussion is also included in [171, Chapter 7]. We gave a condensed summary of the general approach in [232]. For more on the role of elliptic and hyperelliptic curves in cryptography (including the relevance of the problem of machine computation of zeta functions), the standard first reference is [103], though it is some years behind the frontiers of this fast-moving subject. For instance, it predates the growing areas of *pairing-based cryptography* and *isogeny-based cryptography*; the latter is a candidate construction for *quantum-resistant* public-key cryptography by virtue of not being vulnerable to Shor's algorithm.

An important offshoot of the algorithmic approach to rigid cohomology is the computation of *p-adic path integrals* in the sense of Coleman [107]; these play an important role in finding torsion points and rational points on curves via the *Chabauty–Coleman method* [108] and the *Chabauty–Kim method* [273]. See [34, 38] for the case of a single integral (on hyperelliptic curves and general curves, respectively) and [33] for iterated integrals on hyperelliptic curves. Some discussion of iterated integrals on nonhyperelliptic curves can also be found in [37], where the Chabauty–Kim method (or more precisely the *quadratic Chabauty* construction of Balakrishnan and Dogra [35, 36]; see also [153]) was used to solve a longstanding open case of the problem of finding rational points on modular curves.

Since rigid cohomology admits a form of Poincaré duality, it is also possible to perform computations on the other side of the duality; this amounts to working with *Dwork cohomology* [139, 354, 10], which has certain advantages of its own (especially more uniform behavior under degenerations from smooth to singular varieties). For examples of algorithmic applications of Dwork cohomology, see [287, 376].

For completeness, we point out that some promising approaches to interpreting zeta functions via p-adic analysis do not pass through overtly cohomological constructions. In particular, this is true of Dwork's original proof of the rationality of zeta functions of algebraic varieties over finite fields [138], and of a variation described by Harvey [198].

Appendix C

p-adic Hodge theory

For our last application, we turn to the subject of p-adic Hodge theory. Recall that in Chapter 19, we described a "nonabelian Artin–Schreier" construction, giving an equivalence of categories between continuous representations of the absolute Galois group of a positive characteristic local field on a p-adic vector space and certain differential modules with Frobenius structures. In this appendix, we describe an analogous construction for the Galois group of a mixed-characteristic local field. We also mention a couple of applications of this construction.

After the first edition was published, the landscape of p-adic Hodge theory experienced a tectonic shift due to the introduction of *perfectoid spaces*. We have not updated our description accordingly, with the effect that it may read as somewhat old-fashioned; see the chapter notes for some more modern references.

Hypothesis C.0.1. Throughout this appendix, let K be a finite extension of $\mathbb{Q}_p$, let V be a finite-dimensional $\mathbb{Q}_p$-vector space, and let $\tau \colon G_K \to \mathrm{GL}(V)$ be a continuous homomorphism for the p-adic topology on V.

C.1 A few rings

We begin with the "field of norms" construction of Fontaine and Wintenberger.

Definition C.1.1. Put $K_n = K(\zeta_{p^n})$ and $K_\infty = \cup_n K_n$. Let $F = \mathrm{Frac}\, W(\kappa_K)$ and F' be the maximal subfields of K and K_∞, respectively, which are unramified over $\mathbb{Q}_p$. Put $H_K = G_{K_\infty}$ and $\Gamma_K = G_{K_\infty/K} = G_K/H_K$.

Definition C.1.2. Put $\mathfrak{o} = \mathfrak{o}_{\mathbb{C}_p}$. Let $\tilde{\mathbf{E}}^+$ be the inverse limit of the system

$$\cdots \to \mathfrak{o}/p\mathfrak{o} \to \mathfrak{o}/p\mathfrak{o}$$

460

in which each map is the p-power Frobenius (which is a ring homomorphism). More explicitly, the elements of $\tilde{\mathbf{E}}^+$ are sequences $(x_0, x_1, \dots)$ of elements of $\mathfrak{o}/p\mathfrak{o}$ for which $x_{n+1}^p = x_n$ for all n. In particular, for any nonzero $x \in \tilde{\mathbf{E}}^+$, the quantity $p^n v_p(x_n)$ is the same for all n for which $x_n \neq 0$; we call this quantity $v(x)$, and put conventionally $v(0) = \infty$. Choose $\epsilon = (\epsilon_0, \epsilon_1, \dots) \in \tilde{\mathbf{E}}^+$ with $\epsilon_0 = 1$ and $\epsilon_1 \neq 1$. (This choice is somewhat analogous to the choice of a square root of -1 in $\mathbb{C}$.)

The following observations were made by Fontaine and Wintenberger [169].

Proposition C.1.3. *The following are true.*

(a) *The ring $\tilde{\mathbf{E}}^+$ is a domain in which $p = 0$, with fraction field $\tilde{\mathbf{E}} = \tilde{\mathbf{E}}^+[(\epsilon - 1)^{-1}]$.*

(b) *The function $v \colon \tilde{\mathbf{E}}^+ \to [0, \infty]$ extends to a valuation on $\tilde{\mathbf{E}}$, under which $\tilde{\mathbf{E}}$ is complete and $\mathfrak{o}_{\tilde{\mathbf{E}}} = \tilde{\mathbf{E}}^+$.*

(c) *The field $\tilde{\mathbf{E}}$ is the algebraic closure of $\kappa_K((\epsilon - 1))$. (The embedding of $\kappa_K((\epsilon - 1))$ into $\tilde{\mathbf{E}}$ exists because $v(\epsilon - 1) = p/(p - 1) > 0$.)*

Definition C.1.4. Let $\tilde{\mathbf{A}}$ be the ring of Witt vectors of $\tilde{\mathbf{E}}$, i.e., the unique complete discrete valuation ring with maximal ideal p and residue field $\tilde{\mathbf{E}}$. The uniqueness follows from the fact that $\tilde{\mathbf{E}}$ is algebraically closed, hence perfect. In particular, the p-power Frobenius on $\tilde{\mathbf{E}}$ lifts to an automorphism φ of $\tilde{\mathbf{A}}$. (See the notes for Chapter 14 for further discussion of Witt vectors.)

Definition C.1.5. Each element of $\tilde{\mathbf{A}}$ can be uniquely written as a sum of the form $\sum_{n=0}^{\infty} p^n[x_n]$, where $x_n \in \tilde{\mathbf{E}}$ and $[x_n]$ denotes the *multiplicative lift* of x_n (the unique lift of x_n that has a p^m-th root in $\tilde{\mathbf{A}}$ for all positive integers m); note that $\varphi([x]) = [x^p] = [x]^p$. We may thus equip $\tilde{\mathbf{A}}$ with a *weak topology*, in which a sequence $x_m = \sum_{n=0}^{\infty} p^n[x_{m,n}]$ converges to zero if for each n, $v(x_{m,n}) \to \infty$ as $m \to \infty$. Let $\mathbf{A}_{\mathbb{Q}_p}$ be the completion of $\mathbb{Z}_p[([\epsilon] - 1)^{\pm}]$ in $\tilde{\mathbf{A}}$ for the weak topology; as a topological ring, it is isomorphic to the ring $\mathfrak{o}_{\mathcal{E}}$ defined over the base field $\mathbb{Q}_p$ with its own weak topology. It is also φ-stable because $\varphi([\epsilon]) = [\epsilon]^p$.

Definition C.1.6. Let $\mathbf{A}$ be the completion of the maximal unramified extension of $\mathbf{A}_{\mathbb{Q}_p}$, viewed as a subring of $\tilde{\mathbf{A}}$. Put

$$\mathbf{A}_K = \mathbf{A}^{H_K},$$

where the right side makes sense because $\mathbf{A}$ carries a G_K-action by functoriality. Note that $\mathbf{A}_K$ can be written as a ring of the form $\mathfrak{o}_{\mathcal{E}}$, but with coefficients in F' rather than in $\mathbb{Q}_p$.

Definition C.1.7. For any ring denoted with a boldface **A** so far, define the corresponding ring with **A** replaced by **B** by tensoring over $\mathbb{Z}_p$ with $\mathbb{Q}_p$. For instance, $\tilde{\mathbf{B}} = \tilde{\mathbf{A}} \otimes_{\mathbb{Z}_p} \mathbb{Q}_p$ is the fraction field of $\tilde{\mathbf{A}}$.

C.2 (φ, Γ)-modules

We are now ready to describe the mechanism, introduced by Fontaine, for converting Galois representations into modules over various rings equipped with much simpler group actions.

Definition C.2.1. Recall that V is a finite-dimensional vector space equipped with a continuous G_K-action. Put

$$D(V) = (V \otimes_{\mathbb{Q}_p} \mathbf{B})^{H_K};$$

by the Hilbert–Noether–Speiser theorem, $D(V)$ is a finite-dimensional $\mathbf{B}_K$-vector space, and the natural map $D(V) \otimes_{\mathbf{B}_K} \mathbf{B} \to V \otimes_{\mathbb{Q}_p} \mathbf{B}$ is an isomorphism. Since we have only taken H_K-invariants, $D(V)$ retains a semilinear action of $G_K / H_K = \Gamma_K$; it also inherits an action of φ from $\mathbf{B}$. That is, $D(V)$ is a (φ, Γ)-*module* over $\mathbf{B}_K$, i.e., a finite free $\mathbf{B}_K$-module equipped with semilinear φ and Γ_K-actions which commute with each other. It is also *étale*, which is to say the φ-action is étale (unit-root); as in Definition 19.2.4, this is because one can find a G_K-invariant lattice in V.

Theorem C.2.2 (Fontaine). *The functor D, from the category of continuous representations of G_K on finite-dimensional $\mathbb{Q}_p$-vector spaces to the category of étale (φ, Γ)-modules over $\mathbf{B}_K$, is an equivalence of categories.*

Proof The inverse functor is

$$V = (D(V) \otimes_{\mathbf{B}_K} \mathbf{B})^{\varphi=1}.$$

The argument is similar to the proof of Proposition 19.1.5; see [165]. □

In much the same way that Proposition 19.1.5 was refined by Theorem 19.3.1, Theorem C.2.2 was refined by Cherbonnier and Colmez as follows [84]. The big difference is that no additional restriction on the Galois representations is imposed.

Definition C.2.3. Let $\mathbf{B}_{\mathbb{Q}_p}^{\dagger}$ be the image of the bounded Robba ring $\mathcal{E}^{\dagger}$ under the identification of $\mathcal{E}$ (with coefficients in $\mathbb{Q}_p$) with $\mathbf{B}_{\mathbb{Q}_p}$ sending t to $[\epsilon] - 1$. Let $\mathbf{B}_K^{\dagger}$ be the integral closure of $\mathbf{B}_{\mathbb{Q}_p}^{\dagger}$ in $\mathbf{B}_K$. Again, $\mathbf{B}_K^{\dagger}$ carries actions of φ and Γ_K.

Definition C.2.4. Let $\mathbf{A}^\dagger$ be the set of $x = \sum_{n=0}^\infty p^n [x_n] \in \mathbf{A}$ such that $\liminf_{n\to\infty}\{v(x_n)/n\} > -\infty$. Define

$$D^\dagger(V) = (V \otimes_{\mathbb{Q}_p} \mathbf{B}^\dagger)^{H_K};$$

it is an étale (φ, Γ)-module over $\mathbf{B}_K^\dagger$.

The following is the main result of [84].

Theorem C.2.5 (Cherbonnier–Colmez). *The functor $D^\dagger$, from the category of continuous representations of G_K on finite-dimensional $\mathbb{Q}_p$-vector spaces to the category of étale (φ, Γ)-modules over $\mathbf{B}_K^\dagger$, is an equivalence of categories.*

Remark C.2.6. By Theorem C.2.2, to check Theorem C.2.5 it suffices to check that the base extension functor from étale (φ, Γ)-modules over $\mathbf{B}_K^\dagger$ to étale (φ, Γ)-modules over $\mathbf{B}_K$ is an equivalence. The full faithfulness of this functor is elementary; it follows from Lemma 16.3.2. The essential surjectivity is much deeper; it amounts to the fact that the natural map

$$D^\dagger(V) \otimes_{\mathbf{B}_K^\dagger} \mathbf{B}^\dagger \to V \otimes_{\mathbb{Q}_p} \mathbf{B}^\dagger$$

is an isomorphism. Verifying this requires developing an appropriate analogy to Sen's theory of decompletion; this analogy has been developed into a full abstract Sen theory by Berger and Colmez [51].

A further variant was proposed by Berger [46].

Definition C.2.7. Using the identification $\mathbf{B}_{\mathbb{Q}_p}^\dagger \cong \mathcal{E}^\dagger$, put

$$\mathbf{B}_{\mathrm{rig},K}^\dagger = \mathbf{B}_K^\dagger \otimes_{\mathbf{B}_{\mathbb{Q}_p}^\dagger} \mathcal{R},$$

where $\mathcal{R}$ is the Robba ring over $\mathbb{Q}_p$; the subscript "rig" is short for "rigid", in the sense of rigid analytic geometry. Note that $\mathbf{B}_{\mathrm{rig},K}^\dagger$ admits continuous extensions (for the LF-topology) of the actions of φ and Γ_K. Define

$$D_{\mathrm{rig}}^\dagger(V) = D^\dagger(V) \otimes_{\mathbf{B}_K^\dagger} \mathbf{B}_{\mathrm{rig},K}^\dagger;$$

it is an étale (φ, Γ)-module over $\mathbf{B}_{\mathrm{rig},K}^\dagger$.

Theorem C.2.8 (Berger). *The functor $D_{\mathrm{rig}}^\dagger$, from the category of continuous representations of G_K on finite-dimensional $\mathbb{Q}_p$-vector spaces to the category of étale (φ, Γ)-modules over $\mathbf{B}_{\mathrm{rig},K}^\dagger$, is an equivalence of categories.*

Remark C.2.9. The principal new content in Theorem C.2.8 is that the base extension functor from étale φ-modules over $\mathcal{E}^\dagger$ to étale φ-modules over $\mathcal{R}$ is fully faithful; this follows from Corollary 16.3.4. The essential surjectivity of the functor is almost trivial, since étaleness of the φ-action is defined over the Robba ring by base extension from $\mathcal{E}^\dagger$ (Definition 16.3.1). One needs only check that the Γ_K-action also descends to any étale lattice, but this is similar to Lemma 17.4.2.

C.3 Galois cohomology

Since the functor D and its variants lose no information about Galois representations, it is unsurprising that they can be used to recover basic invariants of a representation, such as Galois cohomology.

Definition C.3.1. Assume for simplicity that Γ_K is procyclic; this only eliminates the case where $p = 2$ and $\{\pm 1\} \subset \Gamma$, for which see Remark C.3.2 below. Let γ be a topological generator of Γ. Define the *Herr complex* over $\mathbf{B}_K$ associated to V as the complex (with the first nonzero term placed in degree zero)

$$0 \to D(V) \to D(V) \oplus D(V) \to D(V) \to 0$$

with the first map being $m \mapsto ((\varphi - 1)m, (\gamma - 1)m)$ and the second map being $(m_1, m_2) \to (\gamma - 1)m_1 - (\varphi - 1)m_2$. (The fact that this is a complex follows from the commutativity between φ and γ.) Similarly, define the Herr complex over $\mathbf{B}_K^\dagger$ or $\mathbf{B}_{\mathrm{rig},K}^\dagger$ by replacing $D(V)$ by $D^\dagger(V)$ or $D_{\mathrm{rig}}^\dagger(V)$, respectively.

Remark C.3.2. To give a more conceptual description, which also covers the case where Γ_K need not be profinite, take the total complex associated to

$$0 \to C^\cdot(\Gamma_K, D(V)) \overset{\varphi-1}{\to} C^\cdot(\Gamma_K, D(V)) \to 0$$

where $C^\cdot(\Gamma_K, D(V))$ is the usual cochain complex for computing Galois cohomology (as in [370, §I.2.2]). One might think of this as the "monoid cohomology" of $\Gamma_K \times \varphi^{\mathbb{Z}_{\geq 0}}$ acting on $D(V)$.

Theorem C.3.3. *The cohomology of the Herr complex computes the Galois cohomology of V.*

Proof For $\mathbf{B}_K$, the desired result was established by Herr [201]. The argument proceeds in two steps. One first takes cohomology of the Artin–Schreier sequence

$$0 \to \mathbb{Q}_p \to \mathbf{B} \overset{\varphi-1}{\to} \mathbf{B} \to 0$$

after tensoring with V. This reduces the claim to the fact that the inflation homomorphisms

$$H^i(\Gamma_K, D(V)) \to H^i(G_K, V \otimes_{\mathbb{Q}_p} \mathbf{B})$$

are bijections; this is proved by adapting a technique introduced by Sen.

For $\mathbf{B}_K^\dagger$ and $\mathbf{B}_{\mathrm{rig},K}^\dagger$, the desired result was established by Liu [293]; this proceeds by comparison with the original Herr complex rather than by imitating the above argument, though one could probably do that also. □

Remark C.3.4. As is done in [201, 293], one can make Theorem C.3.3 more precise. For instance, the construction of Galois cohomology is functorial; there is also an interpretation in the Herr complex of the cup product in cohomology.

Remark C.3.5. One can also use the Herr complex to recover Tate's fundamental theorems about Galois cohomology (finite dimensionality, Euler-Poincaré characteristic formula, local duality). This was done by Herr in [202].

C.4 Differential equations from (φ, Γ)-modules

One of the original goals of p-adic Hodge theory was to associate finer invariants to p-adic Galois representations, so as for instance to distinguish those representations which arose in geometry (i.e., from the étale cohomology of varieties over K). This was originally done using a collection of "period rings" introduced by Fontaine; more recently, Berger's work has demonstrated that one can reproduce these constructions using (φ, Γ)-modules. Here is a brief description of an example that shows the relevance of p-adic differential equations to this study. We will make reference to Fontaine's rings $\mathbf{B}_{\mathrm{dR}}, \mathbf{B}_{\mathrm{st}}$ without definition, for which see [47].

Definition C.4.1. Let $\chi: \Gamma_K \to \mathbb{Z}_p^\times$ denote the cyclotomic character; that is, for all nonnegative integers m and all $\gamma \in \Gamma_K$,

$$\gamma(\zeta_{p^m}) = \zeta_{p^m}^{\chi(\gamma)}.$$

For $\gamma \in \Gamma_K$ sufficiently close to 1, we may compute

$$\nabla = \frac{\log \gamma}{\log \chi(\gamma)}$$

as an endomorphism of $D_{\mathrm{rig}}^\dagger(V)$, using the power series for $\log(1 + x)$. The result does not depend on γ.

Remark C.4.2. If one views Γ_K as a one-dimensional p-adic Lie group over $\mathbb{Z}_p$, then ∇ is the action of the corresponding Lie algebra.

Definition C.4.3. Note that ∇ acts on $\mathbf{B}_{\mathrm{rig},K}^{\dagger}$ with respect to

$$f \mapsto [\epsilon] \log[\epsilon] \frac{df}{d[\epsilon]}.$$

As a result, it does not induce a differential module structure with respect to $\frac{d}{dt}$ on $D_{\mathrm{rig}}^{\dagger}(V)$, but only on $D_{\mathrm{rig}}^{\dagger}(V) \otimes_{\mathbf{B}_{\mathrm{rig},K}^{\dagger}} \mathbf{B}_{\mathrm{rig},K}^{\dagger}[(\log[\epsilon])^{-1}]$. We say that V is *de Rham* if there exists a differential module with Frobenius structure M over $\mathbf{B}_{\mathrm{rig},K}^{\dagger}$ and an isomorphism

$$D_{\mathrm{rig}}^{\dagger}(V) \otimes_{\mathbf{B}_{\mathrm{rig},K}^{\dagger}} \mathbf{B}_{\mathrm{rig},K}^{\dagger}[(\log[\epsilon])^{-1}] \to M \otimes_{\mathbf{B}_{\mathrm{rig},K}^{\dagger}} \mathbf{B}_{\mathrm{rig},K}^{\dagger}[(\log[\epsilon])^{-1}]$$

of differential modules with Frobenius structure.

One then has the following results of Berger [46].

Theorem C.4.4 (Berger). *(a) The representation V is de Rham if and only if it is de Rham in Fontaine's sense, i.e., if*

$$D_{\mathrm{dR}}(V) = (V \otimes_{\mathbb{Q}_p} \mathbf{B}_{\mathrm{dR}})^{G_K}$$

satisfies

$$D_{\mathrm{dR}}(V) \otimes_K \mathbf{B}_{\mathrm{dR}} \cong V \otimes_{\mathbb{Q}_p} \mathbf{B}_{\mathrm{dR}}.$$

(b) Suppose that V is de Rham. If V is semistable in Fontaine's sense, i.e., if

$$D_{\mathrm{st}}(V) = (V \otimes_{\mathbb{Q}_p} \mathbf{B}_{\mathrm{st}})^{G_K}$$

satisfies

$$D_{\mathrm{st}}(V) \otimes_F \mathbf{B}_{\mathrm{st}} \cong V \otimes_{\mathbb{Q}_p} \mathbf{B}_{\mathrm{st}},$$

then there exists M as in Definition C.4.3 which is unipotent. Conversely, if such M exists, then V is potentially semistable, i.e., becomes semistable upon restriction to G_L for some finite extension L of K.

Applying Theorem 20.1.4 then yields the following corollary, which was previously a conjecture of Fontaine [166, 6.2].

Corollary C.4.5 (Berger). *Every de Rham representation is potentially semistable.*

Remark C.4.6. The descriptor "de Rham" is meant to convey the fact that if $V = H_{\mathrm{et}}^i(X \times_K K^{\mathrm{alg}}, \mathbb{Q}_p)$ for X a smooth proper variety over K, then V is de Rham and one can use the aforementioned constructions to recover $H_{\mathrm{dR}}^i(X, K)$ functorially from V (solving Grothendieck's "problem of the mysterious functor"). See [47] for more of the story.

C.5 Beyond Galois representations

The category of arbitrary (φ, Γ)-modules over $\mathbf{B}^{\dagger}_{\mathrm{rig},K}$ turns out to have its own representation-theoretic interpretation; it is equivalent to the category of *B-pairs* introduced by Berger [48]. One can associate "Galois cohomology" to such objects using the Herr complex; it has been shown by Liu [293] that the analogues of Tate's theorems (see Remark C.3.5) still hold. These functors can be interpreted as the derived functors of $\mathrm{Hom}(D^{\dagger}_{\mathrm{rig}}(V_0), \cdot)$ for V_0 the trivial representation [247, Appendix].

One may wonder why one should be interested in (φ, Γ)-modules over $\mathbf{B}^{\dagger}_{\mathrm{rig},K}$ if ultimately one has in mind an application concerning only Galois representations. One answer is that converting Galois representations into (φ, Γ)-modules exposes extra structure that is not visible without the conversion.

Definition C.5.1 (Colmez). We say V is *trianguline* if $D^{\dagger}_{\mathrm{rig}}(V)$ is a successive extension of (φ, Γ)-modules of rank 1 over $\mathbf{B}^{\dagger}_{\mathrm{rig},K}$. The point is that these need not be étale, so V need not be a successive extension of representations of dimension 1.

Trianguline representations have the dual benefits of being both classifiable and also somewhat commonplace. On one hand, Colmez [110] classified the two-dimensional trianguline representations of $G_{\mathbb{Q}_p}$; the classification includes a parameter (the $\mathcal{L}$-invariant) relevant to p-adic L-functions. On the other hand, a result of Kisin [274] shows that the Galois representations associated to classical and overconvergent modular forms are trianguline.

Notes

Our presentation here is largely a summary of Berger's [47], which we highly recommend. For detailed study, the recent notes by Brinon and Conrad for the 2009 Clay Mathematics Institute summer school [71] may be of great value. See also the lecture notes of Fontaine [168].

A variant of the theory of (φ, Γ)-modules was introduced by Kisin [275], using the Kummer tower $K(p^{1/p^n})$ instead of the cyclotomic tower $K(\zeta_{p^n})$. This leads to certain advantages, particularly when studying crystalline representations. For instance, Kisin is able to use his construction to establish some classification theorems for finite flat group schemes and for p-divisible groups, conjectured by Breuil. Kisin's work is based on an earlier paper of Berger [49]; both of these use slope filtrations (as in Theorem 16.4.1) to recover a theorem of Colmez and Fontaine classifying semistable Galois representations in terms of certain linear algebraic data.

After [46] appeared, Fontaine succeeded in giving a direct proof of Corollary C.4.5, not going through p-adic differential equations; see [167].

For generalizations of Corollary C.4.5 to the case where K is a complete discretely valued field with possibly imperfect residue field, see [316, 321].

As mentioned at the start of the appendix, our presentation up until this point has ignored the substantial changes wrought in p-adic Hodge theory since the publication of the first edition of this book (and which as of this writing are still ongoing). These changes can be summarized as a "geometrization" of the subject using *Fargues–Fontaine curves* and *perfectoid spaces*. For comprehensive surveys of the developments based on these ideas, see for example [76] and [366]; see also [160] for the original development of curves in p-adic Hodge theory by Fargues and Fontaine, and [161] for the most recent developments available as of this writing.

References

[1] A. Abbes and T. Saito, Ramification of local fields with imperfect residue fields, *Amer. J. Math.* **124** (2002), 879–920.

[2] A. Abbes and T. Saito, Ramification of local fields with imperfect residue fields, II, *Doc. Math.* Extra Vol. (2003), 5–72.

[3] T.G. Abbott, K.S. Kedlaya, and D. Roe, Bounding Picard numbers of surfaces using p-adic cohomology, in *Arithmetic, Geometry and Coding Theory (AGCT 2005)*, Societé Mathématique de France, 2009, 125–159.

[4] T. Abe, Langlands correspondence for isocrystals and existence of crystalline companion for curves, *J. Amer. Math. Soc.* **31** (2018), 921–1057.

[5] S. Abelard, Counting points on hyperelliptic curves in large characteristic: algorithms and complexity, thesis, Université de Lorraine, 2018, https://tel.archives-ouvertes.fr/tel-01876314.

[6] S. Abelard, P. Gaudry, and P.-J. Spaenlehauer, Improved complexity bounds for counting points on hyperelliptic curves, *Found. Comp. Math.* **19** (2019), 591–621.

[7] S. Abelard, P. Gaudry, and P.-J. Spaenlehauer, Counting points on genus-3 hyperelliptic curves with explicit real multiplication, in *ANTS-XIII: Proceedings of the Thirteenth Algorithmic Number Theory Symposium*, Open Book Series 2, Math. Sci. Pub., 2019, 1–19.

[8] M. Abramowitz and I.A. Stegun, *Handbook of Mathematical Functions with Formulas, Graphs, and Mathematical Tables* National Bureau of Standards Applied Math. Series 55, U.S. Government Printing Office, Washington, D.C., 1964.

[9] A. Adolphson, An index theorem for p-adic differential operators, *Trans. Amer. Math. Soc.* **216** (1976), 279–293.

[10] A. Adolphson and S. Sperber, Exponential sums and Newton polyhedra: cohomology and estimates, *Ann. of Math.* **130** (1989), 367–406.

[11] O. Amini and M. Baker, Linear series on metrized complexes of algebraic curves, *Math. Ann.* **362** (2015), 55–106.

[12] O. Amini, M. Baker, E. Brugallé, and J. Rabinoff, Lifting harmonic morphisms I: metrized complexes and Berkovich skeleta, *Res. Math. Sci.* **2** (2015), 67pp.

[13] O. Amini, M. Baker, E. Brugallé, and J. Rabinoff, Lifting harmonic morphisms II: Tropical curves and metrized complexes, *Algebra and Number Theory* **9** (2015), 267–315.

[14] M. Anderson and J. Watkins, Coherence of power series rings over pseudo-Bezout domains, *J. Alg.* **107** (1987), 187–194.

[15] Y. André, Différentielles non commutatives et théorie de Galois différentielle ou aux différences, *Ann. Sci. École Norm. Sup.* (4) **34** (2001), 685–739.

[16] Y. André, Filtrations de type Hasse-Arf et monodromie *p*-adique, *Invent. Math.* **148** (2002), 285–317.

[17] Y. André, Dwork's conjecture on the logarithmic growth of solutions of *p*-adic differential equations, *Compos. Math.* **144** (2008), 484–494.

[18] Y. André, Slope filtrations, *Confluentes Math.* **1** (2009), 1–85.

[19] Y. André, F. Baldassarri, and M. Cailotto, *De Rham Cohomology of Differential Modules on Algebraic Varieties*, second edition, Progress in Math. 189, Birkhäuser, 2020.

[20] Y. André and L. Di Vizio, *q*-difference equations and *p*-adic local monodromy, *Astérisque* **296** (2004), 55–111.

[21] W. Arveson, *An Invitation to C*-Algebras*, Graduate Texts in Math. 39, Springer, 1976.

[22] M. Asakura, An algorithm of computing special values of Dwork's *p*-adic hypergeometric functions in polynomial time, arXiv:1909.02700v3 (2019).

[23] M. Atiyah, Complex analytic connections in fibre bundles, *Trans. Amer. Math. Soc.* **85** (1957), 181–207.

[24] M.F. Atiyah and I.G. Macdonald, *Introduction to Commutative Algebra*, Addison-Wesley, Reading, 1969.

[25] M. Auslander and O. Goldman, Maximal orders, *Trans. Amer. Math. Soc.* **97** (1960), 1–24.

[26] M. Auslander and O. Goldman, The Brauer group of a commutative ring, *Trans. Amer. Math. Soc.* **97** (1960), 367–409.

[27] M. Baker, An introduction to Berkovich analytic spaces and non-Archimedean potential theory on curves, in [360], 123–174.

[28] M. Baker, S. Payne, and J. Rabinoff, On the structure of non-Archimedean analytic curves, in *Tropical and Non-Archimedean Geometry*, Contemp. Math. 605, Amer. Math. Soc., Providence, 2013, 93–121.

[29] M. Baker, S. Payne, and J. Rabinoff, Nonarchimedean geometry, tropicalization, and metrics on curves, *Alg. Geom.* **3** (2016), 63–105.

[30] M. Baker and R. Rumely, Harmonic analysis on metrized graphs, *Canad. J. Math.* **59** (2007), 225–275.

[31] M. Baker and R. Rumely, *Potential Theory and Dynamics on the Berkovich Projective Line*, Math. Surveys and Monographs 159, Amer. Math. Soc., 2010.

[32] M. Baker and S. Norine, Riemann-Roch and Abel-Jacobi theory on a finite graph, *Adv. Math.* **215** (2007), 766–788.

[33] J. Balakrishnan, Iterated Coleman integration for hyperelliptic curves, in *ANTS-X: Proceedings of the Tenth Algorithmic Number Theory Symposium*, Open Book Series 1, MSP, 2013, 41–61.

[34] J. Balakrishnan, R. Bradshaw, and K.S. Kedlaya, Explicit Coleman integration for hyperelliptic curves, in *Algorithmic Number Theory (ANTS-IX)*, Lecture Notes in Comp. Sci. 6197, Springer-Verlag, 2010, 16–31.

[35] J. Balakrishnan and N. Dogra, Quadratic Chabauty and rational points I: *p*-adic heights (with an appendix by J.S. Müller), *Duke Math. J.* **167** (2018), 1981–2038.

[36] J. Balakrishnan and N. Dogra, Quadratic Chabauty and rational points II: Generalised height functions on Selmer varieties, *Int. Math. Res. Notices* **2020** (2020), article ID rnz362.

[37] J. Balakrishnan, N. Dogra, J.S. Müller, J. Tuitman, and J. Vonk, Explicit Chabauty-Kim for the split Cartan modular curve of level 13, *Annals of Math.* **189** (2019), 885–944.

[38] J. Balakrishnan and J. Tuitman, Explicit Coleman integration for curves, *Math. Comp.* **89** (2020), 2965–2984.

[39] F. Baldassarri, Comparaison entre la cohomologie algébrique et la cohomologie *p*-adique rigide à coefficients dans un module différentiel. I. Cas des courbes, *Invent. Math.* **87** (1987), 83–99.

[40] F. Baldassarri, Comparaison entre la cohomology algébrique et la cohomologie *p*-adique rigide à coefficients dans un module différentiel. II. Cas des singularités régulières à plusieures variables, *Math. Ann.* **280** (1988), 417–439.

[41] F. Baldassarri, Continuity of the radius of convergence of differential equations on *p*-adic analytic curves, *Invent. Math.* **182** (2010), 513–584.

[42] F. Baldassarri and L. Di Vizio, Continuity of the radius of convergence of *p*-adic differential equations on Berkovich analytic spaces, arXiv:0709.2008v3 (2008).

[43] W. Balser, W.B. Jurkat, and D.A. Lutz, Birkhoff invariants and Stokes' multipliers for meromorphic linear differential equations, *J. Math. Anal. Appl.* **71** (1979), 48–94.

[44] A. Beilinson, S. Bloch, and H. Esnault, ϵ-factors for Gauss-Manin determinants, *Moscow J. Math.* **2** (2002), 477–532.

[45] R. Bellovin, *p*-adic Hodge theory in rigid analytic families, *Algebra Number Theory* **9** (2015), 371–433.

[46] L. Berger, Représentations *p*-adiques et équations différentielles, *Invent. Math.* **148** (2002), 219–284.

[47] L. Berger, An introduction to the theory of *p*-adic representations, in *Geometric Aspects of Dwork Theory. Vol I, II*, de Gruyter, Berlin, 2004, 255–292.

[48] L. Berger, Construction de (φ, Γ)-modules: représentations *p*-adiques et *B*-paires, *Algebra and Num. Theory* **2** (2008), 91–120.

[49] L. Berger, Équations différentielles *p*-adiques et (φ, N)-modules filtrés, *Astérisque* **319** (2008), 13–38.

[50] L. Berger, Substitution maps in the Robba ring, arXiv:2012.01904v2 (2020).

[51] L. Berger and P. Colmez, Familles de représentations de de Rham et monodromie *p*-adique, *Astérisque* **319** (2008), 303–337.

[52] V.G. Berkovich, *Spectral Theory and Analytic Geometry over Non-Archimedean Fields*, Math. Surveys and Monographs 33, Amer. Math. Soc., Providence, 1990.

[53] V.G. Berkovich, Étale cohomology for non-Archimedean analytic spaces, *Publ. Math. IHÉS* **78** (1993), 5–161.

[54] V.G. Berkovich, Smooth *p*-adic analytic spaces are locally contractible, *Invent. Math.* **137** (1999), 1–84.

[55] V.G. Berkovich, Smooth *p*-adic analytic spaces are locally contractible, II, in *Geometric Aspects of Dwork Theory, Vol. I, II*, Walter de Gruyter, Berlin, 2004, 293–370.

[56] P. Berthelot, *Cohomologie Cristalline des Schémas de Caractéristique $p > 0$*, Lecture Notes in Math. 407. Springer-Verlag, Berlin, 1974.

[57] P. Berthelot, Géométrie rigide et cohomologie des variétés algébriques de caractéristique p, Introductions aux cohomologies p-adiques (Luminy, 1984), *Mém. Soc. Math. France (N.S.)* **23** (1986), 7–32.

[58] P. Berthelot, Cohomologie rigide et cohomologie rigide à support propre. Première partie, Prépublication IRMAR 96-03, available at http://perso.univ-rennes1.fr/pierre.berthelot/ (retrieved June 2020).

[59] P. Berthelot, Finitude et pureté cohomologique en cohomologie rigide (with an appendix in English by A.J. de Jong), *Invent. Math.* **128** (1997), 329–377.

[60] P. Berthelot, Dualité de Poincaré et formule de Künneth en cohomologie rigide, *C.R. Acad. Sci. Paris* **325** (1997), 493–498.

[61] P. Berthelot, Introduction à la théorie arithmétique des $\mathcal{D}$-modules, Cohomologies p-adiques et applications arithmétiques, II, *Astérisque* **279** (2002), 1–80.

[62] P. Berthelot and A. Ogus, *Notes on Crystalline Cohomology*, Princeton University Press, Princeton, 1978.

[63] R. Bhatia, *Matrix Analysis*, Graduate Texts in Math. 169, Springer-Verlag, New York, 1997.

[64] B. Bhatt and P. Scholze, Prisms and prismatic cohomology, arXiv:1905.08229v2 (2019).

[65] V. Bojković, Riemann-Hurwitz formula for finite morphisms of p-adic curves, *Math. Z.* **288** (2018), 1165–1193.

[66] V. Bojković and J. Poineau, On the number of connected components of the ramification locus of a morphism of Berkovich curves, *Math. Ann.* **372** (2018), 1575–1595.

[67] J.M. Borger, Conductors and the moduli of residual perfection, *Math. Annalen* **329** (2004), 1–30.

[68] S. Bosch, *Lectures on Formal and Rigid Geometry*, Lecture Notes in Math. 2105, Springer, Cham, 2014.

[69] S. Bosch, U. Güntzer, and R. Remmert, *Non-Archimedean Analysis*, Grundlehren der Math. Wiss. 261, Springer-Verlag, Berlin, 1984.

[70] U. Brezner and M. Temkin, Lifting problem for minimally wild covers of Berkovich curves, *J. Alg. Geom.* **29** (2020), 123–166.

[71] O. Brinon and B. Conrad, p-adic Hodge theory, notes from the 2009 Clay Mathematics Institute summer school, available online at https://math.stanford.edu/~conrad/ (retrieved June 2020).

[72] S. Boucksom, C. Favre, and M. Jonsson, Valuations and plurisubharmonic singularities, *Publ. Res. Inst. Math. Sci.* **44** (2008), 449–494.

[73] A. Buium, Differential characters of abelian varieties over p-adic fields, *Invent. Math.* **122** (1995), 309–340.

[74] A. Buium, Arithmetic analogues of derivations, *J. Algebra* **198** (1997), 290–299.

[75] K. Buzzard and F. Calegari, Slopes of overconvergent 2-adic modular forms, *Compos. Math.* **141** (2005), 591–604.

[76] B. Cais (ed.), *Perfectoid Spaces: Lectures from the 2017 Arizona Winter School*, Math. Surveys and Monographs 242, Amer. Math. Soc., 2019, 58–205.

[77] D. Caro, F-isocristaux surconvergents et surcohérence différentielle, *Invent. Math.* **170** (2007), 507–539.

[78] D. Caro, Dévissages des F-complexes de $\mathcal{D}$-modules arithmétiques en F-isocristaux surconvergents, *Invent. Math.* **166** (2006), 397–456.

[79] N.L. Carothers, *A Short Course on Banach Space Theory*, Cambridge University Press, Cambridge, 2004.

[80] X. Caruso, D. Roe, and T. Vaccon, *p*-adic stability in linear algebra, in *ISSAC'15—Proceedings of the 2015 ACM International Symposium on Symbolic and Algebraic Computation*, ACM, New York, 2015, 101–108.

[81] X. Caruso, D. Roe, and T. Vaccon, Characteristic polynomials of *p*-adic matrices, in *ISSAC'17—Proceedings of the 2017 ACM International Symposium on Symbolic and Algebraic Computation*, ACM, New York, 2017, 389–396.

[82] J.W.S. Cassels, *Local Fields*, London Math. Society Student Texts 3, Cambridge University Press, Cambridge, 1986.

[83] W. Castryck and J. Tuitman, Point counting on curves using a gonality preserving lift, *Quart. J. Math.* **69** (2018), 33–74.

[84] F. Cherbonnier and P. Colmez, Représentations *p*-adiques surconvergentes, *Invent. Math.* **133** (1998), 581–611.

[85] B. Chiarellotto and G. Christol, On overconvergent isocrystals and *F*-isocrystals of rank one. *Compos. Math.* **100** (1996), 77–99.

[86] B. Chiarellotto and A. Pulita, Arithmetic and Differential Swan conductors of rank one representations with finite local monodromy, *Amer. J. Math.* **131** (2009), no. 6, 1743–1794.

[87] B. Chiarellotto and N. Tsuzuki, Logarithmic growth and Frobenius filtrations for solutions of *p*-adic differential equations, *J. Inst. Math. Jussieu* **8** (2009), 465–505.

[88] B. Chiarellotto and N. Tsuzuki, Log-growth filtration and Frobenius slope filtration of *F*-isocrystals at the generic and special points, *Doc. Math.* **16** (2011), 33–69.

[89] G. Christol, *Modules Différentiels et Equations Différentielles p-adiques*, Queen's Papers in Pure and Applied Math. 66, Queen's University, Kingston, 1983.

[90] G. Christol, Un théorème de transfert pour les disques singuliers réguliers, Cohomologie *p*-adique, *Astérisque* **119–120** (1984), 151–168.

[91] G. Christol, About a Tsuzuki theorem, in *p-adic Functional Analysis (Ioannina, 2000)*, Lecture Notes in Pure and Appl. Math., 222, Dekker, New York, 2001, 63–74.

[92] G. Christol, Thirty years later, in *Geometric Aspects of Dwork Theory (Volume I)*, de Gruyter, Berlin, 2004, 419–436.

[93] G. Christol, *Le Théorème de Turritin p-adique*, manuscript available at https://webusers.imj-prg.fr/~gilles.christol/courspdf.pdf (retrieved July 2020).

[94] G. Christol and B. Dwork, Effective *p*-adic bounds at regular singular points, *Duke Math. J.* **62** (1991), 689–720.

[95] G. Christol and B. Dwork, Differential modules of bounded spectral norm, in *p-adic Methods in Number Theory and Algebraic Geometry*, Contemp. Math. 133, Amer. Math. Soc., Providence, 1992.

[96] G. Christol and B. Dwork, Modules différentielles sur les couronnes, *Ann. Inst. Fourier* **44** (1994), 663–701.

[97] G. Christol and Z. Mebkhout, Sur le théorème de l'indice des équations différentielles *p*-adiques, *Annals de l'Institut Fourier* **43** (1993), 1545–1574.

[98] G. Christol and Z. Mebkhout, Sur le théorème de l'indice des équations différentielles p-adiques II, *Annals of Math.* **146** (1997), 345–410.

[99] G. Christol and Z. Mebkhout, Sur le théorème de l'indice des équations différentielles p-adiques III, *Annals of Math.* **151** (2000), 385–457.

[100] G. Christol and Z. Mebkhout, Sur le théorème de l'indice des équations différentielles p-adiques IV, *Invent. Math.* **143** (2001), 629–672.

[101] R.C. Churchill and J.J. Kovacic, Cyclic vectors, in *Differential Algebra and Related Topics (Newark, NJ, 2000)*, World Scientific, River Edge, NJ, 2002, 191–218.

[102] D. Clark, A note on the p-adic convergence of solutions of linear differential equations, *Proc. Amer. Math. Soc.* **17** (1966), 262–269.

[103] H. Cohen and G. Frey (eds.), *Handbook of Elliptic and Hyperelliptic Curve Cryptography*, Discrete Math. and its Applications 34, Chapman & Hall/CRC, 2005.

[104] A. Cohen, M. Temkin, and D. Trushin, Morphisms of Berkovich curves and the different function, *Adv. Math.* **303** (2016), 800–858.

[105] R.M. Cohn, *Difference Algebra*, John Wiley & Sons, New York, London, Sydney, 1965.

[106] R.F. Coleman, Dilogarithms, regulators and p-adic L-functions, *Invent. Math.* **69** (1982), 171–208.

[107] R.F. Coleman, Torsion points on curves and p-adic abelian integrals, *Ann. of Math.* **121** (1985), 111–168.

[108] R.F. Coleman, Effective Chabauty, *Duke Math. J.* **52** (1985), 765–770.

[109] M. Collins, On Jordan's theorem for complex linear groups, *J. Group Theory* **10** (2007), 411–123.

[110] P. Colmez, Représentations triangulines de dimension 2, *Astérisque* **319** (2008), 213–258.

[111] B. Conrad, Several approaches to non-archimedean geometry, in [360], 24–79.

[112] B. Conrad, Relative ampleness in rigid geometry, *Ann. Inst. Fourier* **56** (2006), 1049–1126.

[113] F.T. Cope, Formal solutions of irregular linear differential equations, II, *Amer. J. Math.* **58** (1936), 130–140.

[114] E. Costa, D. Harvey, and K.S. Kedlaya, Zeta functions of nondegenerate hypersurfaces in toric varieties via controlled reduction in p-adic cohomology, in *ANTS XIII: Proceedings of the Thirteenth Algorithmic Number Theory Symposium*, Open Book Series 2, Math. Sci. Pub., Berkeley, 2019, 221–238.

[115] E. Costa, K.S. Kedlaya, and D. Roe, Hypergeometric L-functions in average polynomial time, in *ANTS-XIV: Proceedings of the Fourteenth Algorithmic Number Theory Symposium*, Open Book Series 4, Math. Sci. Pub., Berkeley, 2020, 143–159.

[116] E. Costa and E.C. Sertöz, Effective obstruction to lifting Tate classes from positive characteristic, arXiv:2003.11037v1 (2020).

[117] R. Crew, F-isocrystals and p-adic representations, in *Algebraic geometry, Bowdoin, 1985 (Brunswick, Maine, 1985)*, Proc. Sympos. Pure Math., 46, Part 2, Amer. Math. Soc., Providence, 1987, 111–138.

[118] R. Crew, Finiteness theorems for the cohomology of an overconvergent isocrystal on a curve, *Ann. Sci. Éc. Norm. Sup.* **31** (1998), 717–763.

[119] R. Crew, Canonical extensions, irregularities, and the Swan conductor, *Math. Ann.* **316** (2000), 19–37.

[120] N.E. Csima, Newton-Hodge filtration for self-dual F-crystals, *Math. Nachr.* **284** (2011), 1394–1403.

[121] C.W. Curtis and I. Reiner, *Representation Theory of Finite Groups and Associative Algebras*, reprint of the 1962 original, Amer. Math. Soc., Providence, 2006.

[122] A. D'Agnolo, and M. Kashiwara, Riemann-Hilbert correspondence for holonomic D-modules, *Publ. Math. IHÉS* **123** (2016), 69–197.

[123] A.J. de Jong, Homomorphisms of Barsotti-Tate groups and crystals in positive characteristic, *Invent. Math.* **134** (1998), 301–333; erratum, *ibid.* **138** (1999), 225.

[124] A.J. de Jong, Barsotti-Tate groups and crystals, *Proceedings of the International Congress of Mathematicians, Vol. II (Berlin, 1998), Doc. Math.* Extra Vol. II (1998), 259–265.

[125] A.J. de Jong and F. Oort, Purity of the stratification by Newton polygons, *J. Amer. Math. Soc.* **13** (2000), 209–241.

[126] P. Deligne, *Équations Différentielles à Points Singuliers Réguliers*, Lecture Notes in Math. 163, Springer-Verlag, Berlin, 1970.

[127] P. Deligne, Catégories tannakiennes, in *The Grothendieck Festschrift, volume II*, Birkhäuser, Boston, 1990, 111–195.

[128] P. Deligne, La conjecture de Weil, II, *Publ. Math. IHÉS* **52** (1980), 137–252.

[129] P. Deligne and N. Katz (eds.), *Groupes de Monodromie en Géométrie Algébrique II, Séminaire de Géométrie Algébrique du Bois-Marie 1967–1969 (SGA 7 II)*, Lecture Notes in Math. 340, Springer-Verlag, Berlin-New York, 1973.

[130] P. Deligne and J.S. Milne, Tannakian categories, in *Hodge Cycles, Motives and Shimura Varieties*, Lecture Notes in Math. 900, Springer-Verlag, Berlin, 1982.

[131] M. Demazure, *Lectures on p-divisible Groups*, Lecture Notes in Math. 302, Springer-Verlag, New York, 1972.

[132] J. Denef and F. Vercauteren, An extension of Kedlaya's algorithm to hyperelliptic curves in characteristic 2, *J. Cryptology* **19** (2006), 1–25.

[133] L. Di Vizio, Local analytic classification of q-difference equations with $|q| = 1$, *J. Noncommutative Geom.* **3** (2009), 125–149.

[134] T. Dokchitser, V. Dokchitser, C. Maistret, and A. Morgan, Semistable types of hyperelliptic curves, in *Algebraic Curves and their Applications*, Contemp Math. 724, Amer. Math. Soc., Providence, 2019, 73–135.

[135] T. Dokchitser, V. Dokchister, C. Maistret, and A. Morgan, Arithmetic of hyperelliptic curves over local fields, arXiv:1808.02936v2 (2018).

[136] V.G. Drinfeld, Langlands' Conjecture for GL(2) over functional Fields, Proceedings of the International Congress of Mathematicians, Helsinki, 565–574 (1978).

[137] A. Ducros, *La Structure des Courbes Analytiques*, manuscript available at https://webusers.imj-prg.fr/~antoine.ducros/livre.html (retrieved June 2021).

[138] B. Dwork, On the rationality of the zeta function of an algebraic variety, *Amer. J. Math.* **82** (1960), 631–648.

[139] B. Dwork, On the zeta function of a hypersurface, *Publ. Math. IHÉS* **12** (1962), 5–68.

[140] B. Dwork, *p*-adic cycles, *Publ. Math. IHÉS* **37** (1969), 27–115.

[141] B. Dwork, On *p*-adic differential equations, II. The *p*-adic asymptotic behavior of solutions of ordinary linear differential equations with rational function coefficients, *Ann. of Math.* (2) **98** (1973), 366–376.

[142] B. Dwork, On *p*-adic differential equations, III. On *p*-adically bounded solutions of ordinary linear differential equations with rational function coefficients, *Invent. Math.* **20** (1973), 35–45.

[143] B. Dwork, On *p*-adic differential equations. IV. Generalized hypergeometric functions as *p*-adic analytic functions in one variable. *Ann. Sci. Éc. Norm. Sup.* **6** (1973), 295–315.

[144] B. Dwork, Bessel functions as *p*-adic functions of the argument, *Duke Math. J.* **41** (1974), 711–738.

[145] B. Dwork, *Lectures on p-adic Differential Equations*, Grundlehren der math. Wissenschaften 253, Springer-Verlag, New York, 1982.

[146] B. Dwork, On the uniqueness of Frobenus operator on differential equations, in *Algebraic Number Theory*, Adv. Stud. Pure Math. 17, Academic Press, Boston, MA, 1989, 89–96.

[147] B. Dwork, *Generalized Hypergeometric Functions*, Clarendon Press, Oxford, 1990.

[148] B.M. Dwork, On exponents of *p*-adic differential modules, *J. reine angew. Math.* **484** (1997), 85–126.

[149] B. Dwork, G. Gerotto, and F. Sullivan, *An Introduction to G-Functions*, Annals of Math. Studies 133, Princeton University Press, Princeton, 1994.

[150] B. Dwork and P. Robba, On ordinary linear *p*-adic differential equations, *Trans. Amer. Math. Soc.* **231** (1977), 1–46.

[151] B. Dwork and P. Robba, Effective *p*-adic bounds for solutions of homogeneous linear differential equations, *Trans. Amer. Math. Soc.* **259** (1980), 559–577.

[152] B. Edixhoven, Point counting after Kedlaya, course notes (2006) available at https://www.math.leidenuniv.nl/~edix/oww/mathofcrypt/carls_edixhoven/kedlaya.pdf (retrieved June 2020).

[153] B. Edixhoven and G. Lido, Geometric quadratic Chabauty, arXiv:1910.10752v3 (2020).

[154] D. Eisenbud, *Commutative Algebra with a View Toward Algebraic Geometry*, Graduate Texts in Math. 150, Springer-Verlag, New York, 1995.

[155] A.-S. Elsenhans and J. Jahnel, Point counting on K3 surfaces and an application concerning real and complex multiplication, *LMS J. Comput. Math.* **19** (2016), 12–28.

[156] P. Enflo, A counterexample to the approximation problem in Banach spaces, *Acta Math.* **130** (1973), 309–317.

[157] A. Escassut, Algèbres d'éléments analytiques en analyse non archimedienne, *Indag. Math.* **77** (1974), 339–351.

[158] X. Faber, Topology and geometry of the Berkovich ramification locus for rational functions, I, *Manuscripta Math.* **142** (2013), 439–474.

[159] X. Faber, Topology and geometry of the Berkovich ramification locus for rational functions, II, *Math. Ann.* **356** (2013), 819–844.

[160] L. Fargues and J.-M. Fontaine, Courbes et fibrés vectoriels en théorie de Hodge *p*-adique, *Astérisque* **406** (2018).

[161] L. Fargues and P. Scholze, Geometrization of the local Langlands correspondence, arXiv:arXiv:2102.13459v2 (2021).

[162] G. Faltings, Coverings of *p*-adic period domains, research announcement (2007) available at http://www.mpim-bonn.mpg.de/preprints/ (retrieved June 2020).

[163] C. Favre and M. Jonsson, *The Valuative Tree*, Lecture Notes in Math. 1853, Springer-Verlag, Berlin, 2004.

[164] C. Favre and M. Jonsson, Valuations and multiplier ideals, *J. Amer. Math. Soc.* **18** (2005), 655–684.

[165] J.-M. Fontaine, Représentations *p*-adiques des corps locaux, I, in *The Grothendieck Festschrift, Vol. II*, Progress in Math. 87, Birkhäuser, Boston, 1990, 249–309.

[166] J.-M. Fontaine, Représentations *p*-adiques semi-stables, Périodes *p*-adiques (Bures-sur-Yvette, 1988), *Astérisque* **23** (1994), 113–184.

[167] J.-M. Fontaine, Représentations de de Rham et représentations semi-stables, Orsay preprint number 2004-12, available online at https://www.imo.universite-paris-saclay.fr/~biblio/pub/2004/ (retrieved June 2020).

[168] J.-M. Fontaine and Y. Ouyang, *Theory of p-adic Galois Representations*, available online at http://staff.ustc.edu.cn/~yiouyang/ (retrieved July 2020).

[169] J.-M. Fontaine and J.-P. Wintenberger, Le "corps de normes" de certaines extensions algébriques de corps locaux, *C.R. Acad. Sci. Paris Sér. A-B* **288** (1979), A367–A370.

[170] J. Fresnel and M. Matignon, Produit tensoriel topologique de corps valués, *Canad. J. Math.* **35** (1983), 218–273.

[171] J. Fresnel and M. van der Put, *Rigid Analytic Geometry and its Applications*, Progress in Mathematics 218, Birkhäuser, Boston, 2004.

[172] W. Fulton, *Intersection Theory*, second edition, Springer-Verlag, Berlin, 1998.

[173] W. Fulton, Eigenvalues, invariant factors, highest weights, and Schubert calculus, *Bull. Amer. Math. Soc. (N.S.)* **37** (2000), 209–249.

[174] P. Gaudry and É. Schost, Genus 2 point counting over prime fields, *J. Symb. Comp.* **47** (2012), 368–400.

[175] I.M. Gelfand, M.M. Kapranov, and A.V. Zelevinsky, *Discriminants, Resultants and Multidimensional Determinants*, reprint of the 1994 edition, Birkhäuser, Boston, 2008.

[176] R. Gérard and A.M. Levelt, Invariants mesurant l'irrégularité en un point singulier des systèmes d'équations différentielles linéaires, *Ann. Inst. Fourier (Grenoble)* **23** (1973), 157–195.

[177] R. Gerkmann, Relative rigid cohomology and deformation of hypersurfaces, *Int. Math. Res. Papers* **2007** (2007), article ID rpm003 (67 pages).

[178] F.Q. Gouvêa, *p-adic Numbers: An Introduction*, third edition, Universitext 223, Springer, Cham, 2020.

[179] P. Griffiths and J. Harris, *Principles of Algebraic Geometry*, Wiley Interscience, New York, 1978.

[180] A. Grothendieck, La théorie de Fredholm, *Bull. Soc. Math. France* **84** (1956), 319–384.

[181] A. Grothendieck, Sur la classification des fibrés holomorphes sur la sphère de Riemann, *Amer. J. Math.* **79** (1957), 121–138.

[182] A. Grothendieck, On the de Rham cohomology of algebraic varieties, *Publ. Math. IHÉS* **29** (1966), 95–103.

[183] A. Grothendieck, *Revêtements Étales et Groupe Fondamental, Séminaire de Géométrie Algébrique du Bois-Marie 1960–1961 (SGA 1)*, Lecture Notes in Math. 224, Springer-Verlag, Berlin, 1971.

[184] A. Grothendieck, Le groupe de Brauer: I. Algébres d'Azumaya et interprétations diverses, in *Séminaire Bourbaki, Vol. 9*, Exposé 290, Soc. Math. France, Paris, 1995, 199–219.

[185] W. Gubler, J. Rabinoff, and A. Werner, Skeletons and tropicalizations, *Adv. Math.* **294** (2016), 150–215.

[186] W. Gubler, J. Rabinoff, and A. Werner, Tropical skeletons, *Ann. Inst. Fourier, Grenoble* **67** (2017), 1905–1961.

[187] B. Guennebaud, Sur une notion de spectre pour les algébres normées ultramétriques, PhD thesis, Université de Poitiers, 1973; available at http://www.ihes.fr/~gabber/GUENN.pdf (retrieved April 2017).

[188] R.M. Guralnick and M. Lorenz, Orders of finite groups of matrices, in *Groups, Rings and Algebras*, Contemp. Math. 420, Amer. Math. Soc., Providence, 2006, 141–161.

[189] H. Hahn, Über die nichtarchimedische Größensysteme, in *Gesammelte Abhandlungen I*, Springer, Vienna, 1995.

[190] U. Hartl, Period spaces for Hodge structures in equal characteristic, *Annals of Math.* **173** (2011), 1241–1358.

[191] U. Hartl, On a conjecture of Rapoport and Zink, *Invent. Math.* **193** (2013), 627–696.

[192] U. Hartl, On period spaces for p-divisible groups, *C.R. Math. Acad. Sci. Paris* **346** (2008), 1123–1128.

[193] U. Hartl and R. Pink, Vector bundles with a Frobenius structure on the punctured unit disc, *Compos. Math.* **140** (2004), 689–716.

[194] R. Hartshorne, *Residues and Duality*, Lecture Notes in Math. 20, Springer, New York, 1966.

[195] R. Hartshorne, On the de Rham cohomology of algebraic varieties, *Publ. Math. IHÉS* **45** (1975), 5–99.

[196] R. Hartshorne, *Algebraic Geometry*, Graduate Texts in Math. 52, Springer-Verlag, New York, 1977.

[197] D. Harvey, Kedlaya's algorithm in larger characteristic, *Int. Math. Res. Notices* **2007** (2007), article ID rnm095 (29 pages).

[198] D. Harvey, Computing zeta functions of arithmetic schemes, *Proc. London Math. Soc.* **111** (2015), 1379–1401.

[199] M. Hazewinkel, Gauss-Manin connection, entry in *Encyclopedia of Mathematics*, third edition, Kluwer, 2001; available online at https://encyclopediaofmath.org/wiki/Gauss-Manin_connection.

[200] M. Hazewinkel, Witt vectors, I, in *Handbook of Algebra. Vol. 6*, Handb. Algebr. 6, Elsevier/North-Holland, Amsterdam, 2009, 319–472.

[201] L. Herr, Sur la cohomologie galoisienne des corps p-adiques, *Bull. Soc. Math. France* **126** (1998), 563–600.

[202] L. Herr, Une approche nouvelle de la dualité locale de Tate, *Math. Ann.* **320** (2001), 307–337.

[203] H. Hironaka, Resolution of singularities of an algebraic variety over a field of characteristic zero: I, *Annals of Math.* **79** (1964), 109–203.

[204] H. Hironaka, Resolution of singularities of an algebraic variety over a field of characteristic zero: II, *Annals of Math.* **79** (1964), 205–326.

[205] W.V.D. Hodge and D. Pedoe, *Methods of Algebraic Geometry, Vol. III* (reprint of the 1954 original), Cambridge University Press, Cambridge, 1994.

[206] R.T. Hoobler, The differential Brauer group, arXiv:1003.1421v1 (2010).

[207] A. Horn, On the eigenvalues of a matrix with prescribed singular values, *Proc. Amer. Math. Soc.* **5** (1954), 4–7.

[208] A. Horn, Eigenvalues of sums of Hermitian matrices, *Pacific J. Math.* **12** (1962), 225–241.

[209] R.A. Horn and C.R. Johnson, *Topics in Matrix Analysis*, Cambridge Univ. Press, Cambridge, 2011.

[210] E. Hrushovski, F. Loeser, and B. Poonen, Berkovich spaces embed in Euclidean spaces, *Enseign. Math.* **60** (2014), 273–292.

[211] R. Huber, A generalization of formal schemes and rigid analytic varieties, *Math. Z.* **217** (1994), 513–551.

[212] R. Huber, *Étale Cohomology of Rigid Analytic Varieties and Adic Spaces*, Aspects of Math., Springer, Wiesbaden, 1996.

[213] R. Huber, Swan representations associated with rigid analytic curves, *J. reine angew. Math.* **537** (2001), 165–234.

[214] H. Hubrechts, Point counting in families of hyperelliptic curves, *Found. Comp. Math.* **8** (2008), 137–169.

[215] H. Hubrechts, Memory efficient hyperelliptic curve point counting, *Int. J. Number Theory* **7** (2011), 203–214.

[216] L. Illusie, Crystalline cohomology, in *Motives (Seattle, WA, 1991)*, Proc. Sympos. Pure Math., 55, Part 1, Amer. Math. Soc., Providence, 1994, 43–70.

[217] N. Jacobson, *Basic Algebra I*, second edition, W.H. Freeman, New York, 1985.

[218] A. Joyal, δ-anneau et vecteurs de Witt, *C.R. Math. Rep. Acad. Sci. Canada* **7** (1985), 177–182.

[219] A. Joyal, δ-anneau et λ-anneaux, II, *C.R. Math. Rep. Acad. Sci. Canada* **7** (1985), 227–232.

[220] T. Kajiwara, Tropical toric geometry, in *Toric Topology*, Contemp. Math. 460, Amer. Math. Soc., Providence, 2008, 197–207.

[221] I. Kaplansky, Maximal fields with valuations, *Duke Math. J.* **9** (1942), 303–321.

[222] K. Kato, Swan conductors for characters of degree one in the imperfect residue field case, in *Algebraic K-theory and algebraic number theory (Honolulu, HI, 1987)*, Contemp. Math. 83, Amer. Math. Soc., Providence, 1989, 101–131.

[223] N.M. Katz, Nilpotent connections and the monodromy theorem: applications of a result of Turrittin, *Publ. Math. IHÉS* **39** (1970), 175–232.

[224] N.M. Katz, Slope filtration of F-crystals, Journées de Géométrie Algébrique de Rennes (Rennes, 1978), Vol. I, *Astérisque* **63** (1979), 113–163.

[225] N.M. Katz, Local-to-global extensions of representations of fundamental groups, *Ann. Inst. Fourier (Grenoble)* **36** (1986), 69–106.

[226] N.M. Katz, On the calculation of some differential Galois groups, *Invent. Math.* **87** (1987), 13–61.

[227] N.M. Katz, A simple algorithm for cyclic vectors, *Amer. J. Math.* **109** (1987), 65–70.

[228] N.M. Katz and T. Oda, On the differentiation of de Rham cohomology classes with respect to parameters, *J. Math. Kyoto Univ.* **8** (1968), 199–213.

[229] K.S. Kedlaya, Counting points on hyperelliptic curves using Monsky-Washnitzer cohomology, *J. Ramanujan Math. Soc.* **16** (2001), 323–338; errata, *ibid.* **18** (2003), 417–418.

[230] K.S. Kedlaya, A p-adic local monodromy theorem, *Annals of Math.* **160** (2004), 93–184.

[231] K.S. Kedlaya, Full faithfulness for overconvergent F-isocrystals, in *Geometric Aspects of Dwork Theory (Volume II)*, de Gruyter, Berlin, 2004, 819–835.

[232] K.S. Kedlaya, Computing zeta functions via p-adic cohomology, *Algorithmic Number Theory (ANTS VI)*, Lecture Notes in Comp. Sci. 3076, Springer-Verlag, 2004, 1–17.

[233] K.S. Kedlaya, Local monodromy of p-adic differential equations: an overview, *Int. J. Number Theory* **1** (2005), 109–154; errata at https://kskedlaya.org/papers/.

[234] K.S. Kedlaya, Slope filtrations revisited, *Documenta Math.* **10** (2005), 447–525; errata, *ibid.* **12** (2007), 361–362.

[235] K.S. Kedlaya, Frobenius modules and de Jong's theorem, *Math. Res. Lett.* **12** (2005), 303–320; errata at https://kskedlaya.org/papers/.

[236] K.S. Kedlaya, Finiteness of rigid cohomology with coefficients, *Duke Math. J.* **134** (2006), 15–97.

[237] K.S. Kedlaya, Fourier transforms and p-adic "Weil II", *Compos. Math.* **142** (2006), 1426–1450.

[238] K.S. Kedlaya, Swan conductors for p-adic differential modules, I: A local construction, *Algebra and Number Theory* **1** (2007), 269–300.

[239] K.S. Kedlaya, Semistable reduction for overconvergent F-isocrystals, I: Unipotence and logarithmic extensions, *Compos. Math.* **143** (2007), 1164–1212.

[240] K.S. Kedlaya, The p-adic local monodromy theorem for fake annuli, *Rend. Sem. Math. Padova* **118** (2007), 101–146.

[241] K.S. Kedlaya, Slope filtrations for relative Frobenius, *Astérisque* **319** (2008), 259–301.

[242] K.S. Kedlaya, Semistable reduction for overconvergent F-isocrystals, II: A valuation-theoretic approach, *Compos. Math.* **144** (2008), 657–672.

[243] K.S. Kedlaya, p-adic cohomology: from theory to practice, in [360], 190–219.

[244] K.S. Kedlaya, Effective p-adic cohomology for cyclic cubic threefolds, in *Computational Algebraic and Analytic Geometry*, Contemp. Math. 572, Amer. Math. Soc., 127–171.

[245] K.S. Kedlaya, Semistable reduction for overconvergent F-isocrystals, III: Local semistable reduction at monomial valuations, *Compos. Math.* **145** (2009), 143–172.

[246] K.S. Kedlaya, p-adic cohomology, in *Algebraic Geometry: Seattle 2005*, Proc. Symp. Pure Math. 80, Amer. Math. Soc., 2009, 667–684.

[247] K.S. Kedlaya, Some new directions in p-adic Hodge theory, *J. Théorie Nombres Bordeaux* **21** (2009), 285–300.

[248] K.S. Kedlaya, Swan conductors for p-adic differential modules, II: Global variation, *J. Institut Math. Jussieu* **10** (2011), 191–224.

[249] K.S. Kedlaya, Semistable reduction for overconvergent F-isocrystals, IV: Local semistable reduction at nonmonomial valuations, *Compos. Math.* 147 (2011), 467–523; errata in [255].

[250] K.S. Kedlaya, Good formal structures for flat meromorphic connections, I: Surfaces, *Duke Math. J.* **154** (2010), 343–418; erratum, *ibid.* **161** (2012), 733–734.

[251] K.S. Kedlaya, Good formal structures for flat meromorphic connections, II: Excellent schemes, *J. Amer. Math. Soc.* **24** (2011), 183–229.

[252] K.S. Kedlaya, Effective p-adic cohomology for cyclic cubic threefolds, in *Computational Algebraic and Analytic Geometry*, Contemp. Math. 572, Amer. Math. Soc., 2012, 127–171.

[253] K.S. Kedlaya, Nonarchimedean geometry of Witt vectors, *Nagoya Math. J.* **209** (2013), 111–165.

[254] K.S. Kedlaya, Reified valuations and adic spectra, *Res. Number Theory* **1** (2015), 1–42.

[255] K.S. Kedlaya, Local and global structure of connections on nonarchimedean curves, *Compos. Math.* **151** (2015), 1096–1156; errata in [269].

[256] K.S. Kedlaya, On commutative nonarchimedean Banach fields, *Doc. Math.* **23** (2018), 171–188.

[257] K.S. Kedlaya, Some ring-theoretic properties of $\mathbf{A}_{inf}$, in *p-adic Hodge Theory*, Springer, 2020.

[258] K.S. Kedlaya, Sheaves, stacks, and shtukas, in [76], 58–205.

[259] K.S. Kedlaya, Convergence polygons for connections on nonarchimedean curves, in *Nonarchimedean and Tropical Geometry*, Simons Symposia, Springer Nature, 2016, 51–98.

[260] K.S. Kedlaya, Good formal structures on flat meromorphic connections, III: Irregularity and turning loci, arXiv:1308.5259v4 (2019); to appear in *Publ. RIMS* (Kashiwara 70th birthday issue).

[261] K.S. Kedlaya, Weil cohomology in practice, lecture notes (2019) available at `http://kskedlaya.org/papers/`.

[262] K.S. Kedlaya, Notes on isocrystals, arXiv:1606.01321v5 (2018).

[263] K.S. Kedlaya, Étale and crystalline companions, I, arXiv:1811.00204v3 (2020).

[264] K.S. Kedlaya, Étale and crystalline companions, II, arXiv:2008.13053v1 (2020).

[265] K.S. Kedlaya, Frobenius modules over multivariate Robba rings, arXiv:1311.7468v3 (2020).

[266] K.S. Kedlaya, Frobenius structures on hypergeometric equations, arXiv:1912.13073v2 (2021).

[267] K.S. Kedlaya, D. Litt, and J. Witaszek, Tamely ramified morphisms of curves and Belyi's theorem in positive characteristic, arXiv:2010.01130v1 (2020).

[268] K.S. Kedlaya and R. Liu, Relative p-adic Hodge theory, II: Imperfect period rings, arXiv:1602.06899v3 (2019).

[269] K.S. Kedlaya and A. Shiho, Corrigendum: Local and global structure of connections on nonarchimedean curves, *Compos. Math.* **153** (2017), 2658–2665.

[270] K.S. Kedlaya and J. Tuitman, Effective bounds for Frobenius structures, *Rend. Sem. Mat. Padova* **128** (2012), 7–16.

[271] K.S. Kedlaya and Liang Xiao, Differential modules on p-adic polyannuli, *J. Inst. Math. Jussieu* **9** (2010), 155–201; erratum, *ibid.* **9** (2010), 669–671.

[272] R. Kiehl, Der Endlichkeitssatz für eigentliche Abbildungen in der nichtarchimedischen Funktiontheorie, *Invent. Math.* **2** (1967), 191–214.

[273] M. Kim, The non-abelian (or non-linear) method of Chabauty, in *Noncommutative Geometry and Number Theory*, Aspects Math. E37, Friedr. Vieweg, Wiesbaden, 2006, 179–185.

[274] M. Kisin, Overconvergent modular forms and the Fontaine-Mazur conjecture, *Invent. Math.* **153** (2003), 373–454.

[275] M. Kisin, Crystalline representations and F-crystals, in *Algebraic Geometry and Number Theory*, Progress in Math. 253, Birkhäuser, Boston, 2006, 459–496.

[276] U. Köpf, Über eigentliche familien algebraischer varietäten über affinoiden räumen, Schriftenreihe Math. Inst. Münster, 2. Serie. Heft 7 (1974).

[277] R.E. Kottwitz, Isocrystals with additional structure, *Comp. Math.* **56** (1985), 201–220.

[278] R.E. Kottwitz, Isocrystals with additional structure. II, *Comp. Math.* **109** (1997), 255–339.

[279] R.E. Kottwitz, On the Hodge-Newton decomposition for split groups, *Int. Math. Res. Notices* **26** (2003), 1433–1447.

[280] M. Krasner, Prolongement analytique uniform et multiforme dans le corps valués complets, *Les Tendances Géom. en Algèbre et Théorie des Nombres*, CNRS, Paris, 1966, 97–141.

[281] M. Krein, A principle of duality for bicompact groups and quadratic block algebras (Russian), *Doklady Akad. Nauk. SSSR* **69** (1949), 725–728.

[282] F.-V. Kuhlmann, Elimination of ramification I: The generalized stability theorem, *Trans. Amer. Math. Soc.* **362** (2010), no. 11, 5697–5727.

[283] W. Krull, Allgemeine Bewertungstheorie, *J. für Math.* **167** (1932), 160–196.

[284] L. Lafforgue, Chtoucas de Drinfeld et correspondance de Langlands, *Invent. Math.* **147** (2002), 1–241.

[285] S. Lang, Algebraic groups over finite fields, *Amer. J. Math.* **78** (1956), 555–563.

[286] A.G.B. Lauder, Deformation theory and the computation of zeta functions, *Proc. London Math. Soc.* **88** (2004), 565–602.

[287] A.G.B. Lauder and D. Wan, Counting points on varieties over finite fields of small characteristic, in *Algorithmic Number Theory*, MSRI Pub. 44, Cambridge Univ. Press, Cambridge, 2008, 579–612.

[288] G. Laumon, *Cohomology of Drinfeld Modular Varieties I*, Cambridge Studies in Advanced Math. 41, Cambridge University Press, Cambridge, 1996.

[289] M. Lazard, Les zéros d'une fonction analytique d'une variable sur un corps valué complet, *Publ. Math. IHÉS* **14** (1962), 47–75.

[290] A.H.M. Levelt, Jordan decomposition for a class of singular differential operators, *Ark. Mat.* **13** (1975), 1–27.

[291] B. Le Stum, *Rigid Cohomology*, Cambridge Tracts in Math. 172, Cambridge University Press, Cambridge, 2007.

[292] J. Lindenstrauss and L. Tzafriri, *Classical Banach Spaces*, Lecture Notes in Math. 338, Springer, 1973.

[293] R. Liu, Cohomology and duality for (φ, Γ)-modules over the Robba ring, *Int. Math. Res. Notices* **2007**, article ID rnm150 (32 pages).

[294] M. Loday-Richaud, Stokes phenomenon, multisummmability and differential Galois groups, *Ann. Inst. Fourier (Grenoble)* **44** (1994), 849–906.

[295] F. Loeser, Exposants p-adiques et théorèmes d'indice pour les équations différentielles p-adiques (d'après G. Christol et Z. Mebkhout), Séminaire Bourbaki, Vol. 1996/97, *Astérisque* **245** (1997), 57–81.

[296] A. Loewy, Über einen Fundamentalsatz für Matrizen oder lineare homogene Differentialsysteme, Sitzungsberichte der Heidelberger Akademie der Wissenschaften, 5. Abhandlung (1918), 1–36.

[297] J. Lubin, Nonarchimedean dynamical systems, *Compositio Math.* **94** (1994), 321–346.

[298] E. Lutz, Sur l'équation $y^2 = x^3 + Ax + B$ sur les corps p-adiques, *J. reine angew Math.* **177** (1937), 238–247.

[299] B. Malgrange, Sur les points singuliers des équations différentielles, *Enseign. Math.* **20** (1974), 147–176.

[300] B. Malgrange, *Équations Différentielles à coeffcients polynomiaux*, Progress in Math. 96, Birkhäuser, Basel, 1991.

[301] B. Malgrange, Connexions méromorphes 2: Le réseau canonique, *Invent. Math.* **124** (1996), 367–387.

[302] Yu. I. Manin, The theory of commutative formal groups over fields of finite characteristic (Russian), *Usp. Math.* 18 (1963), 3–90; English translation, *Russian Math. Surveys* **18** (1963), 1–80.

[303] S. Manjra, A note on non-Robba p-adic differential equations, *Proc. Japan Acad. Ser. A* **87** (2011), 40–43.

[304] A. Marmora, Irrégularité et conducteur de Swan p-adiques, *Doc. Math.* **9** (2004), 413–433.

[305] M. Matignon and M. Reversat, Sous-corps fermés d'un corps valué, *J. Algebra* **90** (1984), 491–515.

[306] S. Matsuda, Local indices of p-adic differential operators corresponding to Artin-Schreier-Witt coverings, *Duke Math. J.* **77** (1995), 607–625.

[307] S. Matsuda, Katz correspondence for quasi-unipotent overconvergent isocrystals, *Comp. Math.* **134** (2002), 1–34.

[308] S. Matsuda, Conjecture on Abbes-Saito filtration and Christol-Mebkhout filtration, in *Geometric Aspects of Dwork Theory. Vol. I, II*, de Gruyter, Berlin, 2004, 845–856.

[309] Z. Mebkhout, Analogue p-adique du théorème de Turrittin et le théorème de la monodromie p-adique, *Invent. Math.* **148** (2002), 319–351.

[310] T. Mochizuki, Good formal structure for meromorphic flat connections on smooth projective surfaces, in T. Miwa et al. (eds.), *Algebraic Analysis and Around*, Advanced Studies in Pure Math. 54, Math. Soc. Japan, 2009, 223–253.

[311] T. Mochizuki, Wild harmonic bundles and wild pure twistor D-modules, *Astérisque* **340** (2011).

[312] T. Mochizuki, The Stokes structure of a good meromorphic flat bundle, *J. Inst. Math. Jussieu* **10** (2011), 675–712.

[313] P. Monsky, Formal cohomology, II: The cohomology sequence of a pair, *Ann. of Math.* **88** (1968), 218–238.

[314] P. Monsky, Formal cohomology, III: Fixed point theorems, *Ann. of Math.* **93** (1971), 315–343.

[315] P. Monsky and G. Washnitzer, Formal cohomology, I, *Ann. of Math.* **88** (1968), 181–217.

[316] K. Morita, Crystalline and semi-stable representations in the imperfect residue field case, *Asian J. Math.* **18** (2014), 143–157.

[317] D. Mumford, J. Fogarty, F. Kirwan, *Geometric Invariant Theory*, third edition, Ergebnisse der Math. 34, Springer-Verlag, Berlin, 1994.

[318] J.R. Munkres, *Topology*, second edition, Prentice Hall, Saddle River, 2000.

[319] M. Nagata, *Local Rings*, Interscience Tracts in Pure and Applied Math. 13, John Wiley & Sons, New York, 1962.

[320] J. Nicaise, Berkovich skeleta and birational geometry, in *Nonarchimedean and Tropical Geometry*, Simons Symposia, Springer, Cham, 2016, 173–194.

[321] S. Ohkubo, The *p*-adic monodromy theorem in the imperfect residue field case, *Algebra Number Theory* **7** (2013), 1977–2037.

[322] S. Ohkubo, A note on logarithmic growth Newton polygons of *p*-adic differential equations, *Int. Math. Res. Notices* **2014**, paper ID rnu017.

[323] S. Ohkubo, On differential modules associated to de Rham representations in the imperfect residue field case, *Algebra Number Theory* **9** (2015), 1881–1954.

[324] S. Ohkubo, On the rationality and continuity of logarithmic growth filtration of solutions of *p*-adic differential equations, *Adv. Math.* **308** (2017), 83–120.

[325] S. Ohkubo, A note on logarithmic growth of solutions of *p*-adic differential equations without solvability, *Math. Res. Lett.* **26** (2019), 1527–1557.

[326] S. Ohkubo, Logarithmic growth filtrations for (φ, ∇)-modules over the bounded Robba ring, *Compos. Math.* **157** (2021), 1265–1301.

[327] M. Ohtsuki, A residue formula for Chern classes associated with logarithmic connections, *Tokyo J. Math.* **5** (1982), 13–21.

[328] F. Oort and T. Zink, Families of *p*-divisible groups with constant Newton polygon, *Doc. Math.* **7** (2002), 183–201.

[329] O. Ore, Theory of non-commutative polynomials, *Annals of Math.* **34** (1933), 480–508.

[330] B. Osserman, The Weil conjectures, in *The Princeton Companion to Mathematics*, Princeton University Press, Princeton, 2008.

[331] A. Ostrowski, Untersuchungen zur arithmetischen Theorie der Körper, *Math. Z.* **39** (1935), 269–404.

[332] S. Pancratz and J. Tuitman, Improvements to the deformation method for counting points on smooth projective hypersurfaces, *Found. Comput. Math.* **15** (2015), 1413–1464.

[333] S. Payne, Analytification is the limit of all tropicalizations, *Math. Res. Lett.* **16** (2009), 543–556.

[334] J. Pila, Frobenius maps of abelian varieties and finding roots of unity in finite fields, *Math. Comp.* **55** (1990), 745–763.

[335] J. Poineau and A. Pulita, The convergence Newton polygon of a *p*-adic differential equation II: Continuity and finiteness on Berkovich curves, *Acta Math.* **214** (2015), 357–393.

[336] J. Poineau and A. Pulita, Continuity and finiteness of the radius of convergence of a p-adic differential equation via potential theory, *J. reine angew. Math.* **707** (2015), 125–147.

[337] J. Poineau and A. Pulita, The convergence Newton polygon of a p-adic differential equation III: global decomposition and controlling graphs, arXiv:1308.0859v1 (2013).

[338] J. Poineau and A. Pulita, The convergence Newton polygon of a p-adic differential equation IV: local and global index theorems, arXiv:1309.3940v2 (2014).

[339] E. Pons, Modules différentiels non solubles. Rayons de convergence et indices, *Rend. Sem. Mat. Univ. Padova* **103** (2000), 21–45.

[340] B. Poonen, Maximally complete fields, *Enseign. Math.* (2) **39** (1993), 87–106.

[341] S. Priess-Crampe and P. Ribenboim, Differential equations over valued fields (and more), *J. reine angew. Math.* **576** (2004), 123–147.

[342] A. Pulita, Frobenius structure for rank one p-adic differential equations, in *Ultrametric functional analysis*, Contemp. Math. 384, Amer. Math. Soc., Providence, 2005, 247–258.

[343] A. Pulita, Rank one solvable p-adic differential equations and finite Abelian characters via Lubin-Tate groups, *Math. Ann.* **337** (2007), 489–555.

[344] A. Pulita, Small connections are cyclic, arXiv:1407.3761v1 (2014).

[345] A. Pulita, The convergence Newton polygon of a p-adic differential equation I: Affinoid domains of the Berkovich affine line, *Acta Math.* **214** (2015), 307–355.

[346] J.-P. Ramis, Théorèmes d'indices Gevrey pour les équations différentielles ordinaires, *Mem. Amer. Math. Soc.* **48** (1984).

[347] P. Ribenboim, *The Theory of Classical Valuations*, Springer-Verlag, New York, 1999.

[348] J.F. Ritt, *Differential Algebra*, Colloq. Pub. XXXIII, Amer. Math. Soc., New York, 1950.

[349] P. Robba, On the index of p-adic differential operators. I, *Ann. Math.* **101** (1975), 280–316.

[350] P. Robba, Lemmes de Hensel pour les opérateurs différentiels. Application à la réduction formelle des équations différentielles, *Enseign. Math.* (2) **26** (1980), 279–311.

[351] P. Robba, On the index of p-adic differential operators. II, *Duke Math. J.* **43** (1976), 19–31.

[352] P. Robba, Index of p-adic differential operators. III. Application to twisted exponential sums, *Astérisque* **119–120** (1984), 191–266.

[353] P. Robba, Indice d'un opérateur différentiel p-adique. IV. Cas des systèmes. Mesure de l'irrégularité dans un disque. *Ann. Inst. Fourier (Grenoble)*, **35** (1985), 13–55.

[354] P. Robba, Une introduction naïve aux cohomologies de Dwork, *Mém. Soc. Math. France* **23** (1986), 61–105.

[355] A.M. Robert, *A Course in p-adic Analysis*, Graduate Texts in Math. 198, Springer-Verlag, New York, 2000.

[356] E. Robertson and J. O'Connor (eds.), *MacTutor History of Mathematics Archive*, https://mathshistory.st-andrews.ac.uk (retrieved July 2020).

[357] R.T. Rockafellar, *Convex Analysis*, Princeton University Press, Princeton, 1970.

[358] N. Saavedra Rivano, *Catégories Tannakiennes*, Lecture Notes in Math. 265. Springer-Verlag, Berlin, 1972.

[359] C. Sabbah, *Introduction to Stokes Structures*, Lecture Notes in Math. 2060, Springer, Heidelberg, 2013.

[360] D. Savitt and D. Thakur (eds.), *p-adic Geometry: Lectures from the 2007 Arizona Winter School*, Univ. Lecture Series 45, Amer. Math. Soc., 2008.

[361] W.H. Schikhof, *Ultrametric Calculus: An Introduction to p-adic Analysis*, Cambridge Studies in Advanced Math. 4, Cambridge University Press, Cambridge, 1984.

[362] W. Schmid, Variation of Hodge structure: the singularities of the period mapping, *Invent. Math.* **22** (1973), 211–319.

[363] P. Schneider, *Nonarchimedean Functional Analysis*, Springer-Verlag, Berlin, 2002.

[364] R. Schoof, Elliptic curves over finite fields and the computation of square roots mod p, *Math. Comp.* **44** (1985), 483–494.

[365] P. Scholze, Perfectoid spaces, *Publ. Math. IHÉS* **116** (2012), 245–313.

[366] P. Scholze and J. Weinstein, *Berkeley Lectures on p-adic Geometry*, Annals of Math. Studies 207, Princeton University Press, Princeton, 2020.

[367] E. Schörner, Ultrametric fixed point theorems and applications, in *Valuation Theory and its Applications, Vol. II (Saskatoon, SK, 1999)*, Fields Inst. Communications 33, Amer. Math. Soc., 2003, 353–359.

[368] J.-P. Serre, Géométrie algébrique et géométrie analytique, *Ann. Inst. Fourier.* **6** (1956), 1–42.

[369] J.-P. Serre, *Local Fields*, Graduate Texts in Math. 67, Springer-Verlag, New York, 1979.

[370] J.-P. Serre, *Galois cohomology*, Corrected reprint of the 1997 English edition, Springer Monographs in Math., Springer-Verlag, Berlin, 2002.

[371] E.C. Sertöz, Computing periods of hypersurfaces, *Math. Comp.* **88** (2019), 2987–3022.

[372] Y. Sibuya, Stokes phenomena, *Bull. Amer. Math. Soc.* **83** (1977), 1075–1077.

[373] J.H. Silverman, *The Arithmetic of Elliptic Curves*, second printing, Graduate Texts in Math. 106, Springer-Verlag, New York, 1991.

[374] M.F. Singer and M. van der Put, *Galois Theory of Difference Equations*, Lecture Notes in Math. 1666, Springer-Verlag, Berlin, 1997.

[375] M.F. Singer and M. van der Put, *Galois Theory of Linear Differential Equations*, Grundlehren der Math. Wiss. 328, Springer-Verlag, Berlin, 2003.

[376] S. Sperber and J. Voight, Computing zeta functions of nondegenerate hypersurfaces with few monomials, *LMS J. Comput. Math.* **16** (2013), 9–44.

[377] T.A. Springer, *Linear Algebraic Groups*, second edition, Birkäuser, Boston, 1991.

[378] The Stacks Project Authors, *Stacks Project*, http://stacks.math.columbia.edu (retrieved Feb 2020).

[379] G.G. Stokes, On the discontinuity of arbitrary constants which appear in divergent developments, *Trans. Cambridge Phil. Soc.* **10** (1857), 106–128.

[380] Y. Sugiyama and S. Yasuda, Belyi's theorem in characteristic two, *Compos. Math.* **156** (2020), 325–339.

[381] T. Tannaka, Über den Dualitätsatz der nichtkommutativen topologischen Gruppen, *Tôhoku Math. J.* **45** (1938), 1–12.

[382] J. Tate, p-divisible groups, in *Proceedings of a Conference on Local Fields (Driebergen, 1966)*, Springer-Verlag, Berlin, 1967, 158–183.

[383] M. Temkin, Stable modification of relative curves, *J. Alg. Geom.* **19** (2010), 603–677.

[384] M. Temkin, Metric uniformization of morphisms of Berkovich curves, *Adv. Math.* **317** (2017), 438–472.

[385] T. Terasoma, Confluent hypergeometric functions and wild ramification, *J. Algebra* **185** (1996), 1–18.

[386] J.-B. Teyssier, Skeletons and moduli of Stokes torsors, *Ann. Sci. Éc. Norm. Sup.* **52** (2019), 337–365.

[387] J.-B. Teyssier, Higher dimensional Stokes structures are rare, in *A Panorama of Singularities*, Contemp. Math. 742, Amer. Math. Soc., Providence, RI, 2020, 201–217.

[388] J.-B. Teyssier, Moduli of Stokes torsors and singularities of differential equations, to appear in *J. Eur. Math. Soc.* (2021).

[389] A. Thuillier, Théorie du potentiel sur les courbes en géométrie analytique non archimédienne. Applications à la theorie d'Arakelov, thesis, Université de Rennes 1, 2005.

[390] A. Thuillier, Géométrie toroïdale et géométrie analytique non archimédienne. Application au type d'homotopie de certains schémas formels, *Manuscripta Math.* **123** (2007), 381–451.

[391] H.L. Turrittin, Convergent solutions of ordinary linear homogeneous differential equations in the neighborhood of an irregular singular point, *Acta Math.* **93** (1955), 27–66.

[392] N. Tsuzuki (as T. Nobuo), The overconvergence of morphisms of étale φ-∇-spaces on a local field, *Compos. Math.* **103** (1996), 227–239.

[393] N. Tsuzuki, Finite local monodromy of overconvergent unit-root F-isocrystals on a curve, *Amer. J. Math.* **120** (1998), 1165–1190.

[394] N. Tsuzuki, The local index and the Swan conductor, *Comp. Math.* **111** (1998), 245–288.

[395] N. Tsuzuki, Slope filtration of quasi-unipotent overconvergent F-isocrystals, *Ann. Inst. Fourier (Grenoble)* **48** (1998), 379–412.

[396] N. Tsuzuki, Morphisms of F-isocrystals and the finite monodromy theorem for unit-root F-isocrystals, *Duke Math. J.* **111** (2002), 385–419.

[397] N. Tsuzuki, On base change theorem and coherence in rigid cohomology, Kazuya Kato's fiftieth birthday, *Doc. Math.* **2003** Extra Vol., 891–918.

[398] N. Tsuzuki, Minimal slope conjecture of F-isocrystals, arXiv:1910.03871v2 (2019).

[399] J. Tuitman, Counting points on curves using a map to $\mathbf{P}^1$, *Math. Comp.* **85** (2016), 961–981.

[400] J. Tuitman, Counting points on curves using a map to $\mathbf{P}^1$, II, *Finite Fields Appl.* **45** (2017), 301–322.

[401] J. Tuitman, Computing zeta functions of generic projective hypersurfaces in larger characteristic, *Math. Comp.* **88** (2019), 439–451.

[402] M. van der Put, The cohomology of Monsky and Washnitzer, Introductions aux cohomologies p-adiques (Luminy, 1984), *Mém. Soc. Math. France* **23** (1986), 33–59.

[403] A.C.M. van Rooij, *Non-Archimedean Functional Analysis*, Monographs and Textbooks in Pure and Applied Math. 51, Marcel Dekker, New York, 1978.

[404] M. Vaquié, Valuations, in *Resolution of singularities (Obergurgl, 1997)*, Progr. Math. 181, Birkhäuser, Basel, 2000, 539–590.

[405] V.S. Varadarajan, Linear meromorphic differential equations: a modern point of view, *Bull. Amer. Math. Soc.* **33** (1996), 1–42.

[406] W. Wasow, *Asymptotic Expansions for Ordinary Differential Equations*, reprint of the 1976 edition, Dover, New York, 1987.

[407] C.A. Weibel, *An Introduction to Homological Algebra*, Cambridge Studies in Advanced Math. 38, Cambridge University Press, Cambridge, 1994.

[408] H. Weyl, Das asymptotische Verteilungsgesetz der Eigenwerte linearer partieller Differentialgleichungen (mit einer Anwendung auf die Theorie der Hohlraumstrahlung), *Math. Ann.* **71** (1912), 441–479.

[409] H. Weyl, Inequalities between the two kinds of eigenvalues of a linear transformation, *Proc. Nat. Acad. Sci. USA* **35** (1949), 408–411.

[410] L. Xiao, Non-archimedean differential modules and ramification theory, thesis, Massachusetts Institute of Technology, 2009.

[411] L. Xiao, On ramification filtrations and p-adic differential equations, I: the equal characteristic case, *Algebra Number Theory* **4** (2010), 969–1027.

[412] L. Xiao, On ramification filtrations and p-adic differential equations, II: mixed characteristic case, *Compos. Math.* **148** (2012), 415–463.

[413] L. Xiao, On the refined ramification filtrations in the equal characteristic case, *Alg. Number Theory* **6** (2012), 1579–1667.

[414] S. Ye, A group-theoretic generalization of the p-adic local monodromy theorem, arXiv:2004.01224v2 (2020).

[415] P.T. Young, Radii of convergence and index for p-adic differential operators, *Trans. Amer. Math. Soc.* **333** (1992), 769–785.

[416] H. Zassenhaus, Über eine Verallgemeinerung des Henselschen Lemmas, *Arch. Math. (Basel)* **5** (1954), 317–325.

[417] I.B. Zhukov, A approach to higher ramification theory, in *Invitation to higher local fields (Münster, 1999)*, Geom. Topol. Monographs 3, Geom. Topol. Publ., Coventry, 2000, 143–150.

[418] I.B. Zhukov, On ramification theory in the case of an imperfect residue field (Russian), *Mat. Sb.* **194** (2003), 3–30; translation, *Sb. Math.* **194** (2003), 1747–1774.

[419] T. Zink, On the slope filtration, *Duke Math. J.* **109** (2001), 79–95.

Index of notation

To compute the sort index of a symbol, read left to right (subscripts before superscripts), retaining only digits, English letters (in any font), Greek letters (α = "alpha", etc.), and the symbols † = "dagger", $'$ = "prime", $*$ = "star", $\times$ = "times", $\vee$ = "vee", $\wedge$ = "wedge". Ties are broken by order of first appearance.

$|A|$ (norm of a matrix)
   archimedean case, 57
   nonarchimedean case, 63
$\mathbb{A}_K$ (Berkovich affine line), 401

χ (cyclotomic character), 467
$\chi(f, V)$ (index of a linear transformation), 131
$\chi(\mathbb{P}_K \setminus Z)$ (Euler characteristic), 429
$\mathbb{C}_p$ (completed algebraic closure of $\mathbb{Q}_p$), 28
$D^\dagger(V)$
   equal characteristic, 344, 345
   mixed characteristic, 465
Δf (Laplacian), 410
δU (boundary of an unbounded affinoid subset), 422
$\mathrm{Diag}(\sigma_1, \ldots, \sigma_n)$ (diagonal matrix), 56
$\mathrm{disc}(f, r)$ (discrepancy), 215
$D^\dagger_{\mathrm{rig}}(V)$, 465
$D(V)$
   equal characteristic, 341
   mixed characteristic, 464

$\mathcal{E}$ (completion of $\mathfrak{o}_K((t)) \otimes_{\mathfrak{o}_K} K$), 166
$\tilde{\mathcal{E}}$ (completed maximal unramified extension of $\mathcal{E}$), 341
$\mathcal{E}^\dagger$ (overconvergent series), 282
$\tilde{\mathcal{E}}^\dagger$ (bounded extended Robba ring), 304
$[E : F]$ (degree of a field extension), 27
$\mathcal{E}_L$ (unramified extension of $\mathcal{E}$), 341
$\mathcal{E}^\dagger_L$ (unramified extension of $\mathcal{E}^\dagger$), 343
$\mathcal{E}_\varphi$ (perfection of $\mathcal{E}$), 289
$\mathcal{E}^\dagger_\varphi$ (perfection of $\mathcal{E}^\dagger$), 289
$\mathrm{Ext}(\cdot, \cdot)$ (Yoneda extension group), 85

F^{alg} (algebraic closure), 46
F_ρ (completion of $K(t)$ for $|\cdot|_\rho$), 165
F'_ρ (completion of $K(t^p)$ for $|\cdot|_{\rho^p}$), 182

F''_ρ (completion of $K((t-1)^p - 1)$ for $|\cdot|_{\rho^p}$), 193
F^{sep} (separable closure), 46
$|F^\times|$ (multiplicative value group), 17

Γ_M (controlling graph), 428
$G_{E/F}$ (Galois group), 46
G_F (absolute Galois group), 46
Γ_K (in p-adic Hodge theory), 462
H^0, H^1 (cohomology)
   of differential modules, 80
   of difference modules, 263
H_K (in p-adic Hodge theory), 462
$\mathcal{H}(x)$ (complete residue field), 399

$IR(V)$ (intrinsic radius), 167
$\mathrm{irr}(V)$ (irregularity of a differential module), 124
$\mathrm{irr}_z(M)$ (irregularity at a point), 132

K (complete nonarchimedean field), 143
K (complete discretely valued field), 261
$K\langle \alpha/t, t]\!]_0$ (bounded series on an annulus), 145
$K\langle \alpha/t, t/\beta \rangle$ (series on a disc/annulus), 144
$K\langle \alpha/t, t/\beta \}$ (series on a half-open annulus), 151
$K\langle \alpha/t, t/\beta]\!]_{\mathrm{an}}$ (analytic elements on a half-open annulus), 152
$K[\![\alpha/t, t/\beta]\!]_{\mathrm{an}}$ (analytic elements on an open annulus), 152
κ_F (residue field), 17
$K\langle t/\beta \rangle$ (series on a disc), 145
$K\{t/\beta\}$ (series on an open disc), 151
$K[\![t/\beta]\!]_0$ (series with bounded coefficients), 145
$K[\![t/\beta]\!]_{\mathrm{an}}$ (analytic elements on a disc), 152
$K[\![t]\!]_\delta$ (series with logarithmic growth), 333

Subject index

Boldfaced page numbers indicate definitions of terms or statements of results.

Printed in the United States
by Baker & Taylor Publisher Services